SEVEN DAY LOAN

This book is to be returned on
or before the date stamped below

1 3 DEC 1999	2 7 NOV 2000
1 0 MAY 2000	
	2 3 MAY 2001
1 6 MAY 2000	
	2 3 OCT 2001
1 2 OCT 2000	
	1 4 NOV 2001
2 6 OCT 2000	
	2 0 MAY 2002
1 3 NOV 2000	

UNIVERSITY OF PLYMOUTH

PLYMOUTH LIBRARY

Tel: (01752) 232323
This book is subject to recall if required by another reader
Books may be renewed by phone
CHARGES WILL BE MADE FOR OVERDUE BOOKS

Hybrid Zones and the Evolutionary Process

Hybrid Zones and the Evolutionary Process

EDITED BY

RICHARD G. HARRISON

Cornell University
Ithaca, New York

New York   Oxford
OXFORD UNIVERSITY PRESS
1993

Oxford University Press

Oxford   New York   Toronto
Delhi   Bombay   Calcutta   Madras   Karachi
Kuala Lumpur   Singapore   Hong Kong   Tokyo
Nairobi   Dar es Salaam   Cape Town
Melbourne   Auckland   Madrid

and associated companies in
Berlin   Ibadan

Library of Congress Cataloging-in-Publication Data
Hybrid zones and the evolutionary process / edited by
Richard G. Harrison.
p.   cm.   Includes bibliographical references and index.
ISBN 0-19-506917-X
1. Hybrid zones—Congresses.   2. Hybridization—Congresses.
3. Evolution (Biology)—Congresses.   I. Harrison, Richard G.
(Richard Gerald), 1945–   .
QH421.H83   1993
575—dc20   92-28327

9 8 7 6 5 4 3 2 1

Printed in the United States of America
on acid-free paper

Preface

Hybrid zones (intergrade zones) have figured prominently in the evolutionary biology literature for more than a century. The decades between 1970 and 1990 witnessed a maturation of hybrid zone theory and a veritable explosion of empirical data derived from long-term, multidisciplinary studies of many hybrid zones. This volume represents an attempt to summarize the current state of knowledge. In so doing, it hopefully provides a stimulus for future research.

Most of the chapters included in the book have evolved from presentations given as part of an invited symposium ("Hybrid Zones and the Evolutionary Process") at the Fourth International Congress of Systematic and Evolutionary Biology (ICSEB) in College Park, Maryland, held in July 1990. The organizers of the Congress not only invited me to organize a symposium on hybrid zones but provided logistic and financial support, both of which were critical to the success of the full-day symposium and the half-day workshop that followed. Of the participants in the symposium, all but one were able to contribute to this book. In addition, I solicited chapters from evolutionary biologists who were not part of the original symposium. These contributions focus on problems and groups of organisms not well represented in the presentations at ICSEB. They make the book a more representative sample of the hybrid zone literature. Each of the chapters has been carefully reviewed and (in most cases) extensively revised. I very much appreciate the willingness of colleagues to provide careful reviews of these chapters. I am particularly indebted to the contributors, who have been remarkably patient during the prolonged gestation period of this book. I have been a demanding editor, and they have been wonderfully responsive to my requests.

The book is divided into two major parts. Part I includes four chapters that examine some of the major conceptual and practical issues associated with the study of hybrid zones. Chapter 1 provides a historical perspective on hybridization and hybrid zones and an overview of major issues; and Chapters 2–4 examine genetic analyses of hybrid zones, the evidence for reinforcement, and the nature and consequences of introgression in plants. Part II includes a series of case studies. These studies represent summaries of long-term research programs that have focused on single hybrid zones or on clusters of hybrid zones within particular groups of organisms. If this book were a complete compendium of hybrid zone research, many other well-studied hybrid zones would certainly have been included. I apologize to those hybrid zone researchers whose work is not represented here and hope that they will understand that it was time and space constraints (not the quality of their research) that led me to limit the dimensions of this book.

Finally, I wish to express my appreciation to many colleagues and students (past and present) who have shared my enthusiasm for studying hybrid zones, bolstering my confidence that a program of research on hybrid zones is a worthwhile endeavor. I am also grateful to the National Science Foundation Systematic Biology Program, which has generously funded my own hybrid zone research over the past 15 years.

Ithaca, N.Y.   R. G. H.
July 1992

Contents

Contributors

Michael L. Arnold
Department of Genetics
University of Georgia
Athens, Georgia 30602

Nicholas H. Barton
Institute of Cell, Animal and
    Population Biology
Division of Biological Sciences
University of Edinburgh
Edinburgh EH9 3JT
United Kingdom

Bobby D. Bennett
Department of Biological Sciences
P.O. Box 599
Arkansas State University
State University, Arkansas 72467

Nelida Contreras
Molecular Evolution and Systematics
    Group
Research School of Biological Sciences
Australian National University
Canberra, A.C.T. 2601, Australia

Katherine S. Gale
Department of Zoology
University of Oxford
South Parks Road
Oxford OX1 3PS
United Kingdom

Fran Groeters
Molecular Evolution and Systematics
    Group
Research School of Biological Sciences
Australian National University
Canberra, A.C.T. 2601 Australia

Richard G. Harrison
Section of Ecology and Systematics
Corson Hall
Cornell University
Ithaca, New York 14853

Godfrey M. Hewitt
School of Biological Sciences
University of East Anglia
Norwich NR4 7TJ
United Kingdom

Daniel J. Howard
Department of Biology
New Mexico State University
Las Cruces, New Mexico 88003

Bert Kohlmann
E.A.R.T.H.
Apartado 4442-1000
San Jose, Costa Rica

James Mallet
Galton Laboratory
Department of Genetics and Biometry
University College London
4 Stephenson Way
London NW1 2HE
United Kingdom

Adam Marchant
Molecular Evolution and Systematics
  Group
Research School of Biological Sciences
Australian National University
Canberra, A.C.T. 2601, Australia

William S. Moore
Department of Biological Sciences
Wayne State University
Detroit, Michigan 48230

James L. Patton
Museum of Vertebrate Zoology
University of California
Berkeley, California 94720

Jeff T. Price
Department of Biological Sciences
Wayne State University
Detroit, Michigan 48230

Loren H. Rieseberg
Rancho Santa Ana Botanic Garden
1500 N. College Ave.
Claremont, California 91711

Jeremy B. Searle
Department of Biology
University of York
Heslington, York, YO1 5DD
United Kingdom

David D. Shaw
Molecular Evolution and Systematics
  Group
Research School of Biological Sciences
Australian National University
Canberra, A.C.T. 2601, Australia

Jacek M. Szymura
Department of Comparative Anatomy
Jagiellonian University
Krakow, Poland

Jonathan F. Wendel
Department of Botany
Iowa State University
Ames, Iowa 50011

1

HYBRID ZONE PATTERN
AND PROCESS

Most evolutionary change occurs slowly or sporadically; as a consequence, direct observation of change over time often yields limited information about the process of evolution. Although it is not difficult to document changes in the frequencies of particular phenotypes or genotypes within populations (microevolutionary changes), the splitting and subsequent divergence of entire lineages (speciation) and the origin of evolutionary novelties are rarely observed. Process is most often inferred from pattern; evolutionary biologists devote much of their time and energy to characterizing and interpreting patterns of variation within and among populations, species, and higher taxa.

Natural hybridization and narrow hybrid zones are striking patterns of variation that have consistently attracted the attention of evolutionary biologists. Hybridization is common in plants and can often be recognized by the presence of localized "hybrid swarms." Debates about the evolutionary significance of hybridization in plants have a long history. Are hybrid populations an important source of evolutionary novelties, or are they evolutionary dead-ends? Under what circumstances are hybrids produced? Should hybrid populations be recognized as distinct species? Although many evolutionary botanists believe that hybridization and introgression are important sources of new variation, the issues are far from resolved.

In animals, hybrid zones often appear as abrupt discontinuities between differentiated groups of populations that are themselves relatively homogeneous over large areas. These discontinuities cry out for an explanation. How did they arise, and why do they persist? Do they represent the coming together, in secondary contact, of populations that have differentiated in allopatry (perhaps populations "on the way" to being good species), or do they arise in a continuous series of populations in response to selection gradients? Are hybrid zones stable or transient? If they are stable, what evolutionary forces are acting to maintain them? If transient, what will be their fate? Will the differentiated populations fuse to form a single descendant lineage, or will interactions within the hybrid zone lead to continuing differentiation and ultimately to speciation?

The four chapters that comprise Part I provide a review of major conceptual and practical concerns of evolutionary biologists studying hybrid zones. Harrison proposes working definitions of "hybrid" and "hybrid zone" and briefly summarizes major issues in hybrid zone research and their relation to traditional themes in evolutionary biology. The chapter by Barton and Gale is an introduction to methods for genetic analysis of hybrid zones. From cline width, cline shape, and patterns of linkage disequilibrium, they show how one can infer the strength of natural selection and the genetic architecture of population (species) differences. In Chapter 3, Howard reviews the arguments for and against the process of reinforcement—the evolution of prezygotic barriers to gene exchange in response to selection against hybrids. Based on a thorough literature review, he suggests that reproductive character displacement, a pattern of variation predicted by the reinforcement model, is not uncommon. Direct evidence for the process of reinforcement, however, is far more difficult to obtain. Finally, Rieseberg and Wendel summarize a voluminous literature on the consequences of hybridization and introgression in plant populations, with emphasis on the recent application of molecular (DNA) markers. With an array of diagnostic genetic markers, it becomes possible to resolve questions about the ancestry of putative hybrid populations.

1

Hybrids and Hybrid Zones: Historical Perspective

RICHARD G. HARRISON

In most groups of animals, diversification over time is portrayed as a series of dichotomous branching events, suggesting that reticulate evolution (hybridization) occurs rarely. This view has been espoused by most evolutionary biologists. According to Mayr (1963), "the evolutionary importance of hybridization seems small in the better-known groups of animals."

In contrast, examples of natural hybridization are common among plants, and many plant species appear to be interconnected by limited gene exchange (Stebbins, 1950, 1959; Grant, 1981). Indeed, botanists use the special term "syngameon" (Lotsy, 1925) to refer to the "most inclusive unit of interbreeding in a hybridizing species group" (Grant, 1981). Stebbins (1959) suggested that hybridization between distinct forms (species or subspecies) is "the rule in flowering plants" and urged that particular attention be given to examples of sympatric closely related species that do not hybridize. Plant hybrid zones tend to be diffuse (not geographically well defined) and are often characterized by local hybrid swarms. In many instances, hybridization appears to occur at ecotones or boundaries between different habitats.

Despite the supposed rarity of animal hybrids in nature, hybridization has been a major focus of studies in animal evolution (Barton and Hewitt, 1985, 1989; Harrison, 1990). For more than a century systematists and evolutionary biologists have struggled to reconcile patterns of variation within hybrid or intergrade zones with concepts of species and subspecies and with proposed models of divergence and speciation. Much of the attention has been devoted to the study of discrete hybrid zones, which often represent abrupt discontinuities between taxa that are themselves relatively homogeneous over large geographic areas. In fact, such discontinuities may be far more common than has generally been believed, with the ranges of many animal species subdivided by hybrid zones (Hewitt, 1988, 1989).

Presumably as a consequence of their different experiences with hybridization in natural populations, botanists and zoologists have developed rather different views of the "evolutionary role" of hybridization. Beginning with Lotsy (1916), botanists have emphasized the evolutionary consequences of hybridization and have suggested that hybridization plays a "key role in race formation" (Grant, 1981). Through polyploidy or introgression, "the products of hybridization can form new evolutionary lines that are isolated from the ancestral types and are therefore free to evolve in new directions" (Stebbins, 1959). Thus botanists view hybridization as a creative force, not only an important source of new variation in plants but also a source of new species.

Zoologists have been more reluctant to recognize hybridization as a catalyst of evolutionary change or innovation. If animal hybrid zones have been directly implicated in evolutionary process, it is as sites where premating barriers are perfected in response to selection against hybrids of reduced fitness. In recent years, zoologists have most often treated hybrid zones as "windows on the evolutionary process" (Harrison, 1990) or as "natural laboratories" (Hewitt, 1988; Barton and Hewitt, 1989) in which to explore the operation of evolutionary forces, the nature of barriers to gene exchange, and the genetic differences responsible for these barriers. Empirical work on animal hybrid zones has stimulated the development of an important body of theory that addresses fundamental questions about both adaptation and speciation.

WHAT IS A HYBRID?

The possibility of crosses between distinct species intrigued the earliest students of natural history, who sometimes invoked hybridization as an explanation for the existence of unusual creatures, both real and mythical (Zirkle, 1941). Although records of crosses between species and varieties have been documented from the sixteenth and seventeenth centuries (Zirkle, 1932, 1934), a modern literature on hybridization can be traced to the systematic studies of plant hybrids carried out by Kolreuter and Linnaeus during the mid-eighteenth century (Stebbins, 1959; Grant, 1981). Nineteenth century evolutionary biologists wrote extensively about hybridization, although they focused on experimental studies and the results of plant and animal breeding (Darwin, 1896; Wallace, 1889). Of particular concern was the distinction between "varieties" and "species." A commonly held view was that crosses between varieties produced *mongrel* offspring that were perfectly fertile, whereas crosses between species produced sterile *hybrid* offspring. Therefore from the perspective of most nineteenth century naturalists, sterility was a criterion for species status and hybrids were by definition sterile. Darwin (1872) was uncomfortable with the prevailing view that species were "specially endowed with sterility in order to prevent their confusion"; he concluded that, although first crosses and hybrids were often sterile, this rule was not universal. He then pointed out the circularity of arguing for the fertility of crosses between varieties as opposed to the sterility of crosses between species.

> It may be urged, as an overwhelming argument, that there must be some essential distinction between species and varieties, inasmuch as the latter, however, much they differ from each other in external appearance, cross with perfect fertility, and yield perfectly fertile offspring. With some exceptions, presently to be given, I fully admit this is the rule. But the subject is surrounded by difficulties, for, looking to varieties produced under nature, if two forms hitherto reputed to be varieties be found in any degree sterile together, they are at once ranked by most naturalists as species.

The futility of proclaiming the existence of discrete classes of crosses was clearly argued in the section on "hybridism" in the 11th Edition of the *Encyclopedia Brittanica.*

"Hybridism therefore grades into mongrelism, mongrelism into cross-breeding, and cross-breeding into normal pairing, and we can say little more than that the success of the union is more unlikely the further apart the parents are in natural affinity" (Mitchell, 1911). Nonetheless, among evolutionary biologists, the notion that "hybrid" should apply only to offspring from *inter*specific crosses persisted well into the twentieth century.

A contrasting view (encountered in the genetics and plant and animal breeding literature) is that hybrid simply refers to the offspring of genetically distinct parents. Such a definition seems appropriate for experimental crosses in which parents are selected for their difference(s) in one or more traits. In most sexually reproducing species, however, every individual in each generation of a natural population is a unique genotype and thus every individual of the next generation would be a "hybrid." Defined in this way, the term "hybrid" would not describe a restricted class of individuals.

Is there a middle ground between hybrids as the products of interspecific crosses and hybrids as offspring of any pair of genetically distinct individuals? Mayr (1963) emphasized that definitions become far more complex when typological thinking is abandoned and hybridization is viewed in the context of variation within and between natural populations. He defined hybridization as "the crossing of individuals belonging to two unlike natural populations that have secondarily come into contact." The meaning of "unlike" was not made clear, but Mayr stressed that populations at some time in the past must have been isolated from one another. Stebbins (1959) proposed that hybridization be defined as the "crossing between individuals belonging in separate populations which have different adaptive norms." In this category he included not only crosses between individuals belonging to different species but also secondary contact between conspecific populations that had diverged in allopathy. His intent appears to have been to define a threshold amount of divergence between parental individuals beyond which their offspring should be considered hybrids. One might imagine that this threshold could be measured by the depth of a valley that separates the two populations on an adaptive landscape (i.e., that the "adaptive norms" of Stebbins' definition correspond to peaks on a landscape).

An operational definition, which does not require subjective determination of whether populations are "unlike" or "have different adaptive norms," is that hybridization is "the interbreeding of individuals from two populations, or groups of populations, which are distinguishable on the basis of one or more heritable characters" (Harrison, 1990; modified from Woodruff, 1973). Although the parents of a hybrid need differ in only one heritable trait, they must be drawn from populations that are diagnosably distinct for that character. Such populations might be called different species by proponents of a phylogenetic species concept (Cracraft, 1983, 1989; Nixon and Wheeler, 1990). The advantages of this definition are that (1) consistent application does not depend on reaching agreement on a single species concept (an unlikely scenario); (2) it is not subject to the necessarily arbitrary assignment of populations to particular taxonomic categories (e.g., races or subspecies); and (3) it does not require judgments about relative fitness of hybrids or differences between parental types in "adaptive norms."

There are also disadvantages. As defined above, the term "hybrid" refers only to the F_1 progeny of a single type of cross. Furthermore, the definition is based on observed differences *between populations* and is useful only when the parental populations are fixed for alternative alleles (diagnosably distinct). Many evolutionary biologists apply the term "hybrid" to F_1 offspring plus the set of all backcross, F_2, and so on individuals that might be found in cases of natural hybridization. If hybrid is equated with an individual of mixed ancestry, hybrids are produced from crosses *within populations,* and the definition given above does not strictly apply. In this case, a hybrid would be any individual heterozygous (intermediate) for at least one of the markers that are diagnosably distinct between the parental populations or fixed for alternative parental alleles at different loci. Such a definition is not always practical, especially if the parental populations are distinguished on the basis of morphology or behavior (rather than "well-behaved" genetic markers). Anderson and Hubricht (1938) emphasized that the term hybrid becomes a problem when several (many) generations of backcrossing have occurred.

> The F_1 is clearly entitled to the term hybrid, but among the progeny of its first cross back to the parent there will be a number of individuals which resemble that species very closely indeed, and each successive backcross will increase the percentage of these indistinguishable or almost indistinguishable mongrels. After a few back-crosses most of the individuals cannot be distinguished by morphological means from the pure species.

Because many "hybrid zones" contain few F_1 individuals, it is perhaps appropriate to allow the term "hybrid" to be used in the sense of mixed ancestry and to specify "F_1 hybrid" when intending to limit discussion to that single class of individuals of mixed ancestry.

WHAT IS A HYBRID ZONE?

Hybrid zones occur when genetically distinct groups of individuals meet and mate, resulting in at least some offspring of mixed ancestry (Barton and Hewitt, 1989; Harrison, 1990). The definition is intentionally broad and includes situations ranging from sporadic or occasional hybridization between species that are broadly sympatric (perhaps associated with different habitats or resources) to narrow zones of hybridization between taxa with effectively parapatric distributions. In some cases the outcome is a "hybrid swarm" (a diverse array of recombinant types). In other situations, only F_1 offspring (in addition to parental types) are found. The definition does not depend on either knowledge of the history of the interaction or an understanding of the evolutionary forces acting to maintain it. Furthermore, it makes no attempt to discriminate on the basis of the geography of hybridization.

Most previous definitions have been more restrictive. Mayr (1942) equated zones of hybridization with regions of secondary intergradation, and many reviews have focused exclusively on zones thought to derive from secondary contact between previously isolated populations (Heiser, 1973; Moore, 1977; Rising, 1983; Littlejohn and Watson, 1985). Yet hybrid zones are often recognized simply by the existence of concordant (or parallel) clines. Clines are nothing more than character gradients (Huxley, 1938), and even steep clines may be the product of selection within and among a continuous series of populations (primary intergradation). In such cases intermediate forms (intergrades) are not necessarily the product of hybridization following secondary contact. Endler (1977) argued that "a steep cline should not be assumed to be a hybrid zone unless there is some evidence for increased variability of fitness and morphology in the steepest part of the cline compared to flatter portions, and beyond that due to mixing and other random effects." Unfortunately, these data are not available for most "hybrid zones." Given the difficulty of inferring history (primary intergradation versus secondary contact) from current patterns of variation (Endler, 1977), it is perhaps best to avoid definitions that depend on knowledge of history.

Some hybrid zone definitions have included an explicit assumption about the dynamics of the zone. For example, Barton and Hewitt (1981) defined a hybrid zone as a "narrow cline maintained by some sort of hybrid unfitness" but have since abandoned this definition in favor of a broader one (Barton and Hewitt, 1985). It seems to be preferable to use "hybrid zone" in a broad sense and to introduce other terms to specify particular subsets. Thus "tension zone" (Key, 1968; Barton and Hewitt, 1985) describes those situations in which a balance between dispersal and selection against individuals of mixed ancestry maintains narrow hybrid zones.

The geography of natural hybridization is highly variable. Many animal hybrid zones represent steep multilocus clines that occur at the junctions of the ranges of two distinct taxa (usually considered subspecies or species). These zones are often remarkably narrow, especially when compared with the widespread distributions of the parental types [e.g., see reviews of hybrid zones in mammals (Chs. 11 and 12) and in the grasshoppers *Chorthippus* and *Caledia* (Chs. 6 and 7)]. There is an extensive literature (both theoretical and empirical) dealing with these phenomena (Barton and Hewitt, 1985, 1989). However, hybrid zones are not always narrow (e.g., the flicker hybrid zone reviewed in Ch. 8); observed width is clearly related to the dispersal ability of the organism and to the strength of selection (see Ch. 2).

The internal structure of hybrid zones can be complex, often reflecting a patchy distribution of habitats and resources. Examples of "mosaic hybrid zones" have been documented in a variety of animals, and many plant hybrid zones seem to be of this type as well (Harrison and Rand, 1989). In these cases, simple cline models may not be an accurate representation of observed patterns of variation; it depends on the dispersal ability of the organism relative to the spatial scale of environmental heterogeneity (Harrison, 1990). There are also examples of broadly sympatric species that hybridize (occasionally or extensively) and yet remain distinct. The ranges of the sunflowers *Helianthus annuus* and *H. petiolaris* now overlap broadly. These species hybridize to produce localized hybrid swarms but elsewhere retain their identities (Heiser, 1947). The butterflies *Colias philodice* and *C. eurytheme* occur together throughout much of the United States, and a complete range of intermediates may be found at many localities (Hovanitz, 1943). Nonetheless, the two species show no evidence of fusing. In both cases, ecological differentiation appears to play an important role in maintaining the distinctness of the parental types.

MAJOR ISSUES IN HYBRID ZONE RESEARCH

Many evolutionary biologists have viewed hybrid zones as active sites of evolutionary change, either as sources of new recombinant types (new species) or as localities in which selection against hybridization leads to strong prezygotic barriers to gene exchange (i.e., less hybridization). For others, hybrid zones are primarily of interest as natural laboratories in which genetic and ecological interactions between differentiated populations can be examined. Understanding the causes and consequences of these interactions provides insights into a range of important problems in evolutionary biology. Here I outline the relation of hybrid zones to other major issues in evolutionary biology and briefly summarize important questions in hybrid zone research. Many of these topics are discussed in detail in the chapters of this book.

Hybrid Zones and Species Concepts. The narrow hybrid zones characteristic of many animals and the localized hybrid swarms found in many plants can be disconcerting phenomena for systematists and evolutionary biologists. What labels should be applied to hybridizing taxa or to the products of natural hybridization? There appear to be two distinct issues: (1) Under what circumstances should local hybrid populations be recognized as new species, distinct from either of the parental populations? (2) Are hybridizing taxa simply races (or perhaps subspecies) because they hybridize, or are they "good species" because allopatric populations of the parental types remain distinct? The answers to these questions clearly depend on what one means by a "species."

Hybrid zones are not easily accommodated within the framework of the "biological species concept," in which the criterion for species status is reproductive or genetic isolation. The entities that are joined (or subdivided) by hybrid zones cannot simply be catalogued as either conspecific or as belonging to different species. In fact, the barriers that limit genetic exchange between hybridizing taxa are often semipermeable (Key, 1968; Harrison, 1986); i.e., there is considerable variance in the extent to which alleles at different loci introgress. This observation has prompted the suggestion that species may need to be defined gene by gene (Barton and Hewitt, 1981). Hybrid zones pose equally serious problems for the recognition concept of species (Paterson, 1985), which defines species as groups of individuals sharing a specific mate recognition system (SMRS). Proponents of the recognition concept (e.g., Lambert et al., 1987; Masters et al., 1987) argued that it is less "relational" than the biological species concept (which they have termed the "isolation concept"), but in practice it is not the case. The boundaries between groups of individuals that share an SMRS are blurred by hybridization and backcrossing, and identification of discrete classes may be impossible.

Phylogenetic species concepts (Cracraft, 1983, 1989; Nixon and Wheeler, 1990) do not rely on reproductive continuity (or discontinuity) as a necessary criterion for defining species and therefore have certain advantages. Species are "the smallest aggregation of populations . . . diagnosable by a unique combination of character states in comparable individuals" (Nixon and Wheeler, 1990). However, hybrid zones may contain individuals in which the "unique combinations of character states" are broken up and recombined. The status of the parental (diagnosably distinct) populations is again ambiguous; "undoubtedly differences of opinion will exist as to the best way to deal with such situations" (Nixon and Wheeler, 1990).

Hybrid zones are clearly a nuisance for those in search of a static species definition in which all individuals can be neatly catalogued as belonging to one species or another. If we overlook this slight inconvenience, however, we find that hybrid zones provide a wealth of information about possible states and degrees of divergence between populations that may be "on the way" to being full species.

Hybrid Zone Origins. Questions about hybrid zone origins have been debated by evolutionary biologists since the late nineteenth century. What historical events and evolutionary forces account for the current distribution of genotypes and phenotypes? Two contrasting scenarios have been proposed. One (perhaps the more popular) considers hybrid zones to be the result of secondary contact between populations that have differentiated in allopatry (e.g., Chapman, 1892; Mayr, 1942). Alternatively, hybrid zones may arise in situ in direct response to spatially varying selection pressures (environmental gradients) (Endler, 1977). These contrasting views are evident in the ornithological literature from 100 years ago. Concerning subspecies of the grackle *Quiscalus aeneus,* Chapman (1892) wrote:

> A question has arisen . . . concerning the manner in which their intergradation is accomplished. Is one bird an imperfectly differentiated offshoot of the other, and are the connecting intergrades geographical intermediates, or have we here two distinct species whose intergradation is due to interbreeding where the confines of their respective habitats adjoin? In other words, the question is one of geographical variation versus hybridization. . . .

During the same year, Allen (1892) reviewed the facts pertaining to intergradation in flickers *(Colaptes)* in the central United States (see Ch. 8) and concluded that all evidence tended to support the "startling hypothesis of hybridization on a grand scale [as opposed to geographic variation in direct response to environment] . . . to account for the occurrence of birds presenting ever-varying combinations of the characters of the two species."

Considerable controversy still surrounds the question of hybrid zone origins and explanations for intergradation between distinct forms. For many years the prevailing view was that hybrid zones invariably reflect secondary contact, a view consistent with the belief that geographic isolation is a prerequisite for differentiation and speciation. Endler (1977) argued persuasively that inference of origins simply from current patterns of variation was risky at best, primarily because both primary intergradation and secondary contact can produce identical patterns of variation. The result has been that evolutionary biologists are now far more cautious in inferring process from pattern. The persistent debate over hybrid zone origins clearly reflects fundamental disagreements about the power of natural selection to lead to differentiation (or the efficacy of gene flow in maintaining homogeneity).

Hybrid zone origins likewise figure prominently in discussions of the relative importance of current ecology (environmental selection) and historical processes (dispersal, vicariance) in determining patterns of variation. Finally, the possibility that hybrid zones (steep clines) are the result of differentiation in situ has been inextricably linked to arguments about the frequency of parapatric speciation events. This linkage is somewhat tenuous, however, because even though a hybrid zone may have arisen by secondary contact, the differences between hybridizing forms may have appeared earlier in evolutionary history (see Ch. 9). The geography of speciation is often obscured by subsequent changes in distribution due to dispersal or local extinction, or both (Hewitt, 1988; Harrison, 1990).

It is often suggested that hybrid zones are a direct consequence of habitat disturbance or environmental change (Anderson, 1948; Hubbs, 1955). Human disturbance certainly has led to the breakdown of ecological isolation in many cases. Moreover, many existing hybrid zones can be explained by invoking Pleistocene and post-Pleistocene range contractions and expansions in the temperate zone (Hewitt, 1989; see Ch. 6) and the tropics (Prance, 1982; but see Ch. 9).

Dynamics of Stable Hybrid Zones. Do hybrid zones represent equilibrium situations? If so, what forces maintain distinct parental types in the face of persistent hybridization? Many hybrid zones appear to be relatively stable, maintained by a balance between dispersal and selection. This balance may involve selection against hybrids or other individuals of mixed ancestry independent of the environment. Such tension zones can form anywhere but tend to move to regions with low population density or barriers to dispersal. Hybrid unfitness is evident in many of the well-studied hybrid zones (see Chs. 6, 7, 11) and is often thought to be characteristic of hybrid zones involving certain types of chromosome rearrangement (see Ch. 12). Alternatively, fitness may be habitat-dependent. Hybrid zones form at habitat boundaries or along steep environmental gradients if different genotypes (species) have higher fitness in different environments (see Chs. 5 and 8).

As detailed in Chapter 2, careful analysis of patterns of clinal variation, linkage disequilibrium, and introgression within and adjacent to hybrid zones can provide important insights into the strength of selection, dispersal rates, and genetic architecture of differences between the hybridizing taxa. The term "genetic architecture" refers to the number, phenotypic effects, and linkage relations of the genes responsible for observed differences. Ultimately, genetic analysis of hybrid zones leads to a better appreciation of the processes involved in the origin of adaptations and the origin of species.

Hybrid Zone Fates. If hybrid zones are transient, what are the likely outcomes? One possibility is that hybrid zones represent secondary contact and neutral diffusion, that the differentiated populations will fuse, yielding a single, possibly polymorphic, species. Alternatively, hybrid zones might represent the "wave of advance" of a superior competitor, resulting in the eventual extinction of one of the two hybridizing taxa. In fact, fusion and extinction are not mutually exclusive outcomes because the fusion of two taxa might involve either selective or random extinction of alleles from each of the parental types (Harrison, 1990). If certain recombinant genotypes produced by hybridization and backcrossing persist as local "hybrid swarms" or "stabilized introgressants," the product of fusion may be considered a distinct species (see Chs. 4 and 5).

One of the most controversial issues surrounding hybrid zones is whether they are sites of "reinforcement"—the evolution of prezygotic barriers to gene exchange in response to selection against hybrids. A mode of speciation originally championed by Dobzhansky (1940, 1941), the "reinforcement model" has met with considerable criticism in recent years (Paterson, 1978, 1982; Butlin, 1987, 1989). In Chapter 3, Howard reviews the issues surrounding this debate and the evidence for reproductive character displacement.

Finally, selection within hybrid zones can lead to the weakening (rather than the strengthening) of barriers to gene exchange. Selection may favor those variants that show the least reduction in viability and fertility when crossed with either of the parental types. Such a scenario has been proposed to explain patterns of variation in a *Clarkia* hybrid zone (Bloom, 1976) and in chromosomal hybrid zones in the shrew *Sorex araneus* (Searle, 1986; see Ch. 12).

Causes and Consequences of Introgression. Hybridization and backcrossing to one or both of the parental types can result in incorporation of alleles from one taxon into the gene pool of the other. Anderson and Hubricht (1938) coined the term "introgressive hybridization" to describe this phenomenon. Attempts to identify and characterize patterns of introgression constitute an important component of the hybrid zone literature (see Ch. 4). Differential introgression is characteristic of some animal hybrid zones (Harrison, 1990; see also Chs. 6 and 7, but see Ch. 10 for a case of concordant clines). Not only do markers introgress to different extents, but there is sometimes a fundamental asymmetry in the direction of introgression that may reflect recent movement of the hybrid zone.

Introgressive hybridization can lead to the production of recombinant genotypes that have properties different from those of either of the parents. Anderson (1948, 1953) argued that introgression is an important source of new variation (more important than mutation) and that variants produced in this way are most likely to succeed in disturbed or changing environments. Whether hybridization and introgression are creative forces in evolutionary change remains unresolved; the strongest proponents come from the botanical community because there are numerous examples of apparent hybrid swarms or populations of stabilized introgressants in the plant literature (Anderson and Stebbins 1954; see also Chs. 4 and 5).

CONCLUSIONS

From these considerations it is clear that studies of hybrid zones and the consequences of hybridization (both natural and experimental) have played an important role in developing our understanding of evolutionary process. Many of the most important issues (e.g., hybrid zone origins, reinforcement) have been debated for years but remain unresolved. However, the decades of the 1970s and 1980s witnessed the development and application of new techniques that have yielded the high resolution genetic markers needed to answer questions about both evolutionary history and current population structure. Over the same time period, hybrid zone theory has matured, allowing tests of alternative hypotheses and providing guidance in sampling design and experimental manipulation. It is clearly an appropriate time to review what we now know and to anticipate what lies ahead.

REFERENCES

Anderson, E. 1948. Hybridization of the habitat. Evolution 2:1–9.

Anderson, E. 1953. Introgressive hybridization. Biol. Rev. 28:280–307.

Anderson, E., and Hubricht, L. 1938. Hybridization in *Tradescantia*. III. The evidence for introgressive hybridization. Am. J. Botany 25:396–402.

Anderson, E., and Stebbins, G. L. 1954. Hybridization as an evolutionary stimulus. Evolution 8:378–388.

Barton, N. H., and Hewitt, G. M. 1981. Hybrid zones and speciation. In W. R. Atchley and D. S. Woodruff, eds., Evolution and Speciation.

Cambridge: Cambridge University Press, pp. 109–145.

Barton, N. H., and Hewitt, G. M. 1985. Analysis of hybrid zones. Annu. Rev. Ecol. Syst. 16:113–148.

Barton, N. H., and Hewitt, G. M. 1989. Adaptation, speciation and hybrid zones. Nature 341:497–503.

Bloom, W. L. 1976. Multivariate analysis of the introgressive replacement of *Clarkia nitens* by *Clarkia speciosa polyantha* (Onagraceae). Evolution 30:412–424.

Butlin, R. K. 1987. Speciation by reinforcement. Trends Ecol. Evol. 2:8–13.

Butlin, R. K. 1989. Reinforcement of premating isolation. In D. Otte and J. Endler, eds., Speciation and its Consequences. Sunderland, MA: Sinauer Associates, pp. 158–179.

Chapman, F. M. 1892. A preliminary study of the grackles of the subgenus *Quiscalus.* Bull. Am. Mus. Natur. Hist. 4:1–19.

Cracraft, J. 1983. Species concepts and speciation analysis. Curr. Ornithol. 1:159–187.

Cracraft, J. 1989. Speciation and its ontology: the empirical consequences of alternative species concepts for understanding patterns and processes of differentiation. In D. Otte and J. Endler, eds., Speciation and its Consequences. Sunderland, MA: Sinauer Associates, pp. 28–59.

Darwin, C. 1872. The Origin of Species, 6th ed. London: John Murray.

Darwin, C. 1896. The Variation of Animals and Plants Under Domestication. New York: D. Appleton.

Dobzhansky, T. 1940. Speciation as a stage in evolutionary divergence. Am. Naturalist 74:312–321.

Dobzhansky, T. 1941. Genetics and the Origin of Species. New York: Columbia University Press.

Endler, J. A. 1977. Geographic Variation, Speciation, and Clines. Princeton, NJ: Princeton University Press.

Grant, V. 1981. Plant Speciation. New York: Columbia University Press.

Harrison, R. G. 1986. Pattern and process in a narrow hybrid zone. Heredity 56:337–349.

Harrison, R. G. 1990. Hybrid zones: windows on evolutionary process. Oxford Surv. Evol. Biol. 7:69–128.

Harrison, R. G., and Rand, D. M. 1989. Mosaic hybrid zones and the nature of species boundaries. In D. Otte and J. Endler, eds., Speciation and its Consequences. Sunderland, MA: Sinauer Associates, pp. 111–133.

Heiser, C. B., Jr. 1947. Hybridization between the sunflower species, *Helianthus annuus* and *H. petiolaris.* Evolution 1:249–262.

Heiser, C. B., Jr. 1973. Introgression reexamined. Bot. Rev. 39:347–366.

Hewitt, G. M. 1988. Hybrid zones—natural laboratories for evolutionary studies. Trends Ecol. Evol. 3:158–167.

Hewitt, G. M. 1989. The subdivision of species by hybrid zones. In D. Otte and J. Endler, eds., Speciation and Its Consequences. Sunderland, MA: Sinauer Associates, pp. 85–110.

Hovanitz, W. 1943. Hybridization and seasonal segregation in two races of a butterfly occurring together in two localities. Biol. Bull. 85:44–51.

Hubbs, C. L. 1955. Hybridization between fish species in nature. Syst. Zool. 4:1–20.

Huxley, J. 1938. Clines: an auxiliary taxonomic principle. Nature 142:219.

Key, K. H. L. (1968). The concept of stasipatric speciation. Syst. Zool. 17:14–22.

Lambert, D., Michaux, B., and White, C. S. 1987. Are species self-defining? Syst. Zool. 36:196–205.

Littlejohn, M. J., and Watson, G. F. (1985). Hybrid zones and homogamy in Australian frogs. Annu. Rev. Ecol. Syst. 16:85–112.

Lotsy, J. P. 1916. Evolution by Means of Hybridization. The Hague: M. Nijhoff.

Lotsy, J. P. 1925. Species or linneon. Genetica 7:487–506.

Masters, J. C., Rayner, R. J., McKay, I. J., Potts, A. D., Nails, D., Ferguson, J. W., Weissenbacher, B. K., Allsopp, M., and Anderson, M. L. 1987. The concept of species: recognition versus isolation. S. Afr. J. Sci. 83:534–537.

Mayr, E. 1942. Systematics and the Origin of Species. New York: Columbia University Press.

Mayr, E. 1963. Animal Species and Evolution. Cambridge: Belknap Press.

Mitchell, P. C. 1910. Hybridism. Encylopaedia Brittanica, 11th Ed. Cambridge, England: The University Press.

Moore, W. S. 1977. An evaluation of narrow hybrid zones in vertebrates. Q. Rev. Biol. 52:263–277.

Nixon, K. C., and Wheeler, Q. D. 1990. An amplification of the phylogenetic species concept. Cladistics 6:211–223.

Paterson, H. E. H. 1978. More evidence against speciation by reinforcement. S. Afr. J. Sci. 74:369–371.

Paterson, H. E. H. 1982. Perspective on speciation by reinforcement. S. Afr. J. Sci. 78:53–57.

Paterson, H. E. H. 1985. The recognition concept of species. In E. Vrba, ed., Species and Speciation. Transvaal Museum, Pretoria, pp. 21–29.

Prance, G. T., ed. 1982. Biological Diversification in the Tropics. New York: Columbia University Press.

Rising, J. D. 1983. The Great Plains hybrid zones. Curr. Ornithol. 1:131–157.

Searle, J. B. 1986. Factors responsible for a karyotypic polymorphism in the common shrew, *Sorex araneus.* Proc. R. Soc. Lond. Biol. 229:277–298.

Stebbins, G. L. 1950. Variation and Evolution in Plants. New York: Columbia University Press.

Stebbins, G. L. 1959. The role of hybridization in evolution. Proc. Am. Philos. Soc. 103:231–251.

Wallace, A. R. 1889. Darwinism. London: Macmillan.

Woodruff, D. S. 1973. Natural hybridization and hybrid zones. Syst. Zool. 22:213–218.

Zirkle, C. 1932. Some forgotten records of hybridization and sex in plants. J. Hered. 23:433–448.

Zirkle, C. 1934. More records of plant hybridization before Koelreuter. J. Hered. 25:3–18.

Zirkle, C. 1941. The Jumar or cross between the horse and the cow. Isis 33:486–506.

2

Genetic Analysis of Hybrid Zones

NICHOLAS H. BARTON AND KATHERINE S. GALE

When two distinct gene pools meet and produce fertile hybrids, the outcome varies from gene to gene. At some loci a universally favorable allele has been established on one side. Such alleles soon spread through the whole population and hence differences are rarely observed. At other loci different alleles may be favored in different environments or genetic backgrounds; selection maintains these differences in the face of random mixing. At other loci—perhaps at most of those we observe in molecular surveys—different alleles may have been established by chance and may have no appreciable effect on fitness. These differences gradually fade away, at a rate that depends on the strength of selection against introgression at the other loci with which they are associated.

The frequencies of the various genotypes found in a hybrid zone tell us about the overall strength of the selection, the number of genes involved, the rate of individual dispersal, and the ease with which alleles cross from one gene pool into the other. The aim of this chapter is to explain how data on discrete markers and on quantitative traits can be used to estimate such parameters. We illustrate the methods using examples from some of the hybrid zones that are discussed in more detail elsewhere in this book and use computer simulations to show that the estimates do not depend on exactly how selection maintains the differences between the hybridizing populations. Previous reviews have considered the wider questions of what hybrid zones can tell us about species and speciation and what role they themselves might play (Barton and Hewitt, 1985, 1989; Hewitt, 1988; Harrison and Rand, 1989; see also Ch. 1). We concentrate instead on the practical issues involved in the genetic analysis of hybrid zones.

A systematist, whose aim is to classify organisms, sees hybrid zones as boundaries between distinct types. A population geneticist, on the other hand, views them as sets of geographic gradients (i.e., of clines) in allele frequencies or quantitative traits. Both extreme views are misleading. Classification of individuals into parental, F_1, F_2, and backcross types wastes much information and, moreover, depends on which markers are used: an individual who is heterozygous for diagnostic alleles at five loci might be classified as an F_1 and yet be homozygous at the sixth locus. If even a small proportion of hybrids reproduce, all the individuals in the vicinity of the hybrid zone eventually carry introgressed alleles in some of their genes. However, describing a population solely in terms of allele frequencies or the means of quantitative traits also throws away much information. The complete data set consists of the genotypes and phenotypes of each individual in the sample. How can this information be reduced to a manageable but informative state?

Hybrid zones can be described in several ways. We concentrate on the allele frequencies at each locus and the pairwise associations between loci ("linkage disequilibria"). For continuously varying characters, the corresponding measures are the mean, variance, and covariances. Our aim is to explain how data presented in this way lead to estimates of selection and gene flow and to find how far these estimates depend on the details of how selection acts. This population genetic description may not always be the most appropriate. Where reproductive isolation is strong, the population may cluster around parental and F_1 genotypes, so that a classification into various recombinant types becomes more natural. One question which we consider is: At what point does selection become so strong that our methods break down? A third description becomes possible when one has sets of closely linked markers, for example, from DNA sequence data, so that the phylogenetic relation between the various genes sampled at a locus can be reconstructed. The extra information that might come from a set of such "gene trees" is discussed briefly at the end.

MECHANISMS FOR MAINTAINING CLINES

Clines can be maintained in two ways. There might simply be a balanced polymorphism, with an equilibrium that varies from place to place. Provided this equilibrium varies gradually enough, the shape of the cline directly reflects the local environment and has a shape that is independent of how far individuals move. For example, sickle cell anemia varies across Africa with the incidence of malaria: the frequency of the Hb^S allele tracks the relative fitnesses. A special case of such "dispersal-independent" clines has been suggested by Moore (1977), who argued that hybrid zones might be maintained by selection *favoring* hybrids within a narrow region of intermediate habitat.

Most hybrid zones cannot be explained in this way. First, dispersal is only negligible when clines are much wider than a characteristic scale, set by the ratio between the dispersal distance and the square root of the selection coefficient (Slatkin, 1973). Hybrid zones often consist of clines that are much narrower than likely environmental gradients, having widths approaching the individual dispersal range (see Figure 3 in Barton and Hewitt, 1985). Second, if cline shape were determined directly by local selective conditions, one would expect it to vary considerably from place to place. In fact, clines often have similar width and shape across different transects. For example, wherever the two races of the alpine grasshopper *Podisma pedestris* meet, the frequency of the Robertsonian fusion that distinguishes them changes in a sigmoid cline 500–900 meters wide (Barton and Hewitt, 1981a, 1989; Nichols and Hewitt, 1986); exceptions can be accounted for by barriers such as streams or scree. The fire-bellied toads *Bombina bombina* and *Bombina variegata* meet in a long hybrid zone that runs round the Carpathian Mountains and the Danube basin. Belly pattern and diagnostic allozymes change in almost exactly the same way across two transects 200 km apart in southern Poland (Szymura and Barton, 1991); the clines near Zagreb, in Croatia, are somewhat wider (9 km versus 6 km) but have the same form (Szymura, 1988; Ch. 10). Finally, if clines at each locus or for each phenotypic trait were maintained in direct response to the environment, one would not expect them to change in the same way or at the same place: in contrast, almost all hybrid zones consist of a cluster of parallel clines, often involving characters with no obvious functional relation (Barton and Hewitt, 1985, 1989).

There are exceptions to this argument: some hybrid zones are so wide that dispersal can hardly be significant, for example, *Thomomys bottae* in the Sangre de Cristo mountains (Hafner et al., 1983). In some cases, genotype frequencies change together but do track the local environment, for example, the crickets *Gryllus firmus* and *Gryllus pennsylvanicus,* which are associated with different soil types (Rand and Harrison, 1989). The concordance of different characters in such "mosaic" hybrid zones (Harrison and Rand, 1989) may be a relic of secondary contact: one would expect that after prolonged hybridization, only those genes directly selected to fit the relevant environment (or that interact epistatically with directly selected genes) would be associated with that environment. In the absence of a sustained influx, differences in neutral traits would decay over a time inversely proportional to the rate of gene flow between habitats.

Their narrow width, consistent shape, and close concordance suggest that most hybrid zones are maintained by a balance between selection and dispersal: the sharp disjunction that would be produced by selection alone is blurred by the random movement of individuals. Selection could act in many ways. The primary distinction is between adaptation to the external environment (so that different alleles are favored in different places) and selection against hybrids (so that alleles are favored in their own genetic background or when at high frequency, regardless of location). The distinction is important, because it determines how the hybrid zone can move, or, in other words, how the sets of genes that distinguish the hybridizing populations compete with each other. If alleles are selected to fit their native habitat, the hybrid zone must lie at a particular point on an environmental gradient. If, on the other hand, selection acts against hybrids, the hybrid zone can move from place to place: it is then termed a "tension zone" (Key, 1968; Barton and Hewitt, 1985).

In reality, different kinds of selection act in the same hybrid zone. For example, the sharp boundaries between different warning patterns in *Heliconius* butterflies are maintained by Müllerian mimicry. This involves selection against heterozygotes, recombinants, and rare alleles, which all tend to produce patterns that are not recognized as distasteful by predators (Mallet, 1986; Mallet et al., 1990). Selection also acts to favor patterns that are common in other species in the same mimicry ring; it can be seen as an adaptation to the external environment, though the set of hybrid zones in all the mimetic species is still free to move as one.

Fortunately, the mechanism of selection has little effect on the shape of the clines; it is this factor that allows us to make inferences without needing to know just how selection is operating. Moreover, provided that selection is not too strong, cline shape does not depend on the local population structure: gene flow through both a continuous habitat and across a grid of demes or "stepping stones" can be approximated by diffusion (Nagylaki, 1975). The effect of gene flow then depends on a single parameter (σ) defined as the standard deviation of the distance between parent and offspring, measured along a linear axis; it does not depend on the whole distribution of dispersal distances, which would be much harder to measure accurately. (Note that in two dimensions, the standard deviation of the *total* distance moved is $\sqrt{2}\sigma$, as it includes movements in two directions). Variations in σ from place to place can greatly affect cline shape (Karlin & Richter-Dyn, 1976); however, they might be detected by the presence of an obvious barrier to dispersal or by variation from transect to transect.

CLINE WIDTH

A variety of models of selection on single loci or quantitative traits produce clines with similar shapes and with width proportional to the ratio between dispersal and the square root of selection ($\sigma/\sqrt{s}$). It is convenient to define the width of a cline in allele frequencies as the inverse of the maximum slope; for a quantitative trait, the analogous definition is the ratio between the difference between the populations on either side and the maximum slope ($\Delta z/(\partial z/\partial x)$). (Explicit theoretical predictions are usually impossible with other definitions, such as the distance between the 20% and 80% points.) Cline widths can be calculated using the diffusion approximation, which holds when selection is weak; in the models considered below, the population is assumed to be diploid and to mate at random.

To compare the widths of clines maintained in different ways, we must use some common measure of selection. Let s* be the difference in mean fitness between populations at the center (defined as the point where the slope is greatest) and the edge. The widths of the clines illustrated in Figure 2-1 are as follows: selection against heterozygotes (H): $2\sigma/\sqrt{s^*}$; selection favoring different alleles on either side of a sharp ecotone, with no dominance (E): $1.732\sigma/\sqrt{s^*}$; the same, but with dominance (D): $1.782\sigma/\sqrt{s^*}$. Thus for a given dispersal rate and a given reduction in mean fitness, the widths of clines maintained by a balance between dispersal and selection are similar. For the case of neutral mixing (N), the width t generations after the two populations met in an abrupt step is $2.51\sigma\sqrt{t}$, which is again similar if t is analogous to $1/s^*$, the characteristic time scale of selection.

Quantitative traits behave somewhat differently (Q in Fig. 2-1). Suppose that weak stabilizing selection favors one optimum on the left of a boundary and another on the right, and the genetic variance (V_g) is constant (a plausible assumption if many loci are involved). The cline width is then $(\Delta/2\sqrt{V_g})\sigma/\sqrt{s^*}$. It is proportional to the difference in optima (Δ), relative to the genetic standard deviation ($\sqrt{V_g}$) (Slatkin, 1978). In this case, the cline width is more simply related to the loss in mean fitness due to genetic variation around the optimum (L): $w = \sigma/\sqrt{L}$. Similar results hold for clines maintained by disruptive selection on a quantitative character (Rouhani and Barton, 1987). In both cases the width is still proportional to $\sigma/\sqrt{s^*}$.

This robust relation should allow one to infer selection pressures from cline widths and dispersal rates. However, it is not usually practicable. Barton and Hewitt (1985, Fig. 3) plotted cline widths against estimated dispersal rates for 26 hybrid zones. Though most clines had widths within one or two orders of magnitude of the estimated dispersal range, many were much wider, implying that they are maintained by unreasonably weak selection. Some of these clines may in fact be maintained by balancing selection alone and be independent of dispersal. However, a more likely explanation is that dispersal is often grossly underestimated. Many of the estimates used in that survey were based on limited mark-recapture studies, subject to great statistical inaccuracies, and inevitably missing long-distance migrants and juvenile dispersal; some

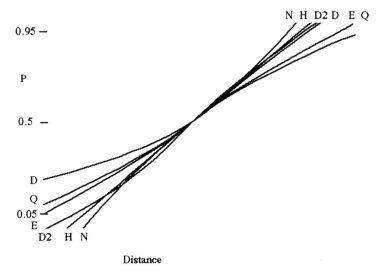

Figure 2-1. Clines maintained by a balance between disperal and selection on a single locus have similar shapes. Allele frequencies are plotted against distance on a logit scale ($\log_e[p/q]$); the clines are scaled so all have the same position and width. Letters refer to different models. **N** = neutral introgression; **H** = heterozygote disadvantage; **E** = an ecotone, with fitnesses $1 + s:1:1 - s$ on the left, and $1 - s:1:1 + s$ on the right; **D** = an ecotone with dominance. Fitnesses are $1:1:1 - s$ on the left, and $1:1:1 + s$ on the right. **D2** refers to the same model, but now the frequency of the recessive homozygote is plotted. **Q** = stabilizing selection on a quantitative trait, with the optimum changing abruptly by Δ at an ecotone.

were not even derived from the same species (e.g., Hafner, 1982). In many examples of apparently wide clines, there is direct or indirect evidence of selection (e.g., compare Hafner, 1982, with Barton, 1982; see also next section).

CLINE SHAPE

The weak dependence of cline shape on the way selection acts is illustrated in Figure 2-1. Clines produced by a variety of models are scaled to the same width and plotted using a logit transformation ($z = \log_e(p/q)$), also known as a logistic transform. It has the advantage that it expands extreme gene frequencies, so a sigmoid curve appears as a straight line. A cline maintained by selection against heterozygotes follows a straight line with slope $(\partial z/\partial x) = 4/w$; other models give clines that approach a straight line on the logit scale. The same method could be used to represent clines in quantitative traits; however, the means of the two parental populations must be known accurately if the transformation is to succeed.

The clearest deviation from this pattern is produced by complete dominance (D). Because selection against rare recessives is ineffective, the frequency of the recessive allele can be high even in regions where it is not favored (Fig. 2-1, left). However, if the frequency of the recessive homozygote is plotted, instead of the frequency of the allele itself, the cline appears much more similar to the others (D2 in Fig. 2-1). A similar pattern is found when positive frequency-dependent selection maintains a cline at a locus with complete dominance (Mallet, 1986).

Interactions among genes can also distort the shape of a set of clines. In the following sections, we investigate a simple model of epistasis, in which the fitness (W) of an individual depends on the fraction of alleles from each of the hybridizing populations, x and $(1 - x)$: $W(x) = 1 - s[4x(1 - x)]^\beta$. When $\beta = 1$, it reduces to a quadratic relation (Fig. 2-2a). When β is large (upper curves in Figure 2-2a: $\beta = 4$ and 16), only individuals with genotype close to x = 50% suffer: a little hybridization hardly reduces fitness. In contrast, when β is small (lower curves in Figure 2-2a: $\beta = \frac{1}{4}$ and $\frac{1}{16}$), a trace of hybridization greatly reduces fitness. With one locus, this model reduces to heterozygote disadvantage: the parameter β then has no effect, because only genotypes with x = 0, 0.5, and 1.0 exist.

With several loci, variation in β changes the fitness of backcross genotypes; one would expect that when β is large—so that backcrosses are almost as fit as pure individuals—the cline would be shallow at the edges (where the unfit genotypes with x = 50% are rare) and steep at the center. Conversely, when β is small, selection between pure and backcross genotypes is strong at the edges and relatively weak near the center, where all hybrids have low fitness. This pattern is seen when large numbers of loci interact (see Figure 2-2c, which shows results for eight loci), though large distortions are seen only when $\beta \gg 1$. However, the type of epistasis has remarkably little effect when just two loci interact (Fig. 2-2b).

The conclusion is that plausible forms of selection are unlikely to distort clines far from a sigmoid cline, which appear as a straight line when plotted on a logit scale. Strong distortions can be produced by epistasis, but only if a moderate to large number of loci interact in their effects on fitness, the effect is such that only individuals close to the central genotype have reduced fitness, and one is observing the loci that are actually under selection. One could produce models of frequency-dependent selection on a single locus, which would produce distorted clines; but, again, implausibly strong nonlinearity is needed.

(a)

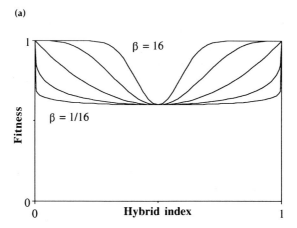

Figure 2-2. (a) In this model, individual fitness is $W(x) = 1 - s(4x[1 - x])^\beta$, where x = the fraction of alleles derived from one of the hybridizing populations. The curves are for $\beta = \frac{1}{16}, \frac{1}{4}, 1, 4,$ and 16.

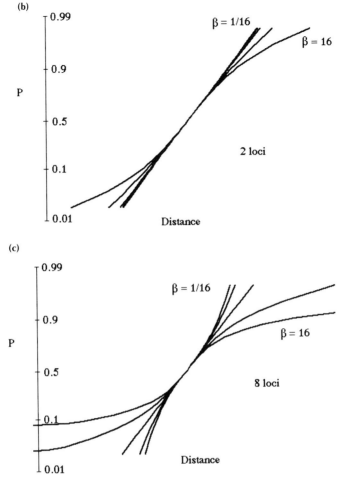

Figure 2-2. *Continued* (**b** and **c**) Note the shapes of clines with two and eight loci, respectively; each shows curves for $\beta = \frac{1}{16}, \frac{1}{4}, 1, 4, 16$ plotted using a logit transform and scaled to have the same width. With two loci (**b**), epistasis has little effect on cline shape. With eight loci (**c**), clines are stepped when $\beta > 1$; however, epistasis still has little effect when $\beta < 1$. Curves for large numbers are similar to those for $n = 8$. Cline shapes were calculated using the diffusion approximation and ignoring linkage disequilibrium (accurate when selection is weak).

LINKAGE DISEQUILIBRIA

The key assumption thus far has been that genes at different loci combine at random. In fact, strong associations are often found in hybrid zones: alleles derived from the same population tend to cluster together in the same individuals. Examples include *Heliconius* (Mallet et al., 1990), *Bombina* (Szymura and Barton, 1986, 1991), *Rana* (Kocher and Sage, 1986), *Uroderma* (Baker, 1981; Barton, 1982), *Gryllus* (Rand and Harrison, 1989), and *Caledia* (Shaw et al., 1985). Such associations are also revealed by correlations between unrelated traits that are unlikely to be explained by the pleiotropic effects of a shared set of genes: for example, in hybrid populations of *Bombina* in southern Poland, there is a correlation between genotype at diagnostic enzyme loci, mating call, and belly pattern (Sanderson et al., 1991; Szymura and Barton, 1991). In this section, we show that these associations can be used, in conjunction with cline widths, to estimate both dispersal and selection. Moreover, they can produce large distortions in cline shape that reflect a genetic barrier to gene flow and can be used to estimate the total selection that maintains all the differences between the hybridizing populations.

Associations between alleles at different loci are referred to as linkage disequilibria. This term is unfortunate, as in the examples cited above the genes involved are not linked, and the hybrid zones may well be in an equilibrium. However, we follow common practice in using the term while bearing in mind these potential confusions.

It is remarkable to find strong associations between alleles at unlinked loci. Such associations should halve in every generation and so must continually be replenished. The most obvious explanation is that linkage disequilibria are produced by the influx of parental combinations of genes into the center of the hybrid zone. To a good approximation, the strength of disequilibrium can be calculated by balancing the rate of influx against the rate of recombination (r). The details are explained in Appendix 1, using a simple model of exchange between two demes. In a continuous population, in which dispersal is approximated by diffusion, the same method can be used, giving (Barton, 1986)

$$D = \frac{\sigma^2}{r} \frac{\partial p}{\partial x} \frac{\partial u}{\partial x} = \frac{\sigma^2}{r w_p w_u} \tag{1}$$

at the center. This value is found after recombination and before diploid individuals migrate; the value after migration would be greater by a factor $(1 + r)$. Linkage disequilibrium is proportional to the product of the gradients at the two loci ($\partial p/\partial x$, $\partial u/\partial x$); at the center they are (by definition) the inverse of the cline widths (w_p, w_u), giving the second formula.

We have tested this prediction against simulations; details can be found in Appendix 2. Figure 2-3 shows results for hybrid zones maintained by heterozygote disadvantage or epistasis, with eight unlinked loci. The observed linkage disequilibrium is compared with the value expected from the cline width predicted by weak selection theory (straight line) and with that expected from the cline width actually seen in simulations (thinner curve). There is good agreement with the prediction from the observed width, even when selection is strong and even when there is epistasis: the values in the two graphs are similar. Thus dispersal rates could be estimated from observed linkage disequilibria, without the need to know just how the clines are maintained. When selection is strong, linkage disequilibria cause the clines to become narrower than expected and, in turn, to generate more disequilibrium. This phenomenon accounts for the discrepancy between the different predictions in Figure 2-3. When selection is weak, and the clines are maintained by heterozygote disadvantage, disequilibrium is somewhat weaker than expected. This may be because random drift scatters the clines to different locations, weakening their interaction. The effect is not seen with epistasis, as the latter tends to pull the clines together. Similar comparisons with just two loci show a similar pattern, though the confidence limits are wider, and there is less steepening of the clines with strong selection.

(a)

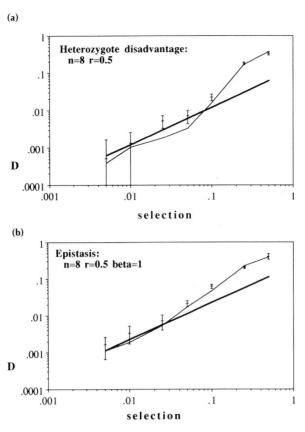

Figure 2-3. Comparison between observed linkage disequilibrium and that expected from a balance between dispersal and recombination. (**a**) Results for selection against heterozygotes, (**b**) Epistasis with $\beta = 1$, i.e., $W = 1 - 4x(1 - x)$. There were 40 demes, each of $2n = 50$ haploid individuals; selection acts on eight unlinked genes. The disequilibrium (D) at the center is plotted on a log scale, against the strength of selection at each locus (s). Haploid gametes migrate after meiosis and before selection; disequilibrium is measured immediately after meiosis, and so the predicted value is $m(1 - m)/rw^2$ (see Appendix 1). The straight line shows the prediction, given the width calculated using a weak selection approximation $w = \sqrt{8m/s}$ for heterozygote disadvantage (**a**) and $w = \sqrt{64m/15s}$ for epistasis (**b**). The thinner line shows the prediction made from the actual width observed in the simulations. Deviations between the two lines are due to differences between observed and expected widths. Observed disequilibria are shown with 95% confidence intervals; data were accumulated from generations 100 to 400 in each run and recorded every 20 generations (apart from $s = 0.005$, for which generations 400 to 1000 were used).

Estimating Linkage Disequilibria: Maximum Likelihood. Estimation of linkage disequilibria is not as simple as estimation of allele frequencies. Because the double heterozygotes PU/QV and PV/QU cannot be distinguished, one cannot simply count up the contributions of different gametic types. Moreover, information from several loci must be combined to give reliable estimates, which is difficult when disequilibria are strong, as estimates from different pairs of loci are not then independent.

The most efficient and elegant method is to use likelihood (Edwards, 1972; Hill, 1974). Consider data from just two loci. Given some value for the pairwise disequilibrium and assuming that gametes combine at random, one can calculate the expected genotype frequencies. The probability that the observed data would have been produced by a population with these frequencies is termed the likelihood of the given disequilibrium. The best estimate is that which gives the observed sample with greatest probability. In other words, it is the maximum likelihood estimate. The relative plausibility of other values is proportional to their likelihood and can be represented by a graph of log likelihood against disequilibrium (a "support curve") or, more compactly, by rejecting values that are less than $\approx e^{-2} = 0.135$ as likely. We refer to this procedure as defining a "2-unit support limit"; with large samples, it corresponds to defining a 95% confidence limit.

The method of maximum likelihood is the most efficient in that with large samples it gives the estimate with smallest variance around the true value. It is also flexible in that it can readily be modified to allow for dominance and epistasis (e.g., Mallet et al., 1990). However, there is no explicit formula for the best estimate, and analysis of data from samples taken after dispersal and from many loci is not straightforward. With random mating, genotype frequencies are in multilocus Hardy-Weinberg proportions immediately after reproduction. However, dispersal generates deviations from Hardy-Weinberg, as well as linkage disequilibria, which distort estimates based on the assumption of random combination. The ideal solution is to set up an explicit model of the hybrid zone that would predict genotype frequencies as a function of dispersal rate and so could be used to calculate the likelihood. Simulations of a three-locus system, however, suggest that deviations from Hardy-Weinberg are small, making this model unnecessary (Mallet and Barton, 1989).

Difficulties increase when one has data from many loci. Again, the ideal solution would be to set up an explicit model, which would predict the complete frequencies of all multilocus genotypes: allele frequencies pairwise disequilibria, three-way disequilibria, and so on. This method would take an inordinate amount of computing; in practice, maximum-likelihood estimates from separate pairwise analyses must be combined. One cannot simply take the average of the separate estimates, as each may be substantially biased. The obvious approach is to use the total log likelihood, summed over all pairwise analyses, as a function of some common value of the pairwise linkage disequilibrium. There are two difficulties here. First, different pairwise estimates are not independent, so one cannot simply sum the log likelihoods. This problem may not be serious if disequilibria are weak, as with large samples from a population in linkage equilibrium estimates are uncorrelated. The second problem is that, if allele frequencies at different loci vary, one expects linkage disequilibria to vary. Indeed, the maximum possible disequilibrium depends on the allele frequencies. A rough solution is to estimate a standardized measure, $R = D/\sqrt{pquv}$. It must lie between -1 and $+1$ and can be thought of as the correlation between the states of the two loci.

Estimating Linkage Disequilibria: Variance in Hybrid Index.

A simpler method, which is adequate in most cases, is to derive the average linkage disequilibrium from the variance in a "hybrid index." The variance is inflated if alleles that increase the hybrid index tend to be found together; when many loci are involved, the bulk of the variance may be caused by linkage disequilibria (Bulmer, 1980).

Suppose that two populations differ in a set of traits (z_i). They might be quantitative traits or Mendelian markers; in the latter case, the three genotypes in a diploid are labeled $z_i = 0$, 1, or 2. Data from n loci or traits can be summarized by a hybrid index, $z = \sum_{i=1}^{n} \alpha_i z_i$. Here we choose to scale this index so that it runs from 0 for one population to 1 for the other. If there are n diagnostic marker genes, the appropriate weighting would be $\alpha_i = (1/2n)$, so that z is just the proportion of alleles derived from one population rather than the other. Assuming Hardy Weinberg proportions, the variance of z is:

$$\text{var}(z) = \sum_{i,j=1}^{n} \alpha_i \alpha_j \, \text{cov}(z_i, z_j)$$

$$= \sum_{i=1}^{n} \alpha_i^2 \, \text{var}(z_i) + \sum_{i \neq j} \alpha_i \alpha_j \, \text{cov}(z_i, z_j)$$

(2a)

This expression has two components: the first due to variation in each contribution to the index [$\text{var}(z_i)$] and the second to covariance between different contributions. Where the index is based on discrete Mendelian markers, these covariances are due to linkage disequilibria: $\text{cov}(z_i, z_j) = 2D_{i,j}$, and $\text{var}(z_i) = 2p_i q_i$. (The factor of 2 arises because z_i is the sum over two copies of i'th gene). With the scaling $\alpha_i = (1/2n)$, we have:

$$\text{var}(z) = \sum_{i=1}^{n} 2\alpha_i^2 p_i q_i + \sum_{i \neq j} 2\alpha_i \alpha_j D_{ij}$$

$$= \frac{1}{2n}\left(\bar{z}(1 - \bar{z}) - \text{var}(p) \right) + \frac{1}{2}\left(1 - \frac{1}{n}\right)\overline{D}$$

(2b)

Here, $\overline{D}$ = average pairwise linkage disequilibrium; $\bar{z}$ = average of the hybrid index; and $\text{var}(p) = 1/n\sum (p_i - \bar{p})^2$ = variance of allele frequency across the n loci. Because the variance of z and the individual allele frequencies can easily be calculated, Eq. 2b gives a straightforward way of estimating linkage disequilibria. Approximate confidence limits can be found by using the critical points of the $F_{n-1,\infty}$ distribution to set limits on var(z) and ignoring uncertainty in the allele frequencies.

We illustrate this method with a sample of 351 *Bombina* taken from the center of the hybrid zone in southern Poland and scored for five diagnostic loci [sample Kopanka 2 (1981) in Szymura and Barton, 1986] (Fig. 2-4). The mean hybrid index is $\bar{z} = 0.486$, and the variance in allele frequency across loci is var(p) = 0.005. One can find the component due to heterozygosity at individual loci from the first term in Eq. 2b: $(1/10)(0.486 \times 0.514 - 0.005) = 0.0245$. The actual variance is 0.0397, which is significantly greater ($F_{356,\infty} = 1.62$, P = 10^{-12}). The excess variance (0.0397 $- 0.0245 = 0.0152$) is due to linkage disequilibria: $0.0152 = (1/2)(1-1/5)\overline{D}$, so that $\overline{D} = 0.038$. Because the critical points of the $F_{356,\infty}$ distribution are at 0.858 and 1.153, confidence limits on $\overline{D}$ are from 0.025 to 0.054. This estimate compares with a maximum likelihood estimate of $\overline{D} = 0.037$, with support limits between 0.027 and 0.045. The maximum likelihood appears somewhat more powerful, though this may be because of correlations between the likelihoods calculated for each pair of loci.

It would be useful to know more about the statistical properties of estimators of multilocus linkage disequilibrium. Previous discussions have concentrated on just two

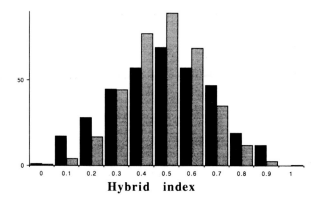

Figure 2-4. Distribution of the hybrid index in a sample of *Bombina* from the center of the hybrid zone in southern Poland (Kopanka 2, 1981; Szymura and Barton, 1986). The figure compares the observed distribution *(solid bars)* with the binomial distribution expected in the absence of linkage disequilibria *(light bars)*. The index is based on five unlinked and diagnostic enzyme loci.

loci or on estimating the random disequilibrium generated by drift (Hill, 1974; Lewontin, 1988; Weir and Cockerham, 1989). A rough comparison of the two methods can be made by examining estimates from repeated subsamples from the actual data (C. MacCallum, pers. comm.). With 2000 replicate samples of 50 individuals (taken with replacement), the standard deviation of the maximum likelihood estimate around the true value is 0.014, which compares with 0.017 for estimates based on the variance in hybrid index. Bias is negligible, and the limits on the estimates are approximately correct; the true value falls within the support limits of the maximum likelihood estimate 93.0% of the time and within the confidence limits based on the hybrid index 97.3% of the time. As expected, the maximum likelihood estimate is more accurate, but the simpler estimate based on the hybrid index performs comparably. Because linkage disequilibria in hybrid zones are often strong and several loci can be sampled, samples of moderate size may suffice. For example, Barton (1982) detected significant disequilibria using data from 25 bats scored for three chromosome rearrangements.

Because analysis and simulations suggest that linkage disequilibrium is primarily maintained by a balance between dispersal and selection, we can work back from the maximum likelihood estimate of $\overline{D} = 0.037$ to find the dispersal rate. These five enzymes are unlinked (Szymura and Farana, 1978) and change across concordant clines with an average width of 6.05 km (Szymura and Barton, 1991). If the sample was taken before dispersal, Eq. 1 would apply and gives a dispersal rate of $\sigma = \sqrt{r\overline{D}w^2} = 0.82$ km gen$^{-1/2}$. If movement occurred before sampling, less dispersal would be required to explain the observed disequilibrium, giving a somewhat lower estimate of $\sigma = \sqrt{r\overline{D}w^2/(1 + r)} = 0.67$ km gen$^{-1/2}$. (Note the units: σ^2 is the variance in distance moved along some axis per generation.) These values are rather lower than the figure of $\sigma = 0.99$ km gen$^{-1/2}$ estimated by Szymura and Barton (1991), which was based on data from many samples across the hybrid zone. The main uncertainty in their estimate arose from random variation in disequilibria between samples, rather than from statistical errors in estimates from each sample.

COVARIANCE BETWEEN QUANTITATIVE TRAITS

Equation 2 can readily be extended to allow linkage disequilibria and hence dispersal rates, to be estimated from the covariance between quantitative traits. Suppose that two traits, z and z*, are each determined by the sum of effects of a number of genes (n, n*, respectively):

$$z = \sum_{i=1}^{n} \alpha_i z_i + E, \ z^* = \sum_{j=1}^{n^*} \alpha_j^* z_j^* + E^*$$

As before, z_i and z_j^* label the effects of the three genotypes at a locus, and take the values 0, 1, or 2. If these traits are determined by different sets of genes and the environmental components (E, E*) are independent, any covariance must be due to linkage disequilibrium:

$$\text{cov}(z, z^*) = \sum_{i=1}^{n} \sum_{j=1}^{n^*} \alpha_i \alpha_j^* \ \text{cov}(z_i, z_j^*)$$

$$= 2 \sum_{i=1}^{n} \sum_{j=1}^{n^*} \alpha_i \alpha_j^* \ D_{ij} \tag{3}$$

Now, if we assume that linkage disequilibria are due to a balance between migration and recombination, substitution from Eq. 1 gives:

$$\text{cov}(z, z^*) = 2 \sum_{i=1}^{n} \sum_{j=1}^{n^*} \alpha_i \alpha_j^* \ \frac{\sigma^2}{r} \frac{\partial p_i}{\partial x} \frac{\partial p_j}{\partial x} \tag{4a}$$

This equation can be rewritten in terms of the gradients of the traits themselves:

$$\text{cov}(z, z^*) = \frac{\sigma^2}{2r} \frac{\partial z}{\partial x} \frac{\partial z^*}{\partial x} \tag{4b}$$

Because we have defined the width of a cline in a quantitative trait as the ratio between the difference on either side and the gradient at the center $\{w = [\Delta z/(\partial z/\partial x)]\}$, we have:

$$\text{cov}(z, z^*) = \frac{\sigma^2}{2r} \frac{\Delta z}{w} \frac{\Delta z^*}{w^*} \tag{4c}$$

Note that this derivation requires only that genetic variation be additive. Different loci might have effects that vary in strength and direction, and alleles need not be fixed on either side.

Fire-bellied and yellow-bellied toads differ in mating call as well as by biochemical markers: the natural logarithm of one component of call, the cycle length, differs by $\Delta z^* = 1.21$ (Sanderson et al., 1991). The cline in cycle length is not significantly different in width from the enzyme clines: $w^* \approx w = 6.05$ km. In the center of the hybrid zone, mating call is correlated with enzyme genotype: the covariance between ln(cycle length) and a hybrid index (z) based on six diagnostic loci and scaled from 0 to 1 is $\text{cov}(z, z^*) = 0.021$. Because it is difficult to imagine that six arbitrarily chosen enzymes have a pleiotropic effect on call, this covariance is likely to be due to linkage disequilibria. Assuming the loci to be unlinked (r = 0.5), the covariance can be explained by a dispersal rate $\sigma = \sqrt{2r \ \text{cov}(z,z^*)} ww^*/\Delta z \Delta z^* = 0.80$ km gen$^{-1/2}$. This value is close to the value estimated from covariances among the enzyme markers. Belly pattern also differs between the taxa and is correlated with both mating call and enzyme genotype to about the same extent. These covariances between unrelated characters support the idea that linkage disequilibria are built up primarily by dispersal, rather than by epistatic selection favoring particular combinations of alleles and traits.

VARIANCE OF QUANTITATIVE TRAITS

Both high heterozygosity and positive linkage disequilibria contribute to the variance in the hybrid index based on enzyme genotype—by definition, an additive and completely heritable trait (Fig. 2-4). In general, both these effects cause an increase in the variance of any quantitative trait in a hybrid zone. In addition, the nongenetic "environmental" variance may be greater if hybrids have reduced developmental stability. Consider a specific example: the variance in log(cycle length) in *Bombina*, which increases from an average of 0.029 outside the hybrid zone to 0.088 in the central Kopanka sample. Some of this increase can be ascribed to linkage disequilibria. To determine how much, suppose for the moment that the difference in trait between the two populations is entirely due to loci that are fixed for alternative alleles in the two races and that change in parallel.[1] Some variance is nongenetic (V_e), and some may be due to loci that are polymorphic in both races (V_{g0}); this amount remains constant across the hybrid zone. At the center, $p_i q_i = \frac{1}{4}$; if linkage disequilibria are generated by dispersal, for a given degree of linkage they are the same for all pairs of loci ($\overline{D}$ on average). Then, from Eq. 2b:

$$\text{var}(z) = \sum_{i=1}^{n} 2\alpha_i^2 \left(\frac{1}{4} - \overline{D} \right) + \sum_{i,j=1}^{n} 2\alpha_i\alpha_j\overline{D} + V_{g0} + V_e \qquad (5a)$$

Because the difference between the races is $\Delta z = 2\Sigma\alpha_i$, the variance at the center is:

$$\text{var}(z) = \frac{\Delta z^2}{2n_e} \left(\frac{1}{4} - \overline{D} \right) + \frac{\Delta z^2}{2} \overline{D} + V_{g0} + V_e \qquad (5b)$$

The increase in variance is due partly to increased heterozygosity (first term) and partly to linkage disequilibria (second term). The effect of heterozygosity decreases inversely with the effective number of loci contributing to the difference between races, defined as $n_e = (\Sigma\alpha_i)^2/(\Sigma\alpha_i^2)$.

In the *Bombina* example, the consistent covariances between different traits suggest that $\overline{D}$ can be estimated from the disequilibrium between diagnostic enzyme markers: $\overline{D} = 0.037$. Hence $\Delta z^2\overline{D}/2 = 0.027$ of the increase in variance of 0.058 can be ascribed to linkage disequilibria. If the remaining increase of 0.032 is due to increased heterozygosity at diagnostic loci, the effective number of loci responsible for the difference in call must be small: $n_e \approx (0.25 - 0.037)/(2 \times 0.032) = 3.3$ loci (San-

---

1. The assumption that alternative alleles are fixed in the two races is not essential. If we take $\overline{D}$ to be the average pairwise disequilibrium between alleles at *diagnostic* loci ($\Delta p = 1$), the disequilibrium between alleles at loci that contribute less to the difference is $D_{ij} = \Delta p_i \Delta p_j \overline{D}$; the second term in Eq. 5b still applies, but the effective number of genes contributing to the trait is now defined as $(\Sigma\alpha_i\Delta p_i)^2/(\Sigma(\alpha_i\Delta p_i)^2)$, and ($\frac{1}{4} - \overline{D}$) is replaced by the average of $[(pq/\Delta p^2) - \overline{D}]$ at the center (the average being weighted by $\alpha\Delta p$).

derson et al., 1991). This is an application of the Castle/Wright/Lande method of estimating gene numbers (Castle, 1921; Lande, 1981), which takes explicit account of
linkage disequilibria. It suffers from the same disadvantages: primarily, sensitivity to
dominance and epistasis, and poor statistical power. Moreover, this particular example should be taken only as an illustration, as the excess variance in mating call may
be due to decreased stability of hybrids. Nevertheless, the calculation shows how measurements of genetic variance in quantitative traits across hybrid zones might give
information about the genetic basis of quantitative variation. In the next sections, we
see how the increase in genetic variance affects the shape of a hybrid zone and the rate
of gene flow between the hybridizing taxa.

LONG-RANGE MIGRATION

If linkage disequilibrium is generated by the diffusion of genes from place to place,
it should be proportional to the gradients in gene frequency [$D_{ij} \approx (\sigma^2/r)(\partial p_i/\partial x)$
$(\partial p_j/\partial x)$]. It should therefore be lower at the edge than at the center. This pattern is
seen with *Bombina;* even after linkage disequilibrium is standardized relative to allele
frequencies ($R = D/\sqrt{pquv}$), it is lower at the edges (Fig. 2-5). However, there is much
more linkage disequilibrium at the edges than would be expected from the shallow
gradients there (dotted line in Fig. 2-5). The problem is that if genes move in a series
of small steps, they would take many generations to diffuse from one side of the hybrid
zone across to the other. Associations between unlinked alleles decrease by half each
generation and so cannot be preserved for this long. These associations could be due
to selection; epistasis produces linkage disequilibrium at a rate proportional to the
product of allele frequencies (pquv) (Barton and Turelli, 1991), and so the standardized disequilibrium should be proportional to $pquv/\sqrt{pquv} = \sqrt{pquv} \approx pq$, as is
observed. However, because linkage disequilibria at the edge of the zone are weak, this

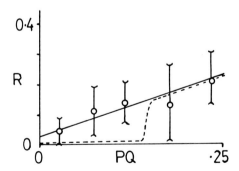

Figure 2-5. Average standardized linkage disequilibrium ($R = D/\sqrt{pquv}$), plotted against the
product of average allele frequencies, $\bar{p}\bar{q}$, across the hybrid zone in *Bombina* in southern Poland.
The edges of the hybrid zone are at the left, and the center is to the right. The dotted line shows
the standardized linkage disequilibrium that would be generated solely by local dispersal; the
solid line is the linear regression through the observed values. (From Szymura and Barton, 1991.
With permission.)

selection would have to act directly on the enzyme loci, which seems unlikely. The most plausible explanation is that a few toads move long distances, thus taking intact sets of genes across the hybrid zone. The observed linkage disequilibria could be accounted for if about one per thousand toads at the edge comes from the other side of the hybrid zone. This hypothesis is consistent with the small fraction of foreign parental and F_1 genotypes: 4 of 1448 toads sampled between 6 and 40 km from the center (Szymura and Barton, 1991). Linkage disequilibrium can thus give information about the distribution of dispersal distances as well as the net rate.

CLINE SHAPE AND BARRIERS TO GENE FLOW

We argued above that selection on one gene, or one quantitative trait, should give a smooth sigmoid cline, which appears as a straight line when plotted on a logit scale (Fig. 2-1). Selection favoring particular combinations of genes can produce a distorted cline, but only with extreme interactions among several loci and with selection that acts directly on the observed loci (Fig. 2-2). However, these arguments neglect the linkage disequilibria that are generated by dispersal and are found in many hybrid zones. Because selection on one locus causes changes at all the loci in linkage disequilibrium with it, the "effective" selection experienced by each locus may be greatly increased; even neutral loci are affected by selection on loci with which they are associated. This increase in selection causes clines to become steeper, which in turn increases the linkage disequilibrium. There is a positive feedback, which produces a sharp step flanked by shallow tails of introgression (Barton, 1983).

This pattern is seen in many hybrid zones, for example, in *Bombina* (Fig. 2-6), *Ranidella* (Blackwell and Bull, 1978), *Uroderma* (Baker, 1981), *Caledia* (Moran et al., 1980), *Mus* (Hunt and Selander, 1973), and in some places *Podisma* (Currie, 1992; Jackson, 1992). Because a stepped pattern can be shown to be statistically significant only if there are many samples from both the center and edges of a continuous transect, it is difficult to judge its prevalence. It is also difficult to know whether stepped clines are caused by linkage disequilibria or, more simply, by either a physical obstacle to gene flow or long-range migration. With *Bombina* these possibilities can be distinguished. There is no evidence of any physical barrier at either of the two transects studied in detail (Fig. 2-6), and the similarity between these transects speaks against this possibility. There are some long-distance migrants, but they are rare and produce F_1 and F_2 progeny that are substantially less fit and so do not contribute significantly to introgression (Szymura and Barton, 1991).

A stepped pattern, in which most change occurs in a narrow cline at the center and yet foreign alleles introgress long distances, reflects a barrier to gene exchange. This situation is true whether it is produced by a physical obstacle or by linkage disequilibria with loci under selection. The strength of the barrier is measured by the ratio between the step in allele frequency and the gradient at the edge [$B = \Delta p/(\partial p/\partial x)$] (Nagylaki, 1976b). This ratio has the dimensions of a distance and can be thought of as the length of unimpeded habitat that would pose the same obstacle to the flow of a neutral allele. For *Bombina,* the barrier is substantial ($B = 51$ km for flow into *B. variegata;* support limits 22–81 km) and could delay neutral alleles for a few thousand generations ($T \approx (B/\sigma)^2$). However, local barriers of this sort cannot much delay the

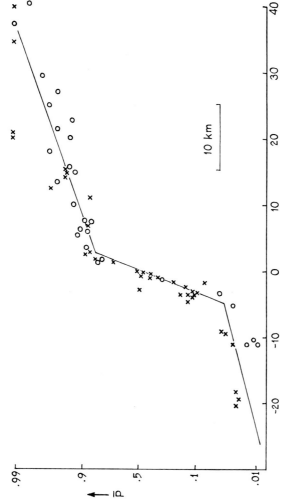

Figure 2-6. Stepped clines seen across two transects through the hybrid zone between *Bombina Bombina* and *B. variegata* in southern Poland (circles = Przemysl; crosses = Krakow). The graph shows the average frequency of *B. variegata* alleles at six diagnostic enzyme loci, plotted on a logit scale. The strength of the barrier to gene flow into *B. variegata* is 51.2 km (2 unit support limits 22–81 km); the ratio between the gradients at the center and at the edge is 0.11 (limits 0.09–0.21). (From Szymura and Barton, 1991. With permission.)

spread of an advantageous allele because once a few copies recombine into the new genetic background they can spread rapidly ($T \approx \log[(B/\sigma)^2 \pi S/2]/2S$, where S = the selective advantage). The observed barrier cannot even account for the presence of fixed allozyme differences between *B. bombina* and *B. variegata*, suggesting that some weak selection must act to maintain these differences ($s \approx 0.37\%$) (Szymura and Barton, 1991).

There is a remarkably general relation between the barrier to gene flow and the net strength of selection maintaining the hybrid zone. Consider clines in loci or traits that are neutral or, at least, are under selection weaker than that induced by linkage disequilibria. The ratio between the gradients at the center and the edge is:

$$\left\{ \frac{(\partial p/\partial x)_{edge}}{(\partial p/\partial x)_{center}} \right\} = w \left\{ \frac{\partial p}{\partial x} \right\}_{edge} = \left\{ \frac{\overline{W}_{center}}{\overline{W}_{edge}} \right\}^{1/r} \qquad (6)$$

where r = the harmonic mean recombination rate between selected and neutral loci; and w = the width of clines at the neutral loci. If the barrier is strong, $\Delta p \approx 1$, and so $B = \Delta p/(\partial p/\partial x)_{edge} \approx w(\overline{W}_{center}/\overline{W}_{edge})^{-1/r}$. The key assumptions when deriving this relation are that selection is weak and that it is not frequency-dependent. Equation 6 then applies exactly to clines maintained by weak selection against hybrids and to a good approximation to clines maintained by adaptation to different environments (Barton, 1986). It can be applied to find the mean fitness of hybrid populations without the need to know just how selection is acting. With *Bombina*, for example, the number of chromosomes and chiasmata implies a mean recombination rate of $r \approx 0.25$; the mean fitness required to explain the observed barrier is 0.58 (support limits 0.54–0.68).

Because Eq. 6 was derived on the assumption that selection is weak, it may not be safe to extrapolate to the strong selection pressures needed to produce strong barriers. We have checked the predictions against simulations of clines maintained by heterozygote disadvantage and by epistasis (Fig. 2-7). Selection acts on 16 loci, spaced evenly along a single chromosome. These loci alternate with 16 neutral marker loci; the recombination rate between adjacent loci is 5%, giving a map length of $31 \times 0.05 = 1.55$ Morgans. This map is much shorter than is usual; however, because the barrier increases with the number of loci but decreases with map length, this choice may give results typical of selection on more loci.

The observed ratio between the gradients at center and edge fits reasonably well with theory (shown by the solid line) when selection is weak. Although there are apparently some significant deviations, the confidence limits shown on the graphs may be too narrow because the frequency of the neutral alleles changes slowly as a result of random drift, so that successive estimates are somewhat autocorrelated. When selection is strong, allele frequencies change abruptly between adjacent demes, and so all demes are close to fixation for one or other parental genotype. Then, as selection increases further, the frequency of introgressing alleles decreases, so the net load on the population remains constant: the mean fitness never declines below about 0.5. Because migration has an effect similar to that of mutation, one would expect, by analogy with the mutation load, that mean fitness would be approximately $1 - m$ and to depend only weakly on the nature of selection. However, we have not been able to find a simple expression for the "migration load"; this lack is to be expected because with epistasis there is no analogous expression for the mutation load in a sexually repro-

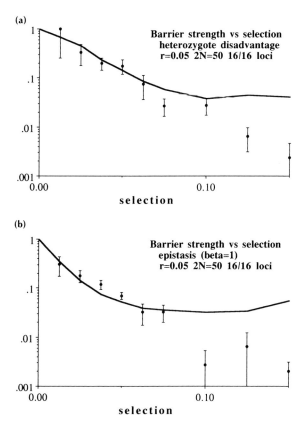

Figure 2-7. Ratio between the gradient in allele frequency at neutral loci at the edge of the hybrid zone and at the center, plotted against selection on each locus: a small ratio indicates a strong barrier to gene flow. (a) Selection acts against heterozygotes (b). There is epistasis, with $\beta = 1$. Observed values are from simulations of 16 selected loci, alternating with 16 neutral markers, with recombination of r = 0.05 between adjacent loci. There are 40 demes (except for s = 0.025, where there are 80). Statistics are calculated every 50 generations from generations 400 to 800; and the slopes on either side are averaged. The solid line gives the predicted ratio, $\overline{W}^{1/r}$; the harmonic mean recombination rate (r) between neutral and selected loci is here 0.1782.

ducing population (Kondrashov, 1988). The barrier to gene flow continues to become stronger with increasing selection against hybrids, even though the mean fitness does not decrease further. With *Bombina,* the observed ratio of gradients is 0.11 (limits 0.09–0.21); in this range, our simulations show that there is good agreement between simulations and theory.

   Even where a barrier to gene flow is primarily caused by a physical obstacle, it may be augmented by selection. With the alpine grasshopper, *Podisma pedestris,* two chromosomal races are separated in most places by smooth sigmoid cline about 800 meters wide. However, at Lac Autier and near Col de la Lombarde, the pattern is sharply stepped (Currie, in prep.; Jackson, 1992) (Figs. 2-8 and 2-9a). Both steps are caused in part by small streams, which impede movement. However, the numbers of marked animals seen moving across the streams are too high to be consistent with the

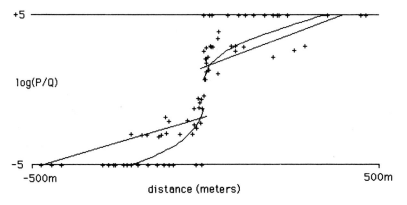

Figure 2-8. Frequency of the chromosomal fusion that distinguishes two races of the grasshopper *Podisma pedestris* across transects at Col de la Lombarde, in the Alpes Maritimes, plotted on a logit scale. (Data courtesy of F. Cox, C. Raboud, and D. Currie; see Jackson, 1991). In most places, the cline is smooth and would give a single straight line on this scale. Here, however, the frequency changes in a sharp step across a stream. This step could simply be due to a physical barrier to dispersal, disrupting a cline maintained by selection against chromosomal heterozygotes (straight lines). However, such change requires a physical barrier an order of magnitude stronger than is measured directly from grasshopper movements. A model in which the physical barrier is augmented by selection against introgressing alleles at 100 loci predicts a further change just near the stream (curved lines). A model including selection on linked loci using direct measures of barrier strength (B = 95 meters) and the width seen in an adjacent, unimpeded transect (w = 600 meters), is substantially more likely than a model with no genetic barrier, the same width, and a strong barrier: B = 950 meters for flow to the left, 1535 meters to the right; $\Delta\log(L)$ = 10.53. It implies a ratio between selection and recombination of S/R = 0.89 (2-unit support limits 0.68–1.24) against introgression to the left and 0.55 (0.45–0.072) against introgression to the right; the net barrier would then be 7400 meters on the left, and 2360 meters on the right.

observed barrier: the strength of the physical barrier to movement is estimated to be 150 meters at Lac Autier, and 95 meters at Col de la Lombarde, whereas the barrier strengths estimated from cline shape are significantly stronger; averaging over movement in either direction, the estimates are 1.70 and 1.24 km, respectively.

The most likely explanation of this discrepancy is that selection against hybrids produces a genetic barrier to gene exchange, which augments the direct physical barrier. Hybrids from laboratory crosses and from nature have viability reduced by about 50% (Barton and Hewitt, 1981b). This finding is supported by a sharp change in frequency immediately adjacent to the stream of Col de la Lombarde (Fig. 2-8). When genes first cross the stream, they find themselves in unfit F_1 and F_2 hybrids, and so are rapidly eliminated. However, as genes recombine into the new genetic background, they experience much weaker selection, primarily due to their own effects on fitness. They can then penetrate well away from the central barrier. This sharp drop adjacent to the physical barrier is statistically significant and implies strong selection against sets of introgressing genes (for details, see Fig. 2-8). At Lac Autier there is no evidence of such a change next to the stream, perhaps because chromosome frequencies here are close to fixation and so give little information. The data are consistent with a weak physical barrier (albeit somewhat stronger than that observed), augmented by selection against hybrids (Fig. 2-9a). It is not yet clear, however, whether the ratio between selection and recombination can be small enough to give a smooth cline in open habitat (as is observed) and yet high enough to reduce significantly the introgression of genes at low frequency across a physical barrier. It depends on the relation between selection against rare foreign alleles relative to the selection on alleles segregating at high frequency in the center of a hybrid zone (Fig. 2-2).

We have derived the reduction of mean fitness in hybrid populations from the size of the step induced in neutral or weakly selected clines. The width of this step gives further information, as it reflects the width of the region in which mean fitness is reduced and hence the width of the clines at the loci under selection. If this region is

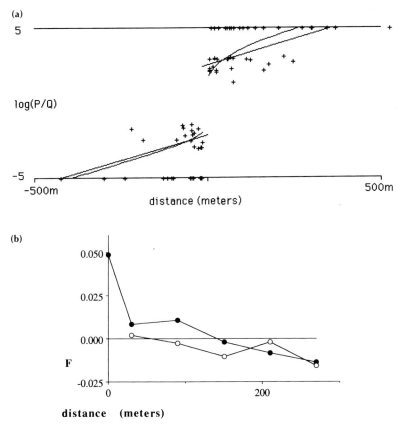

Figure 2-9. (a) Transect at Lac Autier shows a step similar to that shown in Figure 2-8. Here the best fit is to a simple model (straight lines). However, it again requires an unreasonably strong physical barrier (B = 1390 meters for flow to the left, 2050 meters to the right, versus 150 meters measured directly). The data are consistent with a weak physical barrier (lower limits B = 280 meters to the left, 130 meters to the right), augmented by a genetic barrier due to selection on 100 loci (S/R = 0.26 to the left, 0.61 to the right, giving net barriers of 1670 meters on the left and 3870 meters on the right). (b) Covariance between fluctuations in allele frequency at six polymorphic enzyme loci plotted against distance; data are from the hybrid zone in *Podisma pedestris* at Lac Autier. Closed circles = sites on the same side of the stream; open circles = sites on opposite sides. The net barrier estimated from the difference between the two lines is 1.5 km (limits 500 meters to ∞). This finding is consistent with the net barrier strength estimated from the cline in chromosomal fusion (see Figs. 2-8 and 2-9a).

wide relative to the dispersal rate, selection on each locus must be weak; and many loci must be responsible for maintaining the differences between the hybridizing taxa. Consider *Bombina*. The dispersal rate (estimated from linkage disequilibria at the center of the hybrid zone) *is* $\sigma = 0.99$ km gen$^{-1/2}$, and the width of the enzyme clines averages w = 6.05 km. If each cline were maintained by selection against heterozygotes, $w = \sigma\sqrt{8/s}$, and so = 22%. At the center of the hybrid zone, half the individuals would be heterozygous at any one of the n loci, and so the mean fitness would be $\overline{W}$ = $(1 - s)^{n/2} \approx \exp(-ns/2)$. The cline shape gave an estimate of $\overline{W} = 0.58$, implying ns = 1.09, and hence n $\approx$ 5 loci.

This calculation grossly underestimates the number of loci involved because it neglects the strong linkage disequilibria that are known to be present and that greatly steepen clines at multiple loci. This fact can be seen in Figure 2-3, where strong selection produces steeper clines and hence greater disequilibrium than expected from single-locus calculations. The selection coefficient s = 22% is not the selection acting directly on each enzyme locus; rather, it is the total selection acting on all loci in linkage disequilibrium with the observed marker (s $\approx$ Σ $s_i D_{Mi}/p_M q_M$, where M is the marker). To account for these effects, one must assume a particular model of selection. The resulting estimate of gene number is therefore less robust than those of dispersal rate and mean fitness. In the *Bombina* example, selection of 2% must act against heterozygotes at 55 loci to give the observed cline shape and linkage disequilibrium (support limits 26–88). This calculation is supported by the similarity of clines at different genes and for different traits and by the extensive biochemical and morphological differences between *B. bombina* and *B. variegata* (Szymura and Barton, 1991).

This estimate has come from a comparison between the dispersal rate (estimated from linkage disequilibrium) and the width of the region in which mean fitness is reduced (inferred from perturbations to marker loci). The same method was applied by Barton and Hewitt (1981b), using direct measurements of dispersal and viability from the grasshopper *Podisma pedestris*. Viability was reduced over a region much wider than the dispersal range (330 meters, versus 20 meters), implying that selection is spread over many loci (n $\approx$ 75, assuming heterozygote disadvantage). The method could be applied to a wide variety of hybrid zones; however, there are as yet few cases that provide sufficient data on cline shape or fitness components across a continuous transect.

FLUCTUATIONS IN CLINE SHAPE

Information can come from random fluctuations, as well as from the average patterns. Wright (1943) showed that in the "island model," in which each deme draws migrants from a common pool, the standardized variance of allele frequence (F_{st} = var(p)/pq) depends primarily on the number of migrants exchanged per generation (Nm). Populations spread over two dimensions are similar: F_{st} depends mainly on Nm (in a continuum, on neighborhood size, $4\pi\rho\sigma^2$,[2] and only weakly on the selection or mutation

---

2. These parameters are often referred to as measures of "gene flow." This terminology is somewhat misleading, as they are the product of population size and dispersal rates and so describe the relative strengths of random drift and gene flow. The shapes of clines and the degree of linkage disequilibrium do not depend on N or ρ in a dense population and are determined primarily by the measures of gene flow, m and σ^2.

that maintains the polymorphism (Wright, 1943; Slatkin and Barton, 1989). This is because random drift ($\approx 1/N$ or $1/\rho$) and gene flow ($\approx$ m or σ^2) quickly reach a balance, which determines the amount of spatial variation. Wright et al. (1942) applied these results to estimate Nm from the frequencies of recessive lethals in *Drosophila pseudoobscura.* Since that early work, much effort has gone into the estimation of Nm and neighborhood size from molecular data.

The same methods can be used to analyze random fluctuations in cline shape. The variance of allele frequency around a cline depends on the product of density and dispersal in the same way as does the variance around a spatially uniform polymorphism, which has been shown rigorously for a one-dimensional population (Nagylaki, 1978); though the pattern of fluctuations is qualitatively different in two dimensions, it should still be similar for a cline and a uniform polymorphism (Rouhani and Barton, 1987, unpublished data). For example, in a transect across the chromosomal cline of *Podisma,* near Tende in the Alpes Maritimes, the neighborhood size estimated from fluctuations in chromosome frequency is 135 (limits 70–310), compared with an estimate of 335 (265–440) from five polymorphic enzyme loci in the same area (Barton and Hewitt, 1981a; Halliday et al., 1983; unpublished data). There is rather more variation in cline shape than would be expected from fluctuations in uniform polymorphisms.

With *Bombina,* there are data from several concordant clines, and so allele frequency variances can be estimated either from differences in frequency between loci within sites or from fluctuations in the average frequency beween sites (see Szymura and Barton, 1991). After allowing for consistent differences in position and width of clines at different loci, there is little residual variation (F_{st} = 0.015; limits 0.012–0.019). However, there is significantly more variation in average frequency from place to place than would be expected from the close concordance across loci. This is because some sites are consistently more *B. variegata*-like, and others are more *B. bombina*-like. Some correlation between deviations at different loci is to be expected as a result of drift alone, as there is linkage disequilibrium between all loci and covariance between diagnostic metric traits. However, the pattern is too strong to be explained in this way. It could be caused either by sporadic colonization (which might also account for the discrepancy in estimates from *Podisma*) or by selection favoring *B. variegata*-like alleles in *B. variegata*-like habitat, and vice versa.

These possibilities could be distinguished through fluctuations in linkage disequilibria. Recent colonization should generate both strong disequilibrium and concordant deviations in allele frequency. There is a suggestion that this is so for *Bombina.* Among the 33 central populations, linkage disequilibrium (measured by R/pq) (see Fig. 2-5) is correlated with fluctuations in the cline $[(p_{obs} - p_{exp})^2/(p_{exp}q_{exp})]$, though not significantly (Kendall's rank correlation is 0.167; P = 9%). A simpler way of explaining fluctuations would be to look for correlations with relevant environment variables. For example, within the hybrid zone between the crickets *Gryllus firmus* and *G. pennsylvanicus* there is a strong association between soil type and genetic and morphological traits, giving a "mosaic" distribution (Rand and Harrison, 1989).

Variation in allele frequency is usually described by a single statistic, F_{st}, which is then used to estimate Nm or neighborhood size. However, density and dispersal vary from place to place, and so a single measure may be misleading. For example, the original aim of estimating population structure was to determine if random drift could overcome selection and gene flow, enabling a local shift from one adaptive peak to another (Wright, 1931). However, such shifts may be frequent even when the average neighborhood size is high, provided it is small in an appreciable fraction of the species' range. New adaptive peaks can spread if they are either at a strong advantage over the old peak or the sparse populations in which they arise occasionally expand.

Because heterogeneity in population structure can have important consequences, it would be useful to be able to measure it using spatial patterns at polymorphic markers. It would be particularly useful to find ways of detecting hybrid zones and barriers to gene flow. Sokal and his collaborators (1990) have developed a variety of statistical techniques for analyzing spatial patterns and have used them to reconstruct historical changes in European populations from the extensive genetic data available for humans. Sharp changes in gene frequency coincide, in most cases, with language boundaries in Europe.

A more detailed analysis is possible if one has a definite model of the processes responsible for spatial patterning. The variance in frequency of polymorphic alleles, and the covariance between frequencies in different places, depend on the balance between drift and gene flow (through Nm or neighborhood size) and on the balance between gene flow and the force maintaining the polymorphism (through $\sigma/\sqrt{2\mu}$ if the alleles are neutral and are maintained by recurrent mutation, μ). The effect of a barrier on this pattern can be detected: for a given geographic separation, the covariance between sites on the same side of the barrier is greater than that between sites on different sides (Pounds & Jackson, 1981). For *Podisma,* this pattern is found for allele frequencies at six polymorphic enzyme loci, across the hybrid zone at Lac Autier (Fig. 2-9b). Simulations show that a difference of this size is unlikely to arise by chance if there is in fact no barrier; the best estimate of the barrier strength is ≈ 1.5 km, which is consistent with the shape of the chromosomal cline and much stronger than indicated by direct observations of movement across the stream (Currie, in prep.).

ANALYZING GENE TREES

The examples we have used to illustrate our analytical methods have involved morphological, chromosomal, and electrophoretic variation. What more can new molecular techniques tell us about the evolution of hybrid zones? One advantage is that synonymous changes, or variation in noncoding regions, are likely to be neutral, or nearly so, and so provide a reliable baseline against which to compare traits of evolutionary significance. For example, we have argued that weak selection maintains the fixed allozyme differences that distinguish B. bombina from B. variegata; this hypothesis could be tested by comparison with sequences known to be neutral. However, genotype frequencies within hybrid zones may be little affected by weak selection, as is evidenced by the concordance across traits in B. bombina. Having large numbers of diagnostic markers does not help in such cases, as strong selection maintains linkage disequilibria among all the divergent loci. Analysis of any set of markers provides information about the net level of selection and gene flow over the whole genome but not about the markers themselves. If selection is weaker relative to recombination, associations between unlinked loci are weaker; and so more markers become valuable. Disequilibrium may then be detected between linked loci, allowing dispersal and selection to be estimated; and the widths of clines may vary, reflecting different selection on different loci or traits.

There is a great deal of interest at present in mapping the loci responsible for quantitative variation through associations between Mendelian markers and polygenic traits. This approach cannot be applied directly to samples from hybrid zones, as associations may be caused by linkage disequilibrium. However, any associations seen *within* families must be due to linkage. Use of such families has the advantage that it requires only one generation of crossing, rather than two.

Whereas linkage disequilibria may be strong near the center of hybrid zones, making extra markers redundant, they should become weaker as sets of alleles introgress into a new genetic background. If closely linked markers are available, one could date introgressing alleles by finding whether they are still associated with neighboring alleles; the average length of an introgressing block of DNA is exponentially distributed with mean $1/t$ Morgans. Analysis would be most efficient if only those rare individuals found carrying foreign alleles at a chosen locus were sequenced.

Given enough closely linked markers, one can reconstruct the phylogeny that connects all the genes in a sample. If this "gene tree" can be estimated reliably, it is the most appropriate way of summarizing the data. To obtain the same information in the traditional way, one would need to include all the higher-order disequilibria. Techniques for estimating population genetic parameters from phylogenies of alleles are only just being developed. Slatkin and Maddison (1989, 1990) have shown that the numbers of migrants between demes (Nm) can be estimated by regarding spatial location simply as another character, which changes randomly, in the same way that mutation (μ) causes random changes in sequence. Just as the diversity of sequence within a population gives an estimate of $N\mu$, so the diversity of locations among related genes gives an estimate of Nm. Slatkin and Maddison (1989) showed that their method obtains results of accuracy comparable to the results obtained using the standardized variance of allele frequency (F_{st}). However, they argued that phylogenetic information might allow different kinds of gene flow to be distinguished, for example, sporadic extinction and recolonization versus steady dispersal.

In the context of hybrid zones, we can ask whether gene flow between hybridizing taxa could be detected more readily from gene trees. The best understood example is the introgression of mitochondrial DNA from *Mus musculus domesticus* into *M. m. musculus*. Though the morphological, enzyme, and nuclear DNA sequence all change together across the hybrid zone in Jutland, an mtDNA variant derived from *M. m. domesticus* is found in apparently *M. m. musculus* populations in southern Sweden (Ferris et al., 1983, Sage et al., 1990). Mitochondrial DNA does change across this hybrid zone, but this difference involves a more recently derived variant (Vanlerberghe et al., 1988). This pattern suggests that introgression has occurred through an ancient founder event, rather than through persistent introgression across the present hybrid zone.

This example, together with the general finding that mitochondrial DNA introgresses more readily than differences coded by the nuclear genome (Barton and Hewitt, 1989) shows that phylogenetic information from DNA sequences can tell us about the nature of barriers to gene flow. An objection to this approach is that it tells us about the history of only one or a few loci and so cannot give a good estimate of the general rate of gene exchange. An allele that happened to cross the barrier in the distant past can leave many descendants, giving an inflated estimate of gene flow. Ideally, one would like good trees from many loci, together with a theory for interpreting the variation among these trees. A serious practical difficulty is that once one moves away from mitochondrial and chloroplast genomes to nuclear sequences, recombination may make it impossible to reconstruct accurate phylogenies.

The most fruitful application of phylogenetic information may be in disentangling the history of groups of many hybridizing taxa, such as the chromosomal races of mice and shrews in Europe, the warning color races of *Heliconius* butterflies in South America, or the diverse human races. There is a continuum here between taxa so distinct that introgression is negligible, through to populations where clines for different loci or traits are scattered independently. In the latter case, phylogenies of different loci vary, so that though one may be able to draw trees describing the relatedness of populations, these must be an average over many characters—as, for example, with human populations in Europe (Sokal et al., 1990). It is certainly possible to describe the history of such groups; what is not yet clear is whether it is feasible to go beyond that and describe the rates of the various processes responsible for the diversification of populations: the rate of establishment of new "adaptive peaks" and hybrid zones, their rates of expansion and contraction, and the rate at which hybridizing taxa develop into coexisting species.

In this chapter, we have shown how the study of hybridization between pairs of populations can tell us about the processes that keep them distinct and that presumably form the basis for separating full species. We hope that in the future comparable methods will be developed for analyzing data from parapatric groups of taxa, and that these methods will tell us about processes at a higher level.

ACKNOWLEDGMENTS

This work was supported by grants from the N.E.R.C. (GR/6864A) and S.E.R.C. (GR/E/08507) and by the Darwin Trust. We are grateful to C. MacCallum for supplying simulations of the sampling properties of disequilibrium estimates; D. Currie, F. Cox, and C. Rabaud for data on the cline in *Podisma;* and D. Currie for simulations of fluctuations across a barrier.

REFERENCES

Baker, R. J. 1981. Chromosome flow between chromosomally characterized taxa of a volant mammal, *Uroderma bilobatum* (Chiroptera: Phyllostomatidae). Evolution 35:296–305.

Barton, N. H. 1982. The structure of the hybrid zone in *Uroderma bilobatum* (Chiroptera: Phyllostomatidae). Evolution 36:863–866.

Barton, N. H. 1983. Multilocus clines. Evolution 37:454–471.

Barton, N. H. 1986. The effects of linkage and density-dependent regulation on gene flow. Heredity 57:415–426.

Barton, N. H. and Hewitt, G. M. 1981a. A chromosomal cline in the grasshopper *Podisma pedestris.* Evolution 35:1008–1018.

Barton, N. H., and Hewitt, G. M. 1981b. The genetic basis of hybrid inviability between two chromosomal races of the grasshopper *Podisma pedestris.* Heredity 47:367–383.

Barton, N. H., and Hewitt, G. M. 1985. Analysis of hybrid zones. Ann. Rev. Ecol. Syst. 16:113–148.

Barton, N. H., and Hewitt, G. M. 1989. Adaptation, speciation and hybrid zones. Nature 341:497–503.

Barton, N. H., and Turelli, M. 1991. Natural and sexual selection on many loci. Genetics 127:229–255.

Bazykin, A. D. 1973. Population genetic analysis of disrupting and stabilising selection. II. Systems of adjacent populations and populations within a continuous area. Genetika 9:156–166.

Blackwell, J. M., and Bull, C. M. 1978. A narrow hybrid zone between the western Australian frog species, Ranidella insignifera and R. pseud. insignifera: the extent of introgression. Heredity 40:13–25.

Bulmer, M. G. 1980. The Mathematical Theory of Quantitative Genetics. Oxford: Oxford University Press.

Castle, W. E. 1921. An improved method of estimating the number of genetic factors concerned in cases of blending inheritance. Science 54:223.

Crosby, J. L. 1970. The evolution of genetic discontinuity: computer models of the selection of barriers to interbreeding between subspecies. Heredity 25:253–297.

Edwards, A. W. F. 1972. Likelihood. Cambridge: Cambridge University Press.

Ferris, S. D., Sage, R. D., Huang, C.-M., Nielsen, J. T., Ritte, U., and Wilson, A. C. 1983. Flow of mitochrondrial DNA across a species boundary. Proc. Natl. Acad. Sci. USA 80:2290–2294.

Hafner, J. C. 1982. Genetic interaction at a contact zone of Uroderma bilobatum (Chiroptera: Phyllostomatidae). Evolution 36:852–866.

Hafner, J. C., Hafner, D. J., Patton, J. L., and Smith. M. F. 1983. Contact zones and the genetics of differentiation in the pocket gopher Thomomys bottae (Rodentia: Geomyidae). Syst. Zool. 32:1–20.

Halliday, R. B., Barton, N. H., and Hewitt, G. M. 1983. Electrophoretic analysis of a chromosomal hybrid zone in the grasshopper Podisma pedestris. Biol. J. Linn. Soc. 19:51–62.

Harrison, R. G., and Rand, D. M. 1989. Mosaic hybrid zones and the nature of species boundaries. In D. Otte and J. A. Endler (eds.). Speciation and its Consequences. New York: Sinauer Press, pp. 111–133.

Hewitt, G. M. 1988. Hybrid zones: natural laboratories for evolutionary studies. Trends Ecol. Evol. 3:158–167.

Hill, W. G. 1974. Estimation of linkage disequilibrium in randomly mating populations. Heredity 33:229–239.

Hunt, W. G., and Selander, R. K. 1973. Biochemical genetics of hybridisation in European house mice. Heredity 31:11–33.

Jackson, K. S. 1992. The population dynamics of a hybrid zone in the alpine grasshopper Podisma pedestris: an ecological and genetical investigation. Ph.D. thesis, University College London.

Karlin, S., and Richter-Dyn, N. 1976. Some theoretical analyses of migration-selection interaction in a cline: a generalized two-range environment. In S. Karlin and E. Nevo (eds.). Evolutionary Processes and Theory. Orlando, FL: Academic Press, pp. 659–706.

Key, K. H. L. 1968. The concept of stasipatric speciation. Syst. Zool. 17:14–22.

Kimura, M. 1953. "Stepping stone" model of population. Ann. Rep. Natl. Inst. Genet. Jpn. 3:62–63.

Kocher, T. D., and Sage, R. D. 1986. Further genetic analyses of a hybrid zone between Leopard frogs (Rana pipiens complex) in Central Texas. Evolution 40:21–33.

Kondrashov, A. S. 1988. Deleterious mutations and the evolution of sexual reproduction. Nature 336:435–441.

Lande, R. 1981. The minimum number of genes contributing to quantitative variation between and within populations. Genetics 99:541–553.

Lewontin, R. C. 1988. On measures of gametic disequilibrium. Genetics 120:849–852.

Li, W. H., Nei., M. 1974. Stable linkage disequilibrium without epistasis in subdivided populations. Theor. Pop. Biol. 6:173–183.

Malecot, G. 1948. Les Mathematiques d'Heredite [in French]. Paris: Masson et Cie.

Mallet, J. L. B. 1986. Hybrid zones of Heliconius butterflies in Panama and the stability and movement of warning colour clines. Heredity 56:191–202.

Mallet, J. L. B., and Barton, N. H. 1989. Inference from clines stabilized by frequency-dependent selection. Genetics 122:967–976.

Mallet, J. L. B., Barton, N. H., Lamas, G. M., Santisteban, J. C., Muedas, M. M., and Eeley, H. 1990. Estimates of selection and gene flow from measures of cline width and linkage disequilibrium in Heliconius hybrid zones. Genetics 124:921–936.

Moore, W. S. 1977. An evaluation of narrow hybrid zones in vertebrates. Rev. Biol. 52:263–278.

Moran, C., Wilkinson, P., and Shaw, D. D. 1980. Allozyme variation across a narrow hybrid zone in the grasshopper Caledia captiva. Heredity 44:69–81.

Nagylaki, T. 1975. Conditions for the existence of clines. Genetics 80:595–615.

Nagylaki, T. 1976a. The evolution of one- and two-locus systems. Genetics 83:583–600.

Nagylaki, T. 1976b. Clines with variable migration. Genetics 83:867–886.

Nagylaki, T. 1978. Random genetic drift in a cline. Proc. Natl. Acad. Sci. USA 75:423–426.

Nichols, R. A., and Hewitt, G. M. 1986. Population structure and the shape of a chromosomal cline between two races of Podisma pedestris (Orthoptera: Acrididae). Biol. J. Linn. Soc. 29:301–316.

Pounds, J. A., and Jackson, J. F. 1981. Riverine barriers to gene flow and the differentiation of fence lizard populations. Evolution 35:516–528.

Rand, D. M., and Harrison, R. G. 1989. Ecological genetics of a mosaic hybrid zone: mitochondrial, nuclear, and reproductive differentiation of crickets by soil type. Evolution 43:432–449.

Rouhani, S., and Barton, N. H. 1987. Speciation and the "shifting balance" in a continuous population. Theor. Pop. Biol. 31:465–492.

Sage, R. D., Prager, E. M., Tichy, H., and Wilson, A. C. 1990. Mitochondrial DNA variation in house mice, *Mus domesticus* (Rutty). Biol. J. Linn. Soc. 41:105–123.

Sanderson, N., Szymura, J. M., and Barton, N. H. 1991. Variation in mating call across the hybrid zone between the fire-bellied toads *Bombina bombina* and *B. variegata.* Evolution 46:595–607.

Shaw, D. D., Coates, D. J., Arnold, M. J., and Wilkinson, P. 1985. Temporal variation in the chromosomal structure of a hybrid zone and its relationship to karyotype repatterning. Heredity 55:293–307.

Slatkin, M. 1973. Gene flow and selection in a cline. Genetics 75:733–756.

Slatkin, M. 1978. Spatial patterns in the distribution of polygenic characters. J. Theor. Biol. 70:213–228.

Slatkin, M., and Barton, N. H. 1989. A comparison of three indirect methods for estimating average levels of gene flow. Evolution 43:1349–1368.

Slatkin, M., and Maddison, W. P. 1989. A cladistic measure of gene flow inferred from the phylogenies of alleles. Genetics 123:603–613.

Slatkin, M., and Maddison, W. P. 1990. Detecting isolation by distance using phylogenies of alleles. Genetics 126:249–260.

Sokal, R. R., Oden, N. L., Legendre, P., Foster, M.-J., Kim, J., Thomson, B. A., Vauder, A., Harding, R. M., and Barbujani, G. 1990. Genetics and language in European populations. Am. Naturalist 135:157–175.

Szymura, J. M. 1988. Zroznocowanie geograf-

iczne i strefy mieaszancowe kumakow *Bombina bombina* (L.) i *Bombina variegata* (L.) w Europie. Rozprawy habilitacyjne 147. Univ. Jagiellonski [in Polish].

Szymura, J. M., and Barton, N. H. 1986. Genetic analysis of a hybrid zone between the fire-bellied toads *Bombina bombina* and *B. variegata,* near Cracow in Southern Poland. Evolution 40:1141–1159.

Szymura, J. M., and Barton, N. H. 1991. The genetic structure of the hybrid zone between the fire-bellied toads *Bombina bombina* and *B. variegata:* comparisons between transects and between loci. Evolution 45:237–261.

Szymura, J. M., and Farana, I. 1978. Inheritance and linkage analysis of five enzyme loci in interspecific hybrids of toadlets, genus *Bombina.* Biochem. Genet. 16:307–319.

Vanlerberghe, F., Boursot, P., Nielsen, J. T., and Bonhomme, F. 1988. A steep cline for mitochondrial DNA in Danish mice. Genet. Res. (Camb.) 52:185–193.

Weir, B. S. and Cockerham, C. C. 1989. Analysis of disequilibrium coefficients. In W. G. Hill and T. F. C. Mackey (eds.). Evolution and Animal Breeding. Wallingford, Oxon: C.A.B. International, pp. 45–52.

Wright, S. 1931. Evolution in Mendelian populations. Genetics 16:97–159.

Wright, S. 1943. Isolation by distance. Genetics 28:114–138.

Wright, S. 1968. Evolution and the Genetics of Populations. I. Genetic and Biometric Foundations. Chicago: University of Chicago Press.

Wright, S., Dobzhansky, Th., and Hovanitz, W. 1942. Genetics of natural populations. VII. The allelism of lethals in the third chromosome of *Drosophila pseudoobscura.* Genetics 27:363–394.

APPENDIX 1: GENERATION OF LINKAGE DISEQUILIBRIUM BY DISPERSAL

To illustrate how dispersal into a hybrid zone generates associations between alleles, consider two demes (labeled 1 and 2). Two loci segregate in each; one locus carries alleles P and Q at frequencies p and q, and the other carries alleles U and V at frequencies u and v. Associations between alleles are measured by the coefficient of linkage disequilibrium D, which is defined as the difference between the actual frequency of PU gametes, and the frequency expected if the genes combine at random (pu). Gamete frequencies in the two demes immediately after reproduction are as follows.

	Deme 1	*Deme 2*	
QV	$q_1v_1 + D_1$	$q_2v_2 + D_2$	
QU	$q_1u_1 - D_1$	$q_2u_2 - D_2$	(A1)
PV	$p_1v_1 - D_1$	$p_2v_2 - D_2$	
PU	$p_1u_1 + D_1$	$p_2u_2 + D_2$	

We will see that the order of migration, random mating, and recombination have a significant effect on the linkage disequilibrium. The stage in the life cycle at which disequilibrium is measured is also important.

It is simplest to assume, in the first instance, that after meiosis a fraction (m) of the haploid gametes are exchanged between demes; they then combine at random to produce diploid zygotes, which go through meiosis to produce the next generation. The allele frequencies after migration are as follows.

$$p_1^* = (1 - m)p_1 + mp_2 \qquad p_2^* = (1 - m)p_2 + mp_1 \qquad \text{(A2a)}$$

$$u_1^* = (1 - m)u_1 + mu_2 \qquad u_2^* = (1 - m)u_2 + mu_1 \qquad \text{(A2b)}$$

The effect of migration on the frequency of PU gametes is similar:

$$(p_1^* u_1^* + D_1^*) = (1 - m)(p_1 u_1 + D_1) + m(p_2 u_2 + D_2) \qquad \text{(A3a)}$$

$$(p_2^* u_2^* + D_2^*) = (1 - m)(p_2 u_2 + D_2) + m(p_1 u_1 + D_1) \qquad \text{(A3b)}$$

Rearranging, we find the linkage disequilibrium immediately after migration.

$$D_1^* = (1 - m)D_1 + mD_2 + m(1 - m)(p_2 - p_1)(u_2 - u_1) \qquad \text{(A4a)}$$

$$D_2^* = (1 - m)D_2 + mD_1 + m(1 - m)(p_2 - p_1)(u_2 - u_1) \qquad \text{(A4b)}$$

Random mating, followed by recombination, reduces disequilibrium to $D_1^{**} = (1 - r)D_1^*$. The next change in disequilibrium is thus:

$$\Delta D_1 = -rD_1 + m(1 - r)(D_2 - D_1)$$

$$+ m(1 - m)(p_2 - p_1)(u_2 - u_1)(1 - r) \quad \text{(A5a)}$$

$$\Delta D_2 = -rD_2 + m(1 - r)(D_1 - D_2)$$

$$+ m(1 - m)(p_2 - p_1)(u_2 - u_1)(1 - r) \quad \text{(A5b)}$$

The first term describes the reduction in disequilibrium by recombination, the second term the averaging of linkage disequilibria between the two demes, and the third term (the most important) the generation of disequilibrium by the mixing of genomes from populations with different allele frequencies. This model is the same as that analyzed by Li and Nei (1974).

In most cases it is more plausible to assume that diploid individuals migrate, mate at random in their new location, and produce diploid offspring to found the new generation. Because the Wahlund effect causes a deficit in heterozygotes after migration, recombination is less effective at breaking down linkage disequilibrium. The change in disequilibrium (measured in the gamete pool immediately after meiosis) in this case is as follows.

$$\Delta D_1 = -rD_1 + m(1 - r)(D_2 - D_1) + m(1 - m)(p_2 - p_1)(u_2 - u_1) \quad \text{(A6a)}$$

$$\Delta D_2 = -rD_2 + m(1 - r)(D_1 - D_2) + m(1 - m)(p_2 - p_1)(u_2 - u_1) \quad \text{(A6b)}$$

The difference between diploid and haploid migration lies in the third term: with diploid migration, the disequilibrium generated by mixing of different populations is not broken down by recombination in the same generation.

Even in this example, which involves only migration and recombination, the full dynamics are complicated. However, a simple approximation is to assume that the population reaches a "quasiequilibrium" in which linkage disequilibrium generated by migration or other forces is balanced by its loss through recombination. This approach was developed by Nagylaki (1976a), and has been generalized by Barton and Turelli (1991). In this model, the disequilibria in each deme converge; and balancing the first and third terms in Eq. A6 gives:

$$D_1 = D_2 = \frac{m(1 - m)(p_2 - p_1)(u_2 - u_1)}{r} \tag{A7}$$

This method is only approximate, as the population is not in fact in equilibrium. Migration continually erodes the differences between demes. In this case, Eq. A6 can be solved exactly. After the effects of initial conditions have decayed away, the linkage disequilibria converge to:

$$D_1 = D_2 = \frac{m(1 - m)(p_2 - p_1)(u_2 - u_1)}{[r - 4m(1 - m)]} \qquad (r > 4m(1 - m)) \tag{A8}$$

Thus the quasilinkage equilibrium (QLE) approximation of Eq. A7 is accurate if $m \ll r$, in other words, if the effect of recombination is much faster than changes caused by migration.

These values are measured immediately after recombination. If measurements are made on adults, after migration, disequilibria are higher. The QLE approximation is then:

$$D_1 = D_2 = \frac{m(1 - m)(p_2 - p_1)(u_2 - u_1)(1 + r)}{r} \tag{A9}$$

If loci are unlinked, $r = 0.5$, and so disequilibria are 50% greater if measured after migration instead of before. Formulas analogous to Eq. A7 were used in previous analyses (Barton, 1982; Szymura and Barton, 1986, 1991); they apply if measurements are made before dispersal.

Throughout most of this chapter we have concentrated on clines maintained in a stable balance between selection and dispersal. The QLE approximation can still be applied by balancing the disequilibrium generated by selection and dispersal against recombination: it is accurate when selection is weak, and disequilibra are small. [Strictly, results are accurate to order s^2 (Barton, 1986; Barton and Turelli, 1991)]. If one observes neutral markers embedded among selected loci, Eqs. A5 and A6 still apply. However, if the observed loci are themselves selected, and if this selection favors particular combinations of alleles, an additional term must be added. When a pair of clines is maintained purely by epistasis between two loci (i.e., favoring $++$ and $--$) (Bazykin, 1973), numerical calculations show that when selection is weak Eqs. A5 and A6 can underestimate disequilibrium by a factor of two. However, the discrepancy is smaller with other forms of selection and when more loci are involved (Barton, 1983). Mallet and Barton (1989) used simulations to investigate the effect of the order of reproduction, selection, and migration. They found that the weak selection approximations are more accurate when selection follows reproduction, as linkage disequilibria and deviations from Hardy-Weinberg are then lower than they were immediately after migration.

APPENDIX 2: DETAILS OF COMPUTER SIMULATIONS

A fundamental difficulty when simulating multilocus clines is that although we are primarily interested in deterministic processes it is not practicable to simulate populations without including random sampling drift. In principle, one could iterate analytical recurrence relations for the gamete frequencies. However, even with as few as 10 loci, one would need to follow $2^{10} = 1024$ variables, which is barely feasible. If there were no linkage and if all loci had equivalent effects on selection, one could reduce the problem by following the frequencies of gametes carrying 0, 1, ... n "1" alleles; in other words, one could follow the distribution of the hybrid index without needing to keep track of individual genotypes. However, we are interested in the effects of linkage and the interaction between selected and neutral loci. One must therefore make a direct simulation of selection, recombination, and gene flow in a finite population, following the genotypes of each individual. Large numbers of genes can be studied; however, random drift causes substantial fluctuations, necessitating a statistical analysis of the results. A particular difficulty is that we wish to compare simulations with theoretical predictions that are valid for weak selection: yet the effects of drift are most severe when selection is weak.

The simulation was written in Pascal and run on a Macintosh SE/30. A copy is available on request. Because each locus segregates for two alleles (labelled "0" and "1"), each haploid genome (or "haplotype") can be described by a binary number, represented in the computer by a set of 16-bit integers. This method allows us to save memory and time by using built-in binary operations (Crosby, 1970).

Events occur in the following order: movement of haploid gametes, selection, random union of gametes, and meiosis. The population is distributed across a linear chain of demes (Kimura, 1953); a proportion (m) of the gametes from each deme migrates in each generation, divided equally between the two neighbors. This proportion is fixed, and so must equal 0/2N, 1/2N, ...; in all the simulations reported here, we set m = ½ to approximate a continuous spatial distribution. When selection is weak, gamete frequencies change smoothly from place to place. A variety of models of gene flow, including the stepping stone model used here, can then be approximated by diffusion. With this limit, the effect of gene flow depends only on the variance of parent-offspring distance, $\sigma^2 = m$. However, when selection is strong, the details of the dispersal process become important. A two-dimensional model would be more realistic. The effects of drift are qualitatively different in one and two dimensions (Malecot, 1948); however, this problem is not serious here, as we are mainly interested in deterministic processes, which do not depend on the number of dimensions if the hybrid zone is straight.

After migration, haploid gametes unite at random to produce diploids. Selection acts on these diploids and is followed by meiosis. For each haploid offspring, a diploid is chosen as the parent, with probability proportional to its fitness. Fitness may depend on only a subset of the loci, the rest being neutral markers. The offspring is derived from the two parental haplotypes by recombination; genes are arranged along a single chromosome, with probability (r) of crossing-over between adjacent genes.

Several tricks are used to speed up the simulations. Fitness depends only on the number of heterozygous loci, or on the number of "1" as opposed to "0" alleles. A table of fitnesses can therefore be compiled at the beginning of the run. Moreover, because the haplotype is stored as a set of integers, a table listing the number of "1" alleles in each 16-bit integer can be compiled, so the number of "1" bits does not need to be counted for each individual in each generation. Selection of a diploid parent (i.e., a pair of haplotypes is done by setting a table of cumulative fitnesses: for each deme in each generation, the fitness $w(i,j)$ of each possible diploid pair is calculated, and a table $[0, w(1,1), w(1,1) + w(1,2), \ldots, \sum_{i,j=1}^{2N} w(i,j)]$ is set up. A random number is drawn from a uniform distribution between 0 and $\sum_{i,j=1}^{2N} w(i,j)$; the point in the table where this number lies gives the chosen parent.

The population is started with a sharp step. All demes to the left of the midpoint are fixed for "0" alleles, and those on the right are fixed for "1" alleles. Because it is a stochastic simulation, many estimates must be made. Strictly speaking, we should take these estimates from many independent replicates. However, because 100 or more generations must pass before the system settles down, this is not feasible: we therefore record statistics at set intervals, after a warm-up period. To the extent that successive estimates are autocorrelated, confidence intervals are underestimated.

The key statistics are based on the mean and variance of the "hybrid index" $x(0 < x < 1)$. The mean of x is just the mean allele frequency. The variance of x includes two components: the genic variance, due to heterozygosity at individual loci, and the remainder, due to linkage disequilibria (Eq. 2b). Because the genic variance can be calculated from individual allele frequencies, the average disequilibrium can be found without the need to calculate all $n(n - 1)$ pairwise associations.

To test the theory, we must estimate cline width, average linkage disequilibrium and mean fitness at the center, and the strength of the barrier to gene exchange. Because averages must be taken over many generations, they must be calculated automatically. Where selection is weak enough that there is no appreciable barrier effect, cline width is estimated by regressing $\ln(p/q)$ against distance, using the region between $\ln(p/q) = -2$ and $+2$. Here p is the average allele frequency; if the clines at different loci lie in different places, this method overestimates cline width. Where there is an appreciable barrier, the cline is divided into three regions. Because we expect gradients in allele frequency to be proportional to $W^{1/r}$, the center is defined as the region where mean fitness is reduced by at least 5% of its maximum drop. Three linear regressions of $z = \ln(p/q)$ against distance are then taken (excluding the edges, where demes are near fixation, and z is outside the range -5 to $+5$). The gradient in allele frequency is related to the gradients of these regressions by $(dp/dx) = pq(dz/dx)$; allele frequency gradients are calculated by extrapolation to the points where these regressions cross.

We expect linkage disequilibrium to be proportional to $(\partial p/\partial x)^2$; because in simple models the gradient $(\partial p/\partial x)$ is proportional to (pq), the disequilibrium at the center can be estimated by fitting $D = \alpha(pq)^2$, using least-squares. This method gives a better estimate than relying on the maximum D, or the single value nearest the center. However, when the clines are narrow, no demes have $p \approx 0.5$, and so the disequilibrium estimated by regression may be greater than 0.25 (see Fig. 2-3). Mean fitness at the center is estimated in a slightly different way; because we expect cline shape to depend on the actual mean fitness, we use the actual minimum, averaged over all the sampled generations. Except where selection is strong, this method gives a value similar to that estimated by regression of mean fitness on (pq).

The program was tested by comparison with standard population genetic results. The number of crossover events followed a Poisson distribution; the rate of increase of gene frequency variance under drift was equal to pq/2N; linkage disequilibrium decayed at a rate $(1 - r)^t$; migration caused a step to collapse to a cline with width $\sqrt{2\pi\sigma^2 t}$; the decrease in correlation between gene frequency fluctuations under migration, heterozygote advantage, and drift was as predicted by Malecot (1948); and the width and shape of clines maintained by heterozygote disadvantage and epistasis was as expected when selection is weak. These tests give us confidence that, despite its complexity, the program is performing as it should.

3

Reinforcement: Origin, Dynamics, and Fate of an Evolutionary Hypothesis

DANIEL J. HOWARD

Species concepts and definitions have provided fertile ground for disagreements among biologists for more than 100 years (Darwin, 1859; Jordan, 1905; Du Rietz, 1930; Dobzhansky, 1937; Mayr, 1942, 1963, Burma, 1954; Mecham, 1961; Paterson, 1978, 1980, 1981, 1982; Donoghue, 1985; Coyne et al., 1988; Cracraft, 1989; Templeton, 1989). Yet most biologists would agree that in sexually reproducing organisms barriers to gene exchange exist between distinct species, and these barriers can be broken down into two classes: prezygotic and postzygotic. Included among prezygotic barriers are seasonal and habitat differences, long-distance signaling and courtship differences, and differences in genitalic structures. Included among postzygotic barriers are embryo inviability, hybrid inviability, hybrid sterility, and hybrid breakdown (for a more complete description of reproductive barriers see Mayr, 1963, and Dobzhansky, 1970).

Prezygotic barriers prevent individuals from wasting their gametes in the formation of unfit hybrid individuals; hence such barriers are susceptible to enhancement by natural selection (Dobzhansky, 1940; Mayr, 1963). On the other hand, it is difficult to envision how natural selection could operate to strengthen postzygotic barriers such as hybrid inviability or sterility (Darwin, 1859; Mayr, 1963), except under fairly restrictive circumstances (Coyne, 1974). The potential involvement of natural selection in the formation of prezygotic barriers to gene exchange is what sets this group of barriers apart from postzygotic barriers. The process by which prezygotic barriers to gene exchange are improved by natural selection is known as reinforcement.

In recent years the definition of reinforcement has become almost as contentious an issue as the process of reinforcement. Thus it has become necessary to begin a discussion of reinforcement with a clear statement of the process under consideration and to distinguish reinforcement and reproductive character displacement. I define *reinforcement* as the evolution of prezygotic isolating barriers in zones of overlap or hybridization (or both) as a response to selection against hybridization. One of the potential outcomes of this process is reproductive character displacement. *Reproductive character displacement* describes a pattern of greater divergence of an isolating trait in areas of sympatry between closely related taxa than in areas of allopatry. The term *trait* in this definition refers not only to a morphological or a signaling system character but also to the species discrimination ability of individuals.

Butlin (1987, 1989) has argued in favor of different definitions of reinforcement and reproductive character displacement. He considered the process of reinforcement to be limited to situations in which fertile hybrids are actually formed and selected against. He believed that reproductive character displacement, rather than describing a pattern as in my definition, describes the process whereby prezygotic isolating barriers are strengthened between taxa that are already completely isolated by postzygotic barriers. The basis of Butlin's argument is the contention that reinforcement is considered to be a process of speciation; and that when postzygotic isolation is complete, speciation is complete. Therefore reinforcement cannot describe the process that occurs after effective postzygotic isolating barriers are in place.

However, since its inception in Dobzhansky's writings (a topic I return to shortly), reinforcement has generally been considered a model for the evolution of prezygotic isolation during an allopatric speciation event, not a model of speciation. Complete postzygotic isolation merely provides a stronger selection pressure for the evolution of prezygotic isolating barriers. The fundamental aspect of reinforcement, selection against individuals that undergo hybrid matings, is still present. For these reasons, I believe reinforcement should retain its traditionally broad definition, and reproductive character displacement should be limited to a description of pattern, a limitation in keeping with the original exposition of character displacement by Brown and Wilson (1956).

In the remainder of this chapter, I trace the origin of the hypothesis of reinforcement, examine the theoretical arguments in favor of and against the hypothesis, describe empirical evidence that supports the predictions of the hypothesis, develop the idea of hybrid zones as centers of species formation, and set forth some guidelines for the study of reproductive character displacement. I do not address or develop competing hypotheses to explain the evolution of prezygotic isolation, such as sexual selection—not because I regard these competing hypotheses as unimportant but because I want to focus squarely on reinforcement.

ORIGIN OF THE HYPOTHESIS AND DESCRIPTION OF THE PROCESS

The origin of the hypothesis of reinforcement has been a matter of some confusion. In 1966 Grant suggested that selection for reproductive isolation be termed the Wallace effect because early in the development of evolutionary thought, Alfred Russel Wallace (1889) advocated a process of selection for reproductive isolation as an important component of some speciation events. Indeed, Wallace did describe selection for reproductive isolation; but as pointed out by Littlejohn in 1981, Wallace developed a model for the evolution of postzygotic, not prezygotic, barriers to gene exchange. Thus he did not describe the process of reinforcement as we know it today.

Littlejohn (1981) attributed the first development of the hypothesis of reinforcement to the Reverend John Thomas Gulick, a contemporary of Wallace, who wrote some truly insightful papers on the process of speciation at the end of the nineteenth century (Gulick, 1890a,b,c,d, 1891). Based on my own reading of Gulick, I find hints of the hypothesis but no clear statement. Gulick noted that breeding between different species would have harmful effects.

> It needs no experiments to prove that, if the members of a species are impelled to consort only with the members of other species, they will either fail to leave offspring, or their off-spring will fail to inherit the characteristics of the species. [Gulick, 1890b, p. 235]

However, Gulick did not regard these harmful effects as a driving force in the development of barriers to mating. Instead, he attributed the development of reproductive barriers between species to the accumulating effects of the action of the barriers.

> We also see that an endowment which prevents the destruction of the species through the complete isolation of individuals, and which co-operates with migrational instincts in securing dispersal without extinction, may be perfected by the accumulating effects of its own action. [Gulick, 1890b, p. 235]

What Gulick meant by the "accumulating effects of its own action" is that isolation of populations due to behavioral differences or ecological differences leads to further divergence of populations and hence to ever greater isolation. Gulick's emphasis on the importance of isolation, not only behavioral and ecological but also geographic, for the divergence of populations is remarkable for clarity of thought and depth of development; but Gulick did not envision a role for natural selection in the development of barriers to gene exchange.

> I believe that qualities simply producing Segregation can never be accumulated by Natural Selection; . . . [Gulick, 1890b, p. 243]

My reading of the literature indicates that the development of the hypothesis of reinforcement should be attributed to Dobzhansky, although Muller (1940) offered a brief description of the hypothesis at about the same time. The best early statement of the hypothesis is contained in Dobzhansky's 1940 paper, where he argued that geographic isolation leads to the buildup of two separate adaptive complexes uniquely attuned to their own environments. So long as the diverging populations do not come into contact there is no selection for factors that would prevent interbreeding. However, once the populations come back into contact, interbreeding would break up the adaptive complexes of the parental populations, and hybrid offspring would suffer decreased viability or fertility.

> [O]ccurrence of hybridization between races and species constitutes a challenge to which they may respond by developing or strengthening isolating mechanisms that would make hybridization difficult or impossible. Where hybridization jeopardizes the integrity of two or more adaptive complexes, genetic factors which would decrease the frequency or prevent the interbreeding would thereby acquire a positive selective value, even though these factors by themselves might be neutral. [Dobzhansky, 1940, p. 316]

Thus prezygotic isolating barriers evolve after the adaptive divergence of the incipient species, and this evolution occurs primarily in areas where the two taxa overlap in distribution. Dobzhansky's contention that prezygotic barriers to gene exchange evolve at the boundaries between races placed hybrid zones at the center of studies of speciation.

Dobzhansky was enthusiastic about the process he outlined and he plainly thought that it was important in natural populations. Yet as late as 1935 he was of an entirely different opinion with regard to the evolution of barriers to gene exchange.

> Although the mechanisms preventing free and unlimited interbreeding of related forms are as yet little understood, it is already clear enough that a large number of very different mechanisms of this kind are functioning in nature. This diversity of the isolating mechanisms is in itself remarkable and difficult to explain. It is unclear how such mechanisms can be created at all by natural selection, that is what use the organism derives directly from their development. We are almost forced to conjecture that the isolating mechanisms are merely by-products of some other differences between the organism in question, these latter differences having some adaptive value and consequently being subject to natural selection [Dobzhansky, 1935, p. 349]

The dramatic change in Dobzhansky's thinking by 1940 can be traced to his work with Koller on *Drosophila pseudoobscura* and *D. miranda* (Dobzhansky and Koller, 1938). Dobzhansky and Koller discovered that strains of *D. pseudoobscura* from areas of sympatry with *D. miranda* exhibited stronger sexual isolation from the *D. miranda* inhabiting the same region than did strains of *D. pseudoobscura* from outside the zone of overlap. These observations suggested to Dobzhansky that isolating barriers were strengthened in contact zones as a consequence of selection against hybridization.

Despite his enthusiasm for the hypothesis of reinforcement, Dobzhansky (1940) was cautious enough to admit that reproductive isolation occasionally arises as a by-product of isolation and adaptation to different environments. The basic problem, as he saw it, was how frequently and to what extent could isolating barriers be regarded as adaptational by-products arising without the intervention of the process he outlined.

DYNAMICS OF THE HYPOTHESIS

The initial reception of the hypothesis was warm. For example, Mayr wrote approvingly of Dobzhansky's ideas in his 1942 book, at the same time maintaining that species can complete their development in isolation.

Blair (1955) first used the term reinforcement to identify the process described by Dobzhansky. He also provided data from natural populations of anurans that supported the hypothesis.

The first serious assault on the hypothesis came from John Moore in a 1957 paper. Moore raised three objections to the process that have substance.

1. Isolating mechanisms would have selective advantage in the zone but not elsewhere. Therefore they could not flow out of the zone.
2. Hybridization must be frequent if natural selection is to increase the frequencies of genetic differences that reduce crossing. Consequently, as isolating mechanisms become more effective, the ability of selection to augment them steadily decreases.
3. There is no evidence of reinforcement in nature (i.e., there are no examples of the expected outcome, reproductive character displacement).

Moore's paper was influential. In his 1963 classic, *Animal Species and Evolution,* Mayr's attitude toward the process of reinforcement had greatly changed. The change was clearly fueled by Moore's paper as well as by Mayr's well-known aversion to anything that smacked of sympatric speciation. Mayr echoed Moore's objections and added one of his own.

4. Introgression makes the interacting species more similar, weakening isolation rather than strengthening it.

Since 1963 other objections to the hypothesis have been raised. Among them:

5. Gene flow from populations outside the zone swamp the effects of selection within the zone (Bigelow, 1965).
6. Models of hybrid inferiority generally lead to an unstable equilibrium. Therefore extinction of one of the populations or elimination of factors responsible for the hybrid inferiority are more likely than the development of prezygotic isolating mechanisms (Paterson, 1978).
7. Mating systems involve coadaptation between male and female components; hence they lack variation and so are difficult to change (Paterson, 1978).
8. An assortative mating locus is not likely to be polymorphic (W. S. Moore, 1979).
9. In general, hybrid zones are narrow, so few suitable mutations are available (Barton and Hewitt, 1981).
10. Recombination continually breaks down the association between assortative mating genes and the genes causing hybrid unfitness (Barton and Hewitt, 1981).

Most of these objections have been addressed and countered in a debate that has gone on for more than 30 years. It is probably fair to state that the objections to reinforcement are better known to evolutionary biologists than are the refutations of the objections. Therefore I develop the counter-arguments in some detail.

Objections 1, 5, and 9 have as an implicit or explicit assumption the view that contact zones are narrow and that the vast bulk of the two interacting taxa occur outside the zone. Remove this assumption and the arguments carry much less weight. In fact, some hybrid zones are broad (Gartside, 1980; Howard, 1986; Howard and Waring, 1991), which diminishes the influence of parental population gene flow and enhances the possibility that suitable mutations for the evolution of prezygotic reproductive barriers are available. Moreover, many evolutionary biologists now concur that not all genetic divergence takes place in large populations separated by geographic barriers. If genetic divergence occurs in a small peripheral isolate that subsequently expands its range to come into contact with the parental species, the bulk of the daughter species may occur within the contact zone. Thus prezygotic barriers to gene exchange would have an advantage throughout most of the range of the daughter species and could easily spread (Littlejohn, 1981). The same argument would apply in the case of sympatric speciation.

The genetic model that J. A. Moore (1957) had in mind when he voiced objection 2 is unclear. The objection seems to be based on the assumption that species are isolated by multiple barriers that are built up as a result of changes at many gene loci. If this assumption is true, Moore's reservation has merit. However, reviews of the literature on reproductive isolation note that there is a growing amount of evidence that differences in traits responsible for reproductive isolation can be controlled by one or a few loci with major effects (Bush and Howard, 1986; Howard, 1993). Population genetic models of the evolution of reproductive isolation indicate that complete isolation can evolve rapidly when assortative mating is controlled by a single locus (Maynard Smith, 1966; W. S. Moore, 1981).

Objection 4 does present a problem for the reinforcement hypothesis, but introgression as a result of secondary contact does not inevitably lead to fusion of two divergent populations. A model presented by Wilson (1965) indicated that the outcome of contact, fusion or reproductive isolation, is uncertain over a wide range of hybrid fitnesses and frequencies of mate choice error. Thus in most contact zone situations, there may be a "race" between fusion and the evolution of prezygotic isolation with no predetermined winner.

Objection 6 is based on a model that has the following characteristics: Two populations meet in an area of sympatry; individuals mate at random; the populations differ by one chromosome translocation, which causes heterozygote inferiority; there is no migration in or out of the area of sympatry; and the sizes of the populations are not regulated independently (Paterson, 1978). These conditions are not found in most zones of contact. If the two interacting taxa have diverged ecologically, the sizes of the two populations are regulated independently, and a stable polymorphism can result (Maynard Smith, 1966). Alternatively, a stable polymorphism can be maintained by the balance between dispersal and selection against heterozygotes if chromosome frequencies vary spatially (Bazykin, 1969, 1972, 1973; Barton and Hewitt, 1981), which seems likely in a zone of secondary contact. Once stability has been achieved, further divergence and the evolution of reproductive isolation is possible. Finally, many hybrid zones are characterized by a mosaic pattern of variation with clear reversals of gene frequency changes in linear transects running through them. In other words, there is a patchy quality to the hybrid zone with one taxon occurring in one patch and the other taxon occurring in another patch; finally, some patches contain a mix of the two taxa. In most cases, the mosaicism of the zone can be ascribed to distinct habitat requirements of the two interacting taxa and to a patchy distribution of the distinct habitats and intermediate habitats in the zone of overlap (Harrison, 1986; Howard, 1986; Rand and Harrison, 1989). Harrison (1986) has coined the term "mosaic hybrid zone" to describe such zones, and Harrison and Rand (1989) have pointed out that global extinction of one of the interacting taxa is unlikely in such a zone.

Objection 7 is simply not supported by the available evidence. Sexual behavior is variable, often being the most noticeable difference between closely related species (Darwin, 1874), as well as differing among local populations of the same species (Carson, 1986; West-Eberhard, 1983). Moreover, as pointed out by Harrison (1990), isolation among closely related taxa may be achieved by means other than mate recognition, such as by differences in flowering time in plants.

Paterson's (1978, 1981, 1982, 1986) assault on the hypothesis of reinforcement was part of a more general attack on the "isolation concept" of species (Paterson's [1980] term for the biological species concept of Dobzhansky [1937] and Mayr [1942]), a concept that he argued should be replaced by a "recognition concept" of species. Although some of Paterson's criticisms of the isolation concept and its theoretical underpinnings have value, there seem to be more problems than advantages associated with the recognition concept of species (Coyne et al., 1988). Moreover, some of the supposed advantages of the recognition concept are illusory; in particular, contrary to the claims of Paterson, the recognition concept is just as relational as the isolation concept. The delimitation of species and the assignment of individuals to species still involves comparisons between populations and species (Templeton, 1987; Coyne et al., 1988).

Objection 8 is based on the mass-action model of Moore (1979, 1981). The essential feature of this model is that there is a cost associated with assortative mating because an organism with the rarer genotype is less likely to find a mate. Thus the minority allele goes to extinction. Moore (1979) pointed out that an assortative mating locus that satisfied the assumptions of this model would be unlikely to respond to indirect selection pressure resulting from disruptive selection at a second locus. First, such an assortative mating locus is likely to be invariant. Second, the degree of hybrid inferiority would have to be high to overcome the effects of minority disadvantage.

One of the critical questions raised by Moore's paper is whether there is a cost associated with assortative mating. A lack of data makes it difficult to evaluate this question, but one might expect that in animal taxa characterized by widely dispersed individuals or by a lack of long distance mate attraction signals there is likely to be a cost associated with assortative mating. On the other hand, the cost of assortative mating should be reduced or nonexistent in animal taxa that exhibit long distance mate attraction signals and that occur in populations of moderate to high density. It is worth noting again that sexual behavior appears to be highly variable in natural populations.

Another important question raised by Moore's paper is the intensity of selection against hybridization in natural populations. Evolutionary biologists have a tendency to reject models that work only if selection intensities are high; yet we know little about selection intensities in nature. In some of the few studies that have rigorously measured selection against hybridization in natural hybrid zones, selection intensities appear to be high (Woodruff, 1979; Dowling and Moore, 1985; Kocher and Sage, 1986).

The final objection to reinforcement—that recombination breaks down the association between genes causing assortative mating and the genes responsible for hybrid unfitness (Barton and Hewitt, 1981)—is also a criticism leveled at sympatric speciation. Recombination does present a major obstacle to the evolution of reproductive isolation in two allele assortative mating models, but it may be overcome by strong selection against hybridization or by weaker selection combined with some initial nonrandom mating (Felsenstein, 1981). The nonrandom mating may result from different habitat or resource requirements of the two interacting taxa. In contrast to two allele models of assortative mating, recombination does not present a threat to reinforcement when positive assortative mating is attributable to the fixation of the same allele (such as an allele that reduces the migration rate of its bearers) at an assortative mating locus in two subpopulations (Felsenstein, 1981).

At this point in the debate over the importance of reinforcement, it has become clear that the issue cannot be settled by theoretical arguments but by evidence from natural populations. It is precisely in this area that the case for reinforcement appears weakest—until one scratches below the surface.

REPRODUCTIVE CHARACTER DISPLACEMENT

The mantra in the literature since the early 1970s has been that the expected outcome of reinforcement, reproductive character displacement, is a rare phenomenon. The origin of this mantra can be traced to a paper written by Thomas Walker in 1974. Walker's paper is cited in virtually all subsequent papers that comment on the lack of examples of reproductive character displacement. The attention attracted by Walker's work is understandable. Walker examined potential cases of reproductive character displacement among the singing Orthoptera, a group of animals in which a single trait—male calling song—seems to be of overwhelming importance in isolating closely related species. Blair (1964) pointed out that such groups of animals provide the best material for determining whether reinforcement has occurred. The perception of the paper can best be understood from two quotes.

> Walker (1974) examined 254 potential cases of reproductive character displacement in species pairs of gryllids and tettigoniids, and found only five possible cases, none of which was particularly convincing.
> [Littlejohn, 1981, p. 324]

> This hypothesis [reinforcement], once widely accepted, has recently been criticized . . . for a number of theoretical deficiencies and because very few cases of the predicted reproductive character displacement have been reported, in spite of many attempts to find them [Walker (1974)].
> [Phelan and Baker, 1987, p. 206]

The impression left by Littlejohn, Phelan and Baker, and many others who have cited Walker's report is that Walker had adequate data to test whether there was greater divergence of male calling song in areas of sympatry than in areas of allopatry for a large number of pairs of acoustic insects. This impression is wrong.

In fact, Walker did not examine 254 potential cases; he only pointed out that there were 254 species of Gryllidae and Tettigonidae in the United States and Canada, excluding species of western Decticinae and Phaneropterinae.

For the most part, Walker's study consisted of an after-the-fact examination of calling song data gathered during the course of systematic studies. In only one potential case that he reviewed had the data been gathered to test for the presence of reproductive character displacement. This case was a study of *Teleogryllus commodus* and *T. oceanicus* reported by Hill, Loftus-Hills, and Gartside in 1972. Hill et al. turned up no evidence of character displacement in male calling song or female discrimination.

The adequacy of the rest of the data is best summarized in Walker's own words. The following quotation is taken from the discussion section as Walker provided possible explanations for why examples of reproductive character displacement should be lacking in crickets and katydids.

> Too small a sample has been studied intensively to prove that this pattern of variation [reproductive character displacement] is scarce in United States crickets and katydids. In only 22 of 43 presumptive cases have I the minimal data for showing displacement . . . and in only a few of these have I sufficient data to show minor geographic variation. [Walker, 1974, p. 1147]

Walker rejected four potential cases of reproductive character displacement, not because the patterns were incompatible with reproductive character displacement but because the data were not extensive enough. However, if the data were not extensive enough to allow acceptance of reproductive character displacement, they were also not extensive enough to allow rejection.

My comments should not be construed as a criticism of Walker. He did not intend for his report to be the final word on reproductive character displacement in nature. He was well aware of the deficiencies in the data with which he was working, and he ended the report with an appeal for further studies.

If Walker's work is not be the final word on reproductive character displacement, what do the data from natural populations indicate? Is the pattern as rare as suggested by numerous authors? I have examined this question by scouring the literature for studies of reproductive character displacement. I regarded a study as adequate if evidence existed that the trait under analysis sometimes served as a reproductive barrier in the group of animals or plants that were the subject of the investigation and if a statistical analysis of variation in allopatry versus sympatry had been carried out. I suspended the requirement for a statistical analysis only when enough raw data were presented to provide a clear picture of variation in allopatry and in sympatry. Studies deemed adequate by the foregoing criteria were excluded only if evidence of selection against hybridization had been rigorously sought but not found. I instituted the last requirement because contact between populations that have differentiated in allopatry do not automatically initiate the process of reinforcement. Reinforcement is not expected to occur unless there is selection against hybridization. Table 3-1 summarizes the results of my survey of the literature. The pairs of organisms listed are united in having a geographic distribution appropriate for character displacement. The members of the pair occur in a zone of sympatry, and one or both also occur in an area outside the zone. Table 3-1 notes whether there is evidence of selection against hybridization, whether heterospecific matings occur in nature, and of course whether the patterns of variation in one or both species indicate reproductive character displacement, that is, whether there is significantly greater divergence of an isolating trait in areas of sympatry than in areas of allopatry.

My survey of the literature turned up 48 potential cases of reproductive character displacement that had been examined in enough detail to meet the minimal requirements outlined above. Reproductive character displacement was found in 33 of the cases; no reproductive character displacement was found in 15. Cases of reproductive character displacement occur in mammals, birds, amphibians, reptiles, fishes, insects, other invertebrates, and plants. Contrary to the prevalent notion in the literature, the pattern of reproductive character displacement appears to be widespread and frequent.

This finding does not mean that the process of reinforcement is commonplace in nature. It does mean that the pattern of variation expected to be produced by reinforcement is common. Ascertaining whether reinforcement can account for the patterns of reproductive character displacement takes much more work in most cases. This work entails a demonstration that: (1) heterospecific matings occur or probably did occur in nature; (2) selection against hybridization is operating in the field; (3) the displacement detected is perceptible to the opposite sex (where such a determination is relevant); (4) variation is heritable and thus capable of responding to selection; and (5) displacement has not occurred for other reasons, most notably ecological reasons. The first step is necessary so character displacement due to selection against hybridization can be distinguished from character displacement attributable to interference between the mate recognition signals of taxa that do not hybridize. Templeton (pers. comm.) referred to the latter form of character displacement as facilitated reproductive character displacement. The second step is required because one cannot invoke

Table 3-1. Results of Studies of Reproductive Character Displacement

Organisms	Character(s)	Selection Against Hybridization	Heterospecific Matings in Nature	Reproductive Character Displacement	Study
Mammals					
Peromyscus leucopus, P. gossypinus (deer mice)	Mate preference	Unknown	Yes	Yes	McCarley (1964)
Birds					
Aechmophorus occidentalis: light phase, dark phase (western grebe)	Male preference	Unknown	Yes	Yes	Nuechterlein (1981)
Geospiza fulginosa, G. difficilis (ground finches)	Male preference	Unknown	No	Yes	Ratcliff & Grant (1983)
Ficedula hypoleuca, F. albicollis (flycatchers)	Male song	Yes	Yes	Yes	Alatalo et al. (1982; Wallin (1986)
Amphibians and reptiles					
Hyla regilla, H. californiae (frogs)	Mating call	Yes	Yes	No	Ball & Jameson (1966)
Hyla cinera, H. gratiosa (frogs)	Mating call	Unknown	Unknown	No	Asquith et al. (1988)
Hyla chrysoscelis, H. versicolor (frogs)	Mating call	Yes	Unknown	Yes	Ralin (1976, 1977)
Gastrophryne olivacea, G. carolinensis (frogs)	Mating call	Unknown	Yes	Yes	Blair (1955)
Crinia insignifera, C. pseudinsignifera (frogs)	Mating call	Unknown	Yes	No	Littlejohn (1959)
Crinia laevis, C. victoriana (frogs)	Mating call	Yes	Yes	No	Littlejohn et al. (1971)
Rana blairi, R. berlandieri, R. sphenocephala (frogs)	Breeding season	Unknown	Yes	Yes	Hillis (1981)
Litoria paraewingi, northern *L. ewingi* (frogs)	Mating call	Yes	Yes	No	Watson (1972); Watson & Littlejohn (1978)
Litoria ewingi, L. paraewingi (frogs)	Mating call	Yes	Yes	No	Littlejohn & Watson (1983)
Pseudophryne bibroni, P. semimarmorata (frogs)	Mating call	Yes	Yes	No	McDennell et al. (1978)
Bombina bombina, B. variegata (toads)	Mating call	Yes	Yes	No	Szymura & Barton (1986); Sanderson et al. (pers. comm.)
Sceloporus undulatus, elongatus, S. graciosus graciosus (iguanid lizards)	Push-up display	Unknown	No	Yes	Ferguson (1973)
Sceloporus woodi, S. undulatus undulatus (iguanid lizards)	Female preference	Unknown	Yes	No	Jackson (1973)

Table 3-1. Results of Studies of Reproductive Character Displacement (*Continued*)

Organisms	Character(s)	Selection Against Hybridization	Heterospecific Matings in Nature	Reproductive Character Displacement	Study
Fishes					
Etheostoma lepidum, E. spectabile (percid fishes)	Egg receptivity	Yes	No	Yes	Hubbs (1960)
Insects					
Calopteryx maculata, C. aequabilis (damselflies)	Male preference, female wing pigmentation	Yes	Yes	Yes	Waage (1975, 1979)
Allonemobius socius, A. fasciatus (ground crickets)	Male calling song	Yes	Yes	Yes	Howard (1986); Benedix & Howard (1991)
Teleogryllus commodus, T. oceanicus (field crickets)	Male calling song, female preference	Yes	No	No	Fontana & Hogan (1969); Hill et al. (1972)
Laupala (sword-tail crickets)					
L. palola, L. nui	Male calling song	Unknown	Unknown	Yes	Otte (1989)
L. paloto(?), L. nui(?)	Male calling song	Unknown	Unknown	No	Otte (1989)
L. nui, L. tantalis	Male calling song	Unknown	Unknown	Yes	Otte (1989)
Sp. 1, Sp. 4	Male calling song	Unknown	Unknown	No	Otte (1989)
Sp. 1, Sp. 5.	Male calling song	Unknown	Unknown	Yes	Otte (1989)
Sp. 4, Sp. 5	Male calling song	Unknown	Unknown	Yes	Otte (1989)
Sp. 3, Sp. 4	Male calling song	Unknown	Unknown	No	Otte (1989)
Chorthippus parallelus parallelus, C. p. erythropus (grasshoppers)	Mate preference	Yes	Yes	No	Hewitt et al. (1987); Ritchie et al. (1989)
Barytettix psolus, B. paloviridis (grasshoppers)	Aedeagal morphology	Unknown	Yes	Yes	Cohn & Cantrall (1974)
Drosophila paulistorum (fruit flies)					
Amazonian Andean	Male preference	Yes	Unknown	Yes	Dobzhansky et al. (1964); Ehrman (1965)
Amazonian Guianan	Male preference	Yes	Unknown	Yes	Dobzhansky et al. (1964); Ehrman (1965)

Amazonian Orinocan	Male preference	Yes	Unknown	Yes	Dobzhansky et al. (1964); Ehrman (1965)
Andean Guianan	Male preference	Yes	Unknown	Yes	Dobzhansky et al. (1964); Ehrman (1965)
Orinocan Andean	Male preference	Yes	Unknown	Yes	Dobzhansky et al. (1964); Ehrman (1965)
Orinocan Guianan	Male preference	Yes	Unknown	Yes	Dobzhansky et al. (1964); Ehrman (1965)
Centro-American Amazonian	Male preference	Yes	Unknown	No	Dobzhansky et al. (1964); Ehrman (1965)
Centro-American Orinocan	Male preference	Yes	Unknown	Yes	Dobzhansky et al. (1964); Ehrman (1965)
Drosophila mojavensis, D. arizonae (fruit flies)	Male preference	Yes	No	Yes	Wasserman & Koepfer (1977); Ruiz et al. (1990)
Aedes albopictus, A. pseudalbopictus (mosquitoes)	Female preference	Yes	No	Yes	McLain & Rai (1986)
Aedes albopictus, A. seatoi (mosquitoes)	Female preference	Yes	No	Yes	McLain & Rai (1986)
Other invertebrates					
Oronectes rusticus, O. sanbornii (crayfishes)	Size	Yes	Yes	Yes	Butler & Stein (1985); Butler (1988)
Partula suturalis, P. olympia (land snails)	Shell chirality	Yes	Yes	Yes	Murray & Clarke (1980); Johnson (1982)
Partula suturalis, P. mooreana (land snails)	Shell chirality	Yes	No	Yes	Murray & Clarke (1980); Johnson (1982)
Partula suturalis, P. tohiveana (land snails)	Shell chirality	Yes	No	Yes	Murray & Clarke (1980); Johnson (1982)
Plants					
Phlox pilosa, P. glaberrima (phlox)	Color of corolla	Yes	Yes	Yes	Levin & Kerster (1967)
Solanum grayi, S. lymholtzianum (night shades)	Flower size	Yes	Unknown	Yes	Whalen (1978)
Solanum rostratum, S. citrullifolium (night shades)	Flower size	Yes	Unknown	Yes	Whalen (1978)

selection against hybridization to explain a pattern of character displacement if no basis for the selection can be found. The third step is needed to ensure that the displacement results in a strengthening of isolation. The fourth step is an essential part of any research program attempting to determine whether natural selection is the causative agent driving patterns of variation observed in the wild (Endler, 1986). The final step is necessary to eliminate the possibility that habitat differences or some other ecological differences in the zone of sympatry induced the observed character displacement.

At present the most convincing cases for the operation of reinforcement are the damselflies *Calopteryx maculata* and *C. aequabilis* (Waage, 1975, 1979), the crayfishes *Oronectes rusticus* and *O. sanbornii* (Butler and Stein, 1985; Butler, 1988), the land snails *Partula suturalis* and *P. olympia* (Murray and Clarke, 1980; Johnson, 1982), and the *Phlox* species *P. pilosa* and *P. glaberrima* (Levin and Kerster, 1967). In all of the above cases the character displacement is clear, selection against hybridization has been documented, and interspecific matings are known to occur.

One question might be asked of the above survey: Are the results biased? In other words, are studies of reproductive character displacement more likely to be carried out when preliminary evidence suggests a positive result, and are positive results more likely to be collated and published than negative results? These questions are difficult to answer.

One way of eliminating the bias problem is to explore predictions of the reinforcement hypothesis for which a reporting bias would not be expected. One such prediction is that evidence of positive assortative mating would be found in many hybrid zones. There is no reason to expect a reporting bias with regard to patterns of interaction in a hybrid zone.

To test this prediction, I surveyed 172 hybrid zone papers in my files. After eliminating hybrid zones contained in Table 3-1, I identified 37 hybrid zones in which the pattern of interaction between the hybridizing taxa had been characterized on the basis of behavioral, morphological, or genetic data. Twenty-three zones had been studied from the standpoint of electrophoretic or chromosomal characters (or both). Nine zones had been characterized on the basis of morphological characters. Both morphological and genetic characters had been studied in three zones. Actual observations of matings in the field comprised the data from two zones.

Character index analysis was the most widespread method of visualizing the genotypic or phenotypic composition of a hybrid zone population and hence the level of isolation between two taxa. There are a variety of ways of constructing character index profiles, but all lead to the expectation that, with random mating and no differential selection among genotypes, character index scores from a mixed taxa population should cluster in the middle of a histogram. Positive assortative mating or selection against hybrids would lead to a clustering of scores at the two extremes of a histogram. The other common method of characterizing patterns of interaction was the testing of genotypic data for deviations from random mating expectations.

Random mating expectations were fulfilled in 16 of the zones (Table 3-2). In 19 of the zones random mating expectations were not fulfilled. The pattern of interaction was positively assortative. Interactions varied geographically in two of the zones; that is, in some areas of the zone random mating expectations were fulfilled, whereas in

Table 3-2. Patterns of Interaction in Hybrid Zones

Organisms	Pattern of Interaction	Selection Against Hybridization	Study
Mammals			
Mus musculus musculus, M. musculus domesticus (mice)	Random	Unknown	Hunt & Selander (1973)
Mus musculus: CD population, normal population (mice)	Random	Unknown	Spirito et al. (1980)
Peromyscus leucopus: chromosome form 1, chromosome form 2 (deer mice)	Random	Unknown	Stangl (1986)
Spalax ehrenbergi: 58 chromosome form, 60 chromosome form (mole rats)	Random	Unknown	Nevo & Bar-El (1976)
Spalax ehrenbergi: 58 chromosome form, 54 chromosome form (mole rats)	Assortative	Unknown	Nevo & Bar-El (1976)
Spalax ehrenbergi: 58 chromosome form, 52 chromosome form (mole rats)	Assortative	Unknown	Nevo & Bar-El (1976)
Thomomys bottae actuosus, T. b. ruidosae (pocket gophers)	Random	Unknown	Patton et al. (1979)
Thomomys bottae, T. b. connectens, T. b. opulentis (pocket gophers)	Random	Unknown	Smith et al. (1983)
Thomomys bottae: 76 chromosome form, 88 chromosome form (pocket gophers)	Random	Unknown	Hafner et al. (1983)
Thomomys bottae canus, T. townsendii relictus (pocket gophers)	Assortative	Unknown	Thaeler (1968); Patton et al. (1979)
Thomomys b. saxatilis, T. t. relictus (pocket gophers)	Assortative	Unknown	Thaeler (1968)
Thomomys b. pascalis, T. b. mewa (pocket gophers)	Assortative	Unknown	Thaeler (1968)
Thomomys talpoides: 54 chromosome form, 48 chromosome form (pocket gophers)	Random	Unknown	Thaeler (1974)
Thomomys talpoides: 54 chromosome form, 48B chromosome form	Assortative	Unknown	Thaeler (1974)

Table 3-2. Patterns of Interaction in Hybrid Zones (*Continued*)

Organisms	Pattern of Interaction	Selection Against Hybridization	Study
Mammals cont.			
Thomomys talpoides: 46 chromosome form, 48C chromosome form	Assortative	Unknown	Thaeler (1974)
Geomys breviceps, G. bursarius (pocket gophers)	Assortative	Unknown	Cothran & Zimmerman (1985)
Birds			
Vermivora pinus, V. chrysoptera (warblers)	Assortative	Unknown	Gill & Murray (1972)
Icterus galbula galbula, I. g. bullockii (orioles)	Variable	Unknown	Corbin & Sibley (1977); Rising (1983)
Colaptes auratus cafer, C. a. auratus (northern flickers)	Random	No	Moore (1987); Moore & Koenig (1986)
Amphibians and reptiles			
Bufo boreas, B. punctatus (toads)	Assortative	Yes	Feder (1979)
Scaphiopus bombifrons, S. multiplicatus (toads)	Assortative	Yes	Sattler (1985)
Plethodon glutinosus, P. jordani (salamanders)	Variable	Unknown	Highton & Henry (1970)
Eurycea bislineata, E. n. sp. (salamanders)	Assortative	Unknown	Guttman & Karlin (1986)
Desmognathus fuscus, D. ochrophaeus (salamanders)	Assortative	Unknown	Karlin & Guttman (1981)
Bolitoglossa franklini, B. resplendens (salamanders)	Assortative	Unknown	Wake et al. (1980)
Ensatina eschscholtzii eschscholtzii, E. e. klauberi (salamanders)	Assortative	Unknown	Wake et al. (1986)
Plethodon cinereus: glaciated, unglaciated (salamanders)	Random	Unknown	Wynn (1986)
Sceloporus grammicus: fission-6, polymorphic-1 (iguanid lizards)	Assortative	Unknown	Hall & Selander (1973)
Fishes			
Gasterosteus aculeatus: red, black (stickleback fishes)	Random	Yes	McPhail (1969)
Notropis cornutus, N. chrysocephalus (cyprinid fishes)	Assortative	Yes	Dowling & Moore (1985)
Gila orcutti, Hesperoleucus symmetricus (cyprinid fishes)	Assortative	No	Greenfield & Deckert (1973)

Table 3-2. Patterns of Interaction in Hybrid Zones (*Continued*)

Organisms	Pattern of Interaction	Selection Against Hybridization	Study
Insects			
Gryllus pennsylvanicus, G. firmus (field crickets)	Assortative	Yes	Harrison (1983, 1986)
Caledia captiva: Torresian race, Moreton race (grasshoppers)	Random	Yes	Moran (1979); Shaw & Wilkinson (1980)
Chauliognathus pennsylvanicus: long-spot, short-spot (soldier beetles)	Variable	Unknown	McLain (1985)
Heliconius melpomene, H. m. melpomene, H. m. thelxope, H. m. meriana (butterflies)	Random	Unknown	Turner (1971)
Other invertebrates			
Cerion incanum, C. casablancae (land snails)	Random	Unknown	Woodruff & Gould (1987)
Plants			
Pinus contorta, P. banksiana (pines)	Random	Yes	Wheeler & Guries (1987)

other areas of the zone the interaction was positively assortative. Evidence of negative assortative mating was not found in any zone.

Once again, the pattern predicted to result from reinforcement appears to be common in nature. In hybrid zones, interacting taxa often remain distinct rather than forming hybrid swarms. However, although the finding of positive assortment is consistent with the reinforcement hypothesis, it cannot be considered a demonstration that reinforcement is occurring. Assortative mating is not the only explanation for the presence of "pure" species individuals and significant deviations from random mating expectations in a hybrid zone. Alternative explanations include reduced fitness of individuals of mixed ancestry or the presence of migrants from populations outside the hybrid zone. Distinguishing among the alternative explanations requires direct observations of mating, a detailed understanding of the fitness of hybrids in the field, and an assessment of patterns of dispersal. Once positive assortative mating has been established, the factors responsible must be identified, and data on variation in allopatry versus sympatry must be gathered. The only zone listed in Table 3-2 in which positive assortative mating has been clearly established is the long-spot and short-spot *Chauliognathus pennsylvanicus* zone (McLain, 1985). The factors responsible for the positive assortment have not been identified in this zone, but when individuals from allopatric populations are brought together in the laboratory they do not mate assortatively, indicating that barriers to gene exchange have evolved *in situ.* Thus reinforcement seems a likely explanation for the pattern of interaction in the *C. pennsylvanicus* hybrid zone. A great deal of work is necessary to explain the findings for the remainder of the zones in which patterns of variation do not fulfill random mating expectations.

Since the mid-1970s a consensus seems to have been reached that patterns of variation in natural populations do not support an important role for reinforcement in the evolution of prezygotic reproductive barriers. This consensus seems to have emerged from an overreliance on a single study and is not supported by a more sweeping look at the literature.

In fact, reproductive character displacement appears to be a common phenomenon in nature, as do deviations from random mating expectations in hybrid zones. What we still lack are the detailed studies of hybrid zones and of individual cases of reproductive character displacement necessary to determine whether the patterns suggestive of the operation of reinforcement can indeed best be explained by this process. A lack of data, however, is different from negative evidence, a distinction that sometimes gets lost in the arguments over the importance of reinforcement.

I am not the first to contend, after a review of the literature, that evidence from nature is consistent with an important role for reinforcement in the evolution of prezygotic barriers to gene exchange. After a survey of patterns of speciation in *Drosophila,* Coyne and Orr (1989) reported that prezygotic isolation evolves more rapidly than postzygotic isolation in sympatric species pairs but not in allopatric pairs. They pointed out that reinforcement is the best explanation for this pattern.

HYBRID ZONES AS CENTERS OF SPECIES FORMATION

Perhaps the most damaging theoretical criticism of reinforcement as an explanation for the evolution of prezygotic isolating mechanisms was raised by Dobzhansky (1940, p. 318).

> One of the difficulties with the theory lies in explaining how the gene complexes responsible for the isolating mechanisms come finally to permeate the whole bodies of these species. What selective value have the isolating mechanisms for those parts of the species which are not, and never were, exposed to the danger of interbreeding with other species?

This difficulty was seized upon by J. A. Moore in 1957, and Paterson (1982) charged that the criticism remained unanswered.

In fact, Dobzhansky (1940) addressed this difficulty by suggesting that if the genes were selected for in the zone and neutral elsewhere, they might eventually diffuse throughout the species by migration. Caisse and Antonovics (1978) provided theoretical support for Dobzhansky's scenario via a computer simulation model. Other counters to Moore's criticism were outlined earlier in this chapter.

At this point, I want to respond to the problem raised by Dobzhansky with another possibility—the possibility that isolating barriers may not move out of the zone. New species may move out of the zone. Consider the following scenario. Two taxa meet in a contact zone after a long period of geographic isolation. Because of the isolation, one or more traits have diverged sufficiently to partially isolate the two taxa. Moreover, intertaxon matings are costly because hybrid offspring are inviable, sterile, or substantially less fit. The challenge from the related taxon sets up a selection pressure for further divergence in the isolating trait(s) in both taxa. As a result of the pressure, individuals of one or both taxa within the contact zone diverge. If the divergence is great enough, the trait difference may not only serve as a reproductive barrier between taxa inside the contact zone but may isolate daughter populations within the zone from parental populations outside the contact zone. What moves out of the contact zone may be individuals reproductively isolated from the parental stock. Of course, for such movement to be successful, ecological differences must also evolve within the contact zone populations.

Evidence that divergence within a contact zone may reproductively isolate contact zone populations from parental populations comes from work on anurans and *Drosophila*. Littlejohn (1965) reported dramatic divergence in the calling song of *Hyla verreauxi* in a zone of overlap with *H. ewingi*. Later, Littlejohn and Loftus-Hills (1968) demonstrated that the marked differences between the calls of sympatric and allopatric *H. verreauxi* were highly effective as an isolating barrier. Zouros and D'Entremont (1980) demonstrated that females of *Drosophila mojavensis* from the zone of contact with *D. arizonae* discriminated against conspecific males from outside the contact zone. They also provided evidence that the discrimination was a by-product of selection for reproductive isolation between *D. mojavensis* and *D. arizonae* within the contact zone.

The idea that new species may evolve in contact zones can be extended. In a mosaic hybrid zone, populations of one taxon may be semiisolated from each other, and the outcome of interactions with a second taxon may be disparate in different areas. In other words, one trait might diverge in one area and another trait might diverge in another area. Alternatively, the same trait may diverge in different areas but in discordant ways. Thus the challenge from a related species may lead to the development of reproductive barriers not only between populations inside and outside the zone but between different populations within the zone. Contact zones may be centers for species formation.

Evidence consistent with this possibility has been found in a study of reproductive character displacement in iguanid lizards. Ferguson (1973) reported that some aspects of divergence in push-up displays were not concordant in two areas of sympatry between the iguanid lizards *Sceloporus undulatus* and *S. graciosus*. I suspect that more evidence of the foregoing phenomenon will be found when evolutionary biologists take a closer look at geographic variation in reproductively isolating traits within contact zones.

HOW TO STUDY REPRODUCTIVE CHARACTER DISPLACEMENT

Traditionally, reproductive character displacement has been envisioned as a discontinuous step-like pattern, with the divergence of character occurring in both species, beginning abruptly at the borders of the contact zone and persisting throughout the zone (Grant, 1972). However, such a pattern is likely to occur only if gene flow from parental populations is equal throughout the zone, if the selective pressures on the two taxa are similar and invariant throughout the zone, and if the selected trait has the same level of heritable variation in both taxa. These conditions are restrictive and are unlikely to be fulfilled in most zones of overlap.

First, the level of gene flow from parental populations varies throughout the zone. Contact zone populations in closer proximity to parental populations are exposed to a higher level of gene flow than populations farther away. The higher the level of gene flow from parental populations, the less likely the evolution of reproductively isolating trait differences (Bigelow, 1965; Remington, 1968; Howard, 1986). Second, the selective pressure exerted by wasteful hybrid matings varies from population to population within the zone, depending on the relative abundance of the two taxa. The taxon that is more rare is under stronger selection for divergence because individuals of this taxon have a greater chance of encountering and mating with individuals of the wrong taxon (Sawyer and Hartl, 1981; Benedix and Howard, 1991). Finally, the response to selection depends on the level of heritable variation in the isolating trait. The higher the level of heritable variation, the greater the possibility of response to selection (Wade and Arnold, 1980). The level of heritable variation is likely to vary among different taxa.

The first two considerations lead to the expectation that in overlap zones, where the relative abundance of a taxon declines with distance from its parental populations, a clinal pattern of increasing divergence of an isolating trait with increasing distance from the species border will develop. Waage (1979) discovered such a clinal pattern of displacement in the wing color of females of two species of *Calopteryx* damselflies. He related the degree of divergence to the chance of encountering a heterospecific individual.

Mosaic zones of overlap and hybridization represent a more complicated situation. Because of the quiltwork nature of changes in relative abundance, no regular geographic pattern of divergence in an isolating trait would be expected to develop. On a site by site basis, reproductive character displacement is more likely to occur in a taxon where it is less abundant. Distance from the species border is less important in mosaic hybrid zones with large patch sizes because pure species populations occur throughout the zone.

Response by only one species (unilateral reproductive character displacement) is another possible outcome of interactions between two species in an overlap zone. Unilateral reproductive character displacement would be expected to occur when the loss of fitness associated with hybrid mating is greater for one taxon than the other (Butler, 1988) or when one species harbors greater heritable variation in a critically important isolating trait than the other.

The strongest possible falsification of the reinforcement hypothesis for a particular taxon in a zone of overlap occurs when no evidence for the expected outcome of reinforcement—reproductive character displacement—can be found in samples from sites where the displacement is most likely to occur, namely, at sites in which the taxon is relatively rare that are distant from the species border.

One of the consequences of all of the above is that biologists studying reproductive character displacement must take account of relative abundance and distance from the contact zone border when setting up their studies. Another consequence is that pooling of results from different localities within the contact zone should be avoided.

CONCLUSIONS

In a 1978 paper Paterson wondered why authors ever invoke the reinforcement model given that the theoretical objections to the model are strong. He wondered whether the reason could be that reinforcement is needed to justify preconceived ideas on the nature of species. I suggest that the answer to the question is different: Evolutionary biologists keep invoking the model because it seems to explain patterns they find in nature. The eventual fate of the model will be determined by some difficult field and laboratory studies.

ACKNOWLEDGMENTS

I am grateful to Brayton Smoot and Charles Thaeler for their comments on an early draft of this chapter. The writing of this chapter was partially supported by NSF grants BSR 8600429 and BSR 9006484.

REFERENCES

Alatalo, R. V., Gustafsson, L., and Lundberg, A. 1982. Hybridization and breeding success of collared and pied flycatchers on the island of Gotland. Auk 99:285–291.

Asquith, A., Altig, R., and Zimba, P. 1988. Geographic variation in the mating call of the green treefrog *Hyla cinerea*. Am. Midl. Nat. 119:101–110.

Ball, R. W., and Jameson, D. L. 1966. Premating isolating mechanisms in sympatric and allopatric *Hyla regilla* and *Hyla californiae*. Evolution 20:533–551.

Barton, N. H., and Hewitt, G. M. 1981. Hybrid zones and speciation. In W. R. Atchley and D. S. Woodruff (eds.). Evolution and Speciation: Essays in Honor of M. J. D. White. Cambridge: Cambridge University Press, pp. 109–145.

Bazykin, A. D. 1969. Hypothetical mechanism of speciation. Evolution 23:685–687.

Bazykin, A. D. 1972. The disadvantage of heterozygotes in a system of two adjacent populations. Genetika 8:155–161.

Bazykin, A. D. 1973. Population genetic analysis of disruptive and stabilizing selection. Part II. Systems of adjacent populations and populations within a continuous area. Genetika 9:156–166.

Benedix, J. H., Jr., and Howard, D. J. 1991. Calling song displacement in a zone of overlap and hybridization. Evolution 45:1751–1759.

Bigelow, R. S. 1965. Hybrid zones and reproductive isolation. Evolution 19:449–458.

Blair, W. F. 1955. Mating call and stage of speciation in the *Microhyla olivacea-M. carolinensis* complex. Evolution 9:469–480.

Blair, W. F. 1964. Isolating mechanisms and interspecies interactions in anuran amphibians. Q. Rev. Biol. 39:333–344.

Brown, W. L., and Wilson, E. O. 1956. Character displacement. Syst. Zool. 5:49–64.

Burma, B. H. 1954. Reality, existence, and classification: a discussion of the species problem. Madrono 12:193–209.

Bush, G. L., and Howard, D. J. 1986. Allopatric and nonallopatric speciation: assumptions and evidence. In S. Karlin and E. Nevo (eds.). Evolutionary Processes and Theory. Orlando, FL: Academic Press, pp. 411–438.

Butler, M. J. 1988. Evaluation of possibly reproductively mediated character displacement in the crayfishes, *Oronectes rusticus* and *O. sanbornii*. Ohio J. Sci. 88:87–91.

Butler, M. J., IV, and Stein, R. A. 1985. An analysis of the mechanisms governing species replacements in crayfish. Oecologia 66:169–177.

Butlin, R. 1987. Speciation by reinforcement. Trends Ecol. Evol. 2:8–13.

Butlin, R. 1989. Reinforcement of premating isolation. In D. Otte and J. A. Endler (eds.). Speciation and Its Consequences. Sunderland, MA: Sinauer, pp. 158–179.

Caisse, M., and Antonovics, J. 1978. Evolution in closely adjacent plant populations. IX. Evolution of reproductive isolation in clinal populations. Heredity 40:371–384.

Carson, H. L. 1986. Sexual selection and speciation. In S. Karlin and E. Nevo (eds.). Evolutionary Processes and Theory. Orlando, FL: Academic Press, pp. 391–409.

Cohn, T. J., and Cantrall, I. J. 1974. Variation and speciation in the grasshoppers of the Conalcaeini (Orthoptera: Acrididae: Melanoplinae): the lowland forms of western Mexico, the genus *Barytettix*. San Diego Nat. Hist. Mem. 6:1–131.

Corbin, K. W., and Sibley, C. G. 1977. Rapid evolution in orioles of the genus *Icterus*. Condor 79:335–342.

Cothran, E. G., and Zimmerman, E. G. 1985. Electrophoretic analysis of the contact zone between *Geomys breviceps* and *Geomys bursarius*. J. Mamm. 66:489–497.

Coyne, J. A. 1974. The evolutionary origin of hybrid inviability. Evolution 28:505–506.

Coyne, J. A., and Orr, H. A. 1989. Patterns of speciation in *Drosophila*. Evolution 43:362–381.

Coyne, J. A., Orr, H. A., and Futuyma, D. J. 1988. Do we need a new species concept? Syst. Zool. 37:190–200.

Cracraft, J. 1989. Speciation and its ontology: the empirical consequences of alternative species concepts for understanding patterns and processes of differentiation. In D. Otte and J. A. Endler (eds.). Speciation and Its Consequences. Sunderland, MA: Sinauer, pp. 28–59.

Darwin, C. 1859. On the Origin of Species by Means of Natural Selection, or the Preservation of Favoured Races in the Struggle for Life. London: John Murray.

Darwin, C. 1874. The Descent of Man, and Selection in Relation to Sex, 2nd ed. London: John Murray.

Dobzhansky, T. 1935. A critique of the species concept in biology. Philos. Sci. 2:344:355.

Dobzhansky, T. 1937. Genetics and the Origin of Species. New York: Columbia University Press.

Dobzhansky, T. 1940. Speciation as a stage in evolutionary divergence. Am. Naturalist 74:312–321.

Dobzhansky, T. 1970. Genetics of the Evolutionary Process. New York: Columbia University Press.

Dobzhansky, T., Ehrman, L., Pavlovsky, O., and Spassky, B. 1964. The superspecies *Drosophila paulistorum.* Proc. Natl. Acad. Sci. USA 51:3–9.

Dobzhansky, T., and Koller, P. C. 1938. An experimental study of sexual isolation in *Drosophila.* Biol. Zentralbl. 58:589–607.

Donoghue, M. J. 1985. A critique of the biological species concept and recommendations for a phylogenetic alternative. Bryologist 88:172–181.

Dowling, T. E., and Moore, W. S. 1985. Evidence for selection against hybrids in the family Cyprinidae (genus *Notropis*). Evolution 39:152–158.

Du Rietz, G. E. 1930. The fundamental units of biological taxonomy. Svensk Biotanisk Tidskrift 24:333–428.

Ehrman, L. 1965. Direct observation of sexual isolation between allopatric and sympatric strains of the different *Drosophila paulistorum* races. Evolution 19:459–464.

Endler, J. A. 1986. Natural Selection in the Wild. Princeton, NJ: Princeton University Press.

Feder, J. 1979. Natural hybridization and genetic divergence between the toads *Bufo boreas* and *Bufo punctatus.* Evolution 33:1089–1097.

Felsenstein, J. 1981. Skepticism towards Santa Rosalia, or why are there so few kinds of animals? Evolution 35:124–138.

Ferguson, G. W. 1973. Character displacement of the push-up displays of two partially-sympatric species of spiny lizards, *Sceloporus* (Sauria: Iguanidae). Herpetologica 29:281–284.

Fontana, P. G., and Hogan, T. W. 1969. Cytogenetic and hybridization studies of geographic populations of *Teleogryllus commodus* (Walker) and *T. oceanicus* (Le Guillou) (Orthoptera: Gryllidae). Aust. J. Zool. 17:13–35.

Gartside, D. F. 1980. Analysis of a hybrid zone between chorus frogs of the *Pseudacris nigrita* complex in the southern United States. Copeia 1980:56–66.

Gill, F. B., and Murray, B. G., Jr. 1972. Discrimination behavior and hybridization of the blue-winged and golden-winged warblers. Evolution 26:282–293.

Grant, P. R. 1972. Convergent and divergent character displacement. Biol. J. Linn Soc. 4:39–68.

Grant, V. 1966. The selective origin of incompatibility barriers in the plant genus Gilia. Am. Naturalist 100:99–118.

Greenfield, D. W., and Deckert, G. D. 1973. Introgressive hybridization between *Gila orcutti* and *Hesperoleucus symmetricus* (Pisces: Cyprinidae) in the Cuyama River Basin, California. II. Ecological aspects. Copeia 1973:417–427.

Gulick, J. T. 1890a. Divergent evolution and the Darwinian theory. Am. J. Sci. 39:21–30.

Gulick, J. T. 1890b. Divergent evolution through cumulative segregation. J. Linn. Soc. Zool. 20:189–274.

Gulick, J. T. 1890c. Indiscriminate separation under the same environment, a cause of divergence. Nature 42:369–370.

Gulick, J. T. 1890d. Unstable adjustments as affected by isolation. Nature 42:28–29.

Gulick, J. T. 1891. Intensive segregation, or divergence through independent transformation. J. Linn. Soc. Zool. 23:312–380.

Guttman, S. I., and Karlin, A. A. 1986. Hybridization of cryptic species of two lined salamander (*Eurycea bislineata* complex). Copeia 1986:96–108.

Hafner, J. C., Hafner, D. J., Patton, J. L., and Smith, M. F. 1983. Contact zones and the genetics of differentiation in the pocket gopher *Thomomys bottae* (Rodentia: Geomyidae). Syst. Zool. 32:1–20.

Hall, W. P., and Selander, R. K. 1973. Hybridization of karyotypically differentiated populations of the *Sceloporus grammicus* complex (Iguanidae). Evolution 27:226–242.

Harrison, R. G. 1983. Barriers to gene exchange between closely related cricket species. I. Laboratory hybridization studies. Evolution 37:245–251.

Harrison, R. G. 1986. Pattern and process in a narrow hybrid zone. Heredity 56:337–349.

Harrison, R. G. 1990. Hybrid zones: windows on evolutionary process. Oxford Surv. Evol. Biol. 7:69–128.

Harrison, R. G., and Rand, D. M. 1989. Mosaic hybrid zones and the nature of species boundaries. In D. Otte and J. A. Endler, eds. Speciation and its Consequences. Sunderland, MA: Sinauer, pp. 111–133.

Hewitt, G. M., Butlin, R. K., and East, T. M. 1987. Testicular dysfunction in hybrids between parapatric subspecies of the grasshopper *Chorthippus parallelus.* Biol. J. Linn. Soc. 31:25–34.

Highton, R., and Henry, S. A. 1970. Evolutionary interactions between species of North American salamanders of the genus *Plethodon.* Evol. Biol. 3:211–256.

Hill, K. G., Loftus-Hills, J. J., and Gartside, D. F. 1972. Pre-mating isolation between the Australian field crickets *Teleogryllus commodus* and *T. oceanicus* (Orthoptera: Gryllidae). Aust. J. Zool. 20:153–163.

Hillis, D. M. 1981. Premating isolating mechanisms among three species of the *Rana pipiens* complex in Texas and southern Oklahoma. Copeia 1981:312–319.

Howard, D. J. 1986. A zone of overlap and hybridization between two ground cricket species. Evolution 40:34–43.

Howard, D. J. 1993. (in press). Small populations, inbreeding, and speciation. In N. W. Thornhill, ed. The Natural History of Inbreeding and Outbreeding: Theoretical and Empirical Perspectives. Chicago: The University of Chicago Press.

Howard, D. J., and Waring, G. L. 1991. Topographic diversity, zone width, and the strength of reproductive isolation in a zone of overlap and hybridization. Evolution 45:1120–1135.

Hubbs, C. 1960. Duration of sperm function in the percid fishes *Etheostoma lepidum* and *E. spectabile,* associated with sympatry of the parental populations. Copeia 1960:1–8.

Hunt, W. G., and Selander, R. K. 1973. Biochemical genetics of hybridization in European house mice. Heredity 31:11–33.

Jackson, J. F. 1973. The phenetics and ecology of a narrow hybrid zone. Evolution 27:58–68.

Johnson, M. S. 1982. Polymorphism for direction of coil in *Partula suturalis:* behavioural isolation and positive frequency dependent selection. Heredity 49:145–151.

Jordan, D. S. 1905. The origin of species through isolation. Science 22:545–562.

Kocher, T. D., and Sage, R. D. 1986. Further genetic analyses of a hybrid zone between leopard frogs (*Rana pipiens* complex) in central Texas. Evolution 40:21–33.

Levin, D. A., and Kerster, H. W. 1967. Natural selection for reproductive isolation in *Phlox.* Evolution 21:679–687.

Littlejohn, M. J. 1959. Call differentiation in a complex of seven species of *Crinia* (Anura, Leptodactylidae). Evolution 13:452–468.

Littlejohn, M. J. 1965. Premating isolation in the *Hyla ewingi* complex (Anura: Hylidae). Evolution 19:234–243.

Littlejohn, M. J. 1981. Reproductive isolation: a critical review. In W. R. Atchley and D. S. Woodruff, eds. Evolution and Speciation: Essays in Honor of M. J. D. White. Cambridge: Cambridge University Press, pp. 298–334.

Littlejohn, M. J., and Loftus-Hills, J. J. 1968. An experimental evaluation of premating isolation in the *Hyla ewingi* complex (Anura: Hylidae). Evolution 22:659–663.

Littlejohn, M. J., and Watson, G. F. 1985. The *Litoria ewingi* complex (Anura: Hylidae) in south-eastern Australia. VII. Mating-call structure and genetic compatibility across a narrow hybrid zone between *L. ewingi* and *L. paraewingi.* Aust. J. Zool. 31:193–204.

Littlejohn, M. J., Watson, G. F., and Loftus-Hills, J. J. 1971. Contact hybridization in the *Crinia laevis* complex (Anura: Leptodactylidae). Aust. J. Zool. 19:85–100.

Maynard Smith, J. 1966. Sympatric speciation. Am. Naturalist 100:637–650.

Mayr, E. 1942. Systematics and the Origin of Species. New York: Columbia University Press.

Mayr, E. 1963. Animal Species and Evolution. Cambridge, MA: Belknap Press.

McCarley, H. 1964. Ethological isolation in the cenospecies *Peromyscus leucopus.* Evolution 18:331–342.

McDonnell, L. J., Gartside, D. F., and Littlejohn, M. J. 1978. Analyses of a narrow hybrid zone between two species of *Pseudophryne* (Anura: Leptodactylidae). Evolution 32:602–612.

McLain, D. K. 1985. Clinal variation in morphology and assortative mating in the soldier beetle, *Chauliognathus pennsylvanicus* (Coleoptera: Cantharidae). Biol. J. Linn. Soc. 25:105–117.

McLain, D. K., and Rai, K. S. 1986. Reinforcement for ethological isolation in the southeast Asian *Aedes albopictus* subgroup (Diptera: Culicidae). Evolution 40:1346–1350.

McPhail, J. D. 1969. Predation and the evolution of a stickleback *(Gasterosteus).* J. Fish. Res. Bd. Can. 26:3183–3208.

Mecham, J. S. 1961. Isolating mechanisms in anuran amphibians. In W. F. Blair, ed. Vertebrate Speciation. Austin: University of Texas Press, pp. 24–61.

Moore, J. A. 1957. An embryologist's view of the species concept. In E. Mayr, ed. The Species Problem. Washington, DC: American Association for the Advancement of Science, pp. 325–338.

Moore, W. S. 1979. A single locus mass-action model of assortative mating, with comments on the process of speciation. Heredity 42:173–186.

Moore, W. S. 1981. Assortative mating genes selected along a gradient. Heredity 46:191–195.

Moore, W. S. 1987. Random mating in the northern flicker hybrid zone: implications for the evolution of bright and contrasting plumage patterns in birds. Evolution 41:539–546.

Moore, W. S., and Koenig, W. D. 1986. Comparative reproductive success of yellow-shafted, red-shafted and hybrid flickers across a hybrid zone. Auk 103:42–51.

Moran, C. 1979. The structure of the hybrid zone in *Caledia captiva*. Heredity 42:13–32.

Muller, H. J. 1940. Bearings of the 'Drosophila' work on systematics. In J. Huxley, ed. The New Systematics. London: Oxford University Press, pp. 185–268.

Murray, J., and Clarke, B. 1980. The genus *Partula* on Moorea: speciation in progress. Proc. R. Soc. Lond. Biol. 211:83–117.

Nevo, E., and Bar-El, H. 1976. Hybridization and speciation in fossorial mole rats. Evolution 30:831–840.

Nuechterlein, G. L. 1981. Courtship behavior and reproductive isolation between western Grebe color morphs. Auk 98:335–349.

Otte, D. 1989. Speciation in Hawaiian crickets. In D. Otte and J. A. Endler, eds. Speciation and Its Consequences. Sunderland, MA: Sinauer, pp. 482–526.

Paterson, H. E. H. 1978. More evidence against speciation by reinforcement. S. Afr. J. Sci. 74:369–371.

Paterson, H. E. H. 1980. A comment on "mate recognition systems." Evolution 34:330–331.

Paterson, H. E. H. 1981. The continuing search for the unknown and unknowable: a critique of contemporary ideas on speciation. S. Afr. J. Sci. 77:113–119.

Paterson, H. E. H. 1982. Perspective on speciation by reinforcement. S. Afr. J. Sci. 78:53–57.

Paterson, H. E. H. 1986. Environment and species. S. Afr. J. Sci. 82:62–65.

Patton, J. L., Hafner, J. C., Hafner, M. S., and Smith, M. F. 1979. Hybrid zones in *Thomomys bottae* pocket gophers: genetic, phenetic, and ecologic concordance patterns. Evolution 33:860–876.

Phelan, P. L., and Baker, T. C. 1987. Evolution of male pheromones in moths: reproductive isolation through sexual isolation? Science 235:205–207.

Ralin, D. B. 1976. Comparative hybridization of a diploid-tetraploid cryptic species pair of treefrogs. Copeia 1976:191–196.

Ralin, D. B. 1977. Evolutionary aspects of mating call variation in a diploid-tetraploid species complex of treefrogs (Anura). Evolution 31:721–736.

Rand, D. M., and Harrison, R. G. 1989. Ecological genetics of a mosaic hybrid zone: mitochondrial, nuclear, and reproductive differentiation of crickets by soil type. Evolution 43:432–449.

Ratcliffe, L. M., and Grant, P. R. 1983. Species recognition in Darwin's finches (*Geospiza,* Gould). II. Geographic variation in mate preference. Anim. Behav. 31:1154–1165.

Remington, C. L. 1968. Suture zones of hybrid interaction between recently joined biotas. Evol. Biol. 2:321–428.

Rising, J. D. 1983. The progress of Oriole hybridization in Kansas. Auk 100:885–897.

Ritchie, M. G., Butlin, R. K., and Hewitt, G. M. 1989. Assortative mating across a hybrid zone in *Chorthippus parallelus* (Orthoptera: Acrididae). J. Evol. Biol. 2:339–352.

Ruiz, A., Heed, W. B., and Wasserman, M. 1990. Evolution of the *Mojavensis* cluster of cactophilic *Drosophila* with descriptions of two new species. J. Hered. 81:30–42.

Sattler, P. W. 1985. Introgressive hybridization between the spadefoot toads *Scaphiopus bombifrons* and *S. multiplicatus* (Salientia: Pelobatidae). Copeia 1985:324–332.

Sawyer, S., and Hartl, D. 1981. On the evolution of behavioral reproductive isolation: the Wallace effect. Theor. Pop. Biol. 19:261–273.

Shaw, D. D., and Wilkinson, P. 1980. Chromosomal differentiation, hybrid breakdown and the maintenance of a narrow hybrid zone in *Caledia.* Chromosoma 80:1–31.

Smith, M. F., Patton, J. L., Hafner, J. C., and Hafner, D. J. 1983. *Thomomys bottae* pocket gophers of the central Rio Grande Valley, New Mexico: local differentiation, gene flow, and historical biogeography. Occ. Papers Museum Southwestern Biology, University of New Mexico, No. 2, pp. 1–16.

Spirito, F., Modesti, A., Perticone, P., Cristaldi, M., Federici, R., and Rizzoni, M. 1980. Mechanisms of fixation and accumulation of centric fusions in natural populations of *Mus musculus* L. I. Karyological analysis of a hybrid zone between two populations in the central Apennines. Evolution 34:453–466.

Stangl, F. B., Jr. 1986. Aspects of a contact zone between two chromosomal races of *Peromyscus leucopus* (Rodentia: Cricetidae). J. Mamm. 67:465–473.

Szymura, J. M., and Barton, N. H. 1986. Genetic analysis of a hybrid zone between the fire-bellied toads, *Bombina bombina* and *B. variegata,* near Cracow in southern Poland. Evolution 40:1141–1159.

Templeton, A. R. 1987. Species and speciation. Evolution 41:233–235.

Templeton, A. R. 1989. The meaning of species and speciation: a genetic perspective. In D. Otte and J. A. Endler, eds. Speciation and Its Consequences. Sunderland, MA: Sinauer, pp. 3–27.

Thaeler, C. S. 1968. An analysis of three hybrid populations of pocket gophers (genus *Thomomys*). Evolution 22:543–555.

Thaeler, C. S., Jr. 1974. Four contacts between ranges of different chromosome forms of the *Thomomys talpoides* complex (Rodentia: Geomyidae). Syst. Zool. 23:343–354.

Turner, J. R. G. 1971. Two thousand generations of hybridization in a *Heliconius* butterfly. Evolution 25:471–482.

Waage, J. K. 1975. Reproductive isolation and

the potential for character displacement in the damselflies *Calopteryx maculata* and *C. aequabilis* (Odonata: Calopterygidae). Syst. Zool. 24:24–36.

Waage, J. K. 1979. Reproductive character displacement in Calopteryx (Odonata: Calopterygidae). Evolution 33:104–116.

Wade, M. J., and Arnold, S. J. 1980. The intensity of sexual selection in relation to male sexual behavior, female choice, and sperm precedence. Anim. Behav. 28:446–461.

Wake, D. B., Yanev, K. P., and Brown, C. W. 1986. Intraspecific sympatry in a "ring species," the Plethodontid salamander *Ensatina eschscholtzii*, in southern California. Evolution 40:866–868.

Wake, D. B., Yang, S. Y., and Papenfuss, T. J. 1980. Natural hybridization and its evolutionary implications in Guatemalan Plethodontid salamanders of the genus *Bolitoglossa*. Herpetologica 36:335–345.

Walker, T. J. 1974. Character displacement and acoustic insects. Am. Zool. 14:1137–1150.

Wallace, A. R. 1889. Darwinism. An Exposition of the Theory of Natural Selection with Some of Its Applications. London: Macmillan.

Wallin, L. 1986. Divergent character displacement in the song of two allospecies: the pied flycatcher *Ficedula hypoleuca*, and the collared flycatcher *Ficedula albicollis*. Ibis 128:251–259.

Wasserman, M., and Koepfer, H. R. 1977. Character displacement for sexual isolation between *Drosophila mojavensis* and *Drosophila arizonensis*. Evolution 31:812–823.

Watson, G. F. 1972. The *Litoria ewingi* complex (Anura: Hylidae) in south-eastern Australia. II. Genetic incompatibility and delimitation of a narrow hybrid zone between *L. ewingi* and *L. paraewingi*. Aust. J. Zool. 20:423–433.

Watson, G. F., and Littlejohn, M. J. 1978. The *Litoria ewingi* complex (Anura: Hylidae) in South-eastern Australia. V. Interactions between northern *L. ewingi* and adjacent taxa. Aust. J. Zool. 26:175–195.

West-Eberhard, M. J. 1983. Sexual selection, social competition, and speciation. Q. Rev. Biol. 58:155–183.

Whalen, M. D. 1978. Reproductive character displacement and floral diversity in Solanum section *Androceras*. Syst. Bot. 3:77–86.

Wheeler, N. C., and Guries, R. P. 1987. A quantitative measure of introgression between lodgepole and jack pines. Can. J. Bot. 65:1876–1885.

Wilson, E. O. 1965. The challenge from related species. In H. G. Baker and G. L. Stebbins, eds. The Genetics of Colonizing Species. Orlando, FL: Academic Press, pp. 7–24.

Woodruff, D. S. 1979. Postmating reproductive isolation in *Pseudophryne* and the evolutionary significance of hybrid zones. Science 203:561–563.

Woodruff, P. S., and Gould, S. J. 1987. Fifty years of interspecific hybridization: genetics and morphometrics of a controlled experiment on the land snail *Cerion* in The Florida keys. Evolution 41:1022–1045.

Wynn, A. H. 1986. Linkage disequilibrium and a contact zone in *Plethodon cinereus* on the Del-Mar-Va peninsula. Evolution 40:44–54.

Zouros, E., and D'Entremont, C. J. 1980. Sexual isolation among populations of *Drosophila mojavensis:* response to pressure from a related species. Evolution 34:421–430.

4

Introgression and Its Consequences in Plants

LOREN H. RIESEBERG AND JONATHAN F. WENDEL

The role of introgression in plant evolution has been the subject of considerable discussion since the publication of Anderson's influential monograph, *Introgressive Hybridization* (Anderson, 1949). Anderson promoted the view, since widely held by botanists, that interspecific transfer of genes is a potent evolutionary force. He suggested that "the raw material for evolution brought about by introgression must greatly exceed the new genes produced directly by mutation" (1949, p. 102) and reasoned, as have many subsequent authors, that the resulting increases in genetic diversity and number of genetic combinations promote the development or acquisition of novel adaptations (Anderson, 1949, 1953; Stebbins, 1959; Rattenbury, 1962; Lewontin and Birch, 1966; Raven, 1976; Grant, 1981). In contrast to this "adaptationist" perspective, others have accorded little evolutionary significance to introgression, suggesting instead that it should be considered a primarily local phenomenon with only transient effects, a kind of "evolutionary noise" (Barber and Jackson, 1957; Randolph et al., 1967; Wagner, 1969, 1970; Hardin, 1975). One of the vociferous doubters of a significant role of hybridization in plant evolution was Wagner (1969, p. 785), who commented that the "ultimate contributions made by hybrids must be very small or negligible." Wagner's frequently expressed opinion appears to be based on ecological and compatibility arguments, which were encapsulated as follows: "In the rare cases that two well differentiated species happen to be interfertile enough to produce fertile progeny, their hybrids will usually have to fit into some hybrid niche. Such fertile hybrids will therefore tend to be transient, disappearing once the differentiated community returns and the parental species re-occupy their normal habitats" (Wagner, 1970, p. 149).

These divergent opinions primarily reflect differences in how various authors view the relative frequency of particular evolutionary outcomes, with few if any authors advocating the position that introgression *never* plays a significant evolutionary role. Anderson himself expressed uncertainty regarding its importance: "It is premature to attempt any generalizations as to the importance of introgressive hybridization in evolution" (1949, p. 61). Heiser was similarly equivocal, reaching the conclusion that "Introgression does undoubtedly play a role in evolution. . . . It may play a very significant role; but it must be admitted, there is yet no strong evidence to support such a claim" (1973, p. 362).

In this chapter, we ask if any such "strong evidence" has accumulated since the last comprehensive review of plant introgression by Heiser nearly two decades ago and if any new perspectives or issues have been raised by empirical studies. We review evidence bearing on the extent of introgression in plants and its putative consequences. Particular emphasis is placed on insights gained from the application of molecular techniques, which provide the simply inherited genetic markers necessary to address many of the relevant questions but that were unavailable when the subject was last reviewed (see also Rieseberg and Brunsfeld, 1992). It is shown that these molecular data have greatly enhanced our ability to detect and quantify introgression, with perhaps a less dramatic influence on our ability to perceive its evolutionary consequences. We also review recent studies that have provided several insights into patterns and possible mechanisms of plant introgression.

WHAT IS INTROGRESSION?

The term "introgression" has been used to describe a wide range of phenomena, from backcrossing in hybrid swarms and breeding experiments to the exchange of genes between primarily allopatric species. Introgression or "introgressive hybridization" was first defined by Anderson and Hubricht (1938, p. 396) as "the infiltration of germ plasm from one species into another through repeated backcrossing of the hybrids to the parental species." Later authors (e.g., Stebbins, 1959; Heiser, 1973) suggested that the term be restricted to those situations involving the "permanent" addition of genes from one species into another, thereby attempting to draw a distinction between introgression and transient gene flow in local hybrid swarms. This distinction seems to us to be useful despite the difficulties inherent in assessing "permanence" of genetic transfer. Heiser (1973) pointed out that the term introgression need not apply only to results from backcrossing, inasmuch as the particular genetic history is likely to be unknown in most situations and because experimental data indicate that sib-crossing in conjunction with backcrossing is often superior to strict backcrossing for facilitating interspecific gene exchange (Wall, 1970). There is little rationale, in our opinion, for requiring any specific crossing scheme in a definition of introgression, regardless of issues of relative "efficiency." Although introgression was originally restricted to gene flow between species, it has been pointed out (Anderson, 1949; Heiser, 1973; Grant, 1981) that this definition is necessarily arbitrary owing to differences in species concepts among taxonomists. Furthermore, Anderson (1949) suggested that gene flow between intraspecific taxa represents essentially the same phenomenon as introgression between species. We concur and suggest that introgression can appropriately describe gene exchange between species, subspecies, races, or any other set of differentiated population systems. Given these considerations, introgression can be defined as the permanent incorporation of genes from one set of differentiated populations into another, i.e., the incorporation of alien alleles into a new, reproductively integrated population system.

When applying this definition to this review, we found it necessary to exclude from consideration certain evolutionary phenomena. The vast literature on local, ephemeral hybrid swarms is not presented. With rare exception, hybridization leading to polyploidy is not discussed, despite the fact that it is one of the most prominent processes of plant speciation (Stebbins, 1950; Jackson, 1976; Lewis, 1980; Grant, 1981; Levin, 1983). Other than where it pertains to evolutionary issues, the vast literature on artificial hybridization and backcrossing in crop plants is omitted. Our discussion thus focuses on "natural" hybridization and introgression among diploid plants.

HISTORICAL PERSPECTIVE

Botanists have long been fascinated by plant hybridization, and the literature on this subject is voluminous. Excellent summaries of the early literature are given by Roberts (1929) and Heiser (1949a). Two of the earliest papers addressing the subject of introgression are those by DuRietz (1930) and Marsden-Jones (1930). DuRietz (1930) studied populations of *Dacrophyllum, Coprosma,* and *Salix* and noted that sympatric populations of certain species pairs tended to converge in certain morphological features. He attributed this convergence to the "infection" of one species with particular genes from another species. Marsden-Jones (1930) produced artificial hybrids between *Geum urbanum* and *G. rivale* and then backcrossed the hybrids to the parental species. He demonstrated that the backcrosses contained individuals that were almost indistinguishable from the parental species, confirming predictions made by Ostenfald (1928).

That introgression may have evolutionary implications, however, was first recognized by Edgar Anderson (Anderson, 1936a, 1949, 1953). In a series of papers on *Tradescantia* (Anderson and Diel, 1932; Anderson and Woodson, 1935; Anderson, 1936a, Anderson and Hubricht, 1938), Anderson and coworkers provided the first careful experimental studies of introgression, developed methods for its analysis (hybrid indices), introduced the term "introgressive hybridization," and suggested several possible consequences of introgression, including an increase in genetic diversity, the transfer of adaptations, and the development of new adaptations. They also were the first to explicitly recognize the relation between hybridization and the habitat (Anderson and Hubricht, 1938; Anderson, 1948). Anderson (1948) noted that different habitat preferences often form strong barriers to hybridization. Natural or human disturbance may remove these barriers, leading to extensive hybridization. In later papers (Anderson, 1949; Anderson and Gage, 1952), Anderson introduced additional methods for the detection of introgression, including the method of pictorialized scatter diagrams, which were widely used for several decades.

Anderson (1939, 1949, 1953) also introduced the concept of character coherence. Briefly, he suggested that linkage among genes affecting taxonomic characters results in strong correlations among these characters in the offspring of species hybrids. Thus hybrid indices and pictorialized scatter diagrams, in his view, provide an efficient means for analyzing suspected cases of introgression and for distinguishing the morphological results of introgression from those of convergent mutations or retention of ancestral character states. Although the concept of character coherence is still accepted by most plant taxonomists (e.g., Grant, 1981), theoretical and experimental studies (Dempster, 1949; Goodman, 1966) have indicated that factors other than linkage may also be responsible for taxonomic character correlations. For example, selective elimination of recombinant types could result in strong character coherence regardless of linkage. In addition, if introgression is recent, correlations caused by overlapping hybrid and backcross generations can result in character correlations not generated by linkage alone (Goodman, 1966). Thus Goodman (1966) suggested that Andersonian techniques are applicable only in cases where it is known that there is no differential selection among recombinant and nonrecombinant chromosomal types.

This example typifies a problem that pervades the introgression literature; i.e., the supporting evidence often has alternative explanations (Gottlieb, 1972; Heiser, 1973). Many of these alternatives have been recognized for decades. Dobzhansky (1941), for example, recognized the implications presented by convergent morphological evolution. He also suggested the possibility that remnants of the ancestral population from which two species differentiated might have the appearance of hybrids—an early and explicit recognition of symplesiomorphy (retention of primitive character-states). It has also been recognized that primary intergradation could be difficult to distinguish from secondary intergradation. Barber and Jackson (1957), for example, questioned the traditional assumption that a steep cline—an abrupt change in a particular character or group of characters—always results from the merger of two previously differentiated populations. Baker (1947) was skeptical of the use of hybrid indices and other biometric tools in the absence of information regarding the genetic basis of the characters being scored. These and several additional explanations for variation patterns suggestive of introgression were discussed by Gottlieb (1972) and Heiser (1973), including segregation in a polyploid species, inbreeding and selection following hybridization in an autogamous species, the occurrence of hybrid swarms that are no longer in contact with the parental species, and the presence of highly variable F_1 hybrids. Given this panoply of potential problems, it is not surprising that Rieseberg et al. (1988a), following Heiser (1973), considered most putative examples of introgression to be based on circumstantial evidence.

Despite the inherent difficulties with adequately demonstrating introgression, there has been considerable discussion regarding its evolutionary significance. A recurring theme in many early studies of introgression was its importance as a source of genetic variation on which selection could act (Anderson, 1949, 1953). This view seems to have been accepted by many botanists. For example, Stebbins suggested that mutation can never provide enough variability to allow major evolutionary advances to take place: "Genetic recombination must, therefore, be the major source of such variability. . . . This is accomplished most effectively by mass hybridization between populations with different adaptive norms" (1959, p. 248). Rattenbury (1962) attributed the survival of tropical elements in the New Zealand flora to intermittent periods of hybridization (cyclic hybridization) and the production of variable offspring of which some could survive under cooler conditions. Knobloch (1972, p. 97) provided a list of 23,675 natural plant hybrids and states that "although mutation has been given the major role in effecting diversity in the natural world . . . , it is now quite clear to many biologists that the role of hybridization in speciation has been much larger." Likewise, Raven (1976, p. 298) suggested that "the formation of hybrids is a consistent feature of the adaptive system in many, if not most, groups of plants." As mentioned above, this point of view has not met with universal acceptance (Barber and Jackson, 1957; Randolph et al., 1967; Wagner, 1969, 1970; Hardin, 1975).

Many of the "modern" contributions to the study of introgression have been methodological. The use of secondary chemical compounds (e.g., Alston and Turner, 1962; Flake et al., 1978) broadened the database available for the analysis of introgression. More recently, molecular evidence has been applied to the study of introgression, and it has been argued by several authors (e.g., Levin, 1975; Doebley, 1989a; Doebley and Wendel, 1989; Rieseberg and Brunsfeld, 1992) that molecular data are often preferable to morphological data for analyzing ambiguous cases of introgression because of (1) the ready availability of large numbers of independent molecular markers that allow the detection and quantification of even rare introgression; (2) the generally infrequent nonheritable molecular variation (Hillis, 1987); and (3) the apparent selective neutrality of many molecular markers (Kimura, 1982). In contrast, there are often few morphological characters differentiating hybridizing taxa, and these characters are often functionally or developmentally correlated. Moreover, morphological characters typically have an unknown, but presumably complicated, genetic basis, have a nonheritable component that is difficult to estimate, and often converge when exposed to similar selective pressures. Molecular markers also provide the opportunity to monitor both nuclear and cytoplasmic gene flow. Several authors have noted the advantages of employing chromosomally linked molecular markers for elucidating the direction of introgression and distinguishing it from symplesiomorphy and convergence (Avise and Saunders, 1984; Doebley, 1989a; Doebley and Wendel, 1989; Rieseberg et al., 1990a). The use of linked markers greatly increases the potential for the *simultaneous* appearance of multiple markers in an introgressed individual. Clearly, if a putative introgressant possessed multiple, linked markers of a potential hybridizer, the probability that this situation could be attributed to symplesiomorphy or convergence would be minimized.

Although molecular markers clearly provide an important alternative to morphological characters for the study of introgression, there are potential problems, particularly for multigene families such as the nuclear ribosomal RNA gene family (rDNA). Multigene families, which occur as tandem arrays or dispersed throughout the genome, are often subject to "concerted evolution," where sequences within a gene family are corrected against each other by processes such as unequal crossing over, gene conversion, slippage replication, or RNA-mediated exchanges (Drouin and Dover, 1990). If members of a multigene family are used to study introgression, these molecular processes may represent a serious source of error, as the frequency of the introgressed genes in an individual, population, or taxon could be increased or decreased because of this process. It is noteworthy that gene conversion has been experimentally demonstrated for rDNA (Hillis et al., 1991), a gene family often used in studies of introgression (e.g., Arnold et al., 1990a; Rieseberg et al., 1990a). Furthermore, biased gene conversion has been used to account for introgressive patterns of rDNA variation in grasshoppers (Arnold et al., 1988; Marchant et al., 1988), demonstrating the potential evolutionary significance of this phenomenon.

EXTENT OF INTROGRESSION IN PLANTS

To estimate the extent of introgression in nature, it is worthwhile to first examine the frequency of hybridization, as hybridization is a prerequisite to introgression. Naturally occurring interspecific hybrids have been detected in all major groups of plants and in all well-studied floras (Grant, 1981). Although only a small fraction of the plant kingdom has been examined in detail and numerous hybrids go unreported, Knobloch (1972) was able to list 23,675 putative interspecific or intergenetic hybrids. It is more difficult to estimate the frequency of introgression in plants owing to the difficulty of obtaining unambiguous evidence (as discussed in this chapter). Nonetheless, we have attempted to list "noteworthy" cases of introgression in plants (Table 4-1) so as to: (1) provide a reasonably broad introduction to the primary literature; (2) illustrate the breadth of the phenomenon with respect to taxonomic groups and life history features; and (3) summarize the types of empirical evidence used in documenting introgression as well as its proposed consequences. This list is by no means exhaustive, and we have undoubtedly missed many noteworthy examples. In addition, we have included several studies where previously hypothesized examples of introgression have been examined and disproved (e.g., Rieseberg et al., 1988a; Spooner et al., 1991), because these "negative" papers play a role in clarifying the introgression literature.

Table 4-1 lists 165 proposed cases of introgression, many involving more than one species. The examples include one fern *(Trichomanes)* and the full spectrum of seed plant diversity, including three genera in two families of gymnosperms *(Abies, Pinus, Juniperus)* and 82 genera in 40 families of angiosperms. Within angiosperms, approximately 85% of the examples are from dicotyledonous families (34 families, 71 genera), including representatives from all dicot subclasses. Twelve genera in six families of monocots are also represented. Nearly all growth forms, which range from annuals to perennials and trees to herbs, are included. Many pollination syndromes are represented, with wind pollination being rarer than various forms of animal pollination. This finding perhaps runs counter to expectations based on the physical promiscuity inherent in the former, relative to the frequent specificity of the latter. Mating system variation also covers the full spectrum from obligate outcrossers to predominant selfers.

Each study listed in Table 4-1 was evaluated with respect to whether introgression had been rigorously documented in the sense that alternative explanations (see above) were eliminated. In general, we had greater confidence in those studies that employed molecular characters (see above for rationale) or employed numerous morphological, cytological, or chemical characters. Our judgment was that introgression had been documented in 65 cases. Other studies included in Table 4-1 substantiate the existence of hybrid swarms and *may* represent examples of introgression (denoted by "?"). We did not believe, however, that the "permanent addition" of alleles from another species was unambiguously demonstrated in these cases.

In his 1973 review, Heiser distinguished localized introgression, where gene flow extends only a short distance from the area of hybridization, from dispersed introgression, where gene flow is widespread (Fig. 4-1). He suggested that localized introgression is common, but dispersed introgression is rare. Although most of the examples on this list undoubtedly represent instances of localized introgression (denoted by "L"), several cases of dispersed introgression have now been documented as well ("D" in Table 4-1). This distinction, however, is not easily drawn. As pointed out previously (Rieseberg and Brunsfeld, 1991), it is difficult to determine whether apparent patterns of dispersed introgression represent widespread gene flow or result from the establishment and spread of stabilized introgressants (Fig. 4-1). That is, the genetic constitution of introgressed populations occurring away from the area of contact may be the same regardless of the process by which they were derived. Similarly, in cases where localized

Table 4-1. Proposed Examples of Introgression in Plants

Taxon	Evidence[a]	Introgression?[b]	Proposed Consequences[c]	References
Abies borisii-regis	M	?	S	Mattfeld (1930)
Acer saccharophorum/nigrum	M	?	T	Dansereau & Desmarais (1947)
Achillea roseo-alba	M,C	?	S	Ehrendorfer (1959)
Adenostoma fasciculatum	M	?	I,T	Anderson (1954)
Aegilops species	M	?	I,O	Feldman (1965)
Aesculus species	I,M,C	D	I	Hardin (1957); dePamphilis and Wyatt (1989, 1990)
Amaranthus species	M,C	?	B,E	Sauer (1957); Tucker and Sauer (1958)
Apocynum species	M,C	?	I	Anderson (1936b)
Aquilegia formosa/pubescens	M,E	?	I	Grant (1952); Chase and Raven (1975)
Arctostaphylos viscida/patula	I	L	n.a.	Ellstrand et al. (1987)
Argyranthemum sundingii	M	?	S	Brochmann (1987)
Argyranthemum coronopifolium/frutescens	M	L	B	Brochmann (1984)
Argyroxiphium grayanum	P	yes	n.a.	Baldwin et al. (1990)
Ascepias exaltata/syriaca	M	?	I,O	Kephart et al. (1988)
Asclepias speciosa/syriaca	S,I	?	n.a.	Adams et al. (1987)
Asclepias tuberosa	M	?	n.a.	Woodson (1947, 1962); Wyatt and Antonovics (1981)
Aster multiflorus/novae-angliae	M	?	I	Wetmore and Delisle (1939)
Bothriochloa intermedia	M	?	T,C	Harlan and de Wet (1963)
Brassica napus	P,[R]	yes	n.a.	Erickson et al. (1983); Palmer et al. (1983); Palmer (1988); Song et al. (1988)
Bromus pumpellianus/inermis	M	?	T,E,C	Elliott (1949)
Calyptridium monospermum	M	?	R	Hinton (1976)
Carduus nutans/acanthoides	M,I,P,R,S,C	L	n.a.	Warwick et al. (1989)
Ceonothus species	M,C	?	E,T,S	McMinn (1944)
Cercis canadensis	M	?	T,I	Anderson (1953)
Cercocarpus traskiae	M,I	L	B	Rieseberg et al. (1989)
Cinchona species	M	?	T,O	Camp (1948)
Cistus species	M	?	S,E	Dansereau (1941)
Citrullus lanatus/colocynthis	I	L	n.a.	Zamir et al. (1984)
Clarkia bottae	M,C	?	S	Lewis and Lewis (1955)

Species				Reference
Clarkia sect. *Fibula*	P,M	?	S,O	Lewis and Lewis (1955); Sytsma et al. (1990)
Clarkia speciosa ssp. *polyantha*	C,I,M	yes	E	Bloom (1976); Soltis (1985)
Cucurbita species	M,I	?	n.a.	Decker-Walters et al. (1990)
Cucurbita pepo/texana	I	L	n.a.	Decker and Wilson (1987); Kirkpatrick and Wilson (1988)
Cypripedium candidum/pubescens	M,I	L	E,T	Klier et al. (unpublished data)
Delphinium gypsophilum	M,C	?	S	Lewis and Epling (1959)
Diplacus species	C,M,E	?	I,E	Beeks (1962)
Dubautia species	P,S	L	n.a.	Crins et al. (1988); Baldwin et al. (1990)
Elymus species	M	?	I,E	Brown and Pratt (1960)
Elymus glaucus	M,C	?	S	Snyder (1950, 1951); Stebbins (1957)
Epimedium trifoliatobinatum	M,C	?	S	Suzuki (1986)
Eucalyptus risdonii/amygdalina	M	?	D	Potts and Reid (1988)
Fuchsia perscendens	P,[M]	?,	n.a.	Sytsma et al. (1991)
Galium dumosum	M	?	S	Ehrendorfer (1958)
Gaillardia pulchella	I,S	L	n.a.	Heywood and Levin (1984); Heywood (1986)
Geum urbanum/rivale	C,M	?	n.a.	Ravanko (1979)
Gilia species	M,C	?	O,S,E	Grant (1953)
Gilia schilleaefolia	M,C	?	S	Grant (1954)
Gilia cana asp. *speciosa*	M,C	?	E	Grant and Grant (1960)
Gilia capitata ssp.	M,C	?	S,E	Grant (1950)
Gilia latiflora ssp. *davyi*	M,C	?	E	Grant and Grant (1960)
Gilia leptantha ssp. *transversa*	M,C	?	E	Grant and Grant (1960)
Gilia ochroleuca ssp. *vivida*	M,C	?	E	Grant and Grant (1960)
Gossypium bickii	P,[R,I]	yes	S,O	Wendel et al. (1991)
Gossypium aridum	P	yes	n.a.	Wendel and Albert (1991)
Gossypium arboreum	I	yes	I	Wendel et al. (1989)
Gossypium barbadense	I	yes	I,R,T	Percy and Wendel (1990)
Gossypium cunninghamii	P	yes	n.a.	Wendel and Albert (1992)
Gossypium darwinii	I	yes	n.a.	Wendel and Percy (1990)
Gossypium herbaceum	I	yes	I	Wendel et al. (1989)
Gossypium hirsutum	I	yes	R	Percy and Wendel (1990)
Helianthus annuus ssp. *texanus*	P,R,M,C	yes	T,E,C or N	Heiser (1951a); Rieseberg et al. (1990a)
Helianthus annuus	P,M,C	yes	T,E,C or N	Heiser (1947); Rieseberg et al., 1991a; Stebbins and Daly (1961)

Table 4-1. Proposed Examples of Introgression in Plants *Continued*

Taxon	Evidence[a]	Introgression?[b]	Proposed Consequences[c]	References
Helianthus anomalus	P,R,I	yes	O,S	Rieseberg (1991)
Helianthus argophyllus	M,C	?	T,E,C	Heiser (1951b)
Helianthus bolanderi	M,C,[P,R,I,S]	no	T,S	Heiser (1949b); Oliveri and Jain (1977); Rieseberg (1987); Rieseberg et al. (1988a,b)
Helianthus debilia ssp. *cucumerifolius*	P,R,M,C	yes	N	Heiser (1951a); Rieseberg et al. (1990a, 1991b)
Helianthus debilis ssp. *silvestris*	P,[R]	?	N	Rieseberg et al. (1991a)
Helianthus deserticola	P,R,I	yes	O,S	Rieseberg (1991)
Helianthus divaricatus/microcephalus	M	L	T	Heiser (1979)
Helianthus neglectus	P,[R,I]	yes	N	Rieseberg et al. (1990b)
Helianthus paradoxus	P,R,I	yes	O,S	Rieseberg et al. (1990b)
Helianthus petiolaris	P,R,M,C,	yes	T,E,C or N	Heiser (1947); Dorado et al. (1992)
Heuchera hallii	P	D	n.a.	Soltis et al. (1991)
Heuchera micrantha	P	yes	n.a.	Soltis et al. (1991)
Heuchera nivalis	P	D	n.a.	Soltis et al. (1991)
Heuchera parviflora	P	yes	n.a.	Soltis et al. (1991)
Impatiens aurella	M	?	T,S	Ornduff (1967)
Ipomopsis aggregate/tenuituba	M,S	L	n.a.	Grant and Wilken (1988)
Iris species	M,C	?	B	Lenz (1959)
Iris chrysophylla/tenax	S,[M]	L	n.a.	Carter and Brehm (1969)
Iris fulva/hexagona	R,M,C,E,I,N,[P]	D and L	O.T.	Riley (1938); Anderson (1949); Arnold et al. (1990a,b, 1991); Arnold et al. (1991)
Iris nelsonii	I,N	yes	S	Randolph (1965); Arnold et al. (1990b, 1991)
Juniperus ashei/virginianum	M,[S]	no	n.a.	Hall (1952); Flake et al. (1969); Adams and Turner (1970)
Juniperus virginianum/horizontalis	M	?	n.a.	Fassett (1945a,b)
Juniperus virginianum/scopularum	M,S	D	n.a.	Flake et al. (1978)
Juniperus scopularum/horizontalis	M	?	n.a.	Fassett (1945a,b)
Lasthenia burkei	M,C,[I]	?	S	Ornduff (1969, 1976); Crawford and Ornduff (1989)
Lasthenia ferrisiae	M,C	?	S	Ornduff (1966)
Lesquerella densipila/lescurii	C,M	L	n.a.	Rollins and Solbrig (1973)
Lesquerella densipila/stonensis	C,M	L	n.a.	Rollins and Solbrig (1973)

Species				Reference
Lycopersicon esculentum var. esculentum	M,I	yes	n.a.	Rick (1958); Rick et al. (1974)
Lycopersicon chilense	P	?	n.a.	Palmer and Zamir (1982)
Lycopersicon chmielewskii	P	?	n.a.	Palmer and Zamir (1982)
Melandrium dioicum	M,C	?	E	Baker (1948)
Orphrys species	M	no	n.a.	Stebbins and Ferlan (1956)
Orphrys murbeckii	M	?	S	Stebbins and Ferlan (1956)
Oryza species	M,C,I	L	E,T,I	Chu and Oka (1970); Second (1982); Dally and Second (1990); Langevin et al. (1990)
Oxytropis albiflorus	M	?	T	Anderson (1953)
Parthenium argentatum	M	?	T,I	Rollins (1949)
Penstemon spectabilis	M	?	S,O	Straw (1955)
Penstemon clevelandii	M	?	S,O	Straw (1955)
Persea steyermarkii/P. nubigena	R,P	yes	S	Furnier et al. (1990)
Phlox anoena ssp. lightipei	M,C	?	E	Levin and Smith (1966)
Phlox bifida	M	?	I,T	Anderson and Gage (1952)
Phlox drummondii/cuspidata	I,M,C,[S]	L	N	Erbe and Turner (1962); Levin (1967, 1975)
Phlox divaricata ssp. laphamii	M,C	?	E	Levin (1967)
Phlox glaberrima/pilosa	I	L	n.a.	Levin and Schaal (1972)
Phlox maculata ssp. pyramidalis	M,Ph,S	?	E	Hadley and Levin (1969); Levin (1963, 1966)
Phlox pilosa ssp. deamesii	M,C	?	E	Levin and Smith (1966)
Phytolacca species	M	?	S,E	Davis (1985)
Phytolacca species	M	?	T,C	Fassett and Sauer (1950)
Phytolacca americana	M	?	T,I	Anderson (1953)
Pinus contorta/banksiana	M,S,I,[P]	yes	n.a.	Forrest (1980); Critchfield (1985); Wagner et al. (1987); Wheeler and Guries (1987)
Pinus muricata	I	?	n.a.	Millar (1983)
Pisum sativum	P	?	n.a.	Palmer et al. (1985)
Populus fremontii/angustifolia	N	L	T	Keim et al. (1989)
Populus nigra	P,[R]	yes	n.a.	Smith and Sytsma (1990)
Potamogeton Xhaynesii	M,S	?	S	Haynes and Williams (1975); Hellquist and Crow (1986)
Potamogeton ogdenii	M,S	?	S	Hellquist and Hilton (1983)
Potentilla glandulosa ssp. hansenii	M,C	?	E	Clausen et al. (1940); Clausen and Hiesey (1958)
Primula vulgaris/elatior	M	?	n.a.	Valentine (1948)
Purshia tridentata/Cowania stansburyana	M	?	T	Stutz and Thomas (1964)
Purshia glandulosa	M	?	S,T	Stutz and Thomas (1964)

Table 4-1. Proposed Examples of Introgression in Plants *Continued*

Taxon	Evidence[a]	Introgression?[b]	Proposed Consequences[c]	References
Pyrrhopappus species	M,[I]	no	n.a.	Northington (1974); Peterson et al. (1990)
Quercus species	M	?	T,O	Muller (1952); Jensen and Eshbaugh (1976a,b)
Quercus alba	P,[M,R]	yes	S	Hardin (1975); Whittemore and Schaal (1991)
Quercus alvordiana	M	?	n.a.	Tucker (1952)
Quercus douglasii/turbinella	M	?	n.a.	Benson et al. (1967)
Quercus drummondii	M	?	S	Muller (1952)
Quercus dumosa/turbinella	M	?	n.a.	Tucker (1953)
Quercus durata	M,E	?	T	Forde and Farris (1962)
Quercus ganderi	M	?	S	Wolf (1944)
Quercus ilicifolia/marilandica	M	?	N	Stebbins et al. (1947)
Quercus macrocarpa	P,[R]	yes	n.a.	Whittemore and Schaal (1991)
Quercus marilandica/velutina	M	?	n.a.	Cooperrider (1957)
Quercus michauxii	P,[R]	yes	n.a.	Whittemore and Schaal (1991)
Quercus rober/petraea	M	?	n.a.	Rushton (1979)
Quercus stellata	P,[R]	yes	n.a.	Whittemore and Schaal (1991)
Ranunculus species	M,E	?	I	Briggs (1962)
Ranunculus victoriensis	M,E	?	S	Briggs (1962)
Raphanus sativus	M,C	?	T,I	Panetos and Baker (1967)
Sabatia formosa/arenicola	M,I,E	L	n.a.	Bell and Lester (1978)
Salix melanopsis	P,[I]	yes	n.a.	Brunsfeld (1990)
S. taxifolia	P,I,M	yes	S	Brunsfeld (1990)
Salvia mellifera/apiana	M	L	I,N,E	Epling (1947); Meyn and Emboden (1987)
Scaevola gaudichaudiana/mollis	M	?	I	Gillett (1966)
Solanum species	M,S	?	I,C,T,S,A	Hawkins (1962); Johns et al. (1987)
Solanum raphanifolium	M,[R,P]	no	S	Ugent (1970); Spooner et al. (1991)
Solidago rugosa/sempervirens	M	?	I	Goodwin (1937)
Stephanomeria diegensis	I	yes	S	Gallez and Gottlieb (1982)
Stipa californica	M	?	S	Johnson (1962)
Tellima grandiflora	P,[I]	yes	S	Soltis et al. (1991)
Tradescantia occidentalis	M	?	T	Anderson and Hubricht (1938)

Species	Approaches[a]	Introgression?[b]	Consequences[c]	References
Trichomanes species	M	?	n.a.	Bierhorst (1977)
Triticum turgidum	C,P	?	T,S	Gill and Chen (1987)
Typha latifolia/angustifolia	M	?	I	Fassett and Calhoun (1952)
Vaccinium corymbosum	M	?	S	Camp (1945)
Viola species	M,C	?	T,O	Russell (1954); Moore (1959)
Viola cucullata/septentrionalis	M	?	n.a.	Russell (1955)
Wyethia species	M	?	T	Weber (1946)
Zea mays ssp. *mays*/*Z. mays* ssp. *mexicana*	M,P,I	yes	n.a.	Doebley et al. (1987); Doebley and Sisco (1989)
Zea mays ssp. *mays*/*Z. mays* var. *parviglumis*	M,[C,E,I]	no	n.a.	Kato (1976, 1984); Wilkes (1977); Doebley (1984); Doebley et al. (1984, 1987)
Zea perennis	P,[M,I,C]	yes	n.a.	Doebley (1989a)
Zea luxurians/*Z. mays* ssp. *mays*	I	yes	n.a.	Doebley et al. (1984)
Zea diploperennis/*Z. mays* ssp. *mays*	I,[C,E]	yes	n.a.	Doebley (1984); Doebley et al. (1984)

[a]C = cytological or crossing studies; E = ecological; I = isozymes; M = morphological; N = random, nuclear DNA markers; R = ribosomal DNA markers; P = chloroplast DNA; ph = physiological studies; S = secondary compounds. Letters in brackets refer to approaches employed for which no evidence of introgression was observed.

[b]Yes = introgression probably documented; L = local introgression; D = dispersed introgression. A question mark is indicated for cases where, in our judgment, interpretations other than introgression were not adequately ruled out.

[c]B = breakdown of reproductive barriers; D = dispersal mechanism; C = colonization; E = origin of new ecotypes; I = increase in genetic diversity; N = evolutionary noise; O = origin of new species or variety; T = transfer of adaptations; n.a. = no consequences proposed by authors.

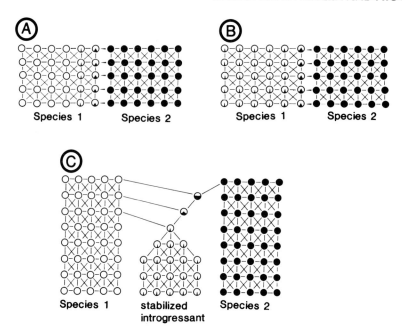

Figure 4-1. Localized introgression, dispersed introgression, and the origin of a stabilized introgressant. Open circles = populations of species 1; closed circles = populations of species 2; black lines = crosses between populations; arrows = direction of introgression. (**A**) Unidirectional localized introgression from species 2 into species 1. (**B**) Unidirectional dispersed introgression. (**C**) Origin of a stabilized introgressant.

introgression is apparent, dispersed introgression may have occurred but in a manner that could not be detected by the methods employed.

More than 90% of the examples listed in Table 4-1 are from temperate regions of the world. This bias undoubtedly reflects the location of scientists who have been interested in the phenomenon. Consequently, it is difficult to extrapolate to tropical plants. Likewise, approximately 25% of the cases listed in Table 4-1 are from California. This disproportionately high figure is due largely to the number of botanists in California, although the species and habitat richness of the California flora may also contribute to the observed high frequency.

A correlation between ecological stability and the occurrence and consequences of hybridization has long been apparent (Anderson and Hubricht, 1938; Dansereau, 1941; Anderson, 1948; Stebbins, 1959; Arnold et al., 1990a). The importance of ecological factors is twofold. First, ecological differences often result in barriers to hybridization (Anderson, 1948). If these barriers are broken down due to natural disturbance (e.g., fire, flood, volcanic activity) or human disturbance, hybridization often follows. Second, the establishment of hybrids and their progeny appears to be aided by relaxed competition in disturbed or open habitats. Thus introgressive hybridization is thought to be promoted by highly disturbed habitats such as those resulting from human activities (e.g., Anderson, 1948; Lenz, 1959). Many examples of introgression listed in Table 4-1 appear to be a direct result of natural or human habitat disturbance (Epling,

1947; Anderson, 1948; Sauer, 1957; Tucker and Sauer, 1958; Lenz, 1959; Stebbins and Daly, 1961; Heiser, 1979; Arnold et al., 1990a,b; Klier et al., 1991).

Although the focus of this chapter concerns introgression between wild species, introgression between wild and domesticated plants is suspected to be common (Anderson, 1949; Harlan, 1965; Heiser, 1973; deWet and Harlan, 1975; Ladizinsky, 1985; Doebley, 1989b) and provides some of the best cases of introgression in plants (Table 4-1). Molecular evidence for introgression has been reported in populations of *Brassica* (Erickson et al., 1983; Palmer et al., 1983; Palmer, 1988; Song et al., 1988), *Chenopodium* (Wilson, 1990b), *Citrullus* (Zamir et al., 1984), *Cucurbita* (Decker and Wilson, 1987; Wilson, 1990a), *Gossypium* (Wendel et al., 1989; Percy and Wendel, 1990; Wendel and Percy, 1990), *Helianthus* (Rieseberg and Seiler, 1990), *Lycopersicon* (Rick et al., 1974; Palmer and Zamir, 1982), *Oryza* (Chu and Oka, 1970; Second, 1982; Langevin et al., 1990), *Pisum* (Palmer et al., 1985), and *Zea* (Doebley et al., 1984, 1987; Doebley, 1989a; Doebley and Sisco, 1989).

It is often thought that introgression between crop plants and their weedy relatives is primarily unidirectional, with gene flow proceeding from the cultigen into the wild or weedy relative (deWet and Harlan, 1975; Ladizinsky, 1985; however, see Doebley, 1989b, for a criticism of this view). Presumably this situation reflects culling of weed × crop hybrids prior to the next planting season. It has also been suggested that introgression from wild plants into domesticated ones has played an important role in the origin and development of the latter (Anderson, 1961; Harlan, 1965; Heiser, 1973; Wilson, 1990b). A complete treatment of this subject is beyond the scope of this review, but some of the complications and types of evidence used may be illustrated by the origin of cytoplasmic male sterility (CMS) in the cultivated sunflower. Stable sunflower CMS (CMS 89) was first discovered in an interspecific cross between *H. petiolaris* and *H. annuus* (Leclercq, 1969). This discovery enabled production of the commercial hybrid sunflower, resulting in dramatic increases in yield over older varieties. Subsequent cpDNA analysis of cultivated and wild populations of *Helianthus* indicated that hybrid sunflower cultivars had the chloroplast genome of *H. annuus,* not that of *H. petiolaris* (Rieseberg and Seiler, 1990), a surprising finding given that CMS 89 is derived from *H. petiolaris* (Leclercq, 1969). Because cpDNA is inherited maternally in *Helianthus* (Rieseberg et al., 1991a), the most likely explanation for the absence of *H. petiolaris* cpDNA genotypes in the hybrid sunflowers was that the source population for CMS 89 was an introgressive population of *H. petiolaris* and *H. annuus.* This hypothesis was confirmed by analysis of seven individuals from the source population of *H. petiolaris* from which CMS 89 was derived; all plants had the morphology and nuclear ribosomal RNA genes of *H. petiolaris* but the cpDNA patterns characteristic of *H. annuus.*

DIFFERENTIAL INTROGRESSION

One of the more surprising results of this survey concerns the relative frequency of cytoplasmic (mitochondria and chloroplast) versus nuclear introgression in plants (Rieseberg and Soltis, 1991). A total of 37 instances of cytoplasmic introgression are reported [all documented using chloroplast DNA (cpDNA) evidence]; of these cases, 29 are considered to represent robust demonstrations (Table 4-1). This number is remarkably high given the little time that cpDNA variation has been amenable to study and the small sample sizes employed in most studies (one to few individuals per taxon). At least half of these examples were completely unexpected based on previous morphological treatments. Furthermore, in 18 of these cases, detailed analyses using isozymes, morphology, or nuclear ribosomal RNA genes (rDNA) failed to detect simultaneous introgression of nuclear genes (e.g., Palmer et al., 1983; Doebley, 1989a; Brunsfeld, 1990; Rieseberg et al., 1990b, 1991b; Smith and Sytsma, 1990; Rieseberg, 1991; Soltis et al., 1991; Wendel et al., 1991; Whittemore and Schaal, 1991).

For many of these cases, the discrepancy between nuclear and cytoplasmic evidence appears to reflect biphyletic or reticulate phylogenetic events. This phenomenon can be illustrated by the unusual evolutionary history of an Australian cotton species, *Gossypium bickii* (Wendel et al., 1991). *Gossypium bickii* is one of three arid zone species included in section *Hibiscoidea* (with *G. australe* and *G. nelsonii*). In contrast to expectations based on the distinctive shared morphology of the group, *G. bickii* possesses a chloroplast genome similar to that of *G. sturtianum* of section *Sturtia* (Fig. 4-2). Yet phylogenetic analysis of allozyme and nuclear rDNA markers indicate that the nuclear genome of *G. bickii* shares a more recent common ancestor with *G. australe* and *G. nelsonii* than it does with *G. sturtianum* (Fig. 4-2). Fifty-eight accessions were examined with reference to nuclear markers, but not a single diagnostic *G. sturtianum* nuclear marker was detected in the nuclear genome of *G. bickii*. Wendel et al. (1991) suggested that these data reflect an ancient hybridization event, with *G. sturtianum* (or a similar species) serving as the maternal parent in a cross with a paternal ancestor in the lineage leading to *G. australe* and *G. nelsonii*. The maternal nuclear genomic contribution may have been eliminated subsequently from the hybrid or its descendants. Lineage sorting is sometimes viewed as an alternative explanation to introgression for nuclear or cytoplasmic discordance, but it is implausible in this situation because of the high sequence divergence values observed between the cpDNAs of *G. sturtianum* and the *G. australe/G. nelsonii* clade.

The differential between cytoplasmic and nuclear gene flow has been addressed in greatest detail by Rieseberg and coworkers in the genus *Helianthus*. Rieseberg et al. (1990a, 1991b) employed cpDNA and nuclear rDNA markers to assess levels of nuclear and cytoplasmic gene flow between *H. annuus* and *H. debilis* ssp. *cucumerifolius* in eastern Texas. Chloroplast DNA and rDNA markers of *H. debilis* ssp. *cucumerifolius* were detected in, respectively, 10 and 16 plants of 154. *H. annuus* plants assayed. Thus levels of nuclear and cytoplasmic gene flow from *H. debilis* ssp. *cucumerifolius* into *H. annuus* were roughly equivalent. In contrast, 193 of the 262 plants of *H. debilis* ssp. *cucumerifolius* surveyed had the cpDNA genotype of *H. annuus*, where only eight individuals had alien rDNA markers. Interspecific flow of chloroplast genotypes from *H. annuus* into *H. debilis* ssp. *cucumerifolius* was approximately 10 times greater than that for nuclear rDNA markers.

Rieseberg et al. (1991b) attributed the asymmetric cpDNA flow to the greater abundance of *Helianthus annuus* in east Texas. Assuming *H. annuus* and *H. debilis* ssp. *cucumerifolius* have roughly equivalent dispersal rates, the greater abundance of *H. annuus* would result in a proportionately greater introduction of *H. annuus* achenes into *H. debilis* ssp. *cucumerifolius* populations than vice versa. The integrity of the nuclear genome would be maintained by selection against foreign nuclear genes coupled with the effects of linkage (below).

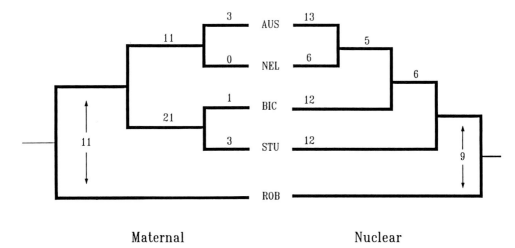

Maternal   Nuclear

Figure 4-2. Discordant maternal and nuclear phylogenetic hypotheses of five species of Australian cotton *(Gossypium)* based on Wagner parsimony analysis. *Gossypium bickii, G. australe,* and *G. nelsonii* are monophyletic with respect to nuclear molecular markers (on right) and morphology (not shown). In contrast, evidence from chloroplast DNAs (on left) indicates a recent, shared common ancestry between *G. bickii* and G. sturtianum, the latter from a different taxonomic section. The maternal phylogeny was derived from restriction site loss/gain mutations in chloroplast DNAs (consistency index = 1.00). The nuclear tree was constructed from restriction site loss/gain mutations in 18S–25S ribosomal DNAs and presence/absence data for allozymes (CI = 0.81). In both trees, character states were polarized with respect to the outgroup species *G. robinsonii.* Arabic numerals indicate the number of nonhomoplasious synapomorphies or autapomorpies along each branch segment. AUS = *G. australe;* NEL = *G. nelsonii;* STU = *G. sturtianum;* BIC = *G. bickii;* ROB = *G. robinsonii.* (From Wendel et al., 1991. With permission.)

Analysis of cpDNA versus rDNA introgression between races of *H. annuus* and *H. petiolaris* from southern California also revealed that the interspecific flow of chloroplast genotypes was more frequent than that of nuclear ribosomal genes (Dorado et al., 1992). Both species are widespread and polytypic, occurring commonly in the western United States and less frequently eastward (Fig. 4-3). In California, *H. annuus* is a common roadside weed, occurring frequently in the central valley and in southern California. *Helianthus annuus* was already present in southern California when the first botanical collections were made and was used by native Americans for various purposes (Heiser, 1949b). Because it does not now occur in natural sites in either central or southern California, it was likely introduced recently by native Americans (Heiser, 1949b). *Helianthus petiolaris* represents an even more recent introduction to southern California, perhaps during the mid-1940s. According to herbarium records (RSA/POM), the first collections were made in 1947, and the species is now rather common. Yet all but four of the 141 individuals (six populations) of *H. petiolaris* analyzed from southern California had the chloroplast genotype of *H. annuus.* In contrast, only two plants had alien rDNA markers (Fig. 4-3). No introgression was observed from *H. petiolaris* into *H. annuus.* It is noteworthy that identification of "authentic" cpDNA genotypes of *H. annuus* and *H. petiolaris* was based on analysis of 51 popu-

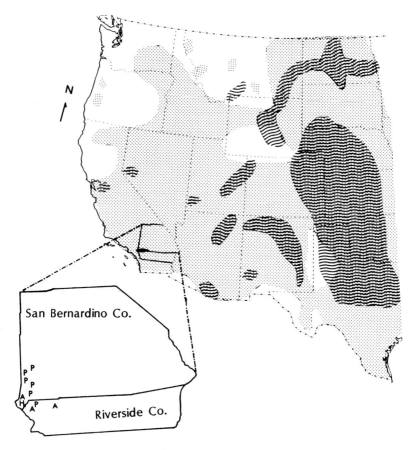

Figure 4-3. Geographic distribution and collection localities of *H. annuus* (dots) and *H. petiolaris* (wavy lines) in the western United States. A = *H. annuus* collection locality; P = *H. petiolaris* collection locality; H = hybrid population locality. (From Dorado et al., 1992. With permission.)

lations (212 individuals) and 12 populations (72 individuals), respectively, from throughout the range of both species (Rieseberg et al., 1991a). Thus misidentification of authentic *H. petiolaris* cpDNA cannot account for the observed situation.

Phylogenetic analysis of *Quercus* populations revealed a similar differential between organellar and nuclear gene flow (Whittemore and Schaal, 1991). A cladogram based on cpDNA data was congruent with geographic distribution rather than morphological species boundaries. Furthermore, the same cpDNA genotype was shared by individuals of two species growing in mixed stands. In contrast, rDNA marker distributions were consistent with morphology. Whittemore and Schaal (1991) suggested that nuclear genes may be exchanged less freely between species of oak than are chloroplast genomes.

Only two instances of nuclear introgression in the absence of cytoplasmic introgression have been reported in plants (Wagner et al., 1987; Arnold et al., 1991). Introgression of morphological features (Critchfield, 1985), terpenoids (Forrest, 1980), and allozymes (Wheeler and Guries, 1987) has been detected between populations of *Pinus contorta* (lodgepole pine) and *Pinus banksiana* (jack pine). In contrast, extensive sampling of sympatric and allopatric populations (including some of the same populations used in the previous studies) with reference to species-specific cpDNA markers revealed no evidence of cytoplasmic introgression (Wagner et al., 1987). Likewise, examination of nuclear markers support a hypothesis of bidirectional nuclear introgression between *Iris fulva* and *I. hexagona* (Arnold et al., 1990a,b, 1991; see Ch. 5). Little cpDNA introgression was observed, however, suggesting that localized and dispersed introgression between these species is largely due to pollen transfer (Arnold et al., 1991).

Cytoplasmic and nuclear gene flow patterns are often different in animals as well, usually with cytoplasmic genes being exchanged more freely (e.g., Ferris et al., 1983; Powell, 1983; Carr et al., 1986; Gyllensten and Wilson, 1987; Tegelstrom, 1987; Marchant, 1988; Aubert and Solignac, 1990). In contrast, several studies have demonstrated roughly equivalent frequencies of nuclear and cytoplasmic gene flow (e.g., Avise et al., 1984; Syzmura et al., 1986; Harrison et al., 1987).

A number of hypotheses have been proposed to explain the different frequencies of cytoplasmic and nuclear gene flow in animals and plants (Barton and Jones, 1983; Powell, 1983; Gyllensten and Wilson, 1987; Aubert and Solignac, 1990; Rieseberg et al., 1991a,b; Wendel et al., 1991; Rieseberg and Soltis, 1991). One possibility is selection against alien nuclear genes but not against cytoplasmic genes (Barton and Jones, 1983; Powell, 1983). Selection against several loci scattered throughout the nuclear genome could greatly reduce overall nuclear gene flow due to linkage (Barton and Bengtsson, 1986; Whittemore and Schaal, 1991).

A second possibility is positive selection for cytoplasmic genes but not nuclear genes (Rieseberg et al., 1991a). Frank (1989) has shown that a native cytoplasm could be largely replaced by an alien one if the latter has a slight fitness advantage conferred by relative ovule success. This process would be promoted by cytoplasmic male sterility (CMS). In this aspect, it is noteworthy that CMS has been observed in a number of interspecific crosses in *Helianthus* (see Rieseberg and Seiler, 1990) and *Gossypium* (Meyer, 1975). Alternatively, differential gene flow could result from differential fitness among cytoplasmic-nuclear combinations (Wendel et al., 1991), whereby natural selection favors a particular alien/native cytoplasmic-nuclear combination.

A third possible mechanism, which may not be wholly separable from CMS, involves a small number of female immigrants of one species deposited into a population of another (Aubert and Solignac, 1990). Male sterility in first generation hybrids and first generation backcrosses could quickly lead to a small population of individuals containing alien cytoplasms. Individuals from subsequent generations would possess nuclear genomes that are increasingly indistinguishable from those of the host population. Similar results might be expected from the introduction of a hybrid, male-sterile propagule into a host population (Gyllensten and Wilson, 1987). This model may be the only one that can account for the *rapid* ($<$ 50 years) replacement of the native *H. petiolaris* cytoplasm with that of *H. annuus* in the Southern California race of *H. petiolaris*. It is noteworthy that the first generation hybrids of *H. annuus* and *H. petiolaris* are sometimes male-sterile due to CMS (Rieseberg, unpublished).

Other mechanisms, such as semigamy, may promote unidirectional cytoplasmic introgression (Wendel et al., 1991). Semigamy is a form of facultative apomixis whereby gamete fusion occurs without nuclear fusion, resulting in progeny that may include maternal haploids, paternal haploids, and chimeric maternal/paternal plants (Turcotte and Feaster, 1967). Wendel et al. (1991) pointed out that semigamy requires neither differential selection for cytotypes nor nuclear genes and can result in the fixation of the nuclear genome of a male donor into a foreign cytoplasm in a single generation. In this context, it may be significant that *Gossypium bickii* represents one of only two reasonably well-documented examples of the complete replacement of a native cytoplasm by an alien one (see also Smith and Sytsma, 1990).

Regardless of the mechanism responsible for cytoplasmic gene flow in the absence of significant nuclear gene flow, the process leads to several implications and conclusions. First, most botanists appear to accept the view that hybridization and introgression are frequent, yet ignore their potential implications for phylogenetic reconstruction [see Funk (1985) and McDade (1990) for notable exceptions]. Ironically, it is precisely this framework, i.e., the estimation of both nuclear and organellar phylogenies, that is responsible for the detection of many cases of introgression, although other evolutionary processes, such as random lineage sorting, can also result in patterns of discordance similar to those resulting from introgression. Nonetheless, the susceptibility of cytoplasmic organelles to introgression, which results in biphyletic organisms, suggests that caution be exercised in the use of organellar sequences for phylogenetic reconstruction. We suggest that an important area of future research is the development of algorithms for phylogeny reconstruction that are designed to account for reticulation.

Second, the degree of similarity between hybridizing species is potentially associated with their ability to exchange cytoplasms. Conversely, the effects of cytoplasmic exchange are more profound for divergent species than for closely related ones. One likely result is CMS, which is usually thought to arise from "incompatibility" between nuclear and cytoplasmic factors. It is also possible that cytoplasmic exchange between divergent species could lead to the formation of a new species as has been proposed for *G. bickii* (Wendel et al., 1991). Cytoplasmic transfer between closely related species generally has not been proposed as a common mechanism for speciation (e.g., Doebley, 1989a; Rieseberg et al., 1991a,b; Soltis et al., 1991; Whittemore and Schaal, 1991).

PROPOSED CONSEQUENCES OF INTROGRESSION

Evidence indicating that introgression is common in plants leads naturally to the question of its evolutionary significance. More specifically, which consequences of introgression, among the several envisioned by Anderson (1949) and others, are evident from empirical data? As indicated in Table 4-1, a number of consequences have been proposed, including increased genetic diversity, transfer of adaptations, origin of adaptations, origin of ecotypes or species, and breakdown or reinforcement of isolating barriers. Introgression has also been suggested as promoting colonization and as a dispersal mechanism (Potts and Reid, 1988). Each of these consequences is of potential interest from an evolutionary perspective. An additional consequence of an entirely different nature, however, emerges from phylogenetic considerations; i.e., introgression, by definition, involves reticulations, thereby rendering the reconstruction of evolutionary histories more difficult, regardless of specific biological consequences. Finally, the possible escape of genetically engineered genes from crop plants through introgression into wild relatives provides an example of a potentially economically and environmentally significant consequence of introgression.

Evidence for each of the consequences listed above is discussed in the following paragraphs. These various consequences are not mutually exclusive; for example, the ability to colonize new habitats is expected to result from introgression of certain necessary adaptations.

Increase in Genetic Diversity. As discussed above, early authors saw introgression as an important source of genetic variation, noting that obviously introgressive populations should exhibit alleles of both parents as well as new single and multilocus genotypes. Thus estimates of genetic diversity calculated from molecular markers and morphological characters are expected to be higher in introgressant populations. Slightly less obvious is the prediction that new alleles may be produced through intragenic recombination or other processes (Golding and Strobeck, 1983). Irrespective of the potential generation of novel alleles, increased genetic diversity has been suggested as possibly altering adaptive potentials of introgressant populations (Lewontin and Birch, 1966).

We are aware of no evidence in the plant literature regarding the generation of novel alleles or "hybrizymes" (Woodruff, 1989) as a consequence of hybridization or introgression, in contrast to the animal literature where hybrizymes have been reported in hybrid zones involving mammals, birds, reptiles, amphibians, and insects (reviewed in Woodruff, 1989). Increased genetic diversity, though, is frequently reported in hybridizing plant populations. Hybridizing populations of white and yellow ladyslipper orchids (Cypripedium), for example, contain considerably higher morphological and allozyme variability than do allopatric populations of either species (Klier et al., 1991). Similarly, large increases in allelic polymorphism (up to 40% greater than the parental taxa) were observed in introgressive populations of three *Aesculus* species (dePamphilis and Wyatt, 1990). A modest increase in allozyme diversity was observed in introgressive populations of jack and lodgepole pines (Wheeler and Guries, 1987). In these and other studies, parallel increases in morphological and allozyme variability are often observed. This is not always the case, however. For example, introgressive populations of *Clarkia speciosa* spp. *polyantha* are no more variable morphologically than their parental taxa (Bloom, 1976), although there is a modest increase in allozyme diversity (Soltis, 1985).

With respect to genetic diversity predictions, it is important to distinguish between *recently* hybridized or introgressant populations (i.e., with recent biparental gene flow) and *stabilized* introgressants that are reproductively isolated from their parental taxa. In the latter, one might expect genetic drift or population bottlenecks to decrease genetic variability relative to progenitor populations, a prediction that has been confirmed for several species of hybrid or introgressant origin. In *Helianthus* (sunflowers), for example, dramatic decreases in overall genetic diversity were observed for three species of hybrid or introgressive origin: *H. anomalus, H. deserticola,* and *H. paradoxus* (Fig. 4-4) Rieseberg et al., 1990b, 1991; Rieseberg, 1991). *Helianthus paradoxus* and *H. deserticola* were polymorphic at only 1 of 17 loci examined, whereas *H. anomalus* was polymorphic at 3 of 17 loci. The number of polymorphic loci was much higher in parental species, ranging from four in *H. annuus* to seven

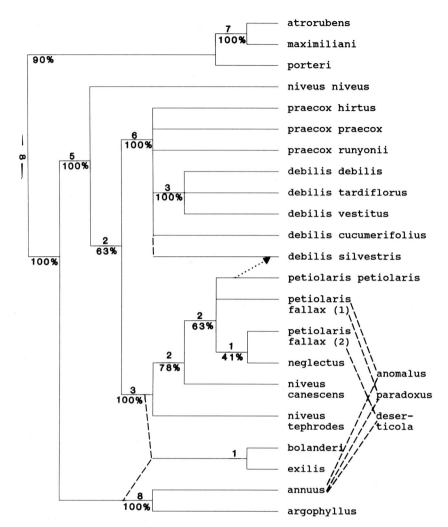

Figure 4-4. Phylogenetic estimate of *Helianthus* section *Helianthus* based on chloroplast DNA and nuclear ribosomal DNA variation. The phylogenetic tree was constructed as follows: (1) six taxa of biphyletic origin were removed from the data matrix: (2) Wagner parsimony analysis was performed on the remaining 19 taxa, resulting in the strict consensus tree presented in this figure; and (3) the six hybrid taxa were then added as parsimoniously as possible to the appropriate nodes of the consensus tree *(dashed lines)* demonstrating reticulate evolution. In the case of *H. debilis* ssp. *silvestris* the dashed lines simply indicate the capture of a foreign cytoplasm; no alien nuclear genes were detected. The consistency of the phylogenetic tree, including reticulations, was 0.85. Taxon designations are given at the ends of the branches, and the number of mutations are given above the branches. Autapomorphies are not shown. Percentages indicate the number of times a monophyletic group occurred in 100 bootstrap samples. (From Rieseberg, 1991. With permission.)

in *H. petiolaris.* Rieseberg et al. (1990b) attributed the low levels of genetic diversity to founder events following hybrid or "recombinational" speciation (Grant, 1981).

Increases in genetic diversity attributable to introgression have also been observed in several domesticated species of *Gossypium* (Wendel et al., 1989; Percy and Wendel, 1990). For example, at least 5 of 10 polymorphic loci in *G. herbaceum* (one of the two species of Old World cultivated cottons) are biallelic rather than monomorphic as a result of introgression (Wendel et al., 1989). A similar, though less extreme, effect is probable with the other Old World cultivated cotton, *G. arboreum;* perhaps 3 or 4 of 11 polymorphic loci are polymorphic owing to introgression (Wendel et al., 1989). Introgression also appears to have contributed to allelic diversity in improved varieties of one of the New World cultivated cottons, *G. barbadense.* Of the 20 polymorphic loci detected in cultivars of *G. barbadense,* nine were polymorphic owing to introgression alone (Percy and Wendel, 1990). Thus introgression appears to be responsible for much of the genetic diversity observed in cultivated cottons.

Transfer of Adaptations. A potentially significant consequence of introgression, transfer of adaptation, has often been proposed, but it is a difficult process to verify experimentally. Even when introgression has been "unambiguously" demonstrated, as with molecular data, it need not imply that adaptively significant genes were also transferred (Rieseberg et al., 1990a). It might be argued that the detection of introgression is synonymous with the transfer of adaptations, as hybrid zone theory predicts that advantageous alleles will cross species barriers more readily than neutral markers (Barton and Bengtsson, 1986). However, the great preponderance of "neutral" molecular markers relative to advantageous alleles leave us to consider this assumption premature. Furthermore, similar selection pressures could produce convergent morphological responses in certain adaptively significant features. It may be a common consequence of similar apomorphic tendencies in related species (Mayr, 1942). Finally, and most problematically, it is difficult to distinguish between transfer of adaptations and simple gene flow even where segregating admixtures contain individuals with morphological features of one species growing in the characteristic environment of the other. This general problem may be illustrated by ladyslipper orchid populations from Iowa (Klier et al., 1991). White and yellow ladyslipper orchid species are ecologically separated: The former occurs in prairies and other open vegetation types, and the latter is restricted to forest understory. Whenever the two species are parapatric in Iowa, they hybridize, resulting in the formation of hybrid swarms in the prairie. Allozyme and morphological data suggest that "yellow" ladyslipper individuals (i.e., morphological "yellows" containing no "white" marker alleles) occur in the prairie. The key question is whether the ability of these "yellows" to survive in the drier, brighter prairie has been transferred from white to yellow ladyslipper via differential selection on genetically recombinant progeny, or establishment of "yellows" in the prairie has been independent of these "white" genes. Distinguishing between these alternatives seems to require both an understanding of the genetic basis for habitat preference and reciprocal transplant studies on a variety of genetically defined parental and hybrid classes.

Despite these experimental difficulties, the literature contains several examples of putative interspecific transfer of adaptations (Table 4-1). Stutz and Thomas (1964), for example, suggested that the lowered palatability to herbivores of certain populations of *Purshia tridentata* has resulted from its acquisition of this characteristic via introgression with *Cowania stansburyana.* The two species are sympatric in Utah, and hybridization is common wherever they come into contact. Although *Cowania* does not occur north of Utah, *Purshia* populations north of Utah express some of *Cowania*'s morphological characters. This situation is thought to have been promoted by some selective advantage for recombinant progeny, such as lowered palatability, which has been observed in a number of *Purshia* populations from the northern portion of its range. However, as pointed out by Heiser (1973), introgression is only one of several plausible explanations for this pattern of variation.

Helianthus provides several examples of the putative transfer of adaptations through introgression. For example, Heiser (1947, 1949b, 1951a,b, 1954, 1965) suggest that the most widespread species, *H. annuus,* was able to expand its range by introgression with native species already locally adapted. In his view, this process resulted in the formation of introgressive races varying toward *H. debilis, H. argophyllus, H. petiolaris,* and perhaps *H. bolanderi.* Molecular evidence provides some support for this interpretation, in that introgression between these species and *H. annuus,* as either the donor or recipient species, has been documented (Rieseberg et al., 1990a,b; Rieseberg, 1991). Likewise, Heiser (1951a) presented evidence suggesting that *H. annuus* was able to invade eastern Texas by "capturing" advantageous alleles of *H. debilis* ssp. *cucumerifolius,* a species already adapted to the area. Rieseberg et al. (1990a, p. 596) provided detailed molecular evidence for introgression in this case but noted that it "does not necessarily prove that the introgression of *H. debilis* ssp. *cucumerifolius* into *H. annuus* was in any way adaptive." In a more recent study, Heiser (1979) examined three hybrid populations of *H. divaricatus* (a species of open habitats) and *H. microcephalus* (a species found in shaded areas) over a 22-year period; considerable hybridity was still evident in one population at the end of this period, probably as a result of continued site disturbance. Although the habitat had become more closed at the other two sites, *H. divaricatus* was dominant at one site and the sole species found at the second site. Heiser suggested that *H. divaricatus* may owe its increased shade tolerance to introgression of genes from *H. microcephalus.*

Similar situations have been suggested for other plant groups. Harlan and deWet (1963) proposed that assimilation of local gene pools by *Bothriochloa intermedia,* a widespread grass species, allowed it to increase its geographic range and ecological tolerance. They introduced the term "compilospecies" to describe those species that, to use their imaginative expression, are "genetically aggressive, plundering related species of their heredities."

Origin of New Adaptations. It has long been evident that hybrids may have characters absent from both parents. Frequently cited examples include (1) bizarre corolla appendages in certain *Antirrhinum* crosses (Hagedoorn, 1921); (2) larger corollas in the hybrid of *Nicotiana langsdoorfi* × *alata* than those observed in either parent (Stebbins, 1966); and (3) novel secondary compounds in *Baptisia* (Alston and Simmons, 1962). Although it is difficult to assign an "adaptive value" to each character, it seems likely that these new characters will have novel adaptive potentials. These novel characters are thought to result from recombination among parental genes leading to novel multilocus genotypes, rather than the generation of new alleles. However, empirical evidence suggests that novel alleles or hybrizymes may also be produced by hybridization. For example, hybrizymes were observed in 19 of the 23 electrophoretic surveys of hybrid zones reviewed by Barton and Hewitt (1985) and have now been reported for most major groups of animals (Woodruff, 1989). Several explanations of this observation have been proposed: increased mutation rates in hybrids, reduced selection, and intragenic recombination between different alleles of the parental species (reviewed in Golding and Strobeck, 1983, and Woodruff, 1989).

These new alleles and genetic combinations may be viewed as the "raw material for evolution" (Anderson, 1949). Experimental evidence that best supports this possibility comes from fruit flies rather than plants. Lewontin and Birch (1966) showed that introgressed populations of *Dacus tryoni* were better able to adapt to higher temperature regimens than populations of either parental species.

Although there are no comparable experimental studies in plants, there is circumstantial evidence for the origin of adaptations through hybridization. Rieseberg (1991) pointed out that the three stabilized hybrid derivatives of *Helianthus annuus* and *H. petiolaris* have notable ecological preferences. *Helianthus anomalus* and *H. deserticola* are xerically adapted sunflowers endemic to desert sand dunes and swales of the southwestern United States, whereas *H. paradoxus* occurs in brackish, saline, marshy areas in west Texas. Both of these habitats are extreme relative to either parent; *H. annuus* occurs in heavy soils that are saturated in the spring but dry by midsummer, and *H. petiolaris* occurs in drier, but not xeric, sandy soils. Riseberg (1991) suggested that these new ecological preferences "provide some support for the view of hybridization as a significant source of genetic novelty upon which selection can act."

Origin of New Types. The evolutionary consequences of hybridization and introgression most frequently emphasized by plant systematists is the generation of new homoploid taxa. In fact, nearly 40% of the examples listed in Table 4-1 propose the origin of a new ecotype or species as a consequence of introgression. These new types are often referred to as stabilized introgressants in the plant literature, i.e., populations that breed true for an alien allele (or alleles). Perhaps the first study to provide evidence that new ecotypes or species could be derived through introgression is that of Dansereau (1941) in the genus *Cistus.* Citing morphological evidence, he suggested that the North African variety of *C. ladaniferus* originated through the introgression of *C. laurifolias* into the typical variety of *C. ladaniferus,* which is restricted to the Iberian peninsula and southern France. *Cistus laurifolias* is sympatric with both varieties, first generation hybrids are often observed where the species occur together, and all the morphological differences that distinguish North African *C. ladaniferus* from typical *C. ladaniferus* are in the direction of *C. laurifolias.* The first detailed evidence for the origin of a new type through introgression was presented by Heiser (1949b) in *Helianthus.* He used morphological, cytological, genetic, and ecological data to postulate that a distinct weedy ecotype of *H. bolanderi,* a native sunflower of California, originated through introgression of genes from the recently introduced *H. annuus* into typical *H. bolanderi.* Ironically, molecular studies (Rieseberg et al., 1988a,b) have failed to confirm the occurrence of introgression in this case. In contrast, molecular evidence Rieseberg et al. (1990a) does support the hypothesized origin of *Helianthus annuus* ssp. *texanus* by the introduction of *H. annuus* into Texas and subsequent introgression of genes from *H. debilis* ssp. *cucumerifolius* into *H. annuus* (Heiser, 1951b).

The origin of stabilized new types through introgression has been suggested for many other groups of plants (Table 4-1). For example, Levin and coworkers postulate an introgressive origin for *Phlox divaricata* ssp. *laphamii* (Levin, 1967), *P. maculata* ssp. *pyramidalis* (Levin, 1963, 1966), *P. pilosa* ssp. *deameii* (Levin and Smith, 1966), and *P. amoena* ssp. *lightipei* (Levin and Smith, 1966). Other genera with large numbers of proposed stabilized introgressants include *Gilia* (Grant, 1950, 1963; Grant and Grant, 1960), *Quercus* (Wolf, 1944; Muller, 1952; Tucker, 1952), and *Salix* (Brunsfeld, 1990). In addition, there are numerous examples where hybrid swarms or populations of stabilized introgressants appear to dominate disturbed or intermediate habitats: *Helianthus annuus/H. bolanderi* (Stebbins and Daly, 1961), *Helianthus divaricatus/H. microcephalus* (Heiser, 1979), *Iris fulva/I. hexagona* (Riley, 1938; Arnold et al., 1990a,b, 1991; see Ch. 5), *Salvia apiana/S. mellifera* (Epling, 1947; Meyn and Emboden, 1987), and *Cypripedium candidum/C. pubescens* (Klier et al., 1991). Many of these populations have been known to exist for more than 40 years, and there is no evidence to suggest that they are transient or ephemeral. Furthermore, in at least one instance (Stebbins and Daly, 1961), plants in the introgressive population appear to have formed a chromosomal sterility barrier isolating them from the parental taxa. It seems likely, therefore, that some of these populations represent the first stage of diploid hybrid speciation.

Models for the origin of new species through hybridization, without a change in chromosome number, have been developed by Grant (1949, 1958) and Stebbins (1957). Grant (1949) suggested that a new species could be derived through hybridization by the formation of an intermediate flower type and subsequent pollination by flower-constant insects. This mode of hybrid speciation, which has been postulated for two species of *Penstemon* (Straw, 1955), would most likely occur if backcrossing of the hybrids to the parents (introgression) is minimal. Alternatively, some other "external" isolating barrier could serve to isolate the hybrids from their parents (Grant, 1981).

A second model for the diploid hybrid origin of a species requires the development of chromosomal sterility barriers between the neospecies and its parents; this process has been termed "recombinational speciation" by Grant (1981). The basic model (Stebbins, 1957; Grant, 1958, 1981) can be summarized as follows: (1) Two parental species are distinguished by two or more separable chromosome rearrangements; (2) their partially sterile hybrid gives rise to new homozygous recombinant types for the rearrangements; and (3) the new recombinant types are fertile *inter se* but at least partially sterile with both parents. This model has been experimentally verified by the synthesis of new "hybrid" species in *Gilia* (Grant, 1966a,b), *Nicotiana* (Smith and Daly, 1959), *Elymus* (Stebbins, 1957), and *Crepis* (Gerrassimova, 1939). Furthermore, at least four wild plant species are now thought to have been derived through this process: *Stephanomeria diegensis* (Gallez and Gottlieb, 1982) and *Helianthus paradoxus, H. anomalus,* and *H. deserticola* (Fig. 4-4) (Rieseberg et al., 1990b; Rieseberg, 1991). In the latter three cases, the chromosomal sterility barriers isolating the hybrid species from their parents are known to be a consequence of their hybrid origin.

These two models clearly are not the only possible means by which reproductive isolation of a nascent hybrid taxon may occur. Other plant species, for which circumstantial evidence suggests a biphyletic origin (Table 4-1), may have been derived by the processes modeled above or by some other unknown mechanism. With respect to some species, however, it seems clear that neither of the above models is appropriate. For example, *Gossypium bickii,* which may have originated through an ancient hybrid speciation event (Wendel et al., 1991), is interfertile with its putative paternal lineage, appears to lack chromosomal rearrangements, and has a floral morphology that is similar to one of its two putative parents.

Reproductive Barriers. Introgression is usually thought to reduce the strength of reproductive isolation barriers and sometimes even lead to the merger of two formerly isolated taxa (e.g., as in the compilospecies concept of Harlan and deWet, 1963). However, one potential outcome of hybridization is actual reinforcement of reproductive isolating mechanisms (see Ch. 3). It usually occurs through reproductive character displacement (e.g., Levin and Kerster, 1967; Levin, 1985; Whalen, 1978), which in the extreme should lead to a cessation of hybridization and therefore of introgression. Although reinforcement of isolating barriers is not generally associated with introgression, there are a few cases where introgression appears to be proceeding in one area of a species range and reproductive barrier reinforcement is taking place in another, e.g., *Calyptridium monospermum* (Hinton, 1976). Reproductive barrier reinforcement appears to have occurred in other plants, but these populations show traces of past introgression events, e.g., *B. barbadense/hirsutum* (Percy and Wendel, 1990).

There is relatively little evidence in the systematic literature for the breakdown of reproductive barriers through introgression, although natural or human disturbance has clearly ameliorated ecological barriers to hybridization and introgression in many genera (e.g., Anderson, 1948; Sauer, 1957; Tucker and Sauer, 1958; Lenz, 1959). Several long-term studies of introgressive populations are especially relevant (Stebbins and Daly, 1961; Hauber and Bloom, 1983; Meyn and Emboden, 1987). Stebbins and Daly (1961) examined an introgressive population of *H. bolanderi* and *H. annuus* over a 9-year period. The two species differ by three or four reciprocal translocations (Heiser, 1949b; Chandler et al., 1986), and first generation hybrids produce 2–10% stainable pollen. A 70–90% increase in pollen viability was observed in introgressive plants by the end of the 9-year period, suggesting a breakdown of chromosomal sterility barriers isolating these species. In a similar study, Meyn and Emboden (1987) compared pollen viabilities for hybrid populations of *Salvia apiana* and *S. mellifera* over a 30-year period and demonstrated that pollen viabilities of introgressed individuals now approach or even equal those of the parental species, whereas first generation hybrids generally have less than 50% stainable pollen. Perhaps the most striking example of the amelioration of chromosomal sterility barriers comes from a 10-year study of a chromosomal hybrid zone between *Clarkia nitens* and *Clarkia speciosa* ssp. *polyantha* (Bloom and Lewis, 1972; Bloom, 1976; Hauber and Bloom, 1983). They demonstrate that hybridization between these two chromosomally differentiated taxa has resulted in the production and establishment of new chromosome arrangements that serve to "genetically link" the parent species. These arrangements have become distributed geographically across the hybrid zone allowing for gene flow between the parental taxa with little loss of fertility.

Rieseberg et al. (1989) suggested that California's rarest tree, *Cercocarpus traskiae,* is in danger of extinction owing to genetic assimilation by its more widespread congener, *C. betuloides.* They pointed out that at least two and possibly as many as four of the seven remaining *C. traskiae* individuals are of hybrid or introgressive origin. Likewise, Brochmann (1984) provided evidence of possible extinction by swamping of *Argyranthemum coronopifolium,* a rare endemic that occurs in two disjunct areas in one of the Canary Islands. The common marguerite, *A. frutescens,* has recently migrated as a weed along new roads into both areas of *A. coronopifolium* and forms extensive hybrid swarms with this species. We are not aware of other examples of plant species in danger of extinction due to introgression, although local extinction may be a frequent phenomenon. For example, *Cypripedium candidum* populations engaged in hybrid swarm formation with *C. pubescens* may no longer contain any "pure" individuals of the former species (Klier et al., 1991). A number of animal species appear to be endangered by introgression, including the cutthroat trout (Allendorf and Leary, 1988), Pecos pupfish (Echelle and Connor, 1989), yellow-crowned parakeet (Taylor, 1975), Seychelles turtle dove (Cade, 1983), Mexican duck (Heusmann, 1974), red wolf (Mech, 1970; Wane and Jenks, 1991), Mojave chub, and Tecopa pupfish (Soltz and Naiman, 1978).

Dispersal Mechanism. Potts and Reid (1988) suggested that hybridization via pollen dispersal may be of evolutionary significance as a means of gene dispersal where seed dispersal is more limited than pollen dispersal. They demonstrated that *Eucalyptus risdonii* is invading the range of *E. amygdalina* but suggested that the rate of colonization may be slowed by limited seed dispersal (Potts and Reid, 1985; Potts, 1986). Gene flow by pollen dispersal is suggested to be more widespread, however. Potts and Reid proposed that *E. risdonii* "appears to be invading favorable habitat islands by gene flow through the more widespread species, *E. amygdalina.*"

Schemske and Morgan (1990) criticized this conclusion based on the "very low frequency of hybridization, potential bias in estimation of distance of interspecific gene flow, lack of information on the extent of intraspecific pollen dispersal distances, and occurrence of comparable levels of hybridization in both species." They suggested that the hybridization observed between *E. amygdalina* and *E. risdonii* might be more realistically viewed as an accidental consequence of the evolutionary history of these two species, rather than as an evolutionary advantage arising from increased dispersal potential.

Phylogenetic and Taxonomic Consequences. Regardless of the adaptive importance of introgression, it is of great concern to practicing taxonomists and phylogeneticists (e.g., McDade, 1990). Introgression blurs taxonomic boundaries, leading to the common phenomenon of taxonomically "difficult groups." Hybridization and introgression have also influenced the development of various species concepts through the challenges they pose for the "biological species concept" (Cracraft, 1983; Donoghue, 1985; Ghiselin, 1987; Templeton, 1989).

Several authors have suggested that phylogenetic studies relying solely on variation patterns in organellar genomes (chloroplast and mitochondria) may be particularly susceptible to errors introduced by introgression (Doebley, 1989a; Furnier et al., 1990; Rieseberg et al., 1990a; Smith and Sytsma, 1990; Rieseberg and Soltis, 1991; Wendel et al., 1991). Three justifications for this conclusion may be offered. First, the maternal inheritance and vegetative segregation of organelles result in organellar genes having an effective population size that is approximately one-fourth that of nuclear genes (Birky et al., 1983), leading to a corresponding increase in the rate of fixation by drift and a decrease in expected gene diversity. Thus the likelihood of maintaining two divergent cpDNAs (native and alien) in a single population over long periods is less than for nuclear genes, with a corresponding decrease in the probability of detecting and correctly diagnosing cytoplasmic introgression (Rieseberg and Soltis, 1991). Second, empirical evidence (discussed above; also see Table 4-1) indicates that nuclear genes may be exchanged less freely between species than organellar genes. Third, putative cases of biphylesis, involving strongly discordant organellar and nuclear phylogenies, are being reported with increasing frequency (Doebley, 1989a; Furnier et al., 1990; Smith and Sytsma, 1990; Rieseberg and Soltis, 1991; Rieseberg et al., 1991a,b; Wendel et al., 1991).

An example of the potential ease with which cytoplasmic introgression could lead to erroneous assessments of monophyly is provided by a detailed study of *Zea* (Doebley, 1989a). In this study, an atypical chloroplast genome was detected in eight individuals from one population ("Piedra Ancha") of *Z. perennis.* Comparison of the atypical cpDNA to other cpDNA types previously described for *Zea* (Doebley et al., 1987) revealed that this genome type was not present in any *Zea* species examined to date. Parsimony analysis indicated that the Piedra Ancha cpDNA genotype shares a most recent common ancestor with the plastid genomes of *Z. mays,* from which it is distinguished by five or six mutations. The cpDNAs from other populations of *Z. perennis,* however, are cladistically sister to *Z. diploperennis,* in accordance with traditional evidence and allozyme data. Doebley suggested that the foreign cytoplasm in the Piedra Ancha population was derived through introgression from some missing taxon. An erroneous phylogenetic conclusion could easily have been reached if Doebley (1989a) had failed to employ a comprehensive sampling strategy, or if he had failed to utilize evidence from nuclear genes.

Zea perennis is tetraploid, but the clade containing the Piedra Ancha plastome type consists solely of diploid species, except of course for the Piedra Ancha population of *Z. perennis.* Thus cytoplasmic introgression may have occurred across ploidy levels. Alternatively, the missing or extinct plastome donor was itself tetraploid.

Rieseberg and Soltis (1991) suggested a number of approaches to avoid erroneous phylogenetic conclusions based on cpDNA data. They included comparisons with phylogenetic hypotheses based on nuclear genes or morphological characters, comprehensive sampling strategies, and methods of data analysis that detect introgression events. At present, the latter consists primarily of conducting separate phylogenetic analysis of data from different genomes (e.g., Fig. 4-2), with subsequent examination of discordant clades. In the future, it may be possible to test for or hypothesize reticulation events using algorithms for phylogeny reconstruction that have been designed with this phenomenon in mind.

Introgression in Crop–Weed Complexes. One of the goals of plant breeding has been utilization of "exotic germplasm" via introgression of "useful" genes from wild relatives into breeding populations of the crop plant. This process traditionally has been restricted to the use of wild relatives with which the crop plant is interfertile, although somatic fusion of sexually incompatible plants is now possible. Advances in molecular biology have made possible the transfer of genes from organisms well outside the traditional crop gene pool through the use of recombinant DNA techniques rather than introgression via sexual processes (Gould, 1988). Thus plant breeders can introduce genes that contribute genetic resistance to specific diseases, herbicide tolerance, pests, and so forth. Although these techniques may undoubtedly lead to improvements in commercial crops, the potential for gene flow between crop plants and their wild or weedy relatives has been demonstrated repeatedly (Table 4-1) providing a possible means of escape for genetically engineered genes. These data thus serve as a cautionary note to those involved in crop plant improvement via biotechnological means (Ellstrand, 1988; Wilson, 1990a). For example, introgression of genetically engineered herbicide resistance genes from crop plants to their weedy relatives could rapidly reduce the commercial value of an herbicide-resistant crop (Ellstrand, 1988).

PROBLEMS

We have reviewed the historical development of ideas regarding introgression in plants, examined the evidence for its extent, and discussed its potential consequences so as to assess its role in plant diversification and speciation. In this section, we discuss several problems that emerged from this review.

A general problem concerning the probability of detecting introgression is evident from the following considerations: (1) The likelihood of hybridization between taxa or differentiated populations, disregarding for the moment issues of opportunity, should, to a first order of approximation, be proportional to their degree of similarity; that is, closely related taxa should hybridize frequently, whereas phylogenetically more distant taxa do so only rarely. (2) The likelihood of detecting introgression should decrease as similarity between hybridizing taxa or populations increases as a consequence of the reduced availability of diagnostic molecular or morphological markers. (3) The likelihood of detecting introgression should decrease (by some unknown function) as the period of time elapsed since introgression increases. Reasons for this situation are threefold. First, the alien contribution is likely to be diluted over time. Second, mutations are likely to occur in the introgressed genes over time, making them difficult to identify. Third, the donor population or species may go extinct or change so dramatically in appearance and genetic constitution that its parental role may not be recognized and its genetic contribution difficult to detect or document.

These considerations have several implications. First, it is likely that a significant proportion of ongoing introgression is cryptic. With respect to evolutionary consequences, we would quote from the epilogue to Anderson's (1949) monograph, *Introgressive Hybridization:* "How important is introgressive hybridization? I do not know. One point seems fairly certain: its importance is paradoxical. The more imperceptible introgression becomes, the greater is its biological significance. . . . Hence our paradox. Introgression is of the greater biological significance, the less is the impact of casual inspection." Second, it is even more likely that a high proportion of historical introgression (i.e., between differentiated taxa that are not currently hybridizing) is cryptic. Third, only a small proportion of ancient introgression events are expected to be experimentally verifiable. In this respect, there is no reason to believe that introgression has not played a major role in the evolution of vascular plants in ancient as well as modern times. Indeed, Epling (1947) argued that the ability to hybridize may actually have been selected for in certain groups as an adaptation to rapidly changing environments (we do not hold this view). This point may have implications for phylogeny estimation at higher taxonomic levels. For example, Stebbins (1950) and Grant (1953) speculated that the lack of clear discontinuities between major evolutionary lineages of plants is a consequence of ancient hybridization events.

In conclusion, the evidence presented in Table 4-1, which indicates that introgression is extensive, may represent a gross underestimate of the true magnitude of the phenomenon in plants. By extension, we are most likely underestimating the significance of introgression in plant evolution. It is also likely that the available empirical evidence is biased toward recent introgression events and toward introgression between more distantly related taxa.

Particular difficulties arise when assessing the evolutionary consequence of ancient introgression events. For example, extensive introgression has been documented in several plant genera, including *Gossypium, Quercus, Helianthus,* and *Salix* (Table 4-1), yet we can only speculate regarding its consequences. Generally, if the origin of a new ecotype or species has been postulated as a consequence of introgression, as with the proposed hybrid origin of *G. bickii* (Wendell et al., 1991) and *Helianthus annuus* spp. *texanus* (Rieseberg et al., 1990), it is unclear whether introgression contributed to the *origin* of the new taxon or it occurred *afterward.* Failure to consider this scenario has frequently led to unjustified conclusions.

A final comment concerns a potentially significant question: Has introgression negatively or positively affected species richness in plants? Table 4-1 demonstrates that most botanists have suggested new ecotypes or species as a result of introgression rather than the breakdown of reproductive barriers or the merger of previously isolated taxa. Early botanists, however, tended to emphasize the potential "creative" role of introgression (e.g., Anderson, 1949; Stebbins, 1950), and it might be argued that their influence largely accounts for the bias observed in Table 4-1. Alternatively, the observed bias may result from the difficulty of demonstrating that a "merged" taxon actually originated from two taxa that are now extinct. In addition, from a mechanistic standpoint, hybridization and introgression may be viewed as a type of genetic recombination. Thus it might be argued that introgression, by expanding the field of recombination (Harrison, 1990), is an evolutionary force contributing to an increase in both genetic and taxonomic diversity. Testing this point of view must await the development of a clearer picture of both the extent of plant introgression and its evolutionary dynamics.

CONCLUSIONS AND FUTURE DIRECTIONS

In the last comprehensive review of plant introgression to date, Heiser (1973, p. 361) concluded that introgression exists but that "most introgression appears to be highly localized. In contrast, dispersed or widespread introgression ... appears to be extremely rare." Empirical evidence (Table 4-1) now available demonstrates that both localized and dispersed introgression occur, and we have indicated several reasons why frequencies of both types are likely to be underestimated. We also now have empirical evidence that introgression may contribute to an increase in genetic diversity, although the evidence for the transfer or origin of adaptations through introgression is less convincing. There are now many examples documenting that (1) populations of introgressants dominate disturbed or intermediate habitats, and (2) these populations are stable and may well contribute to the origin of new species. Furthermore, several diploid plant species are now known to be stabilized hybrid derivatives that were likely derived from ancient hybrid swarms or introgressive populations. Thus in our view the evidence presented here indicates a major role for introgression in evolution, although it is recognized that much remains to be understood.

Molecular methodologies have contributed and will continue to contribute substantially to the detection and quantification of introgression (Rieseberg and Brunsfeld, 1992). Until recently, isozymes were the only molecular markers regularly employed in studies of introgression. A potential flaw of many studies has been the few nuclear markers surveyed, which in most cases range from one (rDNA) to several dozen (isozymes) genes. In the few cases where numerous molecular markers have been employed, little is known about the genomic distribution of successful introgressant alleles or their linkage relations.

These limitations are likely to be alleviated by technological developments that allow a virtually unlimited number of molecular markers to be scored. Restriction fragment analysis of low copy-number nuclear sequences (RFLPs) has been a readily accessible technique (for applications to the study of introgression see Keim et al., 1989; Song et al., 1988). Methodological gains hold the promise for even greater resolution. For example, PCR amplification of random oligonucleotide-primed DNAs (RAPDs) (Williams et al., 1990) may facilitate the detection and screening of numerous, highly variable molecular markers more efficiently than with current RFLP methods (Arnold et al., 1991). In addition to the many advantages afforded by the availability of a large number of markers, detailed genetic linkage maps may be constructed from single segregating progenies (e.g., Bernatsky and Tanksley, 1986; Helentjaris et al., 1986, 1988; Helentjaris, 1987; Tanksley et al., 1988; Michelmore et al., 1989). The advantage of using mapped molecular markers for the study of introgression is that the dynamics of not only individual markers but entire chromosomal segments can be monitored (Doebley and Wendel, 1989; Rieseberg and Brunsfeld, 1991). A significant capability is thus made available, i.e., ascertaining the genomic distribution of introgressed alleles and chromosome segments. For example, introgression may be restricted to a particular set of chromosomes or chromosome segments in a particular population. By establishing linkage relations between adaptively significant traits and molecular markers, the transfer of genetic adaptations can be documented with much greater precision, and hypotheses regarding the permeability of species barriers to advantageous, neutral, and disadvantageous alleles can be tested. This detailed genetic approach has the potential to enhance greatly our understanding of both the mechanisms and adaptive significance of introgression.

These methodological improvements are likely to have the greatest impact when combined with studies of the evolutionary ecology of hybrid and introgressive populations (see Ch. 5). Levin (1979) stressed the importance of experimental manipulations of hybrid swarms and hybrid zones, detailed comparisons of the biology of introgressive and parental plants in different environments, and estimates of fitness (e.g., pollen and seed dispersal) in hybridizing taxa. Similarly, Schemske and Morgan (1990) noted the importance of studying the full range of microevolutionary forces that may affect the direction and magnitude of gene flow. Rieseberg and Brunsfeld (1991) emphasized the need for examining the relative fitness of hybrids, introgressants, and parental plants in a range of environments, as well as studies that document the effects of introgression on isolating barriers. Sophisticated application of the tools of the evolutionary ecologist, when coupled with modern molecular methods, will undoubtedly significantly advance our understanding of introgression.

ACKNOWLEDGMENTS

Both authors' work on introgression has been aided by grants from the National Science Foundation. We thank Peter Bretting, Cristian Brochmann, John Doebley, and Mike Hanson for their careful reviews of the manuscript.

REFERENCES

Adams, R. P., Tomb, A. S., and Price, S. C. 1987. Investigation of hybridization between *Asclepias speciosa* and *A. syriaca* using alkanes, fatty acids, and terpenoids. Biochem. Syst. Ecol. 15:395–399.

Adams, R. P., and Turner, B. L. 1970. Chemosystematic and numerical studies of natural populations of *Juniperus ashei* Bush. Taxon 19:728–750.

Allendorf, F. W., and Leary, R. F. 1988. Conservation and distribution of genetic variation in a polytypic species, the cutthroat trout. Cons. Biol. 2:170–184.

Alston, R. E., and Simmons, J. 1962. A specific and predictable biochemical anomaly in interspecific hybrids of *Baptisia viridis* x *B. leucantha.* Nature 195:825.

Alston, R. E., and Turner, B. L. 1962. New techniques in analysis of complex natural hybridization. Proc. Natl. Acad. Sci. USA 48:130–137.

Anderson, E. 1936a. Hybridization in American Tradescantias. Ann. Missouri Bot. Gard. 23:511–525.

Anderson, E. 1936b. An experimental study of hybridization in the genus *Apocynum.* Ann. Missouri Bot. Gard. 23:159–168.

Anderson, E. 1939. Recombination in species crosses. Genetics 24:668–698.

Anderson, E. 1948. Hybridization of the habitat. Evolution 2:1–9.

Anderson, E. 1949. Introgressive Hybridization. New York: Wiley.

Anderson, E. 1953. Introgressive hybridization. Biol. Rev. 28:280–307.

Anderson, E. 1954. Introgression in *Adenostoma.* Ann. Missouri Bot. Gard. 4:339–350.

Anderson, E. 1961. The analysis of variation in cultivated plants with special reference to introgression. Euphytica 10:79–86.

Anderson, E., and Diehl, D. G. 1932. Contributions to the Tradescantia problem. J. Arnold Arb. 8:213–231.

Anderson, E., and Gage, A. 1952. Introgressive hybridization in *Phlox bifida.* Am. J. Bot. 39:399–404.

Anderson, E., and Hubricht, L. 1938. The evidence for introgressive hybridization. Am. J. Bot. 25:396–402.

Anderson, E., and Woodson, R. W. 1935. The species of *Tradescantia* indigenous to the United States. Contrib. Arnold Arb. 9:1–132.

Arnold, M. L., Bennett, B. D., and Zimmer, E. A.

1990a. Natural hybridization between *I. fulva* and *I. hexagona:* patterns of ribosomal DNA variation. Evolution 44:1512–1521.

Arnold, M. L., Buckner, C. M., and Robinson, J. J. 1991. Pollen mediated introgression and hybrid speciation in Louisiana irises. Proc. Natl. Acad. Sci. USA 88:1398–1402.

Arnold, M. L., Contreras, N., and Shaw, D. D. 1988. Biased gene conversion and asymmetrical introgression between subspecies. Chromosoma 96:368–371.

Arnold, M. L., Hamrick, J. L., and Bennett, B. D. 1990b. Allozyme variation in Louisiana irises: a test for introgression and hybrid speciation. Heredity 65:297–306.

Aubert, J., and Solignac, M. 1990. Experimental evidence for mitochondrial DNA introgression between *Drosophila* species. Evolution 44:1272–1282.

Avise, J. C., Bermingham, E., Kessler, L. G., and Saunders, N. C. 1984. Characterization of mitochondrial DNA variability in a hybrid swarm between subspecies of bluegill sunfish *(Lepomis macrochirus).* Evolution 38:931–941.

Avise, J. C., and Saunders, N. C. 1984. Hybridization and introgression among species of sunfish *(Lepomis):* analysis of mitochondrial DNA and allozyme markers. Genetics 108:237–255.

Baker, H. G. 1947. Criteria of hybridity. Nature 159:221–223.

Baker, H. G. 1948. Stages in invasion and replacement demonstrated by species of *Melandrium.* J. Ecol. 36:96–119.

Baldwin, B. G., Kyhos, D. W., and Dvorak, J. 1990. Chloroplast DNA evolution and adaptive radiation in the Hawaiian silversword alliance (Madiinae, Asteraceae). Ann. Missouri Bot. Gard. 77:96–109.

Barber, H. N., and Jackson, W. D. 1957. Natural selection in action in *Eucalyptus.* Nature 179:1267–1269.

Barton, N., and Bengtsson, B. O. 1986. The barrier to genetic exchange between hybridizing populations. Heredity 56:357–376.

Barton, N. H., and Hewitt, G. N. 1985. Analysis of hybrid zones. Annu. Rev. Ecol. Syst. 16:113–148.

Barton, N. H., and Jones, J. S. 1983. Mitochondrial DNA: new clues about evolution. Nature 306:317–318.

Beeks, R. M. 1962. Variation and hybridization

in southern California populations of *Diplacus* (Scrophulariaceae). Aliso 5:83–122.

Bell, N. B., and Lester, L. J. 1978. Genetic and morphological detection of introgression in a clinal population of *Sabatia* sect. *Campestria* (Gentianaceae). Syst. Bot. 3:87–104.

Benson, L., Phillips, E. A., Wilder, P. A., et al. 1967. Evolutionary sorting of characters in a hybrid swarm. I. Direction of slope. Am. J. Bot. 54:1017–1026.

Bernatsky, R., and Tanksley, S. D. 1986. Toward a saturated linkage map in tomato based on isozymes and random cDNA sequences. Genetics 112:887–898.

Bierhorst, D. W. 1977. Introgression in Trichomanes. Am. J. Bot. 64:1225–1234.

Birky, C. W., Maruyama, T., and Fuerst, P. 1983. An approach to population and evolutionary genetic theory for genes in mitochondria and chloroplasts, and some results. Genetics 103:513–527.

Bloom, W. L. 1976. Multivariate analysis of the introgressive replacement of *Clarkia nitens* by *Clarkia speciosa* subsp. *polyantha.* Evolution 30:412–424.

Bloom, W. L., and Lewis, H. 1972. Interchanges and interpopulational gene exchange in *Clarkia speciosa.* Chromosomes Today 3:268–284.

Briggs, B. 1962. Interspecific hybridization in the *Ranunculus lappaceus* group. Evolution 16:372–390.

Brochmann, C. 1984. Hybridization and distribution of *Argranthemum coronopifolium* (Asteraceae-Anthemideae) in the Canary Islands. Nord. J. Bot. 4:729–736.

Brochmann, C. 1987. Evaluation of some methods for hybrid analysis, exemplified by hybridization in *Argyranthemum* (Asteraceae). Nord. J. Bot. 7:609–630.

Brown, W. V., and Pratt, G. A. 1960. Hybridization and introgression the grass genus *Elymus.* Am. J. Bot. 47:669–676.

Brunsfeld, S. J. 1990. Systematics and evolution in *Salix* section *Longifoliae.* Ph.D. thesis, Washington State University, Pullman.

Cade, T. J. 1983. Hybridization and gene exchange among birds in relation to conservation. In C. M. Schonewald-Cox, S. M. Chambers, B. MacBryde, and W. L. Thomas, eds. Genetics and Conservation: A Reference for Managing Wild Animal and Plant Populations. Menlo Park, CA: Benjamin Cummings, pp. 288–310.

Camp, W. H. 1945. The North American blueberries, with notes on other groups of Vacciniaceae. Brittonia 5:203–275.

Camp, W. H. 1948. Cinchona at high altitudes in Ecuador. Brittonia 6:394–430.

Carr, S. M., Ballinger, S. W., Deer, J. N., Blankenship, L. H., and Bickham, J. W. 1986. Mitochondrial DNA analysis of hybridization between sympatric white-tailed and mule deer in west Texas. Proc. Natl. Acad. Sci. USA 83:9576–9580.

Carter, L. C., and Brehm, B. G. 1969. Chemical and morphological analysis of introgressive hybridization between *Iris tenax* and *I. chrysophylla.* Brittonia 21:44–54.

Chandler, J. M., Jan, C., and Beard, B. H. 1986. Chromosomal differentiation among the *Helianthus* species. Syst. Bot. 11:353–371.

Chase, V. C., and Raven, P. H. 1975. Evolutionary and ecological relationships between *Aquilegia formosa* and *A. pubescens* (Ranunculaceae), two perennial plants. Evolution 29:474–486.

Chu, Y. E., and Oka, H. E. 1970. Introgression across isolating barriers in wild and cultivated *Oryza* species. Evolution 24:344–355.

Clausen, J., and Hiesey, W. M. 1958. Experimental studies on the nature of species. IV. Genetic structure of ecological races. Washington, DC: Carnegie Institute, Publ. 615.

Clausen, J., Keck, D. D., and Hiesey, W. M. 1940. Experimental studies on the nature of species. I. Effect of varied environments on western North American plants. Washington, DC: Carnegie Institute, Publ. 520.

Cooperrider, M. 1957. Introgressive hybridization between *Quercus marilandica* and *Q. velutina* in Iowa. Am. J. Bot. 44:804–810.

Cracraft, J. 1983. Species concepts and speciation analysis. Curr. Ornithol. 1:159–187.

Crawford, D. J., and Ornduff, R. 1989. Enzyme electrophoresis and evolutionary relationships among three species of *Lasthenia* (Asteraceae: Heliantheae). Am. J. Bot. 76:289–296.

Crins, W. J., Bohm, B. A., and Carr, G. D. 1988. Flavonoids as indicators of hybridization in a mixed population of lava-colonizing Hawaiian tarweeds. (Asteraceae: Heliantheae: Madiinae). Syst. Bot. 13:567–571.

Critchfield, W. B. 1985. The later quarternary history of lodgepole and jack pine. Can. J. For. Res. 15:749–772.

Dally, A. M., and Second, G. 1990. Chloroplast DNA diversity in wild and cultivated species of rice (Genus *Oryza,* section Oryza). Cladisticmutation and genetic-distance analysis. Theor. Appl. Genet. 80:209–222.

Dansereau, P. 1941. Etudes sur les hybrides de Cistes. VI. Introgression dans la section Ladanum. Can. J. For. Res. 19:59–67.

Dansereau, P., and Desmarais, Y. 1947. Introgression in sugar maples. II. Am. Midl. Nat. 37:146–161.

Davis, J. I. 1985. Introgression in Central American *Phytolacca* (Phytolaccaceae). Am. J. Bot. 12:1949–1953.

Decker, D. S., and Wilson, H. D. 1987. Allozyme variation in the *Cucurbita pepo* complex: *C. pepo* var. *ovifera* vs. *C. texana.* Syst. Bot. 12:263–273.

Decker-Walters, D. S., Walters, T. W., Posluszny, U., and Kevan, P. G. 1990. Gene flow among annual domesticated species of Cucurbita. Can. J. Bot. 68:782–789.

Dempster, E. R. 1949. Effects of linkage on parental-combination and recombination frequencies in F_2. Genetics 34:272–284.

DePamphilis, C. W., and Wyatt, R. 1989. Hybridization and introgression in buckeyes Aesculus (Hippocastanaceae): a review of the evidence and a hypothesis to explain long-distance gene flow. Syst. Bot. 14:593–611.

DePamphilis, C. W., and Wyatt, R. 1990. Electrophoretic confirmation of interspecific hybridization in Aesculus (Hippocastanaceae) and the genetic structure of a broad hybrid zone. Evolution 44:1295–1317.

DeWet, J. M. J., and Harlan, J. R. 1975. Weeds and domesticates: evolution in the man-made habitat. Econ. Bot. 29:99–107.

Dobzhansky, T. 1941. Genetics and the Origin of Species. New York: Columbia University Press.

Doebley, J. F. 1984. Maize introgression into teosinte—a reappraisal. Ann. Missouri Bot. Gard. 71:1100–1113.

Doebley, J. F. 1989a. Molecular evidence for a missing wild relative of maize and the introgression of its chloroplast genome into Zea perennis. Evolution 43:1555–1558.

Doebley, J. F. 1989b. Isozyme evidence and the evolution of crop plants. In D. E. Soltis and P. S. Soltis, eds. Isozymes in Plant Biology. Portland, OR: Dioscorides Press, pp. 165–191.

Doebley, J., Goodman, M. M., and Stuber, C. W. 1984. Isoenzyme variation in Zea (Gramineae). Syst. Bot. 9:203–218.

Doebley, J., Goodman, M. M., and Stuber, C. W. 1987. Patterns of isozyme variation between maize and Mexican annual teosinte. Econ. Bot. 41:234–246.

Doebley, J., and Sisco, P. H. 1989. On the origin of the maize male sterile cytoplasms: it's completely unimportant, that's why it's so interesting. Maize Genet. Coop. Newslett. 63:108–109.

Doebley, J., and Wendel, J. F. 1989. Application of RFLPs to plant systematics. In T. Helentjaris and B. Burr, eds. Current Communications in Molecular Biology—Development and Application of Molecular Markers to Problems in Plant Genetics. Cold Spring Harbor, NY: Cold Spring Harbor Laboratory, pp. 57–68.

Donoghue, M. 1985. A critique of the biological species concept and recommendations for a phylogenetic alternative. Bryologist 88:172–181.

Dorado, O., Rieseberg, L. H., and Arias, D. 1992. Chloroplast DNA introgression in southern California sunflowers. Evolution 46:566–572.

Drouin, G., and Dover, G. A. 1990. Independent gene evolution in the potato actin gene family demonstrated by phylogenetic procedures for resolving gene conversions and the phylogeny of Angiosperm actin genes. J. Mol. Evol. 31:132–159.

DuRietz, G. E. 1930. The fundamental units of biological taxonomy. Svensk Bot. Tidskr. 24:333–428.

Echelle, A. F., and Conner, P. J. 1989. Rapid geographically extensive introgression after secondary contact between two pupfish species (Cyprinidon, Cyprinidontidae). Evolution 43:717–727.

Ehrendorfer, F. 1958. Ein Variabilitatszentrum als "fossiler" Hybrid-Komplex: Der ost-mediterrane Gallium graceum L.-G. canum Reg.-Formenkreis. Zur phylogenie der Gattung Galium. VI. Oesterr. Bot. Z. 105:229–279.

Ehrendorfer, F. 1959. Differentiation-hybridization cycles and polyploidy in Achillea. Cold Spring Harbor Symp. Quant. Biol. 24:141–152.

Elliott, F. C. 1949. Bromus inermis and B. pumpellianus in North America. Evolution 3:142–149.

Ellstrand, N. C. 1988. Pollen as a vehicle for the escape of engineered genes? Trends Biotechnol. 6:S30–S31.

Ellstrand, N. C., Lee, J. M., Keeley, J. E., and Keeley, S. C. 1987. Ecological isolation and introgression: biochemical confirmation of introgression in an Arctostaphylos (Ericaceae) population. Acta Oecologia/Oecologia Plantarum 8:299–308.

Epling, C. 1947. Natural hybridization of Salvia apiana and S. mellifera. Evolution 1:69–78.

Erbe, L., and Turner, B. L. 1962. A biosystematic study of the Phlox cuspidata-Phlox drummondii complex. Am. Midl. Nat. 67:257–281.

Erickson, L. R., Strauss, N. A., and Beversdorf, W. D. 1983. Restriction patterns reveal single origins of chloroplast genomes in Brassica amphidiploids. Theor. Appl. Genet. 65:201–206.

Fassett, N. C. 1945a. III. Possible hybridization of Juniperus horizontalis and J. scopulorum. Bull. Torrey Bot. Club 72:42–46.

Fassett, N. C. 1945b. IV. Hybrid swarms of J. virginiana and J. horizontalis. Bull. Torrey Bot. Club 72:379–384.

Fassett, N. C., and Calhoun, B. 1952. Introgression between Typha latifolia and T. angustifolia. Evolution 6:367–379.

Fassett, N. C., and Sauer, J. D. 1950. Studies of variation in the weed genus Phytolacca. I. Hybridizing species in northwestern Colombia. Evolution 4:332–339.

Feldman, M. 1965. Further evidence for natural hybridization between tetraploid species of Aegilops sect. Pleionathera. Evolution 19:162–174.

Ferris, S. D., Sage, R. D., Huang, C.-M., Nielson, J. T., Ritte, U., and Wilson, A. C. 1983. Flow of mitochondrial DNA across a species boundary. Proc. Natl. Acad. Sci. USA 80:2290–2294.

Flake, R. H., Urbatsch, L., and Turner, B. L. 1978. Chemical documentation of allopatric introgression in *Juniperus.* Syst. Bot. 3:129–144.

Flake, R. H., von Rudloff, E., and Turner, B. L. 1969. Quantitative study of clinal variation in *Juniperus virginiana* using terpenoid data. Proc. Natl. Acad. Sci. USA 64:487–494.

Forde, M. B., and Farris, D. G. 1962. Effects of introgression on the serpentine endemism of *Quercus durata.* Evolution 16:338–347.

Forrest, G. I. 1980. Geographical variation in the monoterpenes of *Pinus contorta* oleoresin. Biochem. Syst. Ecol. 8:343–359.

Frank, S. A. 1989. The evolutionary dynamics of cytoplasmic male sterility. Am. Nat. 133:345–376.

Funk, V. A. 1985. Phylogenetic patterns and hybridization. Ann. Missouri Bot. Gard. 72:681–715.

Furnier, G. R., Cummings, M. P., and Clegg, M. T. 1990. Evolution of the avocados as revealed by DNA restriction fragment variation. J. Hered. 81:183–188.

Gallez, G. P., and Gottlieb, L. D. 1982. Genetic evidence for the hybrid origin of the diploid plant *Stephanomeria diegensis.* Evolution 36:1158–1167.

Gerassimova, H. 1939. Chromosome alterations as a factor of divergence of forms. I. New experimentally produced strains of *C. tectorum* which are physiologically isolated from the original forms owing to reciprocal translocation. C. R. Acad. Sci. URSS 25:148–154.

Ghiselin, M. 1987. Species concepts, individuality, and objectivity. Biol. Philos. 2:127–143.

Gill, B. W., and Chen, P. D. 1987. Role of cytoplasm-specific introgression in the evolution of the polyploid wheats. Proc. Natl. Acad. Sci. USA 84:6800–6804.

Gillett, G. W. 1966. Hybridization and taxonomic implications in the *Scaveola guadichaudiana* complex of the Hawaiian islands. Evolution 20:506–516.

Golding, G. B., and Strobeck, C. 1983. Increased number of alleles found in hybrid populations due to intragenic recombination. Evolution 37:17–29.

Goodman, M. M. 1966. Correlation and the structure of introgressive populations. Evolution 20:191–203.

Goodwin, R. H. 1937. The cytogenetics of two species of *Solidago* and its bearing on their polymorphy in nature. Am. J. Bot. 24:425–432.

Gottlieb, L. D. 1972. Levels of confidence in the analysis of hybridization in plants. Ann. Missouri Bot. Gard. 59:435–446.

Gould, F. 1988. Evolutionary biology and genetically engineered crops. Bioscience 38:26–33.

Grant, V. 1949. Pollination systems as isolating mechanisms in flowering plants. Evolution 3:82–97.

Grant, V. 1950. Genetic and taxonomic studies in *Gilia.* I. *Gilia capitata.* Aliso 2:239–316.

Grant, V. 1952. Isolation and hybridization between *Aquilegia formosa* and *Aquilegia pubescens. Aliso 2:341–360.*

Grant, V. 1953. The role of hybridization in the evolution of the leafy-stemmed gilias. Evolution 7:51–64.

Grant, V. 1954. Genetic and taxonomic studies in *Gilia.* IV. *Gilia achilleaefolia.* Aliso 3:1–18.

Grant, V. 1958. The regulation of recombination in plants. Cold Spring Harbor Symposia Quant. Biol. 23:337–363.

Grant, V. 1963. The Origin of Adaptations. New York: Columbia University Press.

Grant, V. 1966a. Selection for vigor and fertility in the progeny of a highly sterile species hybrid in *Gilia.* Genetics 53:757–775.

Grant, V. 1966b. The origin of a new species of *Gilia* in a hybridization experiment. Genetics 54:1189–1199.

Grant, V. 1981. Plant Speciation. New York: Columbia University Press.

Grant, V., and Grant, A. 1960. Genetic and taxonomic studies in *Gilia.* XI. Fertility relationships of the diploid cobwebby gilias. Aliso 4:435–481.

Grant, V., and Wilken, D. H. 1988. Natural hybridization between *Ipomopsis aggregata* and *I. tenuituba* (Pomemoniaceae). Bot. Gaz. 149:213–221.

Gyllensten, U., and Wilson, A. C. 1987. Interspecific mitochondrial DNA transfer and the colonization of Scandinavia by mice. Genet. Res. Camb. 49:25–29.

Hadley, E. B., and Levin, D. A. 1969. Physiological evidence of hybridization and reticulate evolution in *Phlox maculata.* Am. J. Bot. 56:561–570.

Hagedoorn, A. C. 1921. Relative value of the processes causing evolution. The Hague: Martinus Nijhoff.

Hall, M. T. 1952. Variation and hybridization in *Juniperus.* Ann. Missouri Bot. Gard. 39:1–64.

Hardin, J. W. 1957. Studies in the Hippocastanaceae. III. Hybridization in *Aesculus.* Rhodora 59:45–51.

Hardin, J. W. 1975. Hybridization and introgression in *Quercus alba.* J. Arnold Abor. 56:336–363.

Harlan, J. R. 1965. The possible role of weed races in the evolution of cultivated plants. Euphytica 14:173–176.

Harlan, J. R., and de Wet, J. M. J. 1963. The compilospecies concept. Evolution 17:497–501.

Harrison, R. G. 1990. Hybrid zones: windows on evolutionary processes. Oxford Surv. Evol. Biol. 7:69–128.

Harrison, R. G., Rand, D. M., and Wheeler, W. C. 1987. Mitochondrial DNA variation in field crickets across a narrow hybrid zone. Mol. Biol. Evol. 4:144–158.

Hawkins, J. G. 1962. Introgression in certain wild potato species. Euphytica 11:26–35.

Hauber, D. P., and Bloom, W. L. 1983. Stability of a chromosomal hybrid zone in the *Clarkia nitens* and *C. speciosa* ssp. *polyantha* complex (Onagraceae). Am. J. Bot. 70:1454–1459.

Haynes, R. R., and Williams, D. C. 1975. Evidence for the hybrid origin of *Potamogeton longiligulatus* (Potamogetonaceae). Mich. Botanist 14:94–100.

Heiser, C. B. 1947. Hybridization between the sunflower species *Helianthus annuus* and *H. petiolaris*. Evolution 1:249–262.

Heiser, C. B. 1949a. Natural hybridization with particular reference to introgression. Bot. Rev. 15:645–687.

Heiser, C. B. 1949b. Study in the evolution of the sunflower species *Helianthus annuus* and *H. bolanderi*. Univ. Calif. Publ. Bot. 23:157–196.

Heiser, C. B. 1951a. Hybridization in the annual sunflowers: *Helianthus annuus* X *H. debilis* var. cucumerifolius. Evolution 5:42–51.

Heiser, C. B. 1951b. Hybridization in the annual sunflowers: *Helianthus annuus* X *H. argophyllus*. Am. Naturalist 85:64–72.

Heiser, C. B. 1954. Variation and subspeciation in the common sunflower, *Helianthus annuus*. Am. Midl. Nat. 51:287–305.

Heiser, C. B. 1965. Sunflowers, weeds and cultivated plants. In H. G. Baker and G. L. Stebbins, eds. The Genetics of Colonizing Species. Orlando, FL: Academic Press, pp. 391–401.

Heiser, C. B. 1973. Introgression re-examined. Bot. Rev. 39:347–366.

Heiser, C. B. 1979. Hybrid populations of *Helianthus divaricatus* and *H. microcephalus* after 22 years. Taxon 28:71–75.

Helentjaris, T. 1987. A genetic linkage map for maize based on RFLPs. Trends Genet. 3:217–221.

Helentjaris, T., Slocum, M., Wright, S., Schaefer, A., and Nienhuis, J. 1986. Construction of genetic linkage maps in maize and tomato using restriction fragment length polymorphisms. Theor. Appl. Genet. 72:761–769.

Helentjaris, T., Weber, D. F., and Wright, S. 1988. Duplicate sequences in maize and identification of their genomic locations through restriction fragment length polymorphisms. Genetics 118:355–363.

Hellquist, C. B., and Crow, G. E. 1986. *Potamogeton* X *haynesii* (Potamogetonaceae); a new species from northeastern North America. Brittonia 38:415–419.

Hellquist, C. B., and Hilton, R. L. 1983. A new species of *Potamogeton* (Potamogetonaceae) from northeastern United States. Syst. Bot. 8:86–92.

Heusmann, H. W. 1974. Mallard-black duck relationships in the Northeast. Wildlife Soc. Bull. 2:171–177.

Heywood, J. S. 1986. Clinal variation associated with edaphic ecotones in hybrid populations of *Gaillardia pulchella*. Evolution 40:1132–1140.

Heywood, J. S., and Levin, D. A. 1984. Allozyme variation in *Gaillardia pulchella* and *G. amblyodon* (Compositae): relation to morphological and chromosomal variation and to geographical isolation. Syst. Bot. 9:448–457.

Hillis, D. M. 1987. Molecular versus morphological approaches to systematics. Annu. Rev. Ecol. Syst. 18:23–42.

Hillis, D. M., Moritz, C., Porter, C. A., and Baker, R. J. 1991. Evidence for biased gene conversion and concerted evolution of ribosomal DNA. Science 251:308–309.

Hinton, W. F. 1976. Introgression and the evolution of selfing in *Calyptridum monospermum* (Portulacaceae). Syst. Bot. 1:95–96.

Jackson, R. C. 1976. Evolution and systematic significance of polyploidy. Annu. Rev. Ecol. Syst. 7:209–234.

Jensen, R. J., and Eshbaugh, W. H. 1976a. Numerical taxonomic studies of hybridization in *Quercus*. I. Populations of restricted area distribution and low taxonomic diversity. Syst. Bot. 1:1–10.

Jensen, R. J., and Eshbaugh, W. H. 1976b. Numerical taxonomic studies of hybridization in *Quercus*. II. Populations with wide area distributions and high taxonomic diversity. Syst. Bot. 1:10–19.

Johns, T., Huaman, Z., Ochoa, C., and Schmiediche, P. W. 1987. Relationship of wild, weed and cultivated potatoes in the *Solanum* X *ajanhairi* complex. Syst. Bot. 12:541–552.

Johnson, B. L. 1962. Amphiploidy and introgression in Stipa. Am. J. Bot. 49:253–262.

Kato, T. A. 1976. Cytological studies of maize. Massachusetts Agric. Exp. Sta. Bull. 635.

Kato, T. A. 1984. Chromosome morphology and the origin of maize and its races. Evol. Biol. 17:219–253.

Keim, P., Paige, K. N., Whitham, T. G., and Lark, K. G. 1989. Genetic analysis of an interspecific hybrid swarm of *Populus*: occurrence of unidirectional introgression. Genetics 123:557–565.

Kephart, S. R., Wyatt, R., and Parrella, D. 1988. Hybridization in North American *Asclepias*. I. Morphological evidence. Syst. Bot. 13:456–473.

Kimura, M. 1982. Molecular Evolution, Protein Polymorphism and Evolutionary Theory. Tokyo: Japanese Scientific Societies Press.

Kirkpatrick, K. J., and Wilson, H. E. 1988. Interspecific gene flow in *Cucurbita*: *C. texana* vs. *C. pepo*. Am. J. Bot. 75:519–627.

Klier, K., Leoschke, M. J., and Wendel, J. F. 1991. Hybridization and introgression in white and yellow ladyslipper orchids (*Cypripedium candidum* and *C. pubescens*). J. Hered. 82:305–319.

Knobloch, I. W. 1972. Intergeneric hybridization in flowering plants. Taxon 21:97–103.

Ladizinsky, G. 1985. Founder effect in crop-plant evolution. Econ. Bot. 39:191–199.

Langevin, S. A., Clay, K., and Grace, J. B. 1990. The incidence and effects of hybridization between cultivated rice and its related weed red rice (*Oryza sativa* L.). Evolution 44:1000–1008.

Leclercq, P. (1969) Une sterilite male cytoplasmique chez le tournesol. Ann. Amelior. Pl. 19:99–106.

Lenz, L. W. 1959. Hybridization and speciation in the pacific coast irises. Aliso 4:237–309.

Levin, D. A. 1963. Natural hybridization between *Phlox maculata* and *P. glaberrima* and its evolutionary significance. Am. J. Bot. 40:714–720.

Levin, D. A. 1966. Chromatographic evidence of hybridization and reticulate evolution in *Phlox maculata*. Am. J. Bot. 53:238–245.

Levin, D. A. 1967. Hybridization between annual species of *Phlox:* population structure. Am. J. Bot. 54:1122–1130.

Levin, D. A. 1967. Variation in *Phlox divaricata*. Evolution 21:92–108.

Levin, D. A. 1975. Interspecific hybridization, heterozygosity, and gene exchange in *Phlox*. Evolution 29:37–51.

Levin, D. A. 1979. Hybridization: An Evolutionary Perspective. Stroudsburg, PN: Dowden, Hutchinson & Ross.

Levin, D. A. 1983. Polyploidy and novelty in flowering plants. Am. Naturalist 122:1–25.

Levin, D. A. 1985. Reproductive character displacement in *Phlox*. Evolution 39:1275–1281.

Levin, D. A., and Kerster, H. W. 1967. Natural selection for reproductive isolation in *Phlox*. Evolution 21:679–687.

Levin, D. A., and Schaal, B. A. 1972. Seed protein polymorphism in *Phlox pilosa*. Brittonia 24:46–56.

Levin, D. A., and Smith, D. E. 1966. Hybridization and evolution in the *Phlox pilosa* complex. Am. Naturalist 100:289–302.

Lewis, H., and Epling, C. 1959. *Delphinium gypsophilum*, a diploid species of hybrid origin. Evolution 13:511–525.

Lewis, H., and Lewis, M. 1955. The genus *Clarkia*. Univ. Calif. Publ. Bot. 20:241–392.

Lewis, W. H., ed. 1980. Polyploidy: Biological Relevance. New York: Plenum Press.

Lewontin, R. C., and Birch, L. C. 1966. Hybridization as a source of variation for adaptation to new environments. Evolution 20:315–336.

Marchant, A. D. 1988. Apparent introgression of mitochondrial DNA across a narrow hybrid zone in *Caledia captiva* species-complex. Heredity 60:39–46.

Marchant, A. D., Arnold, M. L., and Wilkinson, P. 1988. Gene flow across a chromosomal tension zone. I. Relicts of ancient hybridization. Heredity 61:321–328.

Marsden-Jones, E. M. 1930. The genetics of *Geum intermedium* Willd. haud Ehrh. and its back-crosses. J. Genet. 23:377–395.

Mattfeld, J. 1930. Uber Hybridogene Sippen der Tannen. Bibl. Bot. 100:1–84.

Mayr, E. 1942. Systematics and the Origin of Species. New York: Columbia University Press.

McDade, L. 1990. Hybrids and phylogenetic systematics. I. Patterns of character expression in hybrids and their implications for cladistic analysis. Evolution 44:1685–1700.

McMinn, H. E. 1944. The importance of field hybrids in determining species in the genus *Ceonothus*. Proc. Calif. Acad. Sci. 25:323–356.

Mech, L. D. 1970. The Wolf: The Ecology and Behavior of an Endangered Species. Garden City, NY: Natural History Press.

Meyer, V. G. 1975. Male sterility from *Gossypium harknessii*. J. Hered. 66:23–27.

Meyn, O., and Emboden, W. A. 1987. Parameters and consequences of introgression in *Salvia apiana* X *S. mellifera* (Lamiaceae). Syst. Bot. 12:390–399.

Michelmore, R. W., Kesseli, R. V., Hulbert, S. H., Legg, E. J., and Paran, I. 1989. Analysis of lettuce downy mildew using molecular markers. In T. Helentjaris and B. Burr, eds. Current Communications in Molecular Biology—Development and Application of Molecular Markers to Problems in Plant Genetics. Cold Spring Harbor, NY: Cold Spring Harbor Laboratory, pp. 45–50.

Millar, C. V. 1983. A steep cline in *Pinus muriata*. Evolution 37:311–319.

Moore, D. M. 1959. Population studies on *Viola lactea* SM. and its wild hybrids. Evolution 13:318–332.

Muller, C. H. 1952. Ecological control of hybridization in *Quercus:* a factor in the mechanism of evolution. Evolution 6:147–161.

Northington, D. K. 1974. Systematic studies of the genus *Pyrrhopappus* (Compositae, Cichorieae). Spec. Publ. of Museum No. 6. Lubbock, TX: Texas Technical Press.

Oliveri, A. M., and Jain, S. K. 1977. Variation in the *Helianthus exilis-bolanderi* complex. Madrono 24:177–189.

Ornduff, R. 1966. A biosystematic survey of the oldfield genus *Lasthenia*. Univ. Calif. Publ. Bot. 40:1–92.

Ornduff, R. 1967. Hybridization and regional variation in Pacific Northwestern *Impatiens* (Balsaminaceae). Brittonia 19:122–128.

Ornduff, R. 1969. The origin and relationships of *Lasthenia burkei* (Compositae). Am. J. Bot. 56:1042–1047.

Ornduff, R. 1976. Speciation and oligogenic differentiation in *Lasthenia* (Compositae). Syst. Bot. 1:91–96.

Ostenfald, C. H. 1928. The present state of knowledge on hybrids between species of flowering plants. J. R. Hort. Soc. 53:31–44.

Palmer, J. D. 1988. Intraspecific variation and multicircularity in *Brassica* mitochondrial DNAs. Genetics 118:341–351.

Palmer, J. D., Jorgenson, R. A., and Thompson, W. F. 1985. Chloroplast DNA variation and evolution in *Pisum:* patterns of change and phylogenetic analysis. Genetics 109:195–213.

Palmer, J. D., Shields, C. R., Cohen, D. B., and Orton, T. J. 1983. Chloroplast DNA evolution and the origin of amphidiploid *Brassica* species. Theor. Appl. Genet. 65:181–189.

Palmer, J. D., and Zamir, D. 1982. Chloroplast DNA evolution and phylogenetic relationships in *Lycopersicon.* Proc. Natl. Acad. Sci. USA 79:5006–5010.

Panetos, C. A., and Baker, H. G. 1967. The origin of variation in "wild" *Raphanus sativus* in California. Genetica 38:243–274.

Percy, R. G., and Wendel, J. F. 1990. Allozyme evidence for the origin and diversification of *Gossypium barbadense.* Theor. Appl. Genet. 79:529–542.

Peterson, K. A., Elisens, W. J., and Estes, J. R. 1990. Allozyme variation in *Pyrrhopappus multicaulis* and *P. carolinianus* (Asteraceae): relation to mating systems and purported hybridization. Syst. Bot. 15:534–543.

Potts, B. M. 1986. The population dynamics and regeneration of a hybrid zone between *Eucalyptus risdonii* and *E. amygdalina.* Aust. J. Bot. 34:305–329.

Potts, B. M., and Reid, J. B. 1985. Analysis of a hybrid swarm between *Eucalyptus risdonii* Hook. f. and *E. amygdalina* Labill. Aust. J. Bot. 33:543–56.

Potts, B. M., and Reid, J. B. 1988. Hybridization as a dispersal mechanism. Evolution 42:1245–1255.

Powell, J. R. 1983. Interspecific cytoplasmic gene flow: evidence from *Drosophila.* Proc. Natl. Acad. Sci. USA 80:492–495.

Randolph, L. F. 1966. *Iris nelsonii,* a new species of Louisiana iris of hybrid origin. Baileya 14:143–169.

Randolph, L. G., Nelson, I. S., and Plaisted, R. L. 1967. Negative evidence of introgression affecting the stability of Louisiana Iris species. Cornell Univ. Agr. Exp. Stat. Mem. 398:1–56.

Rattenbury, J. A. 1962. Cyclic hybridization as a survival mechanism in New Zealand forest flora. Evolution 16:348–363.

Ravanko, O. 1979. Observations from uniform experimental plot on introgression between *Geum urbanum* and *Geum rivale* in Southwest Finland. Ann. Bot. Fenn. 16:43–49.

Raven, P. H. 1976. Systematics and plant population biology. Syst. Bot. 1:284–316.

Rick, C. M. 1958. The role of natural hybridization in the derivation of cultivated tomatoes of western South America. Econ. Bot. 12:346–367.

Rick, C., Zobel, R. W., and Fobes, J. F. 1974. Four peroxidase loci in red-fruited tomato species: genetics and geographic distribution. Proc. Natl. Acad. Sci. USA 71:835–839.

Rieseberg, L. H. 1987. A reexamination of introgression in *Helianthus.* Ph.D. thesis, Washington State University, Pullman.

Rieseberg, L. H. 1991. Homoploid reticulate evolution in *Helianthus:* evidence from ribosomal genes. Am. J. Bot. 78:1218–1237.

Rieseberg, L. H., Beckstrom-Sternberg, S., and Doan, K. 1990a. *Helianthus annuus* ssp. *texanus* has chloroplast DNA and nuclear ribosomal RNA genes of *Helianthus debilis* ssp. *cucumerifolius.* Proc. Natl. Acad. Sci. USA 87:593–597.

Rieseberg, L. H., Beckstrom-Sternberg, S., Liston, A., and Arias, D. M. 1991a. Phylogenetic and systematic inferences from chloroplast DNA and isozyme variation in *Helianthus* sect. *Helianthus* (Asteraceae). Syst. Bot. 16:50–76.

Rieseberg, L. H., and Brunsfeld, S. 1992. Molecular evidence and plant introgression. In D. E. Soltis, P. S. Soltis, and J. J. Doyle, eds. Molecular Systematics of Plants. New York: Chapman & Hall, pp. 151–176.

Rieseberg, L. H., Carter, R., and Zona, S. 1990b. Molecular tests of the hypothesized hybrid origin of two diploid *Helianthus* species (Asteraceae). Evolution 44:1498–1511.

Rieseberg, L. H., Choi, H. C., and Ham, D. 1991b. Differential cytoplasmic versus nuclear introgression in Helianthus. J. Hered. 82:489–493.

Rieseberg, L. H., and Seiler, G. 1990. Molecular evidence and the origin and development of the domesticated sunflower (*Helianthus annuus* L.). Econ. Bot. 44S:79–91.

Rieseberg, L. H., and Soltis, D. E. 1991. Phylogenetic consequences of cytoplasmic gene flow in plants. Evol. Trends Pl. 5:65–84.

Rieseberg, L. H., Soltis, D. E., and Palmer, J. D. 1988a. A molecular re-examination of introgression between *Helianthus annuus* and *H. bolanderi* (Compositae). Evolution 42:227–238.

Rieseberg, L. H., Soltis, D. E., and Soltis, P. S. 1988b. Genetic variation in *Helianthus annuus* and *H. bolanderi.* Biochem. Syst. Ecol. 16:393–399.

Rieseberg, L. H., Zona, S., Aberbom, L., and Martin, T. D. 1989. Hybridization in the island endemic, Catalina mahogany. Cons. Biol. 3:52–58.

Riley, H. P. 1938. A character analysis of colonies of *Iris fulva, Iris hexagona* var. *giganticaerulea* and natural hybrids. Am. J. Bot. 25:727–738.

Roberts, H. F. 1929. Plant Hybridization before Mendel. New York: Hafner.

Rollins, R. C. 1949. Sources of genetic variation in *Parthenium argentatum* Gray (Compositae). Evolution 3:358–368.

Rollins, R. C., and Solbrig, O. T. 1973. Interspecific hybridization in *Lesquerella.* Contr. Gray Herb. Harvard Univ. 203:3–48.

Rushton, B. S. 1979. *Quercus rober* and *Q. petraea:* A multivariate approach to the hybrid problem. 2. The geographical distribution of population types. Watsonia 12:209–224.

Russell, N. H. 1954.Three field studies of hybridization in the stemless white violets. Am. J. Bot. 41:679–686.

Russell, N. H. 1955. Local introgression between *Viola cucullata* Ait. and *V. septentrionalis* Greene. Evolution 9:436–440.

Sauer, J. 1957. Recent migration and evolution of the dioecious amaranths. Evolution 11:11–31.

Schemske, D. W., and Morgan, M. T. 1990. The evolutionary significance of hybridization in *Eucalyptus.* Evolution 44:2151–2152.

Second, G. 1982. Origin of the genic diversity of cultivated rice (*Oryza* spp.): study of the polymorphism scored at 40 isozyme loci. Jpn. J. Genet. 57:25–57.

Smith, H. H., and Daly, K. 1959. Discrete populations derived by interspecific hybridization and selection in *Nicotiana.* Evolution 13:476–487.

Smith, R. L., and Sytsma, K. J. 1990. Evolution of *Populus nigra* L. (sect. Aigeiros): introgressive hybridization and the chloroplast contribution of *Populus alba* L. (sect. Populus). Am. J. Bot. 77:1176–1187.

Snyder, L. A. 1950. Morphological variability and hybrid development in *Elymus glaucus.* Am. J. Bot. 37:628–636.

Snyder, L. A. 1951. Cytology of inter-strain hybrids and the origin of variability in *Elymus glaucus.* Am. J. Bot. 38:195–202.

Soltis, D. E., Soltis, P. S., Collier, T. G., and Edgerton, M. L. 1991. Chloroplast DNA variation within and among genera of the *Heuchera* group (Saxifragaceae): evidence for chloroplast transfer and paraphyly. Am. J. Bot. 78:1150–1161.

Soltis, P. 1985. Studies of genetic variation in an introgressive complex in *Clarkia* (Onagraceae). Ph.D. thesis, University of Kansas.

Soltz, D. S., and Naiman, R. J. 1978. The natural history of native fishes in the Death Valley system. Serial Publications of the Natural History Museum of Los Angeles County, Science Series 30.

Song, K. M., Osborn, T. C., and Williams, P. H. 1988. *Brassica* taxonomy based on nuclear restriction fragment length polymorphisms (RLFPs). 1. Genome evolution of diploid and amphidiploid species. Theor. Appl. Genet. 75:784–794.

Spooner, D. M., Sytsma, K. J., and Smith, J. F. 1991. A molecular reexamination of diploid hybrid speciation of *Solanum raphanifolium.* Evolution 45:757–763.

Stebbins, G. L. 1950. Variation and Evolution in Plants. New York: Columbia University Press.

Stebbins, G. L. 1957. The hybrid origin of microspecies in the *Elymus glaucus* complex. Cytologia 36S:336–340.

Stebbins, G. L. 1959. The role of hybridization in evolution. Proc. Am. Philos. Soc. 103:231–251.

Stebbins, G. L. 1966. Processes of Organic Evolution. Englewood Cliffs, NJ: Prentice-Hall.

Stebbins, G. L., and Daly, G. K. 1961. Changes in the variation of a hybrid population of *Helianthus* over an eight-year period. Evolution 15:60–71.

Stebbins, G. L., and Ferlan, L. 1956. Population variability, hybridization and introgression in some species of *Ophrys*. Evolution 10:32–46.

Stebbins, G. L., Matzke, E. B., and Epling, C. 1947. Hybridization in a population of *Quercus marilandica* and *Q. ilicifolia.* Evolution 1:79–88.

Straw, R. M. 1955. Hybridization, homogamy and sympatric speciation. Evolution 9:441–444.

Stutz, H. C., and Thomas, L. K. 1964. Hybridization and introgression in *Cowania* and *Purshia.* Evolution 18:183–195.

Suzuki, K. 1986. *Epimedium trifoliatobinatum,* a species derived from hybridization between *E. grandiflorum* and *P. diphyllum.* In K. Iwatsuki, P. H. Raven, and W. H. Bock, eds. Modern Aspects of Species. Tokyo: University of Tokyo Press, pp. 195–209.

Sytsma, K. J., Smith, J. F., and Berry, P. E. 1991. Biogeography and evolution of morphology, breeding systems, and flavonoids in the old world *Fuchsia* sect. *Skinnera* (Onagraceae): evidence from chloroplast DNA. Syst. Bot. 16:257–269.

Sytsma, K. J., Smith, J. F., and Gottlieb, L. D. 1990. Phylogenetics in *Clarkia* (Onagraceae): restriction site mapping of chloroplast DNA. Syst. Bot. 15:280–295.

Syzmura, J. M., Spolsky, C., and Uzzell, T. 1986. Concordant changes in mitochondrial and nuclear genes in a hybrid zone between two frog species (genus *Bombina*). Experimentia 41:1469–1470.

Tanksley, S. D., Bernatsky, R., Lapitan, N. L., and Prince, J. P. 1988. Conservation of gene repertoire but not gene order in pepper and tomato. Proc. Natl. Acad. Sci. USA 85:6419–6423.

Taylor, R. H. 1975. Some ideas on speciation in New Zealand parakeets. Nortornis 22:110–121.

Tegelstrom, H. 1987. Transfer of mitochondrial DNA from the northern red-backed vole (*Clethrionomys rutilus*) to the bank vole (*C. glareolus*). J. Mol. Evol. 24:218–227.

Templeton, A. 1989. The meaning of species and speciation: a genetic perspective. In D. Otte and J. Endler, eds. Speciation and Its Consequences. Sunderland, MA: Sinauer, pp. 3–27.

Tucker, J. 1952. Evolution of the California oak *Quercus alvordiana.* Evolution 6:162–180.

Tucker, J. 1953. The relationship between *Quercus dumosa* and *Quercus turbinella.* Madrono 12:49–60.

Tucker, J. M., and Sauer, J. D. 1958. Aberrant *Amaranthus* populations of the Sacramento-San Joaquin delta, California. Madrono 14:252–261.

Turcotte, E. L., and Feaster, C. V. 1967. Semigamy in Pima cotton. J. Hered. 58:55–57.

Ugent, D. 1970. *Solanum raphanifolium,* a Peruvian wild potato species of hybrid origin. Bot. Gaz. (Crawfordsville) 131:225–233.

Valentine, D. H. 1948. Studies in British primulas. I. Hybridization between primrose and oxlip (*Primula vulgaris* Huds. and *P. elatior* Schreb). New Phytol. 47:111–130.

Wagner, D. B., Furnier, G. R., Saghai-Maroof, M. A., Williams, S. M., Dancik, B. P., and Allard, R. W. 1987. Chloroplast DNA polymorphisms in lodgepole and jack pines and their hybrids. Proc. Natl. Acad. Sci. USA 84:2097–2100.

Wagner, W. H., Jr. 1969. The role and taxonomic treatment of hybrids. Bioscience 19:785–789.

Wagner, W. H., Jr. 1970. Biosystematics and evolutionary noise. Taxon 19:146–151.

Wall, J. R. 1970. Experimental introgression in the genus *Phaseolus.* I. Effect of mating system on interspecific gene flow. Evolution 24:356–366.

Warwick, S. I., Bain, J. F., Wheatcroft, R., and Thompson, B. K. 1989. Hybridization and introgression in *Carduus nutans* and *C. acanthoides* reexamined. Syst. Bot. 14:476–494.

Wayne, R. K., and Jenks, S. M. 1991. Mitochondrial DNA analysis implying extensive hybridization of the endangered red wolf *Canis rufus.* Nature 351:565–568.

Weber, W. A. 1946. A taxonomic and cytological study of the genus *Wyethia,* family Compositae, with notes on the related genus *Balsamorhiza.* Am. Mid. Nat. 35:400–542.

Wendel, J. F., and Albert, V. A. 1992. Phylogenetics of the cotton genus (*Gossypium* L.): character-state weighted parsimony analysis of chloroplast-DNA restriction site data and its systematic and biogeographic implications. Syst. Bot. 17:115–143.

Wendel, J. F., Olson, P. D., and Stewart, J. McD. 1989. Genetic diversity, introgression, and independent domestication of old world cultivated cottons. Am. J. Bot. 76:1795–1806.

Wendel, J. F., and Percy, R. G. 1990. Allozyme diversity and introgression in the Galapagos endemic *Gossypium darwinii* and its relationship to continental *G. barbadense.* Biochem. Syst. Ecol. 18:517–528.

Wendel, J. F., Stewart, J. McD., and Rettig, J. H. 1991. Molecular evidence for homoploid reticulate evolution among Australian species of *Gossypium.* Evolution 45:694–711.

Wetmore, R. H., and Delisle, A. L. 1939. Studies in the genetics and cytology of two species in the genus *Aster* and their polymorphy in nature. Am. J. Bot. 26:1–12.

Whalen, M. D. 1978. Character displacement and floral diversity in *Solanum* section *Androcerus.* Syst. Bot. 3:77–86.

Wheeler, N. C., and Guries, R. P. 1987. A quantitative measure of introgression between lodgepole and jack pines. Can. J. Bot. 65:1876–1885.

Whittemore, A. T., and Schaal, B. A. 1991. Interspecific gene flow in oaks. Proc. Natl. Acad. Sci. USA 88:2540–2544.

Wilkes, H. G. 1977. Hybridization of maize and teosinte in Mexico and Guatemala and the improvement of maize. Econ. Bot. 31:254–293.

Williams, J. G. K., Kubelik, A. R., Livak, K. J., Rafalski, J. A., and Tingey, S. V. 1990. DNA polymorphisms amplified by arbitrary primers are useful as genetic markers. Nucleic Acids Res. 18:6531–6535.

Wilson, H. D. 1990a. Gene flow in squash species. Bioscience 40:449–455.

Wilson, H. D. 1990b. *Quinua* and relatives (*Chenopodium* sect. *Chenopodium* subsect. *Cellulata*). Econ. Bot. 44S:92–110.

Wolf, C. B. 1944. The gander oak, a new hybrid oak from San Diego County, California. Proc. Calif. Acad. Sci. 25:177–188.

Woodruff, D. S. 1989. Genetic anomalies associated with *Cerion* hybrid zones: the origin and maintenance of the new electromorphic variants called hybrizymes. Biol. J. Linn. Soc. 36:281–294.

Woodson, R. E., Jr. 1947. Some dynamics of leaf variation in *Asclepias tuberosa.* Ann. Missouri Bot. Gard. 34:353–432.

Woodson, R. E., Jr. 1962. Butterflyweed revisited. Evolution 16:168–185.

Wyatt, R., and Antonovics, J. 1981. Butterflyweed revisited—spatial and temporal patterns of leaf shape variation in *Asclepias tuberosa.* Evolution 35:525–542.

Zamir, D., Navot, N., and Rudich, J. 1984. Enzyme polymorphism in *Citrullus lanatus* and *C. colocynthis* in Israel and Sinai. Pl. Syst. Evol. 146:163–170.

II

CASE STUDIES OF
HYBRID ZONES

The chapters in this section summarize results and conclusions from selected, long-term research programs on hybrid zones. The taxa represented are a distinctly nonrandom sample of the earth's biota, which in part simply reflects a nonrandom distribution of hybrid zones in nature. For example, none of the studies are of marine organisms, primarily because hybrid zones in the marine realm are rare (although a few clear examples can be found). Many hybrid zones may be cryptic, in the sense that they cannot be recognized easily (or perhaps at all) from examination of morphology and have become apparent only upon detailed analyses of cytogenetic or molecular data. As a result, organisms favored by cytogeneticists may be "overrepresented" in the hybrid zone literature. Several of the examples discussed in this section involve hybrid zones between "chromosomal races" that were not initially recognized as distinct, e.g., in the grasshopper *Caledia captiva* (see Ch. 7) and the shrew *Sorex araneus* (see Ch. 12).

Hybridization is common in many groups of plants, and yet only one botanical example is included here. There are surprisingly few plant hybrid zones that have been examined in detail using a multidisciplinary approach. Botanists have also been slower than zoologists to adopt an explicit population genetic framework as the context in which to analyze examples of natural hybridization. Hybridization in the Louisiana irises is a notable exception, with the early contributions of Riley and Anderson serving as a foundation for the more recent ecological and genetic studies of Arnold and Bennett (see Ch. 5). The latter also clearly illustrate how the use of molecular markers can help to clarify the consequences of hybridization past and present.

There are numerous examples of well-studied animal hybrid zones in terrestial environments, and only a selected subset could be included in this volume. They were chosen to represent a diversity of taxa, habitats, and patterns of variation and to illustrate clearly how hybrid zone research has contributed to our understanding of evolution. Three of the chapters focus on insect examples; four chapters summarize data from studies of vertebrate hybrid zones. Several of these case studies come close to fulfilling M. J. D. White's (*Modes of Speciation,* W. H. Freeman, 1978) criteria for "a fully documented history . . . of speciation"; i.e., they include data on current distributions and inferences about past distributions, detailed analyses of patterns of morphological, chromosomal, and molecular variation, as well as information on dispersal, life history, behavior (mate choice), and habitat and resource requirements.

Each of the contributions addresses at least some of the major issues in hybrid zone research outlined in Chapter 1. Many of the authors confront (albeit briefly) the question of whether hybridizing taxa should be considered distinct species. Patton (see Ch. 11) clearly points out that a major focus in early hybrid zone studies of pocket gophers was to resolve the perplexing question of defining species boundaries in groups of taxa that are primarily distributed allopatrically or parapatrically. Moore and Price (see Ch. 8) advocate a variant of the "cohesion species concept" and conclude that, despite hybridization and introgression, the red-shafted and yellow-shafted flickers should be considered distinct species.

Discussions of hybrid zone origins and of the factors that currently determine the distribution of hybridizing taxa figure prominently in many of the chapters. For hybrid zones in temperate regions, most authors favor a scenario of secondary contact, usually based on consideration of how recent glacial history altered the distributions of plant communities and of the animals associated with them. Among the examples discussed in the following chapters, the hybrid zones for *Chorthippus* (see Ch. 6), *Colaptes* (see Ch. 8), *Bombina* (see Ch. 10), and several species of pocket gophers (see Ch. 11) appear to be examples of secondary contact. Drawing on excellent paleoclimatological data from Europe, Hewitt (see Ch. 6) is able to trace with considerable confidence the historical biogeography of the two subspecies of *Chorthippus* and to date their recent arrival in the Pyrenees. The importance of Pleistocene refugia in the tropics (and subsequent spread from these refugia) as determinants of current distribution patterns remains a contentious issue. Although there has been much support for a model of allopatric divergence in rain forest fragments and subsequent secondary contact, the evidence is perhaps less convincing than in the temperate zone examples. Mallet (see Ch. 9) reviews many of the arguments, using the distributions and hybrid zones of *Heliconius* butterflies as evidence.

Analysis of the current dynamics of hybrid zones also is a common theme of the chapters in this section. Studies of flickers in the Great Plains, *Heliconius* butterflies in Central and South America, grasshoppers in Europe and Australia, and irises in Louisiana all reveal a role for natural selection in determining hybrid zone structure. However, the nature of natural selection varies. Moore and Price (see Ch. 8) argue that the flicker hybrid zone is maintained by "exogenous" selection, that spatially varying selection regimes (environmental gradients) determine both the position of the hybrid zone and (together with dispersal) its dimensions. The two *Iris* species studied by Arnold and Bennett (see Ch. 5) each appear to be better adapted to their own local habitat, and hybrid populations tend to occur at ecotones. Observations and experiments provide convincing evidence for ecological determination of hybrid zone structure. In other cases, narrow hybrid zones are maintained by a balance between dispersal and hybrid unfitness, e.g., for the grasshoppers *Chorthippus* (see Ch. 6) and *Caledia* (see Ch. 7), the toad *Bombina* (see Ch. 10), and many hybrid zones involving karyotypic races of small mammals (see Ch. 12). Hybrid unfitness may be independent of the external environment. The *Chorthippus* hybrid zone arose relatively recently and appears to have remained where it first formed. The situation is rather different for *Caledia*. Although the steep clines for chromosome arrangements are a consequence of selection against hybrids, the current position of the zone appears to be determined by climatic variables, and there is good evidence that the zone has moved a considerable distance over the past 20,000 years. Shaw et al. (see Ch. 7) suggest a possible direct relation between chromosome complement and grasshopper life history (development time) that could provide the final causal link in explaining the distribution of *Caledia* karyotypes.

   Hybrid zones are also natural laboratories for studying the contributions of ecological and behavioral differences to reproductive isolation. Ecological differentiation is found in irises (see Ch. 5), toads (see Ch. 10), and pocket gophers (see Ch. 11); and affiliation with particular habitats influences not only the structure of the hybrid zones but also the probability of encounters between individuals of the hybridizing species. For both *Chorthippus* (see Ch. 6) and *Bombina* (see Ch. 10), the hybridizing taxa differ in components of the male calling song; and for *Chorthippus* there is clear evidence of positive assortative mating.

   None of the case histories reviewed in the following chapters provide evidence for the process of reinforcement. Searle (see Ch. 12) discusses certain situations in which reinforcement is a possible outcome, but the observed patterns of variation in a number of small mammal hybrid zones suggest that selection has not altered the properties of mate recognition systems. Instead, selection appears to have resulted in a reduced frequency of less fit hybrids through the origin of staggered chromosomal clines in which most "hybrids" are "high-fertility simple Robertsonian heterozygotes" rather than "low-fertility complex heterozygotes." Patton (see Ch. 11) raises the possibility of reinforcement in pocket gopher hybrid zones but argues that the properties of these zones make such an outcome unlikely. In the *Chorthippus* hybrid zone, observations and experiments do not suggest an increase in assortative mating within or adjacent to the hybrid zone, although according to Hewitt (see Ch. 6) the "situation seems ideal for reinforcement."

Finally, patterns of introgression have been examined in many of the hybrid zones, often using a variety of markers: morphological, chromosomal, and molecular. For *Bombina,* clines for diagnostic allelic markers at allozyme loci are concordant both within a single transect and among transects (see Ch. 10), but in many other hybrid zones there is strong evidence for differential and sometimes asymmetrical introgression. For *Caledia,* the steep clines for chromosome rearrangements are displaced up to several hundred kilometers from the broader clines seen for a number of different molecular markers (see Ch. 7). In this case, patterns of variation suggest recent hybrid zone movement. Given an approximate date of origin and good estimates of dispersal rates, the *Chorthippus* hybrid zone is wider than expected on the basis of models of clinal variation resulting from secondary contact. Hewitt (see Ch. 6) proposes a more realistic model for the process of range expansion and secondary contact to account for the observed widths.

All of the studies presented in the following chapters have made unique contributions not only to our catalogue of hybrid zone patterns but also to our understanding of evolutionary process. From each case history, we can begin to reconstruct a unique series of historical events and to define an array of evolutionary forces that have combined to produce the patterns that we observe today. Ultimately we are in search of general principles. We hope to estimate the probabilities of particular evolutionary trajectories when differentiated populations interact and to understand the characteristics of organisms and environments that determine these probabilities.

5

Natural Hybridization in Louisiana Irises: Genetic Variation and Ecological Determinants

MICHAEL L. ARNOLD AND BOBBY D. BENNETT

Early studies of natural hybridization were traditionally directed either toward understanding species relationships based on the pattern and extent of natural hybridization or toward understanding the process of hybridization itself, often as a means for unraveling the component parts of speciation (e.g., Wiegand, 1935; Anderson and Stebbins, 1954). Rather than being mutually exclusive, these approaches have been complementary and have led to a greater understanding of both species relationships and the dynamics of hybridization. Many studies concerned with inferring evolutionary process from patterns of natural hybridization have relied on previous, systematic studies to provide the framework on which to evaluate alternate evolutionary hypotheses (e.g., *Aesculus:* Hardin, 1957; dePamphilis and Wyatt, 1989; *Helianthus:* Heiser, 1951; Rieseberg et al., 1990a; *Iris:* Riley, 1938; Anderson, 1949; Arnold et al., 1990a). The reverse relation between these two approaches, in studies of natural hybridization, is also apparent. Thus observations of hybrid genotypes and clinal variation (Huxley, 1938; Endler, 1977) have led to identification and characterization of previously unrecognized taxa (e.g., *Uroderma:* Baker et al., 1975; Baker, 1981; Greenbaum, 1981; *Geomys:* Pembleton and Baker, 1978; Baker et al., 1989; *Caledia:* Shaw et al., 1980).

More recently, a major focus for studies of natural hybridization has been the genetic delineation of zones of contact and the assessment of the impact of natural hybridization on the genetic structure of the hybridizing taxa. In particular, the width of hybrid zones, degree of concordance of character clines, differential penetration of various genetic and morphological markers, appearance of novel genetic elements in the zone of contact, level of disequilibrium between cytoplasmic and nuclear genetic elements, and the association of areas of hybridization with regions of ecological transition or discontinuity have led to inferences concerning the relative contributions of stochastic and deterministic forces to the dynamics of contemporary zones of overlap (Moran and Shaw, 1977; Barton, 1979; Shaw et al., 1980, 1983; Barton and Hewitt, 1981; Barton et al., 1983; Sattler, 1985; Shaw et al., 1985; Stangl, 1986; Arnold et al.,

1987, 1990a,b; Asmussen, 1987; Harrison et al., 1987; Moore, 1987; J. Arnold et al., 1988; Marchant et al., 1988; Bennett, 1989; Dowling et al., 1989; Rand and Harrison, 1989; dePamphilis and Wyatt, 1990; Rieseberg et al., 1990a). Some areas of contact have been termed tension zones to indicate that they have most likely resulted from an overall lower fitness of hybrid genotypes relative to the parental forms (Key, 1968). Other authors have suggested that such hybrid zones might reflect (1) positive selection favoring the hybrid forms in an area of ecological transition or discontinuity (Moore, 1977; Moore and Buchanan, 1985), or (2) the presence of a selection gradient (Endler, 1977). However, the resolution of hybridization events into hybrid zones defined by a pattern of clinal variation is not a universal pattern. Thus Rand and Harrison (1989) have described the concept of a "mosaic" zone of overlap as exemplified by the inter-action of the cricket species *Gryllus firmus* and *G. pennsylvanicus.* Furthermore, their finding of an association of certain genotypic arrays with definable habitat types (soil type) suggests the action of natural selection mediated by ecological parameters (Rand and Harrison, 1989). Such an interdigitation of parental and hybrid forms and the correlation between putative ecological determinants and the parental and hybrid taxa is also characteristic of other cases of natural hybridization. In particular, there are apparent examples where the interaction between hybridizing plant species appears to fit a mosaic rather than a clinal hybrid zone model and where the pattern of interdig-itation is closely associated with ecological discontinuities (e.g., *Baptisia:* Alston and Turner, 1963; *Helianthus:* Heiser et al., 1969).

It is a truism that the biological characteristics of any organism chosen for analysis provide unique opportunities and also place limitations on the questions that can be addressed. However, many plant systems allow detailed experimental manipulation. Furthermore, many of the evolutionary phenomena associated with natural hybrid-ization (e.g., hybrid speciation, syngameons, hybrid swarms, introgression) were his-torically defined or exemplified with reference to plant systems (see Grant, 1981, for review). In the present chapter, we summarize the findings from analyses of a classic example of natural hybridization: the Louisiana iris species complex. There are several factors that make these *Iris* species a model system for studying the process of natural hybridization. First, plant species (in general) allow the identification and monitoring of individuals through time. Second, species-specific molecular and biochemical markers have been identified, screened, and utilized in order to examine the extent to which alleles introgress. Third, findings from experimental ecological and demo-graphic studies have led to the description of biotic and abiotic factors that affect the fitness of *I. fulva, I. hexagona,* and hybrid types and have permitted a test for corre-lations between environmental components and the genetic structure of hybrid pop-ulations. Using the genetic and ecological data, we address questions concerning (1) the structure of contemporary hybrid populations; (2) the evolutionary impact of nat-ural hybridization on this species complex with regard to introgression and hybrid spe-ciation; and (3) the potential for ecological determinants to affect the distribution of *Iris* species and their hybrids and the structure and dynamics of the hybrid "zone."

TAXONOMIC, ECOLOGICAL, GENETIC, AND EVOLUTIONARY STUDIES OF THE LOUISIANA IRISES: A SYNOPSIS

Iris fulva and *I. hexagona* (= *I. hexagona var. giganti-caerulea;* Anderson 1949) come into contact in southern Louisiana when water channels (bayous) derived from the

Mississippi River end in freshwater marshes and swamps. *I. fulva* is associated with shaded, understory areas along the banks of bayou systems deriving from the Mississippi River (Louisiana, Arkansas, Mississippi, Tennessee, Kentucky, Illinois, Missouri; Viosca, 1935; Bennett, 1989). This species has been reported to occur as far east as Georgia (Godfrey and Wooten, 1979), but there are no known specimens east of central Mississippi (N. Henderson, pers. comm.). *I. hexagona* occurs principally in open, freshwater marshes and swamps in Texas, Louisiana, Mississippi, Alabama, Georgia, South Carolina, and Florida (Viosca, 1935; Bennett, 1989; N. Henderson, pers. comm.). In contrast to the former two species, *I. brevicaulis* (= *I. foliosa;* Viosca, 1935) occurs in much drier habitats (unimproved pastures and hardwood forests; Viosca, 1935; Randolph et al., 1967) in Louisiana, Mississippi, Alabama, Arkansas, Tennessee, Oklahoma, Kentucky, Missouri, Illinois, Indiana, Michigan, Wisconsin, and Ontario (N. Henderson, pers. comm.). Although no phylogenetic analysis is currently available that encompasses these three species, they have been placed in series Hexagonae within the section Limniris (Mathew, 1981). A preliminary phylogenetic analysis of *I. fulva, I. hexagona,* and *I. brevicaulis* and a species from a second series *(I. virginica;* Matthew, 1981) suggested a sister-species status for *I. fulva* and *I. hexagona* (Arnold and Hamrick, unpublished data). Furthermore, an analysis of chloroplast DNA RFLP variation present in *I. fulva, I. hexagona,* and *I. brevicaulis* also suggested a closer relation between the first two species.

The first conceptual contribution to studies of natural hybridization, derived from the Louisiana iris species, was their use as a "typical example" of the process of introgressive hybridization (Anderson, 1949). However, prior to Anderson's (1949) seminal work, Viosca (1935), Foster (1937), and Riley (1938) carried out detailed taxonomic, ecological, and cytological investigations. In particular, the data from the morphological analysis of Riley (1938) was used to define introgressive hybridization (Anderson, 1949). Furthermore, the ecological studies of Viosca (1935) represented the first attempt to define the biotic and abiotic factors that might limit the distributions of the Louisiana iris species and their hybrids. The work of each of the early researchers was largely in response to an earlier analysis by Small and Alexander (1931) in which over 80 species of *Iris* were named. Most of these "species" were later demonstrated to be various hybrid segregants and not deserving of specific recognition (Viosca, 1935). However, the findings of Small and Alexander (1931) reflected the extent of natural hybridization between at least three of the *Iris* species: *Iris fulva, I. hexagona,* and *I. brevicaulis.*

Unlike Anderson (1949), Randolph et al. (1967) concluded that natural hybridization between *I. fulva, I. hexagona,* and *I. brevicaulis* had not led to the transfer of genetic material between any of these species, except in areas of sympatry. This conclusion was based on an analysis of morphological characters and chromosome markers in both sympatric and allopatric populations of each of these three species. Thus Randolph et al. (1967) did not detect the presence of morphological characters diagnostic for one species in allopatric populations of alternate species. Furthermore, they did not discover the alternate chromosome markers (Randolph et al., 1961) in any of the allopatric populations. Although these authors failed to find evidence for the process of introgressive hybridization in this species complex, it was concluded that a speciation event had taken place as a result of hybridization between *I. fulva, I. hexagona,* and possibly *I. brevicaulis* (Randolph, 1966). The "new" species, *I. nelsonii,* was defined as having an intermediate morphology, between *I. fulva* and *I. hexagona,* as well as a combination of *I. fulva* and *I. hexagona* marker chromosomes (Randolph, 1966; Randolph et al., 1967). *I. nelsonii* is apparently restricted to extreme southern Louisiana (Randolph, 1966). Significantly, the habitat occupied by *I. nelsonii* was thought to possess characteristics of habitats associated with either *I. fulva* or *I. hexagona* (Randolph, 1966); populations of *I. nelsonii* were found in areas of heavy shade ("*I. fulva* habitat") and fluctuating water levels ("*I. hexagona* habitat") (Randolph, 1966).

NATURAL HYBRIDIZATION IN LOUISIANA IRISES: REPRODUCTIVE ISOLATION, INTROGRESSION, AND HYBRID SPECIATION

Reproductive Isolation. *I. fulva, I. hexagona,* and *I. brevicaulis* differ in chromosome number and rearrangements (*I. fulva, 2n = 42; I. hexagona* and *I. brevicaulis, 2n = 44*) and natural hybrids demonstrate lower pollen fertilities relative to the parental species (Randolph et al., 1967). Data derived from 2 years of experimental hybridization between *I. fulva* and *I. hexagona* also indicate the presence of reproductive isolation and lower fitness associated with interspecific crosses. Thus of 65 total interspecific crosses, only 7 produced seed capsules. This was in contrast with 26 intraspecific crosses of which 10 were successful in producing seed capsules. An average of 27 and 20 seeds per capsule were produced, respectively, for the interspecific and intraspecific crosses. The germination frequencies for the hybrid seeds was 0.135, compared to 0.225 for those from intraspecific crosses. Hence there appears to be selection against hybrid formation prior to seed set and then at the stage of seed germination, which could be due to inhibition of interspecific pollen tube development, hybrid ovule abortion, and/or developmental abnormalities that block embryonic growth and germination. Bennett (1989) and Bennett and Arnold (unpublished data; Fig. 5-1); found that seeds collected from natural hybrid individuals took significantly longer to germinate and had a significantly lower frequency of germination than seeds from either *I. fulva* or *I. hexagona.* These findings and those for the experimental F_1 hybrids are in agreement with the conclusion that hybrid individuals have lower fitness for most

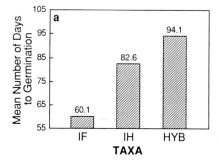 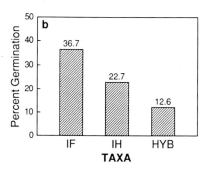

Figure 5-1. Results from the experimental germination of *I. fulva, I. hexagona,* and natural hybrid seeds. (a) Mean number of days to germination. (b) Percent germination. All of the values for both (a) and (b) are significantly different from one another. IH = *I. hexagona:* IF = *I. fulva;* HYB = hybrid seeds.

of the parameters that have been measured (Bennett, 1989; and see below). Despite the selection against interspecific hybrids, largely concurrent flowering times (Fig. 5-2) and common pollinators result in the production of natural hybrids in areas of overlap between *I. fulva* and *I. hexagona* (Riley, 1938; Anderson, 1949; Randolph et al., 1967; Bennett, 1989; Arnold et al., 1990a,b). Furthermore, although there is apparent

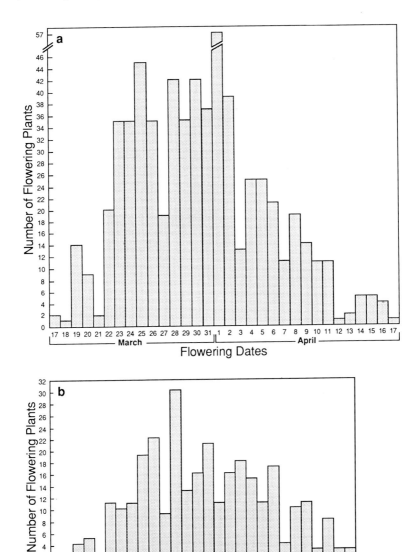

Figure 5-2. Distribution of flowering times for *I. fulva* and *I. hexagona*. (**a**) *I. fulva*. (**b**) *I. hexagona*. Numbers along the x-axis represent the days of 3/17/90 to 4/17/90 *(I. fulva)* and 3/23/90 to 4/19/90 *(I. hexagona)*. Numbers along the y-axis indicate the number of individuals flowering.

asynchrony in the flowering times of *I. brevicaulis* and the other two species, natural hybrid populations between *I. fulva, I. hexagona,* and *I. brevicaulis* have also been reported (Randolph et al., 1967; Arnold et al., 1990b; 1991).

Introgression: Sympatric and Allopatric. Numerous outcomes of the process of hybridization are possible given different levels of reproductive isolation (Mayr, 1963), cohesion (Templeton, 1989), or ecological differentiation between the hybridizing taxa. The various consequences include reinforcement of reproductive barriers through character displacement (Butlin, 1989; see Ch. 3); the creation of an ephemeral, localized hybrid swarm that is maintained in a disturbed habitat (Anderson, 1948); hybrid populations acting as "sinks" for pest organisms and therefore suppressing co-evolutionary responses of pest species to their prey (Whitham, 1989); and the transfer of genetic material from one hybridizing taxon into the other ("introgressive hybridization" or "introgression"; Anderson and Hubricht, 1938).

The process of introgression has also been perceived to have an array of potential evolutionary repercussions (see Ch. 4). Thus the transfer of genetic material between the hybridizing taxa may (1) result in a reduction in reproductive isolation (Grant, 1963) and possibly the fusion of the hybridizing gene pools to form a single, polymorphic taxon; (2) contribute to either an increase (Lewontin and Birch, 1966; Nagle and Mettler, 1969) or decrease (Levin and Bulinska-Radomska, 1988) in relative fitness of the introgressed taxon; (3) facilitate the expansion of the introgressed form into extreme environments (Lewontin and Birch, 1966); (4) offset the effective loss of genetic variation as a consequence of genetic drift (Grant and Grant, 1989); (5) act as a pollen-mediated dispersal mechanism for plant species with limited potential for seed dispersal (Potts and Reid, 1988); or (6) foster speciation (Mecham, 1960; Levin, 1969). Indeed, Stebbins and Daly (1961) have stated that "hybridization between populations having markedly different adaptive properties, particularly if it occurs in a disturbed, changing environment where new ecological niches are likely to be available, is one of the strongest factors bringing about evolutionary change." However, the fertility and viability of the recombinant individuals and the pattern and degree of genetic interaction between these individuals and other hybrid and parental organisms ultimately decide the outcome of hybridization and introgression.

A number of the hypothesized effects of introgression depend on the transfer of genetic material or the spread of the introgressants outside of areas of sympatry/parapatry (Anderson, 1949; Lewontin and Birch, 1966; Potts and Reid, 1988). A measurement of the pattern of variation in areas of sympatry/parapatry and allopatry allows an assessment of the severity of selection against the introduction of genetic material from one form into another (Haldane, 1948). For example, Shaw et al. (1985) and Marchant et al. (1988) found that chromosome rearrangements and highly repeated DNA families (Arnold and Shaw, 1985; Arnold et al., 1986) had not introgressed across a zone of parapatry between two grasshopper subspecies, even though allozyme, ribosomal DNA (rDNA) and mitochondrial DNA (mtDNA) markers were present in allopatric populations up to 450 km from the present-day hybrid zone. Marchant et al. (1988) concluded that this broad pattern of clinal variation for the nuclear and cytoplasmic gene markers implied strong selection against the introgression of the chromosome markers.

The first step in determining the role that introgressive hybridization may have played in the evolution of species groups is the unambiguous detection of gene transfer between hybridizing taxa. Anderson and Hubricht (1938) and Anderson (1949) argued that the most common outcome of natural hybridization is the transfer of genes from one of the hybridizing types to the other. In some cases there appears to be an unequal degree of penetration of different genetic markers between hybridizing taxa (e.g., see Harrison, 1986; Arnold et al., 1987; Marchant et al., 1988). The "semipermeable" nature of the boundary between hybridizing taxa (Key, 1981; Harrison, 1986) is best explained by an interaction of both stochastic and deterministic processes (Bloom, 1976; Shaw et al., 1980; Ferris et al., 1983; Powell, 1983; Arnold et al., 1988, 1990a,b; Marchant et al., 1988; Baker et al., 1989; Rand and Harrison, 1989). In this regard, an analysis of the distribution of genetic markers in sympatric/parapatric and allopatric populations of hybridizing taxa allows inferences concerning historical, ecological, and genetic components that have influenced the present-day distribution of introgressed genes. For example, Bert and Harrison (1988) concluded that (1) time of contact, (2) presence or absence of an ecotone in the region of contact, or (3) a quantitative excess of one species compared to the other were responsible for producing the different structures present in two zones of hybridization between the stone crab species *Menippe mercenaria* and *M. adina*. Although introgressive hybridization may indeed be common, distinguishing introgression simply from observation of pattern is not always easy; and a number of cases of hypothesized introgressive hybridization have been called into question following additional analysis (Randolph et al., 1967; Gibbs, 1968; Birch and Vogt, 1970; Heiser, 1973). In addition, there have been relatively few analyses of natural hybridization involving plant species that have included genetic analyses that would allow a sensitive test for introgressive hybridization (Doyle and Doyle, 1988; Rieseberg et al., 1988, 1990a; dePamphilis and Wyatt, 1989, 1990; Doebly, 1989; Keim et al., 1989; Warwick et al., 1989; Arnold et al., 1990a,b).

In an attempt to test the alternative conclusions of Anderson (1949) and Randolph et al. (1967), with regard to the occurrence of introgression in Louisiana irises, we have identified diagnostic genetic markers for *I. fulva, I. hexagona,* and *I. brevicaulis* (Arnold et al., 1990a,b). Figure 5-3 illustrates the rDNA repeat length differences and the inheritance of the *I. fulva* and *I. hexagona* markers in natural hybrid individuals. Likewise, an analysis of random nuclear DNA sequences has identified a number of diagnostic markers for *I. fulva* and *I. hexagona* that are useful in testing for interspecific gene exchange (Fig. 5-4; Arnold et al., 1991). Finally, in a survey of 50

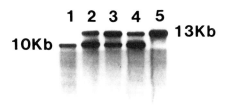

Figure 5-3. Ribosomal repeat units present in *I. fulva* (lane 1), *I. hexagona* (lane 5), and three natural hybrid individuals (lanes 2–4).

a b c d e f g h i j k l m n o p q r

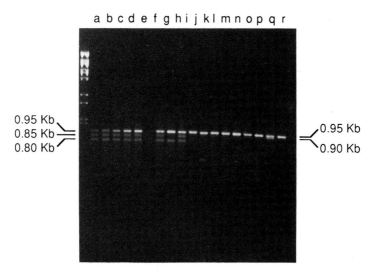

Figure 5-4. RAPD markers for individuals from an *I. fulva* (lanes a–i) and an *I. hexagona* (lanes j–r) population. These markers were generated using a PCR-based method for amplifying random stretches of genomic DNA (Williams et al., 1990).

presumptive isozyme loci, Arnold et al. (1990b) detected several alternate, fixed or nearly fixed loci and numerous unique allozymes for *I. fulva, I. hexagona,* and *I. brevicaulis.* There are three distinct but interrelated phenomena that can be tested by an examination of genetic variation in populations of the various *Iris* species and in putative hybrid populations; they include sympatric introgression (introgression in regions of spatial overlap between the hybridizing taxa), allopatric introgression (introgression of genes of one taxon into an allopatric population of a second taxon), and hybrid speciation (Grant, 1963, 1981; Heiser, 1973; Rieseberg and Brunsfeld, 1991).

We tested for the occurrence of sympatric introgression using the distribution of species-specific genetic markers in an area of overlap between *I. fulva* and *I. hexagona* and in a region of overlap between *I. fulva, I. hexagona,* and *I. brevicaulis.* Figure 5-5 presents the genetic variation found in the Bayou L'ourse hybrid population. This population has been defined as an area of overlap and hybridization between *I. fulva* and *I. hexagona.* The relative proportions of *I. fulva* and *I. hexagona* rDNA, allozyme, and random amplified polymorphic DNA (RAPD; Williams et al., 1990) markers present in 42 plants sampled from this population are illustrated in Figure 5-5 (Arnold et al., 1990a,b, 1991). The proportion of rDNA originating from each species in each of the individuals was determined by densitometric measurements (Arnold et al., 1990a). The relative proportions of the species-specific allozyme and RAPD markers were determined by dividing the number of *I. fulva* allozymes or RAPD bands by the total number of allozymes or bands at the diagnostic loci. The first observation that can be made from these data is that there are multiple genotypes in this population. Because *I. fulva* and *I. hexagona* are highly clonal (Bennett, 1989; and see following discussion), the genetic data indicate the presence of numerous, genetically distinct individuals rather than a limited number of clonal lineages. This finding is in accord with the hypothesis that there is ongoing hybridization and introgression in this pop-

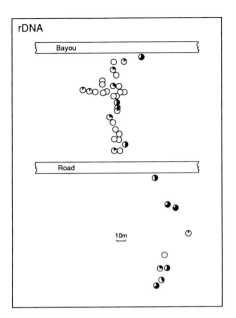

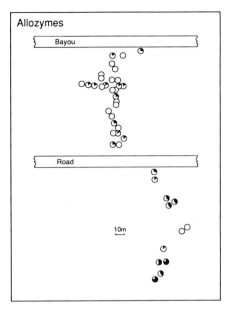

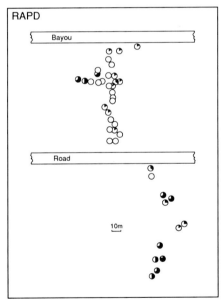

Figure 5-5. Genetic variation in a sample of 42 individuals from the Bayou L'ourse population. Each circle represents an individual plant. The filled-in and open portions of each circle represent the proportion of *I. fulva* and *I. hexagona* markers, respectively. The top-left panel, top-right panel, and bottom panel illustrate the rDNA, allozyme, and RAPD variations, respectively. There are three missing data points (plants) for the rDNA analysis.

ulation (Arnold et al., 1990a,b, 1991). Indeed an analysis of the multilocus genotypes of the 42 individuals, in relation to the expected frequencies of six genealogical classes (*I. fulva, I. hexagona,* F_1, F_2, $B_{1,I,f}$, and $B_{1,I,h}$), resulted in the conclusion that this sample contained a preponderance of advanced generation *"I. hexagona"* backcross individuals (Nason et al., 1992). A comparison of the individual marker systems (rDNA, isozyme, random DNA markers) also demonstrates large differences in the frequencies of the markers in some of the individuals. A pattern of nonconcordance between various genetic and morphological markers in hybrid individuals has been noted in other studies (e.g., Harrison, 1986; Dowling, et al., 1989). Factors that may be important in producing this effect include the loss or retention of certain haplotypes or genotypes due to stochastic processes (Ball et al., 1990), selection against the incorporation of certain genetic components (Shaw et al., 1990), habitat selection (Hamrick and Holden, 1979), and, particularly in the case of rDNA, molecular mechanisms that effect the amplification or loss of specific variants (Dover, 1982; Arnold et al., 1988; Hillis et al., 1991). With regard to the last phenomenon, Arnold et al. (1988) and Marchant et al. (1988) have suggested that the process of biased gene conversion may facilitate the introgression and subsequent amplification of "foreign" rDNA variants in allopatric populations of hybridizing taxa.

A separate analysis, involving a population presumed to have originated from hybridization between *I. fulva, I. hexagona,* and *I. brevicaulis,* involved the examination of isozyme and rDNA variation (Arnold et al., 1990b). Figure 5-6 represents the distribution of allozymes that are diagnostic for either *I. fulva, I. hexagona,* or *I. brevicaulis* in the 48 individuals sampled from this population. It is apparent that there are numerous combinations of the markers, indicative of numerous genotypes. It is also apparent that the three species have not contributed equally to the genetic structure of this hybrid assemblage. In fact there are only five individuals that possess marker allozymes indicative of the genetic contribution of *I. hexagona.* In contrast, there are 35 and 40 individuals, respectively, that possess *I. fulva* or *I. brevicaulis* markers, or both. Preliminary data from an analysis of rDNA repeat length variants also support the predominance of *I. fulva* and *I. brevicaulis* in the genetic makeup of this hybrid population; Arnold et al. (1990b) reported the presence of length variants characteristic of these two species. The distribution and pattern of genetic variation in this population and in the area of overlap between *I. fulva* and *I. hexagona* illustrates the consequences of sympatric introgression. In the *I. fulva* × *I. hexagona* hybrid population, the introgression is biased in that most of the hybrid individuals have apparently resulted from backcrossing into *I. hexagona.* In contrast, individuals from the second area of overlap are characterized by having mainly *I. fulva* or *I. brevicaulis* genetic markers. The stochastic and deterministic components that may have contributed to the genetic structure of the first of these populations are detailed below. For the present discussion it is sufficient to note that there is the opportunity for reticulate evolution resulting from sympatric introgression (Mecham, 1960; Heiser, 1973; Rieseberg and Brunsfeld, 1991); and possibly of even greater importance, these populations represent a potential genetic corridor for the process of allopatric introgression.

Prior to the use of discrete genetic markers, the documentation of allopatric introgression was problematic. Thus Anderson and Hubricht (1938) in their description of the process of introgressive hybridization concluded that the utility of morphological characters for the detection of multiple backcross progeny, and therefore

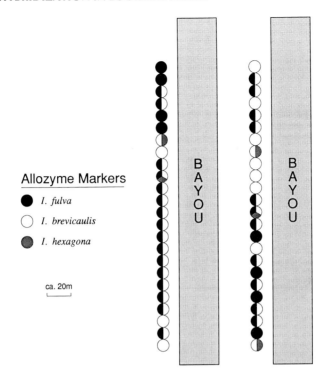

Figure 5-6. Distribution of *I. fulva, I. hexagona,* and *I. brevicaulis* species-specific allozyme markers in 48 *Iris* individuals (Arnold et al., 1990b). Each circle represents an individual plant. The proportion of the circle indicated to be *I. fulva, I. hexagona,* or *I. brevicaulis* is *not* indicative of the proportion of the markers, only their presence or absence. The sampling was carried out along a linear transect, and the individual pictured at the top right-hand corner of this figure is located next to the individual at the lower left-hand corner.

introgressive hybridization, would be limited. In fact, the alternate conclusions of Anderson (1949) and Randolph et al. (1967) concerning introgression between the Louisiana *Iris* species were derived largely from morphological data. It appears therefore that an analysis of multiple, discrete genetic markers represents the most sensitive assay for allopatric introgression (Marchant et al., 1988; Rieseberg and Brunsfeld, 1991; see Ch. 4). In the present instance, data from rDNA, isozyme, and random nuclear DNA studies are available for both sympatric and allopatric populations of *I. fulva, I. hexagona,* and *I. brevicaulis* (Arnold et al., 1990a,b, 1991). As discussed above, these data sets are consistent with a hypothesis of sympatric introgression in the contemporary areas of overlap. Otherwise, the distribution of the alternate markers is restricted to the species for which they are diagnostic, with two notable exceptions. Arnold et al. (1990a) detected the presence of the alternate rDNA repeat length variants in one allopatric population each of *I. fulva* and *I. hexagona.* These populations are characteristic of their respective species with regard to morphological comparisons or habitat observations (Bennett, 1989), although the *I. fulva* population is in an area of apparent ecological succession (Bennett and Arnold, pers. obs.). The closest known populations of the alternate species are located about 10 km and 25 km from the *I.*

fulva and *I. hexagona* populations, respectively. Arnold et al. (1990a) argued that the genetic findings were in accord with the conclusion of Anderson (1949) that these species had been genetically affected by the process of natural hybridization and introgression. Further examinations of the same populations using isozyme markers (Arnold et al., 1990b) and RAPD markers (Arnold et al., 1991) detected the same pattern of genetic variation; alternate markers were present in the same two allopatric populations. In summary, the genetic data support a hypothesis of sympatric introgression between *I. fulva* and *I. hexagona* in one area and between *I. fulva, I. brevicaulis,* and *I. hexagona* in another. Furthermore, these data also indicate that there has been bidirectional, allopatric introgression between *I. fulva* and *I. hexagona.*

Hybrid Speciation. Grant (1981) has suggested that the crucial factor in hybrid speciation is the stabilization of the "breeding behavior" of hybrid individuals. In this regard, Grant (1981) enumerated several mechanisms that might reduce the likelihood of the disruption of hybrid gene complexes by recombination and segregation. These mechanisms included (1) asexual reproduction (apomixis); (2) permanent translocation heterozygosity; (3) permanent odd polyploidy; (4) allopolyploidy; (5) the production of a diploid, hybrid derivative that is reproductively isolated from the parental species by chromosomal barriers (recombinational speciation); and (6) the production of a hybrid type that is isolated by external barriers such as the invasion of a unique habitat. The occurrence of hybrid species has been hypothesized for numerous plant species complexes (Grant, 1963; 1981; Rieseberg et al., 1990a). Analyses of some of these groups have resulted in rejection of the hypothesis of hybridization-mediated speciation, subspeciation, or raciation in some instances (Rieseberg et al., 1988; 1990b; Rieseberg and Brunsfeld, 1991; see Ch. 4) and support for this hypothesis in others (Rieseberg et al., 1990a,b; Rieseberg and Brunsfeld, 1991).

Randolph (1966) named a new species *(I. nelsonii)* in the Louisiana iris complex and argued that this taxon represented a stabilized hybrid derivative of *I. fulva, I. hexagona,* and possibly *I. brevicaulis.* The specific status of *I. nelsonii* was justified on the basis of unique chromosome markers, divergent morphological traits, and its occupation of a habitat that was distinct from any of the putative progenitors (Randolph, 1966). The stability of this taxon was inferred from a high level of pollen viability and levels of morphological variation that did not exceed those present in *I. fulva, I. hexagona,* or *I. brevicaulis* (Randolph, 1966; Randolph et al., 1967). The inference of a hybrid origin for *I. nelsonii* derived from morphological and chromosomal characters that appeared to derive from more than one of the parental species (Randolph, 1966; Randolph et al., 1967). Levels of reproductive isolation between *I. nelsonii* and the other three species and the genetic stability of *I. nelsonii* have not been documented experimentally. However, populations of *I. nelsonii* are characterized by a unique chromosome complement and are highly fertile, as measured by pollen viability. This genetic stability is present even though *I. nelsonii* is endemic to a region of southern Louisiana in which *I. brevicaulis, I. fulva,* and *I. hexagona* occur and hybridize (Randolph et al., 1961; Randolph et al., 1967). The chromosomal markers identified for *I. nelsonii* were compared to the unique marker chromosomes of *I. fulva, I. hexagona,* and *I. brevicaulis* (Randolph et al., 1961). This analysis indicated that *I. nelsonii* possessed at least one of the marker chromosomes characteristic of *I. fulva* and a marker chromosome similar to one found in *I. hexagona* but with a different centromere position (Randolph et al., 1961). None of the marker chromosomes unique to *I. brevicaulis* was apparent in the *I. nelsonii* complement. *I. nelsonii* has also been described as occupying a unique habitat relative to the other three species, being found in areas of low light intensity and deep water (Randolph et al., 1967). Furthermore, it was concluded that *I. nelsonii* possessed a combination of morphological characteristics that were found in *I. fulva* and *I. hexagona.* In addition, the shape of the seed capsule and later flowering time of *I. nelsonii* led to the suggestion that *I. brevicaulis* had also been involved in the formation of this hybrid derivative (Randolph, 1966).

To test the hypothesis that *I. nelsonii* represents a derivative from natural hybridization between *I. fulva, I. hexagona,* and *I. brevicaulis,* Arnold et al. (1990b, 1991) have examined the same markers that were used in the analysis of introgressive hybridization. If *I. nelsonii* is a hybrid derivative, it should possess a combination of the genetic markers that characterize the putative parental taxa (*I. fulva, I. hexagona,* or *I. brevicaulis*). Furthermore, the hypothesis of hybrid speciation would be strengthened if these combinations of genetic markers were also present in contemporary hybrid populations. In this regard, Arnold et al. (1990b) found that the allozyme frequencies present in *I. nelsonii* largely reflected those of *I. fulva,* which is apparent from the relatively high genetic identity between these two species ($I = 0.95$). However, as predicted by the hybrid derivative hypothesis, *I. nelsonii* also shared genetic markers (allozyme or RAPD markers, or both) characteristic for *I. hexagona* and *I. brevicaulis.* Furthermore, similar combinations of the genetic markers found in *I. nelsonii* were also found in a contemporary area of overlap between *I. fulva, I. hexagona,* and *I. brevicaulis* (Fig. 5-6). This observation also supports the contention that *I. nelsonii* is a derivative of introgressive hybridization between these three species.

As discussed above, there are a number of mechanisms that may lead to the stabilization of hybrid species (Grant, 1981). With regard to *I. nelsonii,* a combination of the mechanisms may have played a role in stabilization. First, clonal reproduction by means of a rhizomatous habit is an important aspect of the life history of all of the *Iris* species examined in this study (Bennett, 1989; and see following discussion) and therefore may have facilitated the establishment of *I. nelsonii* subsequent to hybridization. Second, because this species apparently occupies a habitat that is unique in comparison to *I. fulva, I. hexagona,* and *I. brevicaulis* (heavily shaded areas with deep water), an ecological separation between *I. nelsonii* and the parental species may also exist. It is interesting to note that this hybrid derivative was most likely derived through repeated backcrossing into *I. fulva,* as reflected by the high genetic similarity between *I. nelsonii* and *I. fulva* (Arnold et al., 1990b). Lewontin and Birch (1966) have suggested that the transfer of genetic material between hybridizing species may result in an introgressed form that is able to invade an extreme environment. In this regard, the comparison of the genetic structure of an area of overlap between *I. fulva* and *I. hexagona* with environmental parameters indicates that hybrid genotypes are occupying unique habitats with respect to the parental species (see following discussion). These findings support the hypothesis that hybridization between *I. fulva, I. hexagona,* and *I. brevicaulis* has resulted in the production of an introgressed genotype *(I. nelsonii)* that allowed the invasion of an extreme environment. Finally, *I. hexagona, I. fulva, I. brevicaulis,* and *I. nelsonii* are each characterized by unique marker chromosomes

(Randolph et al., 1961) indicating that *I. nelsonii* may be a hybrid derivative that is stable with regard to a "hybrid" chromosomal complement. Arnold et al. (1990b) have argued that the presence of multiple processes for the stabilization of hybrid derivatives should increase the opportunity for hybrid speciation. Such multiple avenues would be available whenever there is any combination of asexual and sexual reproduction, chromosomal differentiation, and open ecological niches associated with hybridizing taxa.

NATURAL HYBRIDIZATION IN LOUISIANA IRISES: ECOLOGICAL DETERMINANTS, SPECIES DISTRIBUTIONS AND THE GENETIC STRUCTURE OF HYBRID POPULATIONS

Several evolutionary models have been formulated that incorporate components thought to be important in determining the genetic structure and outcome of hybrid zones between species (Endler, 1977; Moore, 1977; Barton, 1979; Barton and Hewitt, 1981). Using the terminology of Moore and Buchanan (1985), three of the models may be classified as the "adaptive speciation" hypothesis, the "bounded hybrid superiority" hypothesis, and the "dynamic-equilibrium" hypothesis. The adaptive speciation model predicts that hybridizing taxa will either merge owing to hybridization or become completely reproductively isolated; the outcome would depend on the effectiveness of natural selection against hybrid types. In contrast, the bounded hybrid superiority hypothesis suggests that environmental (ecological) determinants result in a habitat in which the hybrids are superior to the parental types. Within this region the hybrids will outcompete the parental types and the hybrid populations will be maintained. Finally, the dynamic equilibrium model emphasizes the contribution of hybrid inferiority to the genetic structuring of hybrid zones (Barton, 1979). In addition to these models are those that depend on a gradient of selection that maintains clinal variation across the hybrid zone (Endler, 1977). The common denominator of each of these models is the action of natural selection. However, the form of natural selection present in the hybrid populations depends on whether there are interactions between environmental and genetic components, and whether this interaction favors or selects against the hybrid genotype(s).

Ecological Determinants and Species Distributions. A number of biotic and abiotic factors have been hypothesized as determinants of the distribution of the various Louisiana iris species and their hybrids. Most of these putative determinants relate to the aquatic or semiaquatic habitat occupied by the various species. Thus Viosca (1935), Riley (1938), Randolph (1966), and Randolph et al. (1967) argued that the list of major determinants included degree of shading, water content of the soil and the correlated substrate consistency, salinity, pH of water, fluctuation of water level, and degree of interspecific competition. This list was based on field observation and not demographic studies or experimental ecological analyses.

   In a series of experiments designed to test the effects of three of the above factors (shade tolerance, salinity tolerance, and interspecific competitive ability) on the relative fitness of *I. fulva, I. hexagona,* and classes of natural hybrids, Bennett (1989) and Bennett and Grace (1990) discovered significant differences in the response of the two species and their hybrids to the different experimental regimens. In the experiments concerned with the effect of shade tolerance, a hierarchical fitness response (as measured in terms of biomass production) was apparent among the various classes (Bennett and Grace, 1990). Table 5-1 illustrates the effect on overall biomass produced by *I. fulva, I. hexagona,* and two hybrid classes under the medium and high shading regimens. These results indicate that under high shading the fitness of the classes follow the pattern of *I. fulva* > hybrid red > hybrid purple > *I. hexagona;* this degree of shading is found in contemporary hybrid associations (see below). Therefore *I. fulva* shows the least depression in biomass with increased shading, the hybrid classes are intermediate, and *I. hexagona* demonstrates the greatest response to increased shading. It is significant to note that "hybrid red" plants contained, on average, more of the *I. fulva* diagnostic markers. In contrast, the "hybrid purple" class is characterized by individuals that are more similar to *I. hexagona.* Therefore there is a correlation between the proportion of *I. fulva* versus *I. hexagona* genetic material present in a population of individuals and their relative fitness with respect to different levels of shading.

A second parameter tested as a possible abiotic determinant of species distribution was salinity. Using the same classes of individuals (*I. fulva, I. hexagona,* "hybrid red," and "hybrid purple") that were used in the shade experiments, Bennett (1989) showed that there were significant differences between these classes in regard to their responses to the various treatments. In contrast to the response to shade tolerance, however, the salinity data indicate that the relative fitnesses of the various classes are *I. hexagona* > "hybrid red"/"hybrid purple" > *I. fulva.* Therefore although all of the classes were significantly affected by the two highest salt concentrations, *I. hexagona* was the least affected, *I. fulva* was the most affected, and the two hybrid classes were intermediate in their responses (Bennett, 1989).

Finally, an experimental analysis was carried out to measure the relative competitive abilities of *I. fulva, I. hexagona,* and the "hybrid purple" class (Bennett, 1989). The experimental design included growing two plants (representing two of the three classes) together in each container. The biomass obtained in such experiments was then compared to the biomass produced by plants grown in monoculture (control). In addition, Bennett (1989) tested for substrate effects by growing each of the classes in soil collected from *I. hexagona, I. fulva,* and "hybrid purple" populations. Results

Table 5-1. Comparative Shade Tolerance of *I. fulva, I. hexagona,* and Two Hybrid Classes Measured by Biomass Production

Comparison	50% Light Reduction	80% Light Reduction
IF/IH	IF > IH	IF > IH
IF/HP	IF > HP	IF > HP
IF/HR	IF > HR	IF > HR
IH/HP	IH < HP	IH < HP
IH/HR	IH < HR	IH < HR
HP/HR	HP > HR	HP < HR

Source: Data from Bennett and Grace (1990).
IF = *I. fulva;* IH = *I. hexagona;* HR = hybrid red; HP = hybrid purple.
The greater than (>) and less than (<) signs indicate a greater or lesser relative shade tolerance, respectively.

Table 5-2. Relative Interspecific Competitive Performance of *I. fulva, I. hexagona,* and a Hybrid Class

Mixture	Competitive Performance
IH/IF	IH > IF
IH/HP	IH = HP
HP/IF	HP > IF

The results are indicated for mixtures of the taxa, taken two at a time; the results are relative to the results from control experiments where individuals from the same taxon were grown in competition (Bennett, 1989). The greater than (>), less than (<), and equal (=) signs indicate relatively greater, lesser, or equivalent competitive ability, respectively. IF = *I. fulva;* IH = *I. hexagona;* HP = hybrid purple.

from the competition experiment (Table 5-2) suggested that *I. hexagona* and the "hybrid purple" class were equivalent in terms of their competitive ability, and that both of these classes were superior to *I. fulva* in terms of their competitive ability. Once again it is of interest that the relative fitness of the hybrid class is like that of the genotypically most similar parental form (Arnold et al. 1990a,b, 1991). In contrast to the measurable effect on relative fitness caused by competitive interactions, there was no apparent difference that was attributable to the different soil types used in this experiment (Bennett, 1989).

The preceding experiments clearly suggest differences in fitness for *I. fulva, I. hexagona,* and various hybrid classes with regard to each of the three variables tested (shade, salinity, and interspecific competition). Furthermore, as summarized above, numerous authors have reported apparent habitat differences between the Louisiana *Iris* taxa. As an initial test of relative fitness and niche differentiation, Bennett (1989) carried out a 3-year demographic study of *I. fulva, I. hexagona,* and their natural hybrids. This analysis included a single population of *I. fulva,* two populations of *I. hexagona,* and two hybrid populations. Because the Louisiana *Iris* species are long-lived and possess both asexual (rhizomatous) and sexual forms of reproduction, the data derived from this analysis included parameters of survivorship, clonal growth (increase in rhizome length, production of new ramets, number of individuals that produced new ramets) and sexual reproduction (number of flowers produced per flowering ramet, percentage of clonal fragments flowering, number of seed capsules produced, number of seeds per capsule, percentage of seed germination for each population, and seedling establishment; Bennett, 1989). Table 5-3 summarizes the results of this study for each of the parental taxa and the hybrids.

The demographic data indicate a number of trends with respect to the fitness of *I. fulva, I. hexagona,* and the hybrid individuals in their respective habitats. Thus the hybrid individuals that were examined during this study, demonstrated the greatest overall degree of survivorship (Table 5-3). However, for all other measurements the hybrid populations were the same as the least productive parental taxon or were themselves the least productive populations. This is particularly telling in the estimate of reproductive fitness (ERF; Bennett, 1989) calculation that clearly reveals the low level of sexual reproductive capacity of the hybrids, relative to either *I. fulva* or *I. hexagona.* For 6 of the 11 measurements, *I. hexagona* appeared to outperform *I. fulva.* For three

Table 5-3. Fitness Estimates for *I. fulva, I. hexagona,* and Hybrid Populations for Components of Survivorship, Asexual Reproduction, and Sexual Reproduction

Measurement	I. fulva	I. hexagona	Hybrids
Survivorship	+	+ +	+ + +
Rhizome length (increase)	+	+ + +	+
New ramets (no.)	+ +	+ + +	+
Individuals producing new ramets (%)	+ +	+ + +	+
Flowers/flowering ramet (no.)	+ + +	+	+
Percent of flowering individuals	+	+	+
Seed capsules (no.)	+ + +	+ + +	+
Seeds/capsule (no.)	+ + +	+, + + +	+
Seed germination (%)	???	???	???
Seedling establishment (no.)	???	???	???
ERF[a]	43.3	43.2	17.7

The designations are as follows: +, + +, and + + + represent a grade from smallest to greatest production in whatever the trait being measured. If two or more of the taxa have the same symbol (e.g., "Increase in Rhizome Length," + = *I. fulva* and hybrid classes), they are not significantly different in their values for the particular trait.

[a]ERF (estimate of reproductive fitness, Bennett, 1989) = (capsules/inflorescence) × (seeds/capsule) × (% germination).

of the components the two taxa were not significantly different, and of the remaining two *I. fulva* showed the highest productivity and thus the highest inferred fitness of any of the three classes (Table 5-3). It must be remembered that these data indicate performance of *I. fulva, I. hexagona,* and the hybrid individuals in their respective habitats; these results tell us nothing about the relative fitness of the two parental taxa and the hybrids in the alternate habitats. However, given this limitation, it is apparent that the hybrid population of individuals demonstrates a much reduced level of growth and reproductive success in its own habitat compared with that found for the parental taxa in their habitats. The demographic study and the experimental analyses also suggest that *I. fulva,* due to its greater shade tolerance and lower competitive ability, clonal growth, and survivorship, relative to *I. hexagona* and certain hybrid classes, would be excluded from the open marsh habitat and therefore would be restricted to the understory of the bayou margins.

Ecological Determinants and the Genetic Structure of an* I. fulva × I. hexagona *Hybrid Population. Based on the demographic and experimental studies, Bennett (1989) and Bennett and Grace (1990) have argued that the hybrid populations are located in an ecotone between the typical understory habitat of *I. fulva* (margins of bayou systems) and the open marsh habitat characteristic of *I. hexagona* (Fig. 5-7). In this regard, it has already been demonstrated that the components of shading, salinity, and interspecific competition have differential effects on the overall performance of the parental taxa and their hybrids. In addition, another component that has not been examined experimentally, but that may be an important determinant of the distribution of *Iris* taxa, is water depth (Viosca, 1935). If the habitat occupied by the hybrid individuals is an ecotone, and there is selective pressure associated with the changeover

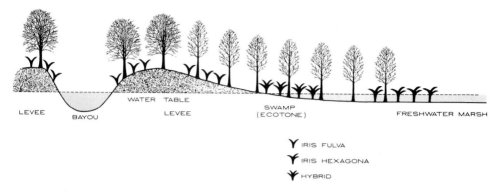

Figure 5-7. Habitat associations for *I. fulva, I. hexagona,* and natural hybrids. This illustration advances the hypothesis that natural hybrids between *I. fulva* and *I. hexagona* occupy an ecotone between the parental habitats. (Adapted from Viosca, 1935.)

between the *"I. fulva"* and the *"I. hexagona"* habitats, there should be demonstrable clinal variation for species-specific genetic markers (Endler, 1977; Heywood, 1986). Arnold et al. (1990a,b, 1991) have examined diagnostic molecular and biochemical markers for 42 individuals collected from a contemporary *I. fulva* × *I. hexagona* hybrid population. We have already discussed the individual genetic marker systems with regard to a test for sympatric introgression. However, for the present discussion, Figure 5-8 represents the multilocus genotypes for each of the 42 individuals; the genotypes are based on the allozyme, ribosomal DNA (rDNA), and random nuclear DNA data (Arnold et al., 1990a,b, 1991). These findings are informative for testing whether the resulting hybrid zone between *I. fulva* and *I. hexagona* fits a clinal (Endler, 1977) or mosaic (Rand and Harrison, 1989) model.

First, it may be asked whether this hybrid population is situated in an area characterized by ecological discontinuity (an ecotone). In this regard, it is important to note that the two regions separated by the roadway are distinguished by two different habitats, and Bennett (1989) considered these areas to be two separate populations. The region between the road and the levee and bayou is an area of cypress swamp characterized by a 33.2 % (± 0.02 %) reduction in light intensity and an average water depth of 27.45 ± 1.17 cm (Bennett and Arnold, unpublished data). However, even this region is not uniform with respect to either shading or water depth owing to the presence of channels and levees associated with the main bayou and light gaps due to fallen cypress trees. Therefore there are parts that are "ecotonal," having a combination of two of the environmental parameters that are associated with either *I. fulva* or *I. hexagona* populations. In contrast, there are regions within this area that approximate an *I. hexagona* habitat type (high light intensity and fluctuating water depth) and other areas that possess an *I. fulva*-like habitat (reduced light intensity and water depth). The latter habitat is particularly apparent near the levee adjacent to the bayou (Fig. 5-8). With respect to the genetic structuring of the cypress swamp region, there are (1) multiple genotypes, rather than a single or a few clonal lineages, and (2) large fluctuations in the proportion of the *I. fulva* and *I. hexagona* markers between individuals (Fig. 5-8; Arnold et al., 1990a,b; and previous discussion). Of the 30 individuals sampled from this subpopulation, seven were found to have an *I. hexagona* multilocus genotype.

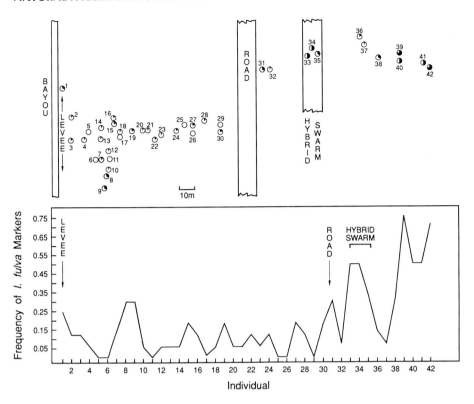

Figure 5-8. Proportions of *I. fulva* and *I. hexagona* genetic markers present in 42 individuals sampled from an area of sympatry between *I. fulva* and *I. hexagona.* The pie diagrams indicate both the proportion of the genetic markers (*I. fulva* = shaded portion, *I. hexagona* = unshaded portion of each circle) and the spatial distribution for each of the individuals. The graph indicates the actual proportion of the markers for each individual but not the spatial distribution.

The remainder of the individuals had various combinations of the marker alleles indicative of advanced hybrid generations (Nason et al., 1992). It would be predicted that *I. fulva*-like individuals would have a higher fitness in more shaded habitats relative to those plants with an *I. hexagona*-like genotype. Furthermore, the competition data (Bennett, 1989) suggest that *I. fulva* would be competitively excluded by *I. hexagona* from areas of elevated light intensities. It is important to note that the fitness component that appears to be most important to the maintenance of the hybrid population is survivorship (Bennett, 1989; and above discussion). Therefore it should be informative to compare the survivorship across years and the multilocus genotypes with the environmental variables associated with individual plants.

The remainder of the sample, located across the roadway, is in an area of man-made disturbance. This region is much drier, with little if any above ground water during the year and with full sunlight except in areas near three *Quercus virginiana* individuals. The genetic data for individuals from this portion of the sample indicate a number of similar trends relative to the subpopulation sampled from the cypress swamp habitat (Fig. 5-8). First, there are no individuals that possess an *I. fulva* multilocus genotype. Second, there are a number of genotypes indicative of multiple genetic lineages. Third, there are widely varying proportions of *I. fulva* and *I. hexagona* genetic markers in the individuals surveyed. However, in contrast to the cypress swamp subpopulation, there are no *I. hexagona* individuals as defined by the genotypic data, and there is an overall increase in the proportion of *I. fulva* genetic markers (Fig. 5-8). Similar to findings from studies of hybrid zones in other organisms (Harrison, 1986; Arnold et al., 1987; Marchant et al., 1988), the pattern of clinal change in the various marker systems (allozymes, rDNA, and RAPD markers) is not identical. Furthermore, there are large differences in the multilocus genotypes between individuals within each of the subpopulations (Fig. 5-5). However, the mean proportion of *I. fulva* markers per individual in the cypress swamp subpopulation and the subpopulation in the disturbed habitat is 0.10 and 0.40, respectively. Arnold et al. (1990a,b, 1991) have argued that the variation in the frequencies of the *I. fulva* and *I. hexagona* diagnostic markers in this sample do not indicate the presence of only a single region of clinal variation. This point was argued because of the large frequency shifts within each of the two subpopulations. Furthermore, these authors suggested that more detailed genetic analysis within this zone might document the presence of multiple clines that reflect the historical contact points between the two species or selection. For example, individual 1, located on the edge of the levee associated with the bayou channel, has a wine-colored flower. Because experimental F₁ hybrids also have wine-colored flowers (Arnold, unpublished data) this finding seems to indicate the presence of a larger proportion of *I. fulva* genomic elements in individual 1 than in the "hybrid purple" phenotype found in the remainder of the cypress swamp (Arnold et al., 1990a,b, 1991). Furthermore, there are additional plants (not sampled in the genetic studies) along this levee that have similar floral coloration to individual 1. There is a decrease in the frequency of *I. fulva* genetic markers between individual 1 and individuals 2–6 (Fig. 5-8). The habitat type occupied by individual 1 has been hypothesized to be a corridor by which *I. fulva* comes into contact with *I. hexagona* (Viosca, 1935). It is possible that the *I. fulva*-like hybrid plants in this area reflect a historical point of contact between these two species, as well as selection for an *"I. fulva"* genotype (Bennett, 1989; and above discussion). The fact that no *I. fulva* individuals, as measured by the multilocus genotypes, have been found along this levee, suggests that this species is being displaced by hybridization with the more numerous *I. hexagona* individuals.

The cline between the cypress swamp habitat and the disturbed habitat across the roadway is explicable on the basis of stochastic or deterministic processes. First, the presence of higher frequencies of the *I. fulva* genetic markers and a wider range of floral coloration in the region of disturbed habitat is suggestive of a historical contact point between this species and *I. hexagona*. The fact that this region is one of extensive man-made disturbance and a much reduced *Iris* population relative to the cypress swamp region (Bennett, 1989; Bennett and Arnold, pers. obs.), suggests that this area is marginal *Iris* habitat. The presence of the wide array of floral phenotypes in this region illustrates the concept of a "hybridized habitat" (Anderson, 1948), resulting from man-made disturbance. In this case the cline might be more indicative of the historical contact of these two species and the relictual nature of the *Iris* individuals in a greatly modified habitat, rather than the influence of selection. The lack of *I. fulva* genotypes among the individuals sampled in this portion of the population (Fig. 5-8) and the lower proportion of *I. fulva* genetic markers (0.40) relative to *I. hexagona* markers suggest that the *I. fulva* genotypes are being displaced in this portion of the hybrid population also.

CONCLUSIONS AND HYPOTHESES

The genetic and ecological data for members of the Louisiana iris species complex suggest the occurrence of a number of evolutionary phenomena that have been mediated by the action of natural hybridization. Figure 5-9 represents the inferred phenomena (sympatric introgression, allopatric introgression, and hybrid speciation) and the hypothesized mechanisms (normalizing selection, directional selection, gene flow) that may have led to these phenomena. First, the pattern of genetic variation within two areas of parapatry is indicative of past and ongoing sympatric introgressive hybridization. Furthermore, the distributions of genotypes in the regions of overlap are not strictly clinal but, rather, reflect an apparent combination of clinal and mosaic characteristics. A comparison of findings from the ecological and genetic analyses suggests that the genetic structuring in at least one of the areas of overlap may be affected by selection due to microhabitat fluctuations. We hypothesize, and the distribution of genotypes support this hypothesis, that this selection has both directional (Dobzhansky, 1970) and normalizing (Waddington, 1957) components. During the initial stages of the hybridization and introgression there would be directional selection in favor of individuals capable of exploiting the ecotonal habitat between the parental taxa. Once the habitat had been invaded, normalizing selection would act to preserve the favored genotypes in contrast to the consequences of further hybridization or recombination.

A second consequence of natural hybridization in this species complex has been the occurrence of allopatric introgression between *I. fulva* and *I. hexagona.* Figure 5-9 indicates two mechanisms by which it may have occurred. It should be noted that both of these scenarios involve an area of sympatric hybridization and introgression; however, in one instance the area of sympatric introgression acts as a corridor for gene flow resulting in allopatric introgression into formerly allopatric regions of the hybridizing taxa. In the second case the area of sympatric introgression actually becomes an allopatric, introgressed population. With the second hypothesis, directional selection resulting in the predominance of genetic material from one of the taxa is hypothesized to be mediated by an environmental change. This habitat change is envisioned to rep-

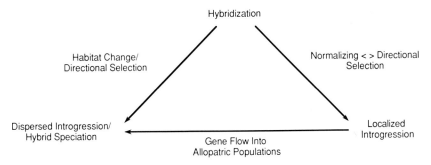

Figure 5-9. Model to explain how natural hybridization may lead to sympatric introgression, allopatric introgression, and hybrid speciation.

resent either succession from a disturbed to a predisturbance habitat type or a response to a directional change in climatic variables. With regard to the Louisiana irises, the bidirectional introgression may reflect both the historical and genetic explanations. Although the allopatric, introgressed populations of *I. hexagona* and *I. fulva* are characterized by ecological variables that are found in other conspecific populations, the *I. fulva* sample comes from a region that is undergoing ecological succession. In contrast, the *I. hexagona* sample comes from an isolated and apparently undisturbed area. A final and equally plausible mechanism for the observed allopatric introgression (not illustrated in Fig. 5-9) would be the direct exchange of genes between allopatric populations of *I. hexagona* and *I. fulva*.

The third evolutionary phenomenon associated with natural hybridization inferred for this species complex is that of hybrid speciation. In accord with the hypothesis of hybrid origin, the genetic analyses discovered a combination of the species-specific markers from *I. fulva, I. hexagona,* and *I. brevicaulis* in individuals of *I. nelsonii.* As indicated in Figure 5-9, the mechanism leading to the production of the hybrid derivative may include either of those hypothesized for the production of allopatric introgression. In this regard, it is significant to note that the area of overlap between *I. fulva, I. hexagona,* and *I. brevicaulis* includes individuals with various combinations of the species-specific markers found in *I. nelsonii.* Therefore the contemporary area of overlap between the three parental taxa may be representative of a similar, intermediate stage that historically led to the origin of the hybrid derivative *I. nelsonii.*

The above phenomena represent important pathways for the evolution of species and species complexes. However, the significance of these phenomena lies not only in their relation to the process of speciation; of equal importance are the phenomena themselves. Sympatric and allopatric introgression and hybrid speciation are being observed in ever-increasing numbers of species complexes owing largely to sensitive methods of detection. The picture that is currently being resolved is one of natural hybridization, with all of its subsequent effects, as a major evolutionary process and not as an evolutionary by-product of imprecise reproductive isolation.

ACKNOWLEDGMENTS

The studies described in this chapter were supported in part by NSF grants BSR-9004242 (M.L.A., B.D.B., and J. L. Hamrick), BSR-9106666 (M.L.A. and J.L. Hamrick), an NSF REU Supplement (M.L.A.), a University of Georgia Faculty Research Grant (M.L.A.), and a grant from the American Iris Society Foundation (B.D.B. and M.L.A.).

REFERENCES

Alston, R. E., and Turner, B. L. 1963. Natural hybridization among four species of *Baptisia* (Leguminosae). Am. J. Bot. 50:159–173.

Anderson, E. 1948. Hybridization of the habitat. Evolution 2:1–9.

Anderson, E. 1949. Introgressive Hybridization. New York: Wiley.

Anderson, E., and Hubricht, L. 1938. Hybridization in *Tradescantia*. III. The evidence for introgressive hybridization. Am. J. Bot. 25:396–402.

Anderson, E., and Stebbins, G. L., Jr. 1954. Hybridization as an evolutionary stimulus. Evolution 8:378–388.

Arnold, J., Asmussen, M. A., and Avise, J. C. 1988. An epistatic mating system model can produce permanent cytonuclear disequilibria in a hybrid zone. Proc. Natl. Acad. Sci. USA 85:1893–1896.

Arnold, M. L., Appels, R., and Shaw, D. D. 1986. The heterochromatin of grasshoppers from the *Caledia captiva* species complex. I. Sequence evolution and conservation in a highly repeated DNA family. Mol. Biol. Evol. 3:29–43.

Arnold, M. L., Bennett, B. D., and Zimmer, E. A. 1990a. Natural hybridization between *Iris fulva* and *I. hexagona:* pattern of ribosomal DNA variation. Evolution 44:1512–1521.

Arnold, M. L., Buckner, C. M., and Robinson, J. J. 1991. Pollen-mediated introgression and hybrid speciation in Louisiana irises. Proc. Natl. Acad. Sci. USA 88:1398–1402.

Arnold, M. L., Contreras, N., and Shaw, D. D. 1988. Biased gene conversion and asymmetrical introgression between subspecies. Chromosoma 96:368–371.

Arnold, M. L., Hamrick, J. L., and Bennett, B. D. 1990b. Allozyme variation in Louisiana irises: a test for introgression and hybrid speciation. Heredity 65:297–306.

Arnold, M. L., and Shaw, D. D. 1985. The heterochromatin of grasshoppers from the *Caledia captiva* species complex. II. Cytological organization of tandemly repeated DNA sequences. Chromosoma 93:183–190.

Arnold, M. L., Shaw, D. D., and Contreras, N. 1987. Ribosomal RNA-encoding DNA introgression across a narrow hybrid zone between two subspecies of grasshopper. Proc. Natl. Acad. Sci. USA 84:3946–3950.

Asmussen, M. A., Arnold, J., and Avise, J. C. 1987. Definition and properties of disequilibrium statistics for associations between nuclear and cytoplasmic genotypes. Genetics 115:755–768.

Baker, R. J. 1981. Chromosome flow between chromosomally characterized taxa of a volant mammal, *Uroderma bilobatum* (Chiroptera: Phyllostomatidae). Evolution 35:296–305.

Baker, R. J., Bleier, W. J., and Atchley, W. R. 1975. A contact zone between karyotypically characterized taxa of *Uroderma bilobatum* (Mammalia: Chiroptera). Syst. Zool. 24:133–142.

Baker, R. J., Davis, S. K., Bradley, R. D., Hamilton, M. J., and Van Den Busshe, R. A. 1989. Ribosomal-DNA, mitochondrial-DNA, chromosomal, and allozymic studies on a contact zone in the pocket gopher, *Geomys.* Evolution 43:63–75.

Ball, R. M., Jr., Niegel, J. E., and Avise, J. C. 1990. Gene genealogies within the organismal pedigrees of random-mating populations. Evolution 44:360–370.

Barton, N. H. 1979. The dynamics of hybrid zones. Heredity 43:341–359.

Barton, N. H., Halliday, R. B., and Hewitt, G. M. 1983. Rare electrophoretic variants in a hybrid zone. Heredity 50:139–146.

Barton, N. H., and Hewitt, G. M. 1981. A chromosomal cline in the grasshopper *Podisma pedestris.* Evolution 35:1008–1018.

Bennett, B. D. 1989. Habitat differentiation of Iris fulva Ker Gawler, *Iris hexagona* Walter, and their hybrids. Ph.D. dissertation, Louisiana State University, Baton Rouge.

Bennett, B. D., and Grace, J. B. 1990. Shade tolerance and its effect on the segregation of two species of Louisiana iris and their hybrids. Am. J. Bot. 77:100–107.

Bert, T. M., and Harrison, R. G. 1988. Hybridization in western Atlantic stone crabs (Genus *Menippe*): evolutionary history and ecological context influence species interactions. Evolution 42:528–544.

Birch, L. C., and Vogt, W. G. 1970. Plasticity of taxonomic characters of the Queensland fruit flies *Dacus tryoni* and *Dacus neohumeralis* (Tephritidae). Evolution 24:320–343.

Bloom, W. L. 1976. Multivariate analysis of the introgressive replacement of *Clarkia nitens* by *Clarkia speciosa polyantha* (Onagraceae). Evolution 30:412–424.

Butlin, R. 1989. Reinforcement of premating isolation. In D. Otte and J. A. Endler, eds. Speciation and Its Consequences. Sunderland, MA: Sinauer.

DePamphilis, C. W., and Wyatt, R. 1989. Hybridization and introgression in buckeyes (*Aesculus:* Hippocastanaceae): A review of the evidence and a hypothesis to explain long-distance gene flow. Syst. Bot. 14:593–611.

DePamphilis, C. W., and Wyatt, R. 1990. Electrophoretic confirmation of interspecific hybridization in *Aesculus* (Hippocastanaceae) and the genetic structure of a broad hybrid zone. Evolution 44:1295–1317.

Dobzhansky, T. 1970. Genetics of the Evolutionary Process. New York: Columbia University Press.

Doebly, J. 1989. Molecular evidence for a missing wild relative of maize and the introgression of its chloroplast genome into *Zea perennis.* Evolution 43:1555–1559.

Dover, G. 1982. Molecular drive: a cohesive mode of species evolution. Nature 299:111–116.

Dowling, T. E., Smith, G. R., and Brown, W. M. 1989. Reproductive isolation and introgression between *Notropis cornutus* and *Notropis chrysocephalus* (family Cyprinidae): comparison of morphology, allozymes, and mitochondrial DNA. Evolution 43:620–634.

Doyle, J. J., and Doyle, J. L. 1988. Natural interspecific hybridization in eastern North American Claytonia. Am. J. Bot. 75:1238–1246.

Endler, J. A. 1977. Geographic Variation, Speciation and Clines. Princeton, NJ: Princeton University Press.

Ferris, S. D., Sage, R. D., Huang, C. M., Nielsen, J. T., Ritte, U., and Wilson, A. C. 1983. Flow of mitochondrial DNA across a species boundary. Proc. Natl. Acad. Sci. USA 80:2290–2294.

Foster, R. C. 1937. A cyto-taxonomic survey of the North American species of *Iris.* Contrib. Gray Herb. 119:3–82.

Gibbs, G. W. 1968. The frequency of interbreeding between two sibling species of *Dacus* (Diptera) in wild populations. Evolution 22:667–683.

Godfrey, R. K., and Wooten, J. W. 1979. Aquatic and Wetland Plants of the Southeastern United States: Monocotyledons. Athens: University of Georgia Press.

Grant, B. R., and Grant, P. R. 1989. Evolutionary Dynamics of a Natural Population: The Large Cactus Finch of the Galapagos. Chicago: University of Chicago Press.

Grant, V. 1963. The Origin of Adaptations. New York: Columbia University Press.

Grant, V. 1981. Plant Speciation. New York: Columbia University Press.

Greenbaum, I. F. 1981. Genetic interactions between hybridizing cytotypes of the tent-making bat *(Uroderma bilobatum)*. Evolution 35:306–321.

Haldane, J. B. S. 1948. The theory of a cline. J. Genet. 48:277–284.

Hamrick, J. L., and Holden, L. R. 1979. Influence of microhabitat heterogeneity on gene frequency distribution and gametic phase disequilibrium in *Avena barbata*. Evolution 33:521–533.

Hardin, J. W. 1957. Studies in the Hippocastanaceae. IV. Hybridization in *Aesculus*. Rhodora 59:185–203.

Harrison, R. G. 1986. Pattern and process in a narrow hybrid zone. Heredity 56:337–349.

Harrison, R. G., Rand, D. M., and Wheeler, W. C. 1987. Mitochondrial DNA variation in field crickets across a narrow hybrid zone. Mol. Biol. Evol. 4:144–158.

Heiser, C. B., Jr. 1951. Hybridization in the annual sunflowers: *Helianthus annuus* X *H. debilis* var. *cucumerifolius*. Evolution 5:42–51.

Heiser, C. B. 1973. Introgression re-examined. Bot. Rev. 39:347–366.

Heiser, C. B., Jr., Smith, D. M., Clevenger, S. B., and Martin, W. C., Jr. 1969. The North American sunflowers *(Helianthus)*. Mem. Torrey Bot. Club 22:1–218.

Heywood, J. S. 1986. Clinal variation associated with edaphic ecotones in hybrid populations of *Gaillardia pulchella*. Evolution 40:1132–1140.

Hillis, D. M., Moritz, C., Porter, C. A., and Baker, R. J. 1991. Evidence for biased gene conversion in concerted evolution of ribosomal DNA. Science 251:308–310.

Huxley, J. 1938. Clines: an auxiliary taxonomic principle. Nature 142:219–220.

Keim, P., Paige, K. N., Whitham, T. G., and Lark, K. G. 1989. Genetic analysis of an interspecific hybrid swarm of *Populus:* occurrence of unidirectional introgression. Genetics 123:557–565.

Key, K. H. L. 1968. The concept of stasipatric speciation. Syst. Zool. 17:14–22.

Key, K. H. L. 1981. Species, parapatry, and the morabine grasshoppers. Syst. Zool. 30:425–458.

Levin, D. A. 1969. The challenge from a related species: a stimulus for saltational speciation. Am. Naturalist 103:316–322.

Levin, D. A., and Bulinska-Radomska, Z. 1988. Effects of hybridization and inbreeding on fitness in *Phlox*. Am. J. Bot. 75:1632–1639.

Lewontin, R. C., and Birch, L. C. 1966. Hybridization as a source of variation for adaptation to new environments. Evolution 20:315–336.

Marchant, A. D., Arnold, M. L., and Wilkinson, P. 1988. Gene flow across a chromosomal tension zone. I. Relics of ancient hybridization. Heredity 61:321–328.

Matthew, B. 1981. The Iris. New York: Universe Books.

Mayr, E. 1963. Animal Species and Evolution. Cambridge, MA: Belknap Press.

Mecham, J. S. 1960. Introgressive hybridization between two southeastern treefrogs. Evolution 14:445–457.

Moore, W. S. 1977. An evaluation of narrow hybrid zones in vertebrates. Q. Rev. Biol. 52:263–277.

Moore, W. S. 1987. Random mating in the northern flicker hybrid zone: implications for the evolution of bright and contrasting plumage patterns in birds. Evolution 41:539–546.

Moore, W. S., and Buchanan, D. B. 1985. Stability of the northern flicker hybrid zone in historical times: implications for adaptive speciation theory. Evolution 39:135–151.

Moran, C., and Shaw, D. D. 1977. Population cytogenetics of the genus *Caledia* (Orthoptera: Acridinae). III. Chromosomal polymorphism, racial parapatry and introgression. Chromosoma 63:181–204.

Nagle, J. J., and Mettler, L. E. 1969. Relative fitness of introgressed and parental populations of *Drosophila mojavensis* and *D. arizonensis*. Evolution 23:519–524.

Nason, J. D., Ellstrand, N. C., and Arnold, M. L. 1992. Patterns of hybridization and introgression in populations of oaks, manzanitas and irises. Am. J. Bot. 79:101–111.

Pembleton, E. F., and Baker, R. J. 1978. Studies of a contact zone between chromosomally characterized populations of *Geomys bursarius*. J. Mamm. 59:233–242.

Potts, B. M., and Reid, J. B. 1988. Hybridization as a dispersal mechanism. Evolution 42:1245–1255.

Powell, J. R. 1983. Interspecific cytoplasmic gene flow in the absence of nuclear gene flow: evidence from *Drosophila*. Proc. Natl. Acad. Sci. USA 80:492–495.

Rand, D. M., and Harrison, R. G. 1989. Ecological genetics of a mosaic hybrid zone: mitochondrial, nuclear, and reproductive differentiation of crickets by soil type. Evolution 43:432–449.

Randolph, L. F. 1966. *Iris nelsonii,* a new species of Louisiana iris of hybrid origin. Baileya 14:143–169.

Randolph, L. F., Mitra, J., and Nelson, I. S. 1961. Cytotaxonomic studies of Louisiana irises. Bot. Gaz. 123:125–133.

Randolph, L. F., Nelson, I. S., and Plaisted, R. L. 1967. Negative evidence of introgression affecting the stability of Louisiana Iris species. Cornell Univ. Agr. Exp. Station Mem. 398:1–56.

Rieseberg, L. H., Beckstrom-Sternberg, S., and Doan, K. 1990a. *Helianthus annuus* ssp. *texanus* has chloroplast DNA and nuclear ribosomal RNA genes of *Helianthus debilis* ssp. *cucumerifolius*. Proc. Natl. Acad. Sci. USA 87:593–597.

Rieseberg, L. H., and Brunsfeld, S. J. 1991. Molecular evidence and plant introgression. In D. E. Soltis, P. S. Soltis, and J. J. Doyle, eds. Plant Molecular Systematics, New York: Chapman and Hall, pp 151–176.

Rieseberg, L. H., Carter, R., and Zona, S. 1990b. Molecular tests of the hypothesized hybrid origin of two diploid *Helianthus* species (Asteraceae). Evolution 44:1498–1511.

Rieseberg, L. H., Soltis, D. E., and Palmer, J. D. 1988. A molecular reexamination of introgression between *Helianthus annuus* and *H. bolanderi* (Compositae). Evolution 42:227–238.

Riley, H. P. 1938. A character analysis of colonies of *Iris fulva, Iris hexagona* var. *giganticaerulea* and natural hybrids. Am. J. Bot. 25:727–738.

Sattler, P. W. 1985. Introgressive hybridization between the spadefoot toads *Scaphiopus bombifrons* and *S. multiplicatus* (Salientia: Pelobatidae). Copeia 1985:324–332.

Shaw, D. D., Coates, D. J., Arnold, M. L., and Wilkinson, P. 1985. Temporal variation in the chromosomal structure of a hybrid zone and its relationship to karyotypic repatterning. Heredity 55:293–306.

Shaw, D. D., Marchant, A. D., Arnold, M. L., Contreras, N., and Kohlmann, B. 1990. The control of gene flow across a narrow hybrid zone: a selective role for chromosomal rearrangement? Can. J. Zool. 68:1761–1769.

Shaw, D. D., Moran, C., and Wilkinson, P. 1980.

Chromosomal reorganization, geographic differentiation and the mechanism of speciation in the genus *Caledia*. In R. L. Blackman, G. M. Hewitt, and M. Ashburner, eds. Insect Cytogenetics. Symposium of the Royal Entomological Society of London No. 10. Oxford: Blackwell, pp. 171–194.

Shaw, D. D., Wilkinson, P., and Coates, D. J. 1983. Increased chromosomal mutation rate after hybridization between two subspecies of grasshoppers. Science 220:1165–1167.

Small, J. K., and Alexander, E. J. 1931. Botanical interpretation of the iridaceous plants of the gulf states. Contrib. N. Y. Bot. Gard. 327:325–357.

Stangl, F. B., Jr. 1986. Aspects of a contact zone between two chromosomal races of *Peromyscus leucopus* (Rodentia: Cricetidae). J. Mamm. 67:465–473.

Stebbins, G. L., and Daly, K. 1961. Changes in the variation pattern of a hybrid population of *Helianthus* over an eight-year period. Evolution 15:60–71.

Templeton, A. R. 1989. The meaning of species and speciation: a genetic perspective. In D. Otte and J. A. Endler, eds. Speciation and Its Consequences. Sunderland, MA: Sinauer.

Viosca, P., Jr. 1935. The irises of southeastern Louisiana: a taxonomic and ecological interpretation. Bull. Am. Iris Soc. 57:3–56.

Waddington, C. H. 1957. The Strategy of the Genes. London: George Allen & Unwin.

Warwick, S. I., Bain, J. F., Wheatcroft, R., and Thompson, B. K. 1989. Hybridization and introgression in *Carduus nutans* and *C. acanthoides* reexamined. Syst. Bot. 14:476–494.

Whitham, T. G. 1989. Plant hybrid zones as sinks for pests. Science 244:1490–1493.

Wiegand, K. M. 1935. A taxonomist's experience with hybrids in the wild. Science 81:161–166.

Williams, J. G. K., Kubelik, A. R., Livak, K. J., Rafalski, J. A., and Tingey, S. V. 1990. DNA polymorphisms amplified by arbitrary primers are useful as genetic markers. Nucleic Acids Res. 18:6531–6535.

6

After the Ice: *Parallelus* Meets *Erythropus* in the Pyrenees

GODFREY M. HEWITT

The Pleistocene Ice Ages greatly modified the geographic distributions of most organisms, with changes in distributions particularly marked in the temperate regions, including Europe and North America for which we have a growing amount of information. At the last glacial maximum in western Europe (18,000–20,000 BP) the Arctic ice sheet extended down across England and North Germany, with separate ice sheets on the Alps, Pyrenees, and other high mountains scattered across from Iberia to Transylvania. There was tundra vegetation in France and largely treeless steppe over much of Spain and southern Europe (Huntley and Birks, 1983; Huntley, 1988). Today in western Europe tundra conditions are found only in northern Norway and the Kola Peninsula in northwestern Russia.

GLACIATION AND GRASSHOPPERS

The short-horned meadow grasshopper *Chorthippus parallelus* is distributed across western Europe and is reported from eastern Europe and Asia. In Spain it is found on the high Sierras, and its range extends north to Scotland and southern Scandinavia. Two distinct subspecies are recognized (Reynolds, 1980): *C. p. parallelus,* which is distributed through most of Europe, and *C. p. erythropus,* which replaces it in Iberia. *Erythropus* was distinguished by its red hind tibiae, its more structured courtship song, and its higher number of pegs in the stridulatory file (which are scraped across the forewings to produce song). After the initial identification a number of further differences have been found including more morphological, song, and behavioral characters as well as chromosomal and molecular ones (Butlin and Hewitt, 1985a,b; Gosalvez et al., 1988; Ritchie, 1990).

In the Pyrenees this grasshopper reaches about 2100 meters in altitude and a hybrid zone between *parallelus* and *erythropus* runs along the spine of the mountain range (Fig. 6-1). The hybrid zone probably formed following the last ice melt. At the western end the two subspecies meet on a broad front, whereas in the high Pyrenees they meet only in cols below 2100 meters. They hybridize readily, but F_1 males from

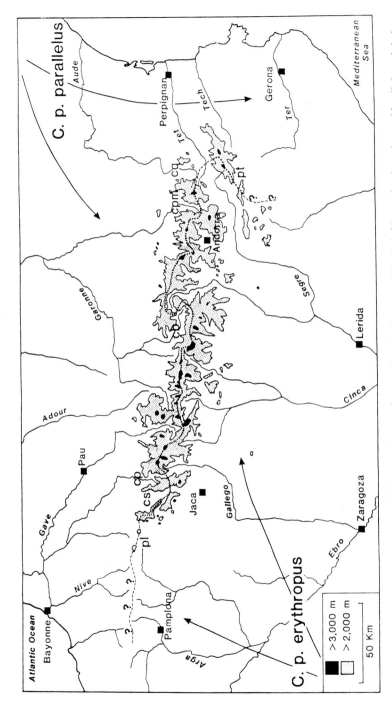

Figure 6-1. Contact between *C. p. parallelus* and *C. p. erythropus* along the Pyrenees. The hybrid zone has been located in all the cols below 2000 meters; pl = Puerto de Larrau; cs = Col de Somport; cp = Col de Portalet; cb = Col de Beret; cpm = Col de Puymorens; cq = Col de la Quillane; pt = Puerto de Tosas. Its exact course at the extreme west and east ends is uncertain at present.

crosses between pure *parallelus* and pure *erythropus* show severe testis dysfunction and are sterile (Hewitt et al., 1987). Such hybrid unfitness demonstrates that the hybrid zone is a tension zone (Key, 1968, 1981; Barton and Hewitt, 1985, 1989). Interestingly, the F_1 females are fertile and appear fully fit—an example of Haldane's rule—so that gene flow can occur through backcrosses (BXs) and subsequent crosses.

During the Pleniglacial ($>$ 18,000 BP) this grasshopper probably survived in the southern limits of Europe and the Balkans. By 13,000 BP the climate had warmed considerably, and the pollen data show that trees rapidly extended their ranges over Europe, particularly *Betula* and *Pinus*. About 11,000 BP there was a severe reversal of this trend, called the Younger Dryas, which was noticeable in the north and west where recently established trees disappeared and glaciers advanced again. This cold snap lasted only about 500–1000 years and by 10,000 BP the warming and rapid advance of vegetation had recommenced. Grasshoppers such as *Chorthippus parallelus* would have expanded from their glacial refugia in southern Europe and the Balkans as climate and food allowed. This expansion was probably rapid, as *C. parallelus* reached Britain, and the English Channel formed around 8000 BP, cutting off invasion by further species that now occur in northern France. The Pyrenean ice cap would have retreated during this period and the vegetation become suitable for grasshoppers first in the foothills, then at the lower ends of the range, and finally in the high cols. Interestingly, the course of the hybrid zone between *parallelus* and *erythropus* maps along the highest part of mountains more accurately than does the political boundary, which argues strongly for the advance of the two subspecies from France and Spain up the sides to meet as conditions and the vegetation improved. It also implies that the zone has not moved far from its original position in the high passes and meadows where the ice last disappeared.

COLS AND CLINES

This hybrid zone has been studied in detail in two high passes in particular, Col de Portalet (1762 meters) in the Western Pyrenees and Col de la Quillane (1713 meters) in the Eastern Pyrenees (Fig. 6-1).

We have scored a range of morphological and song characters from series of collections running through the cols from pure *parallelus* in France to pure *erythropus* in Spain. These characters change clinally (Butlin and Hewitt, 1985a,b; Ritchie, 1988; Butlin, 1989; Butlin et al., 1991). This clinal pattern in the field and the inheritance of the characters in laboratory crosses clearly indicate that they are largely polygenically determined (Butlin and Hewitt, 1988). If these various characters are combined into a multivariate analysis, the number of stridulatory pegs on the male's hind femur contributes most to the first canonical variate, which explains some 70% of the variation. This character is thus a good indicator of the zone's position and its general shape (Fig. 6-2). Interestingly, the clines at Col de Portalet are wider than at Col de la Quillane (Table 6-1) and possible reasons for this will be discussed later. Looking at each col separately the various characters show clines of different widths. For example, at Quillane the cline width for number of pegs (np) is 4.2 km with the center of the cline resting close to the barrage at Lac de Matemale, whereas for echeme interval (Fig. 6-3) the width is 1.4 km and for syllable length (Fig. 6-3) it is 19.5 km, with the centers of these clines calculated to be, respectively, 0.7 and 2.0 km north of the barrage. Our data for

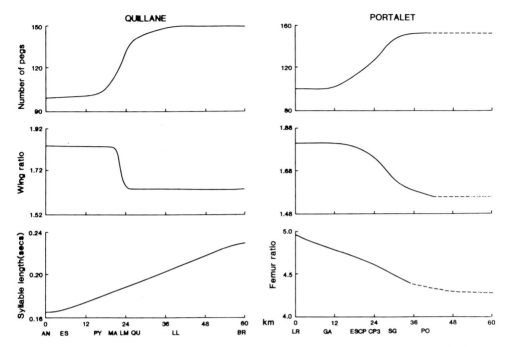

Figure 6-2. Clinal change through Col de la Quillane and Col de Portalet in number of stridulatory pegs on the hind femur and wing ratio of males. Wider clines are shown at Quillane for song syllable length and at Portalet for female femur ratio. Key populations are located by letters, e.g., AN = Aunat; LL = Llo; LR = Laruns; PO = Polituara. (See Butlin et al., 1991, for further details.)

Table 6-1. Widths of Clines for Some Subspecific Characters at Col de la Quillane and Col de Portalet and the Position of Their Center Relative to That of the Cline for Peg Number

	Quillane		Portalet	
Character (Males)	Width (km)	Center (km)	Width	Center
Peg number	4.2 ± 0.57	0	9.0 ± 1.42	0
Peg rows length	8.0 ± 1.86	1.0 S	5.9 ± 1.60	2.6 S
Wing ratio	1.3 ± 0.57	0.3 N	5.6 ± 2.49	3.8 N
Femur ratio	4.2 ± 1.70	2.5 N	10.6 ± 3.15	2.5 N
Pronotum ratio (males)	0.45 ± 1.60	2.5 N	0.37 ± 0.67	2.6 S
Pronotum ratio (females)	1.04 ± 0.83	1.7 N	6.42 ± 2.28	6.6 S
Song echeme interval	1.4 ± 0.65	0.7 N	Wider?	
Courtship index	2.1 ± 2.28	6.3 S	~ 14	
Song syllable length	19.5 ± 3.29	2.0 N	Wider?	
Esterase-2	15 (approx.)		Wider?	
First canonical variate for 11 male morphological measures	3.46 ± 0.423	0.6 N	9.71 ± 1.38	0.4 S

The cline parameters were calculated from a tanh model using 1100 individuals from 22 populations for Portalet and 500 individuals from 15 populations for Quillane (see Butlin, 1989; Butlin, et al., in prep.).

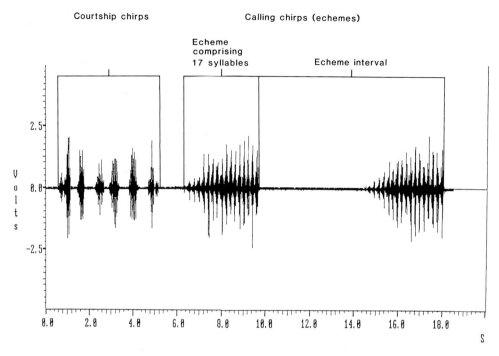

Figure 6-3. Courtship and calling song components of *Chorthippus parallelus* (Acrididae). Echeme length and echeme interval are indicated. Syllable length is calculated as echeme length/ number of syllables. The voltage scale represents the amplitude of the second after passing through a filter.

allozyme differences are not so comprehensive, but clines for alleles of *esterase 2* and *hexokinase* have widths in the region of 15–20 km.

Given that postglacial warming allowed the subspecies to meet at the col some 9000 BP, it seems likely that during the subsequent hybridization and mixing alleles at different loci have introgressed to different degrees depending on the selection against heterozygotes for those loci controlling each character and initially also on linkage relations and selection on the linked loci. Where the heterozygote at a locus is unfit, that cline is narrow. At some loci the alleles may be selectively equivalent in both genetic backgrounds (neutral), and they may have introgressed further. The various character clines are broadly coincident in the col, and there are a number of reasons why this situation should be (Barton and Hewitt, 1985; Hewitt, 1985, 1989). Probably the strongest reason is secondary contact; all the gene differences began introgressing together around 9000 BP at the same place in the col. It is enhanced by epistasis between loci; the alleles within each of the two subspecies' genomes may well be coadapted, so that recombinants and segregants are less fit. An ecotonal change or simply a population density trough may also concentrate some allelic differences in one area. Interestingly, a laboratory crossing program indicated that there may be sex linkage or maternal effects for several characters and dominance, epistasis, and genotype/ environment interactions (Butlin and Hewitt, 1988). Some of these factors, such as epistasis and linkage, would promote coincidence; but one might expect that dominance and environmental differences could displace some cline centers.

It is possible that part of the variation in cline position is caused by these effects, but the overriding feature is how close these clines are to the col; i.e., they are broadly coincident. Furthermore, the greatest width is only 20 km, which is not large compared with the species range of thousands of kilometres. Despite 9000 years of dispersal, hybridization, and recombination, there is apparently no real exchange of genetic material between the subspecies. There are many more hybrid zones with similar dimensions of width compared to the species range, which poses a worry for the taxonomist trying to define species. The biological species concept applied strictly allows for no gene exchange between full species; thus hybrid zones are between subspecies because there is hybridization producing F_1, and backcross offspring (Mayr, 1963; Barton and Hewitt, 1985). The problem is that this situation is not apparently affecting the genotypes in the main body of each of the taxa, just blurring the edges in a few places. In cases such as *C. parallelus,* the independence of the genomes during evolution seems ensured, certainly over thousands of years. An alternative view of a species is the coadaptive species concept (Mayr, 1963; Barton and Hewitt, 1985), which envisages a cluster of phenotypes stable to invasion by foreign genes because each genome is internally coadapted. Where a species complex is subdivided by hybrid zones, the component genomes may well be internally coadapted, but the independence of the genomes does not depend on this point. It may simply be due to the small diffusive gene flow through the zone, not to selection. Thus genome independence would be for geographic rather than genetic reasons.

One might argue that following postglacial secondary contact some alleles have introgressed large distances into the other subspecies, but they have not been recognized. With *C. parallelus* the allozymes, morphology, and song of *erythropus* do not show any obvious impact through France and Great Britain (Dagley, 1988), but it does not exclude the introgression of alleles at some loci. To test this proposition thoroughly one would need to examine long transects for many nuclear DNA markers. This question seems to be important in any hybrid zone, but care is needed when interpreting the results, as apparent introgression may be produced by different processes. For example, after secondary contact and the formation of a hybrid zone, neutral alien alleles may diffuse by dispersal into the abutting genome's range; if they are advantageous, they spread faster and further. On the other hand, the expansion of allopatric races and subspecies following the last glaciation may well have produced small, temporarily isolated hybrid populations before the main bodies of the races met. In such a mixed pioneer population, a reassorted and recombined genome could survive and subsequently spread with range expansion. A number of possible cases already exist, viz., *Vandiemenella viatica* (Hewitt, 1979), *Mus musculus/domesticus* (Gyllensten and Wilson, 1987; Vanlerberghe et al., 1988), *Caledia captiva* (Marchant et al., 1988; Shaw et al., 1988), with more being currently researched. The details of population genetics and biogeography of these samples vary, but all depend on sparse patchy population structure and environmentally induced range changes.

NUCLEOLAR ORGANIZERS: POSITION AND PENETRATION

Although the clines for character differences between *C. p. parallelus* and *C. p. erythropus* are broadly coincident on the ridge of the Pyrenees, we have noted that small but significant variation exists among the exact cline centers. A particularly clear example has come to light for the nucleolar organizing regions (NORs), which comprise repeated copies of ribosomal DNA located in particular regions on the chromosomes of this species. After investigation with a range of staining and fluorescent techniques, C-banding with giemsa and acridine orange revealed NOR sites, and the nucleoli they produce can be seen with silver staining (Golsalvez et al., 1988). *C. p. parallelus* has NORs on the L2 and L3 and X chromosomes, whereas *C. p. erythropus* has them on the L2 and L3 chromosomes only (Fig. 6-4), clearly indicating a considerable difference in the arrangement of the ribosomal DNA sequences in the two subspecies. When the population frequency of the XNOR is determined by C-banding, the change from *parallelus* type (+XNOR) to *erythropus* (−XNOR) occurs several kilometers down from the cols on the Spanish side (Fig. 6-5) (Ferris et al., 1993). Both Col de Portalet and Col de la Quillane show this noncoincidence, but it is more distinct at Quillane because the morphological and song character cline centers are more tightly clustered. The XNOR cline appears to be narrow in both places. At Portalet it is less than 1 km in width; only one mixed population has been found so far, and that had some recombinant X chromosomes (Ferris et al., 1993).

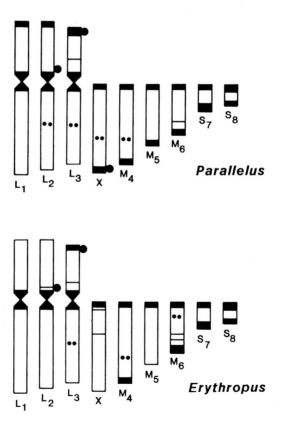

Figure 6-4. Karyograms of two subspecies of *C. parallelus* as described by C-banding, Giemsa, acridine orange, and silver staining (Gosalvez et al., 1988). The nucleolar organizing regions are identified by the nucleolus (external round black spot) each produces; *C. p. parallelus* has an X-NOR. The smaller black dots are produced by C-banding. (Bella et al., in prep.).

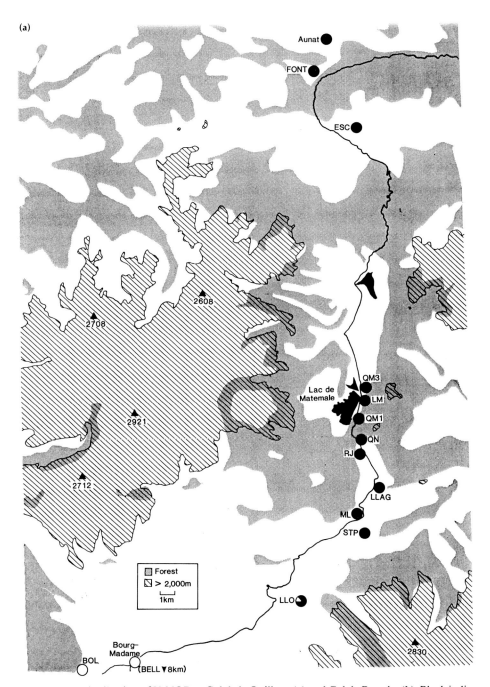

Figure 6-5. Distribution of X-NOR at Col de la Quillane (**a**) and Col de Portalet (**b**). Black indicates the proportion of males with an X NOR C-band site on the X chromosome. The zone center for morphological characters is indicated (by an arrow), along with altitude and forest. Population location names are as in Table 6-2.

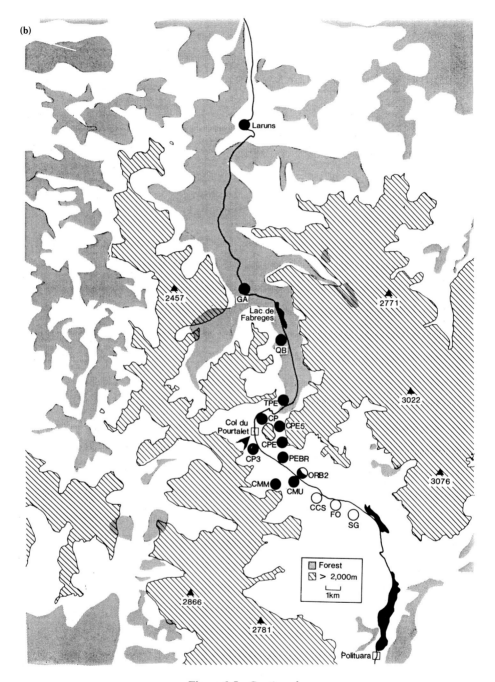

Figure 6-5. *Continued*

The song and morphological character clines provide a measure of the change from a *parallelus* genome to an *erythropus* one, and the XNOR chromosome marker appears to have penetrated some distance into the *erythropus* genome. At Portalet it is displaced as far as the cline center of the most displaced morphological character on the Spanish side, female pronotum ratio, indicating that some introgression has reached this far. However, the XNOR changeover occurs on the tails of most other clines and, the genome must be largely *erythropus* here. Its narrowness indicates selection against heterozygotes for the XNOR or recombinants for the X chromosomes. At Quillane the XNOR changeover is less accurately determined, but it is displaced some distance beyond the morphological and song clines and may well be in nearly pure *erythropus* genome.

Cellular expression of the NOR to produce a nucleolus has been measured by silver staining (Gosalvez et al., 1988). Interestingly, the activity of the XNOR varies through the cols (Table 6-2). At both Portalet and Quillane it is active and visible in all male meiocytes in *parallelus*-like populations but becomes progressively less active as the genome becomes *erythropus*-like. The change in activity of this rDNA locus follows the morphological and song clines and suggests that it functions and is expressed proportionate to the *parallelus* content of the genome. The character clines at Portalet are wider and less coincident near the center, mirrored by the expression of the XNOR. However, the tails of the XNOR expression at Quillane require further study. The copy number and type of rDNA sequences may well be different between *parallelus* and *erythropus*. Certainly the presence of three NORs in one and two NORs in the other indicates a different chromosomal distribution and causes unnatural

Table 6-2. Some Populations on Transects Through the Col de la Quillane and Col de Portalet Giving the Percentage of Meiotic Prophase Cells Per Individual Expressing Three Nucleoli

	Quillane				Portalet		
Label	Location	% Three Nucleoli	km	Label	Location	% Three Nucleoli	km
AN	Aunat	97.6	0	GA	Gabas	98.9	0
FONT	Fontanès d.s.	96.8	7	NQB	N. Quebe Bottes	83.3	2.2
ESC	Escouloubre	93.8	13.2	TPE	Tourmont Peyrelue	82.9	8.0
QM3	Quillane-Matemale3	93.1	23.4	CP	Col de Portalet	75.5	8.7
LM	Lac de Matemale	83.8	24.1	CPE5	Col de Peyrelue5	70.4	9.3
QM1	Quillane-Matemale1	56.1	25.7	CPE	Col de Peyrelue	65.2	11.0
QN	Col de la Quillane	49.1	26.3	CPE2	Col de Peyrelue2	66.7	12.1
RJ	Refuge de Jour	29.0	28.0	CP3	Col de Portalet3	70.4	12.2
LLAG	La Llagonne	11.5	33.0	PEBR	Peyrelue Bridge	53.4	12.7
ML	Mont Louis	20.1	35.3	CMUP	Corral de las Mulas up	48.9	14.0
STP	St. Pierre d.f.	25.2	36.0	ORB2	Old Road Bridge2	35.7	14.4
LLO	Llo	12.3	38.5	CCS2	Cable Car Station2	22.5	16.0
BOL	Bolvir	10.9	50.0	FO	Formigal	6.7	17.4
BELL	Bellver	6.1	57.0	SG	Sallent de Gallego	6.3	19.0

C. p. parallelus has NORs on the X, L2, and L3 chromosomes, whereas *C. p. erythropus* has only the L2 and L3 NORs. They show variable expression through the cols.

mixtures and numbers of rDNA cistrons in various hybrids. It may be this situation that causes the proposed unfitness of hybrids for the XNOR.

Why is the XNOR displaced to the Spanish side? One possibility is suggested after considering the nucleolar composition of hybrids. If we assume that there is an optimal mode of rDNA copy number in a *Chorthippus parallelus* genome, this number is then divided between two sites in *C. p. erythropus* and among three sites in *C. p. parallelus,* giving six NORs in a diploid female *parallelus* and five NORs in a diploid male *parallelus;* diploid male and female *erythropus* have four NORs. F_1 hybrid males containing the XNOR of *parallelus* have more than the modal rDNA copy number, whereas those with the *erythropus* X have less. As more complex hybrids are produced, this bias generally applies. If it is less debilitating to carry more than the modal number of rDNA copies than it is to carry fewer, it gives a selective advantage to the XNOR chromosome. Clearly such hypothesis rests on the actual distribution of rDNA copies and the effect on fitness of variant copy numbers. Results from *Drosophila* suggest that extra NORs may be tolerated more readily than deficiencies. In *Drosophila* the "bobbed" mutants have their rDNA sequences reduced below a threshold level of some 130 copies and show a delayed development, reduced variability and fertility, and visible effects on bristle and cuticular morphology. However, mechanisms of compensation can occur that magnify the copy number in tissues and in some cases increase it in the germ line over several generations toward a normal "equilibrium" value (Ritossa and Scala, 1969). Thus an initially marked effect on fitness in hybrid *C. parallelus* may be modified in subsequent generations (see later).

HYBRIDS AND HOMOGAMY

A useful approach to considering what might have happened after secondary contact following the Ice Age is through studies of laboratory hybridization. There is a significant degree of assortative mating among pure *C. p. parallelus* and pure *C. p. erythropus* when females are given a choice (Ritchie et al., 1989; Butlin and Ritchie, 1991). Nevertheless, females mate several times in nature, and some mated with males of both subspecies. We have set up a program of controlled crosses to examine inheritance and fitness in *parallelus-erythropus* hybridization, and some of the results are most surprising.

One series of crosses involved mating individual pure females with single males of both pure subspecies sequentially. Thus a *parallelus* female was mated successfully first with a *parallelus* male and then with an *erythropus* male (P.PE). Another *parallelus* female was mated sequentially by single *erythropus* and *parallelus* males (P.EP), and so on (E.EP and E.PE) replicated a number of times. The embryos from the eggs produced by these females were C-banded, as seen with giemsa and acridine orange (Gosalvez et al., 1988) to determine their paternity; the chromosome banding differs between the two subspecies. The results show several departures from straightforward inheritance (Bella et al., 1992). Overall there is an excess of pure over hybrid progeny, when equality might be expected; and it is much more marked when *parallelus* is the mother (Table 6-3). There is, in fact, little mortality in young prediapause grasshopper embryos up to the stage they are scored, so although some of this preponderance of pure progeny may be due to differential mortality of early hybrid embryos, there is not enough mortality to explain it all. We are left with evidence of homogamy.

Table 6-3. PEPE Crosses[a]

Cross	Pure Male	Pure Female	Hybrid Male	Hybrid Female	Total Pure	Total Hybrid	Total Male	Total Female
PPE	23	7	1	15	30	16	25	22
PEP	35	16	6	1	51	10	42	20
Subtotal P-	58	23	7	19	81	26	67	42
EEP	18	3	10	18	21	28	31	21
EPE	18	9	7	8	27	15	30	17
Subtotal E-	36	12	17	26	48	43	61	38
Total	94	35	24	45	129	69	128	80

[a]For example, PPE = *parallelus* female crossed first with *parallelus* male and then with *erythropus* male to give progeny scored as embryos by chromosome banding.

Two other major effects are apparent in these results (Table 6-3). First, the degree of homogamy (pure/hybrid progeny) is greater when the second male to mate is the same subspecies as the female; and second, there is a strong sex ratio bias in favor of males (1.54). All three effects have interesting implications for the historical dynamics of this hybrid zone.

When *parallelus* met *erythropus* in these cols around 9000 BP, *parallelus* females mated to both *parallelus* and *erythropus* males would have tended to produce more pure offspring than *erythropus* females similarly mated. It is a form of genetic selection in favor of *parallelus* gametes and increases the frequency of *parallelus* and causes its advance. Furthermore, these pure offspring would have been more fit than the hybrids, as the F_1 males would have shown complete testis dysfunction. Consequently, more fertile *parallelus* offspring should have been produced, other things being equal between the two subspecies (e.g., viability, mating ability, fertility, fecundity). This effect would be exacerbated by the male biased sex ratio, as more *parallelus* males would be pure and hence fertile—tending to drive the *parallelus* genome forward.

However, females can produce semifertile backcross progeny and they in turn can cross and backcross, so that over several generations a mixing of the genomes would occur by segregation and recombination. This situation would tend to separate the loci causing this drive and dissipate its force. It may even destroy it, as the loci are likely to be epistatic in their effects on such a complex phenotype. Nevertheless, if the genes for homogamy are associated with a particular chromosome region, that chromosome region may continue to be driven. Could this explain the displacement of the XNOR region of *parallelus*? The probable strong selection against X chromosome recombinants, as evidenced by the narrow cline between *parallelus* and *erythropus* X chromosome banding, would hold such loci on the X chromosome in linkage disequilibrium for some time, reducing the effects of recombination. On the other hand, one might have expected the XNOR to have moved farther from the putative contact point on the Col around 9000 BP if all or most of the homogamy loci were linked with it, which argues for it having only some of them. Homogamy itself is an interesting phenomenon, and we have evidence for it from another hybrid zone in *Podisma pedestris* (Hewitt et al., 1987). Because in the *C. parallelus* experiments the female received sperm from both types of male and because the eggs are fertilized, laid, and analyzed with no resorption and little loss, the homogamy must result from some form of sperm recognition, preference, or competition—be it of chemical or physical nature.

HYBRIDS AND HALDANE'S RULE (HYBRID FITNESS AND DYSFUNCTION)

As already noted, laboratory F_1 hybrids between pure *C. p. parallelus* and *C. p. erythropus* collected from outside the hybrid zone show severe testis dysfunction in the males while the females are apparently normal, an example of Haldane's rule (Hewitt et al., 1987). Backcrosses of these F_1 females to parental males produce fertile progeny, but the males are still affected, showing reduced follicle size in the testis and somewhat disorganized sperm maturation and processing. Our crossing program is examining this point in more detail, and results have demonstrated a distinct bimodality in follicle length (a reliable measure of follicle size and overall spermatogenic organization) among males from backcrosses of all four types (Virdee, 1991). Because the F_1 female contains both haploid complements it is tempting to think that these long and short (good and bad) follicled males differ by a single factor received from their mother that does or does not match with their father's genome.

We may ask what the consequences of postglacial contact and hybridization might have been for testis function. Obviously, there would have been considerable male sterility as F_1s and backcrosses were produced. If this severe reduction in hybrid fitness is determined by a few genes, we would expect a narrow hybrid zone to develop for this character, and the underlying gene clines would be narrow and steep. The width would also be determined by the insect's dispersal in this region (Barton and Hewitt, 1985); and if only one gene was the cause, the zone would be 120 meters wide from our estimates of dispersal. If 10 genes were involved, it would be about 380 meters wide, and for 100 genes about 1700 meters on a simple additive fitness model (see Barton and Hewitt, 1981, for the mathematics of multiple gene zones). If several genes are involved, mating and recombination over many generations produce whole ranges of multiple heterozygotes, and the simple chance of generating an F_1 hybrid genotype in the zone is of the order of $(\frac{1}{2})^n$, where n is the number of genes contributing to the hybrid testis dysfunction. For 10 genes this number is 9.8×10^{-4}, and the probability of at least a backcross level of heterozygosity is less than 0.62 in the center of the zone. We might not then expect to find obvious signs of a zone of testis dysfunction unless four or fewer genes were involved. This zone would be less than 300 meters wide and require more detailed sampling to identify it.

Our sampling to date has not revealed such a zone for testis dysfunction (Fig. 6-6), (Ritchie and Hewitt, 1993). In the center of the zone for morphological characters at Portalet these samples were 200 meters apart, whereas at Quillane they were less frequent in some parts. Toward the edges of the zone the samples were also wider apart. It seems unlikely that we have missed a narrow oligogenic dysfunction zone, unless it is well displaced from the morphological center. It seems likely that it is multigenic.

One approach to examining this further is that of "bracket crossing." At Portalet a series of five reciprocal crosses were made between the *parallelus* and *erythropus* sides of the Col, progressing from outside the hybrid zone in toward the center. If the testis dysfunction zone is less than about 2–3 km wide, it should be located between

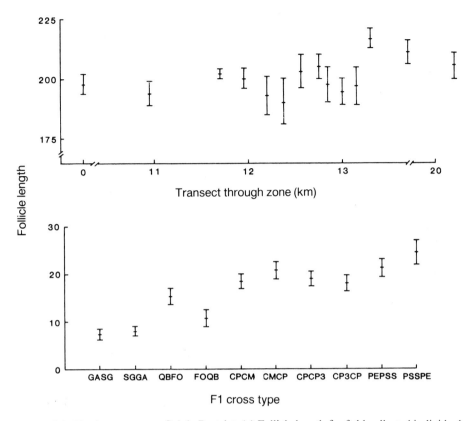

Figure 6-6. Testis structure at Col de Portalet. (a) Follicle length for field collected individuals (data from Ritchie, 1988). (b) Follicle length in F_1 males from crosses bracketing the Col. On the left, GASG is an F_1 between two pure populations (Fig. 6-5), and crosses are closer together progressing to the right (data from S.R. Virdee, pers. comm.).

two of these populations; if it is wider we expect a more gradual change over two or more crosses. The results show a fairly progressive change from severe dysfunction in the widest cross to normality in the crosses closest to the col (Fig. 6-6) (Virdee, 1991), which argues for polygenic control. Another approach we are currently pursuing is that of "crossing in," where a pure anchor population is crossed to each of a series of transect populations traversing the col to a pure population on the other side of the zone (Virdee, 1991). This approach should allow a more accurate location and description of the shape of the cline for testis dysfunction alleles.

Although the apparent lack of a zone of testis dysfunction in the field can be explained by invoking a polygenic basis so that hybrids heterozygous at enough loci rarely occur, there is another possibility—that of amelioration (Hewitt, 1990; Ritchie and Hewitt, 1991). Here alleles that modify the detrimental effects of heterozygosity and thereby increase fitness are selected in the zone (Endler, 1977; Barton and Hewitt, 1981; Searle, 1986; Sanderson, 1989). In addition to having normal testis structure and function, the few hybrid populations so far examined do not show any reduction in a number of fitness characters, such as numbers hatching, time to hatching, development rate, or survival to adult. Neither is there any increase in their developmental asymmetry as measured on hind-femur length and width, the number of stridulatory pegs, and peg row length—for which both legs were compared (Ritchie, 1988; Ritchie and Hewitt, 1991). However, although amelioration is an attractive hypothesis, it should be emphasized that at present these results can simply be explained by a model of polygenic control with dispersal and mixing of alleles in the zone following secondary contact and hybridization. Critical experiments are needed to distinguish these hypotheses.

MATING AND MODIFICATION

C. p. parallelus and *C. p. erythropus* have noticeable differences in male calling song and courtship song (Butlin and Hewitt, 1985b); and when mixed together, individuals from pure populations show significant assortative mating (Ritchie et al., 1989). Evidence reveals that mating involves olfactory cues as well (Ritchie, 1990); and with the morpological differences already described and other behavioral differences, it demonstrates clearly different mate recognition systems for the two subspecies. Because the F_1 hybrids and backcross offspring show such severe male sterility, when they met after the Ice Age there would have been a distinct selective advantage to assortative mating producing fit parental type progeny. This scenario is classic for speciation by reinforcement (see Ch. 3). It has been argued that such selection should lead to a divergence in mate recognition systems, enhancing isolating mechanisms until full species are formed (Wallace, 1889; Mayr, 1963; Dobzhansky, 1970), which has been widely accepted as a major mode of speciation. However, a number of voices have been raised concerning the likelihood of reinforcement actually occurring (Moore, 1957; Paterson, 1978; Templeton, 1981; Barton and Hewitt, 1985; Butlin, 1985, 1987, 1989; Spencer et al., 1986; Sanderson, 1989; see Ch. 3). Although computer simulations can produce an effect, the conditions under which the effect occurs are limited; strong selection and few highly linked genes are necessary for any progress, and within a hybrid zone gene flow from the parental populations tends to swamp the process. Furthermore, when examined carefully (Butlin, 1985, 1987, 1989), there is little empirical evidence that supports reinforcement.

If reinforcement had occurred in the 9000 generations since the *parallelus* genome began hybridization with that of *erythropus* in the Pyrenees, we might expect to find that assortative mating had increased in the zone in response to selection to reduce hybrid unfitness. It would perhaps take the form of an inverse cline where assortment for the parental type increased strongly on the shoulders of the zone with a sharp reverse in the middle. We might also expect that some of the characters comprising the mate recognition system would show some such pattern. However, as has been pointed out, song characters such as syllable length, echeme interval, and courtship index and morphological measures of stridulatory pegs, leg, and body characters all describe relatively smooth simple (tanh) clines. Using virgin individuals in a "bracket pairing" of populations, the level of assortment decreased between populations as one approached the center of the zone. The pattern was different for nonvirgin females, and in one region near the zone there was increased assortment, making the result equivocal (Ritchie et al., 1989). A more elegant and robust experimental test of expectations under reinforcement involved testing individual females from a series of populations traversing the hybrid zone for assortative mating when offered a choice of males, one from a pure *C. p. parallelus* stock and one from pure *C. p. erythropus* stock (Butlin and Ritchie, 1991). Such a design has several advantages: The character measured is actual mating; the males are from the same populations for each female, but not from the female's own population; and the individuals are unrestrained and able to use all their mate recognition system. Because all individuals are laboratory-reared, this experiment should demonstrate genetic differences in preference. The pattern of female preference revealed by this experiment is best fitted by a simple step cline, with a sudden change from *parallelus* to *erythropus* over about 1 km. This situation occurs in a region of dense woodland near Col de la Quillane where grasshopper density is low and suggests relatively strong selection on this character. However, on the *parallelus* side of the zone, there is an indication of higher homogenic preference in a small group of samples, although the pattern is not statistically different from that of a simple cline. Once again it provides no good evidence and only limited support for reinforcement, even though the situation seems ideal for reinforcement and the experiment pertinent and exhaustive. Other zones with different structures should be examined critically.

When postglacial contact occurred there was hybridization and mixing of alleles at gene loci controlling the mate recognition system. There may well have been selection against hybrid MRSs as well as dysfunctional testes, and it could be that the outcomes for MRS are the same as suggested for testis dysfunction, i.e., either dilution of the effect through recombination of a polygenic system or some form of genetic modification of the MRS so that hybrid individuals functioned effectively. The first is the more parsimonious and favored for that reason, but it is possible that some modification could have occurred, perhaps on a local scale. Testing this possibility requires a different type of experiment.

More generally, there are many hybrid zones already known and probably many more in existence, showing a wide range of characteristics (Hewitt, 1988). The lack of evidence for reinforcement from these zones, even after detailed studies such as these on *Chorthippus,* argues that the process is not important here. Of course it could be that on occasion it occurs rapidly and two species are formed with distinct mate recognition systems. If this situation occurs we would expect to see close sibling species with partially overlapping ranges, the overlap depending on ecological compatibility, how long ago speciation occurred, and how fast they can intersperse. Unfortunately, this pattern could also occur if they met secondarily having already speciated in allopatry, possibly by divergence during repeated range contractions into refugia (Hewitt, 1989). This point raises a corollary question as to just how mate recognition systems do change, particularly as there is likely to be strong stabilizing selection operating (Paterson, 1985; Butlin, 1989). Perhaps the most likely pathway is through a change in the community of organisms on which or among which the species in question lives so that mating in a different way, at a different time or place, or using different signals is more efficient, and selection for change does occur (e.g., Diehl and Bush, 1989; Tauber and Tauber, 1989; Rice and Salt, 1990; Wood and Keese, 1990).

DENSITY AND DISPERSAL

The clines for some characters through this hybrid zone are fairly narrow (1 km), whereas others are rather wide (15–20 km). The width of the zone at Portalet is generally wider than at Quillane. Narrow clines can be explained by selection against heterozygotes, recombinants, or ecotypes (Barton and Hewitt, 1985, 1989), but the wider ones suggest neutral diffusion. Using Endler's (1977) diffusion equation for neutral clines [$T = 0.35 (w/d)^2$, where T = the time since contact, w = cline width (1/max slope), and d = dispersal (standard deviation of parent offspring distances)], and a dispersal estimate of some 30 meters per generation, a cline of 20 km would take some 155,000 years to produce. During the 9000 years since the end of the Ice Age dispersal of 30 meters per generation would produce a cline of 4.8 km. Clearly, such figures depend on three things: the dispersal measures, time estimates and the environment in which the cline has formed. First, our dispersal estimate seems robust in magnitude (Virdee and Hewitt, 1990); two experimental designs gave similar answers, but the shape of the distribution would bear further examination. Furthermore, dispersal experiments in another flightless alpine grasshopper, *Podisma pedestris,* using four designs in different places, also gave similar estimates for this species, around 20 meters per generation (Barton and Hewitt, 1982; Nichols, 1984; Mason, 1988). Second, the time estimate of 9000 years also seems robust for these high cols. The zone could not have formed until after the Younger Dryas. Significantly, it would require a dispersal of around 120 meters per generation to produce a 20-km cline in 9000 years. An odd long distance migrant may travel this far, but for the whole population to show a standard deviation of parent offspring distances four times larger than measured is a paradox. Third, the spatial structure of the grasshopper population, as determined by the environment, may influence dispersal and gene flow, which is a more complex consideration.

We have approached the role of the environment through an analysis of grasshopper habitat at Portalet using a series of 4×4 meter quadrats to sample the pasture vegetation. Distinct vegetation types were identified from a classification analysis (TWINSPAN) of these data (Virdee and Hewitt, 1990). It allows us to correlate grasshopper density with plant species assemblages and thereby identify and rank the types of habitat where *Chorthippus parallelus* occurs. In general, the highest grasshopper densities are found in moist flushes and grassy patches near streams that support marsh marigold, self heal, and sedges. The grasshopper can be found regularly up to about 2100 meters, and suitable habitat is found right through the col from Sallent de Gallego to Gabas and beyond. However, the Spanish side of the mountains is drier, which becomes noticeable as one descends from the col toward Sallent de Gallego. The French side becomes much more wooded as one descends, and here grasshoppers are found only in large clearings with suitable vegetation. Now the spatial scale of these flushes, streams, and meadows is relatively large compared with the dispersal distance we have measured. Consequently, migration and gene flow probably occur along and through this irregular network and patchwork of streams and meadows, having consequences for the rate of gene flow through the col (Endler, 1977).

In order to gain a better overall picture of the habitat distribution across the full 30 km width of the zonal transition through the Pyrenees, we have begun processing satellite images from LANDSAT to discriminate suitable vegetation for *C. parallelus* (Virdee, 1991). It involves using sites on the satellite image where the vegetation cover is known from our field work to identify the wavelength compositions that correlate with these different habitats. It is a good predictor of grasshopper density and clearly has many other academic and applied possibilities. Computer modeling can help quantify flow in such vegetational patchworks as they are determined (Endler, 1977).

The particular advantage of the satellite imaging is that large tracts of mountains can be assessed that would be a superhuman task on foot.

For the present we can only guess at just how fast gene flow may be, but a series of spatial bottlenecks would be expected to reduce the current gene flow compared with a more continuous distribution of favorable habitat. The history of pasturage in the high cols is ancient, and the northern slopes such as Val d'Ossau are still heavily forested despite considerable medieval clearing. It seems unlikely, therefore, that overall dispersal and migration is greater than our estimate for thousands of years, although it may well have been less before clearing.

PIONEERS OR PHALANXES

One time when a combination of vegetational conditions and dispersal could have produced a zone such as that observed, which is wider than expected, was when the cols were first colonized by grasshoppers following the retreat of the ice. It seems unlikely that the postglacial hybridization was a simple collision of two dense advancing waves of grasshoppers. More likely, the progress of each subspecies up the mountains was a series of advances and retreats—similar to glacier ends today—with small pockets of pioneers and refugees. When the col itself was finally invaded, it is likely to have been by longer-distance migrants who set up founder colonies that spread to fill the virgin pastures.

We have modeled different types of advance and mixing by computer simulation (Nichols and Hewitt, 1993). Let us first consider individuals carrying different neutral alleles meeting as either two fully stocked advancing fronts or through long-range migrants. One set of simulations used an area of 80×40 demes; a certain proportion of each deme dispersed (e.g., $\frac{1}{20}$) and had a probability of moving distance x proportionate to $1/\sqrt{x}$. The numbers in the next generation in each deme were determined by a logistic equation $[N_{t+1} = N_t + r(K - N_t)/K]$, where the rate of increase was r = 0.9 and the carrying capacity was K = 20 individuals (Fig. 6-7). When the two advancing genomes meet on a broad front as relatively well established populations, a steep cline forms and slowly spreads. When the two genomes are initially some distance apart and make contact through their long-distance dispersers, however, a much wider cline is established that continues to spread slowly once the population numbers reach carrying capacity. What happens in this latter case is that long-distance migrants set up a colony in unoccupied territory that then grows and spreads until it meets other pioneer colonies and the space is filled. These pioneers come from both genomes, and so homozygous patches of both alleles are interspersed over a wider area in a sort of temporary mosaic.

Obviously one can envisage a range of scenarios involving different inherent forms of insect dispersal and different rates and forms of vegetation advance that effectively modify dispersal and mixing. The general point is that much wider clines can be established for neutral alleles by the sort of colonization processes described than by simple contact and diffusion. Most probably the clines for *C. parallelus* were produced under such conditions, and such conditions may well explain the widths of hybrid zones for other species that seem rather wide.

If the two alleles making secondary contact produce heterozygote disadvantage, the results are somewhat different. Initially, the mixing is similar to the neutral alleles

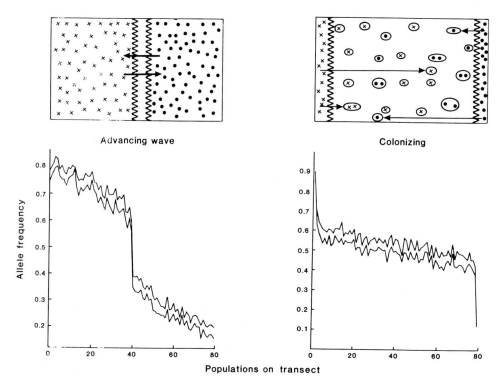

Figure 6-7. Two types of secondary contact and their consequences for zone width. In the first, contact is made by two advancing waves containing different alleles; in the second, colonizing migrants set up localized interspersed populations of each allele. After 1000 generations the cline produced by advancing waves is stepped and much narrower than that produced by colonizers. The clines drawn are each the maximum and minimum of 10 replicate simulations (Nichols and Hewitt, 1993).

model, with a broader cline produced by pioneer migrants. However, as the population numbers increase and migrants find themselves in foreign territory, producing unfit hybrid progeny, the tension zone crystallizes out to its equilibrium dimensions. The latter are determined by a balance between recent selection and dispersal, regardless of the initial establishment 9000 generations ago. An exception is where dispersal is low (Nm < 1), because colonies of one or other homozygote that were set up by the initial invasion can persist as patches behind their own tension zones in the face of weak dispersal. In detail it would be a mosaic of the two forms where the overall allele frequency formed a broad hybrid zone. We have some evidence for such an effect caused by low density in the alpine grasshopper *Podisma pedestris* (Nichols and Hewitt, 1986), and it may play a part in mosaic zones such as those for *Gryllus firmus* and *G. pennsylvanicus* (Harrison and Rand, 1989). However, in the latter case habitat suitability and patchiness seem to be important in determining the mosaic, and it is important to distinguish the two causes.

In the foregoing discussion a number of factors have been identified that could produce a broader hybrid zone than expected on a simple model, i.e., reinforcement, patchy distribution, and interspersed colonization. First, it seems unlikely that reinforcement is involved here; there is a narrow simple cline for preference and no evidence for its bimodality in populations in the zone (Ritchie and Hewitt, 1991). Second, although the environment is patchy, it could not be described as a mosaic zone because the clines are fairly smooth and there is no sign of differential habitat preference. Indeed, the patchy distribution of suitable vegetation may explain why many of the clines at Quillane are much narrower than at Portalet; there are areas of dense forest across Col de la Quillane, so grasshoppers are limited to occasional small pockets in the center of the zone, whereas the Col de Portalet is treeless. The density trough caused by this vegetation at Quillane with its reduction of dispersal could produce a narrower hybrid zone. Third, there is secondary contact with interspersed colonization, which certainly is a possibility. If it is to explain the wider zone at Portalet, it implies a rather different postglacial history for the two cols.

POSTGLACIAL PALYNOLOGY

Information on the formation of this zone following the Ice Age comes from a number of sources, including physical and biological paleontology and archeology, as well as deduction from our understanding of its present structure. Its position in the high cols along the crest of the range is strong evidence for each subspecies advancing up its side of the mountains as the ice cap retreated, finally meeting as suitable vegetation reached into the cols from both sides. The analysis of pollen from cores taken in the Pyrenean region allows reconstruction and dating of this vegetation expansion, and fortunately there has been some excellent work in both the Quillane and Portalet areas (Jalut, 1974; Jalut et al., 1988; Jalut and Vernet, 1989; Reille, 1990a,b, pers. comm.). The core from the peat bog of La Borde is particularly useful, as it comes from an altitude of 1660 meters about 4 km southwest of Col de la Quillane (1713 meters) (Fig. 6-8). It shows a marked reversal and fall in temperature around 10,000 BP, with an increase in steppe plants such as *Artemisia* and Poaceae and a sharp reduction in *Betula* and *Pinus.* A number of other cores in the region also record this cold spell, providing firm evidence for the Younger Dryas period. Accelerator carbon measurements are in hand to date these events more precisely, but it is not likely that *C. parallelus* was able to inhabit this region until after this period when the climate warmed markedly in the Preboreal. Hence the date of estimated contact is 9000 BP.

There has been some debate as to the existence of the Younger Dryas in southern Europe, with some arguing that it was a northern phenomenon. Data such as that of Reille (1990a,b) show that it occurred in the high Pyrenees, and recent cores in the Western Pyrenees from much lower altitude indicate that it occurred there also (Reille, pers. comm.). Sites near Olat (Girona) at 440 meters altitude near the eastern end of the Pyrenees do not provide a clear sequence through this period, but there is some evidence for a rise in *Artemisia* and Poaceae in some cores (Perez-Obiol, 1988). Further south in Spain from Padul near Granada (785 meters), a detailed analysis shows that the 13,000 BP amelioration was pronounced, and the 10,000 BP Younger Dryas pollen spectra indicate dryness with perhaps a less strong cold reversal (Pons and Reille, 1988).

All three of these regions are relevant to the paleobiogeography of *C. parallelus.* The regions near Granada were probably a refuge for the species in the Ice Age. The

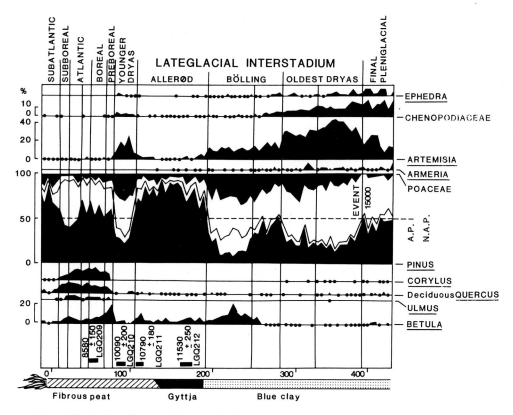

Figure 6-8. Pollen diagram of the relative frequencies of major species found in a core from La Borde *(Pyrénées orientales)* at 1660 meters close to Col de la Quillane. The cold reversal of the Younger Dryas around 10,500 BP is clearly evident. (Simplified from Reille, 1990a. With permission and gratitude.)

course of the zone in the eastern and western ends of the Pyrenees appears somewhat contorted on present data. At the eastern end of the Pyrenees the zone bends back southwest at the head of the Tet Valley (Fig. 6-1) and runs into Spain across the Puerto de Tosas and along the Sierra del Cadi. We are not yet sure where it ends. This distribution suggests that *C. p. parallelus* reached the eastern foothills first, advancing from France to the northeast, and then advanced into Spain until it met *C. p. erythropus* coming the other way. At the western end in the Basque country our collections suggest that the zone may be wider and run farther north of the divide in places. Perhaps *erythropus* reached this end of the Pyrenees first and advanced into France. In fact, this pattern would fit with a southwestern Iberian refugium during the Ice Age, as suggested by pollen evidence of deciduous trees such as *Quercus*. Obviously we need more collections of grasshoppers and more palynology in these regions.

As well as the large reversal of the Younger Dryas, the ameliorating climate had smaller oscillations of perhaps 200 years and 20 years, and subfossil beetle remains are better indicators of rapid changes (Coope, 1977; Atkinson et al., 1987). It seems likely that the expansion of the vegetation and the grasshoppers up to the cols involved advances and retreats. Such "touch and go" at the col would have enhanced the mixing of pioneer populations of both subspecies over the col and helped broaden the cline. Also we should not forget that at this time humans were active hunting deer around the Pyrenees and up the valleys as the ice retreated (Bahn, 1985). There are many artifacts to witness this occurrence, including cave paintings and carved bone; and humans may have crossed the cols before the grasshoppers. There is also an ancient tradition of seasonal herding over the Pyrenees—transhumance—and there is cereal pollen recorded from near Col de la Quillane around 6400 BP. Then, as now, they may have unwittingly carried grasshoppers.

PLEISTOCENE PERSPECTIVE

It is pertinent to ask where *C. parallelus* might have been during the Pleniglacial around 18,000 BP, and how it reached its present distribution. Refugia for deciduous trees, indicative of a climate and vegetation suitable for these grasshoppers, are thought to have survived in southwestern Iberia, southern Italy, and the Balkans (Huntley, 1988). They may each have contained a different form of *C. parallelus. Erythropus* spread to fill Iberia, but *parallelus* could have expanded from southern Italy or the Balkans. In a closely related species group containing *Chorthippus biguttulus, C. brunneus,* and *C. mollis,* work on their songs (Ragge, 1987) shows that there are different song types (= species?) in the three regions, with a complex geographic distribution. This species group is more dissected than was deduced from the morphological descriptions, and *C. parallelus* may prove to be in the same situation.

Such expansions of range have some interesting genetic properties (Hewitt, 1989). First, as the species spreads into new, favorable territory, those populations at the leading edge contribute most pioneer migrants and their genomes set up the new colonies. The main body of the species, once established, largely stays where it is. If a mutant arises at the leading edge, it has a greater chance of fixation because of genetic drift under these founding conditions. A negatively heterotic mutant could establish a territory of homozygotes ahead of the main body. When the two variants met as population sizes increased, a hybrid zone could form. The hybrid zone would effectively prevent the parental genome of the main body from expanding further, whereas the new mutant could continue to do so. This population would spread until another mutation at the leading edge did likewise. There are a number of parapatrically subdivided species in which such a process could have played a part, including the chromosomal taxa of the shrew *Sorex* (Searle, 1988), the mouse *Mus* (Corti et al., 1986), and the grasshoppers *Vandiemenella* and *Warramaba* (White, 1978). Of course other processes may well have been important in these cases. Expansion from the leading edge may also tend to produce overall homozygosity because of genetic drift, and there is some evidence of less protein heterozygosity in species from higher latitudes that expanded there after the Ice Age (Sage and Wolf, 1986). Furthermore, the climate was oscillating somewhat during this postglacial warming, which could strengthen the founder effect in the leading edge of the advance. We need more information on the genetic makeup of species across their range in these regions with which to address these possibilities.

For species such as *Chorthippus parallelus* that have spread north during the postglacial period, the old refugial areas are now too warm. In Spain the species is found only on the high Sierras, and similarly in Italy and the Balkans. When the climate reverses with the next Ice Age, it is these southerly isolates that will expand and advance south into the refugia. The northern population and genomes will probably become extinct. During the Pleistocene generally temperatures have been lower than now, with Ice Age conditions prevailing and only occasional warm interstadia (Kukla, 1981; Gribbin, 1989; Guiot et al., 1989). The last time it was as warm as now was 120,000 BP, and only a few times before that over the last 2 million years. Consequently, *Chorthippus parallelus* had a southerly distribution for most of the Pleistocene and made occasional forays north to form hybrid zones. This process of repeated range expansion and contraction could well have accumulated genetic differences to produce stronger hybrid zones each time (Hewitt, 1989). Such scenarios probably apply to many organisms in these latitudes and may well be important in genome divergence and speciation.

ACKNOWLEDGMENTS

It is a particular pleasure to acknowledge the help and friendship of my several collaborators on this *Chorthippus* hybrid zone: Roger Butlin, Colin Ferris, Richard Nichols, Michael Ritchie, and Sonia Virdee from Britain, and Jose Bella, Jaime Gosalvez, Carmen Lopez-Fernandez, and Jose-Miguel Rubio from Spain. I am also grateful to the British Council (Acciones Integradas), EEC, NERC, and SERC for funding various parts of it. Without these funds this study would not have been possible.

REFERENCES

Atkinson, T. C., Briffa, K. R., and Coope, G. R. 1987. Seasonal temperatures in Britain during the past 22,000 years, reconstructed using beetle remains. Nature 325:587–592.

Bahn, P. G. 1985. Pyrenean Prehistory. Westminster, England: Aris & Phillips.

Barton, N. H., and Hewitt, G. M. 1981. Hybrid zones and speciation. In W. R. Atchley and D. S. Woodruff, eds. Evolution and Speciation: Essays in honour of M. J. D. White. Cambridge: Cambridge University Press, pp. 341–359.

Barton, N. H., and Hewitt, G. M. 1982. A measurement of dispersal in the grasshopper *Podisma pedestris*. Evolution 35:1008–1018.

Barton, N. H., and Hewitt, G. M. 1985. Analysis of hybrid zones. Ann. Rev. Ecol. Syst. 16:113–148.

Barton, N. H., and Hewitt, G. M. 1989. Adaptation, speciation and hybrid zones. Nature 341:497–503.

Bella, J. L., Butlin, R. K., Ferris, C., and Hewitt, G. M. 1992. Asymmetrical homogamy and unequal sex ratio from reciprocal mating order crosses between *Chorthippus parallelus* subspecies. Heredity 68:345–352.

Butlin, R. K. 1985. Speciation by reinforcement. In J. Gosalvez, C. Lopez-Fernandez, and C. Garcia de la Vega, eds. Orthoptera 1. Madrid: Fundacion Ramon Areces, pp. 84–113.

Butlin, R. K. 1987. Speciation by reinforcement. Trends Ecol. Evol. 2:8–13.

Butlin, R. K. 1989. Reinforcement of premating isolation. In D. Otte and J. A. Endler, eds. Speciation and Its Consequences. Sunderland, MA: Sinauer, pp. 158–179.

Butlin, R. K., and Hewitt, G. M. 1985a. A hybrid zone between *Chorthippus parallelus parallelus* and *Chorthippus parallelus erythropus* (Orthoptera: Acrididae): morphological and electrophoretic characters. Biol. J. Linn. Soc. 26:269–285.

Butlin, R. K., and Hewitt, G. M. 1985b. A hybrid zone between *Chorthippus parallelus parallelus* and *Chorthippus parallelus erythropus* (Orthoptera: Acrididae): behavioural characters. Biol. J. Linn. Soc. 26:287–299.

Butlin, R. K., and Hewitt, G. M. 1988. Genetics of behavioural and morphological differences between parapatric subspecies of *Chorthippus parallelus* (Orthoptera: Acrididae). Biol. J. Linn. Soc. 33:233–248.

Butlin, R. K., and Ritchie, M. G. (1991). Variation in female mate preference across a grasshopper hybrid zone. J. Evol. Biol. 4:227–240.

Butlin, R. K., Ritchie, M. G., and Hewitt, G. M. 1991. Comparisons among morphological characters and between localities in the *Chorthippus parallelus* hybrid zone (Orthoptera: Acrididae). Phil. Trans. Roy. Soc. Lond. B 334:297–308.

Coope, G. R. 1977. Fossil coleopteran assemblages as sensitive indicators of climatic change during the Devensian (last) cold stage. Phil. Trans. Roy. Soc. Lond. B 280:313–340.

Corti, M., Capanna, E., and Estabrook, G. F. 1986. Micro-evolutionary sequences in house mouse chromosomal speciation. Syst. Zool. 35:163–175.

Dagley, J. R. 1988. Population differentiation in the grasshopper *Chorthippus parallelus* (Orthoptera: Acrididae): a study of the mate recognition system. Ph.D. thesis, University of East Anglia, Norwich.

Diehl, S. R., and Bush, G. L. 1989. The role of habitat preference in adaptation and speciation. In D. Otte and J. A. Endler, eds. Speciation and Its Consequences. Sunderland, MA: Sinauer, pp. 345–365.

Dobzhansky, Th. 1970. Genetics of the Evolutionary Process. New York: Columbia University Press.

Endler, J. A. 1977. Geographic Variation, Speciation and Clines. Princeton, NJ: Princeton University Press.

Ferris, C., Rubio, J. M., Gosalvez, J. and Hewitt, G. M. 1993. One way introgression of subspecific sex chromosome marker in a hybrid zone. Heredity (submitted).

Gosalvez, J., Lopez-Fernandez, C., Bella, J. L., Butlin, R. K., and Hewitt, G. M. 1988. A hybrid zone between *Chorthippus parallelus parallelus* and *Chorthippus parallelus erythropus* (Orthoptera: Acrididae): chromosomal differentiation. Genome 30:656–663.

Gribbin, J. 1989. The end of the ice ages? New Scientist 17 June: 48–52.

Guiot, J., Pons, A., Beaulieu, J. L., and Reille, M. 1989. A 140,000 year continental climate reconstruction from two European pollen records. Nature 338:309–313.

Gyllensten, U., and Wilson, A. C. 1987. Interspecific mitochondrial DNA transfer and the colonization of Scandinavia by mice. Genet. Res. Camb. 49:25–29.

Harrison, R. G., and Rand, D. M. 1989. Mosaic hybrid zones and the nature of species boundaries. In D. Otte and J. A. Endler, eds. Speciation and Its Consequences. Sunderland, MA: Sinauer, pp. 111–133.

Hewitt, G. M. 1979. Animal Cytogenetics III. Orthoptera. Stuttgart: Gebrüder Borntraeger.

Hewitt, G. M. 1985. The structure and maintenance of hybrid zones—with some lessons to be learned from alpine grasshoppers. In J. Gosalvez, C. Lopez-Fernandez, and C. Garcia de la Vega, eds. Orthoptera I. Madrid: Fundacion Ramon Areces, pp. 15–54.

Hewitt, G. M. 1988. Hybrid zones—natural laboratories for evolutionary studies. Trends Ecol. Evol. 3:158–167.

Hewitt, G. M. 1989. The subdivision of species by hybrid zones. In D. Otte and J. A. Endler, eds. Speciation and Its Consequences. Sunderland, MA: Sinauer, pp. 85–110.

Hewitt, G. M. 1990. Divergence and speciation

as viewed from an insect hybrid zone. Can. J. Zool. 68:1701–1715.

Hewitt, G. M., Butlin, R. K., and East, T. M. 1987. Testicular dysfunction in hybrids between parapatric subspecies of the grasshopper *Chorthippus parallelus.* Biol. J. Linn. Soc. 31:25–34.

Hewitt, G. M., Nichols, R. A., and Barton, N. H. 1987. Homogamy in a hybrid zone in the alpine grasshopper *Podisma pedestris.* Heredity 59:457–466.

Huntley, B. 1988. Glacial and Holocene vegetation history: Europe. In B. Huntley and T. Webb III, eds. Vegetation History. Dordrecht: Kluwer, pp. 341–383.

Huntley, B., and Birks, H. J. B. 1983. An Atlas of Past and Present Pollen Maps for Europe. Cambridge: Cambridge University Press.

Jalut, G. 1974. Evolution de la végétation et variations climatiques, durant les quinze derniers millénaires dans l'extrémité orientale des Pyrénées. Ph.D. thesis, University of Toulouse.

Jalut, G., Andrieu, V., Delibrias, G., Fontugne, M., and Pages, P. 1988. Palaeoenvironment of the Valley of Ossau (Western French Pyrenees) during the last 27,000 years. Pollen Spores 30:357–394.

Jalut, G., and Vernet, J. L. 1989. La végétation du pays de Sault et de ses marges depuis 15,000 ans: réinterprétation des données palynologiques et apports de l'anthracologie. In Pays de Sault: Espaces, Peuplement Populations. Paris: Editions du CNRS, pp. 23–41.

Key, K. H. L. 1968. The concept of stasipatric speciation. Syst. Zool. 17:14–22.

Key, K. H. L. 1981. Species, parapatry and the morabine grasshoppers. Syst. Zool. 30:425–458.

Kukla, G. J. 1981. Pleistocene climates on land. In A. Berger, ed. Climatic Variations and Variability: Facts and Theories (NATO-ASI, Series C, Vol. 72). London: D. Reidel P. C., pp. 207–232.

Mason, P. 1988. Reproductive isolation between the races of the grasshopper *Podisma pedestris.* Ph.D. thesis, University of East Anglia, Norwich.

Marchant, A. D., Arnold, M. L., and Wilkinson, P. 1988. Gene flow across a chromosomal tension zone. 1. Relics of ancient hybridization. Heredity 61:321–328.

Mayr, E. 1963. Animal Species and Evolution. Cambridge: Belknap Press of Harvard University Press.

Moore, J. A. 1957. An embryologists' view of the species concept. In E. Mayr, ed. The Species Problem. Washington, DC: AAAS, pp. 325–338.

Nichols, R. A. 1984. The ecological genetics of a hybrid zone in the alpine grasshopper *Podisma pedestris.* Ph.D. thesis, University of East Anglia, Norwich.

Nichols, R. A., and Hewitt, G. M. 1986. Population structure and the shape of a chromosomal cline between two races of *Podisma pedestris* (Orthoptera: Acrididae). Biol. J. Linn. Soc. 29:301–316.

Nichols, R. A., and Hewitt, G. M. 1993. The genetic consequences of long distance dispersal during colonization. Heredity (submitted).

Paterson, H. E. H. 1978. More evidence against speciation by reinforcement. S. Afr. J. Sci. 74:369–371.

Paterson, H. E. H. 1985. The recognition concept of species. In E. S. Vrba, ed. Species and Speciation. Transvaal Museum Monograph No. 4. Pretoria: Transvaal Museum, pp. 21–29.

Perez-Obiol, R. 1988. Histoire tardiglaciare et holocène de la végétation de la région volcanique d'Olot (N. E. Péninsule Ibérique). Pollen Spores 30:189–202.

Pons, A., and Reille, M. 1988. The Holocéne and Upper Pleistocene pollen record from Padul (Granada, Spain): a new study. Paleogeogr. Palaeoclimatol. Palaeoecol. 66:243–263.

Ragge, D. 1987. Speciation and biogeography of some southern European Orthoptera, as revealed by their songs. In B. Baccetti, ed. Evolutionary Biology of Orthopteroid Insects. Chichester: Ellis Horwood, pp. 418–426.

Reille, M. 1990a. La tourbière de La Borde (Pyrénées orientales, France): un site clé pour l'étude du Tardiglaciaire sud-européen. C. R. Acad. Sci. Paris 310:823–829.

Reille, M. 1990b. Leçons de Palynologie et d'Analyse Pollinique. Paris: Editions du CNRS.

Reynolds, W. J. 1980. A re-examination of the characters separating *Chorthippus montanus* and *C. parallelus* (Orthoptera: Acrididae). J. Nat. Hist. 14:283–303.

Rice, W. R., and Salt, G. W. 1990. The evolution of reproductive isolation as a correlated character under sympatric conditions: experimental evidence. Evolution 44:1140–1152.

Ritchie, M. G. 1988. A Pyrenean hybrid zone in the grasshopper *Chorthippus parallelus* (Orthoptera: Acrididae): descriptive and evolutionary studies. Ph.D. thesis, University of East Anglia, Norwich.

Ritchie, M. G. 1990. Does song contribute to assortative mating between subspecies of *Chorthippus parallelus* (Orthoptera: Acrididae)? Anim. Behav. 39:685–691.

Ritchie, M. G., Butlin, R. K., and Hewitt, G. M. 1989. Assortative mating across a hybrid zone in *Chorthippus parallelus* (Orthoptera: Acrididae). J. Evol. Biol. 2:339–352.

Ritchie, M. G., and Hewitt, G. M. 1993 (in press). Outcomes of negative heterosis. In J. Masters and H. Spencer, eds. Mate Recognition and Speciation. Cambridge: Cambridge University Press.

Ritossa, F. M., and Scala, G. 1969. Equilibrium variations in the redundancy of rDNA in *Drosophila melanogaster.* Genetics 61 (suppl.):305–317.

Sage, R. D., and Wolff, J. O. 1986. Pleistocene glaciation, fluctuating ranges, and low genetic variability in a large mammal *(Ovis dalli).* Evolution 40:1092–1095.

Sanderson, N. 1989. Can gene flow prevent reinforcement? Evolution 43:1223–1235.

Searle, J. B. 1986. Factors responsible for a karyotypic polymorphism in the common shrew, *Sorex araneus.* Proc. R. Soc. Lond. Biol. 229:277–298.

Searle, J. B. 1988. Karyotypic variation and evolution in the common shrew, *Sorex araneus.* In P. E. Brandham, ed., Kew Chromosome Conference III. Norwich: HMSO, pp. 97–107.

Serrano, L., Ferris, C., Lopez-Fernandez, C., and Hewitt, G. M (in prep.). Distribution and expression of X-NOR in *Chorthippus parallelus* hybrid zone in the Pyrenees.

Shaw, D. D., Marchant, A. D., Arnold, M. L., and Contreras, N. 1988. Chromosomal rearrangements, ribosomal genes and mitochondrial DNA: contrasting patterns of introgression across a narrow hybrid zone. In P. E. Brandham, ed. Kew Chromosome Conference III. Norwich: HMSO, pp. 121–129.

Spencer, H. G., McArdle, B. H., and Lambert, D. M. 1986. A theoretical investigation of speciation by reinforcement. Am. Naturalist 128:241–262.

Tauber, C. A., and Tauber, M. J. 1989. Sympatric speciation in insects: perception and perspective. In D. Otte and J. A. Endler, eds. Speciation and Its Consequences. Sunderland, MA: Sinauer, pp. 307–344.

Templeton, A. R. 1981. Mechanisms of speciation—a population genetics approach. Ann. Rev. Ecol. Syst. 12:23–48.

Vanlerberghe, F., Boursot, P., Nielsen, J. T., and Bonhomme, F. 1988. A steep cline for mitochondrial DNA in Danish mice. Genet. Res. Camb. 52:185–193.

Virdee, S. R. 1991. Ecological and genetical determinants of a hybrid zone in *Chorthippus parallelus.* Ph.D. thesis, University of East Anglia, Norwich.

Virdee, S. R., and Hewitt, G. M. 1990. Ecological components of a hybrid zone in the grasshopper *Chorthippus parallelus* (Zetterstedt) (Orthoptera: Acrididae). Bol. San. Veg. Plagas 20:299–309.

Wallace, A. R. 1889. Darwinism. London: Macmillan.

White, M. J. D. 1978. Modes of speciation. San Francisco: W. H. Freeman.

Wood, T. K., and Keese, M. C. 1990. Host-plant-induced assortative mating in *Enchenopa* treehoppers. Evolution 44:619–628.

7

Genomic and Environmental Determinants of a Narrow Hybrid Zone: Cause or Coincidence?

DAVID D. SHAW, ADAM D. MARCHANT,
NELIDA CONTRERAS, MICHAEL L. ARNOLD,
FRAN GROETERS, AND BERT C. KOHLMANN

The application of molecular and biochemical analytical techniques to evolutionary problems has revealed that, in most cases, speciation or cladogenesis is associated with cumulative changes to many components of both the nuclear and cytoplasmic genomes. These components include allelic frequencies at single gene loci, multigene families and polygenic systems, the structure, abundance, and distribution of various classes of nontranscribed DNA, and major genomic reorganization by chromosomal rearrangement. The changes are also frequently correlated with varying levels of both premating and postmating reproductive isolation, although any causal relations have yet to be definitively established.

The recognition of numerous examples of taxa that form hybrid zones in nature has provided evolutionary biologists with a unique opportunity to examine in situ the mechanics and dynamics of phyletic divergence and, in particular, to define the genetic determinants of species formation. Thus the principal models of speciation, as put forward by Mayr (1963), Dobzhansky (1970), Carson (1971), and others could now be quantified and appraised for their evolutionary relevance.

The high expectations of hybrid zone analysis to provide answers to the more contentious and speculative aspects of speciation theory have been optimistically reviewed by Hewitt (1988) and Barton and Hewitt (1989). On a more cautionary note, although we can now describe and define, at the finest level, the spatial organization of hybrid zones in terms of the distribution of both cytoplasmic and nuclear DNA components, we still know little of their underlying evolutionary significance in a neo-Darwinian sense. The relevant and important issues highlighted by hybrid zones are their potential to contribute to our understanding of the genetics of adaptation, the evolution of reproductive isolating mechanisms, and the consequences of gene flow between contiguous taxa. It is relatively easy to define species differences in terms of their DNA sequences using molecular analyses, but which of these sequences play or have played a role in adapting these species to their respective environments or in isolating them from their neighbors are questions that await resolution.

If we highlight the genetics of species *formation,* in contrast to species *differences,* neo-Darwinian theory advocates that populations maintain their continuity in evolutionary time by adapting, via their genetic systems, to changing environmental conditions. Thus geographically distinct populations that were derived from the same ancestor but subsequently were subjected to different environmental regimes would be expected to evolve different adaptive norms. A range expansion of two or more such populations may result in an evolutionary confrontation at the interface between the expanding populations. It is implicit in this thesis that the genetic differentiation observed between taxa has arisen primarily by the action of the same genetic forces that operate within a taxon (Lewontin, 1974). Because every component of the genome has been shown to be involved in divergent evolution, the problem of assessing their relative contributions to any selective basis of species formation is a daunting one. Similarly, the role of adaptation by natural selection in this extrapolation from within- to between-population evolution is still far from clear, although it has been emphasized as the "primary determinant of speciation" (Templeton, 1982).

Any hybridization at the zone of overlap between parapatric taxa creates a natural experiment from which we should then be able to "quantify the genetic differences responsible for speciation, to measure the diffusion of genes between divergent taxa, and to understand the spread of alternative adaptations" (Barton and Hewitt, 1989). Unfortunately, it has not proved to be the case, and most analyses of hybrid zones remain descriptive—a major criticism that has plagued speciation studies for years. Similarly, we are still obliged to rely heavily on mathematical models "to measure the genetic basis of reproductive isolation and to understand the spread of different adaptive peaks" (Barton and Hewitt, 1989).

We present a critical review of a hybrid zone between two taxa of the Australian grasshopper *Caledia captiva.* This hybrid zone has been studied from a variety of aspects, including molecular (ribosomal, mitochondrial, and highly repeated DNA sequences) and enzyme electrophoretic analyses of 28 loci, cytology (chromosomal and C-band variation), reproductive compatibility (premating behavior patterns and postmating embryonic viability), life history strategies (variation in development time) and environmental parameters (bioclimate prediction systems).

The chromosomes of these hybridizing taxa are diagnostically differentiated by a series of pericentric rearrangements involving all members of the genome. By comparing the patterns of change of these chromosomal differences across the hybrid zone with those of other cytoplasmic and nuclear markers, we have exposed strikingly different profiles. Chromosomal change from one karyotype to the other occurs abruptly and completely within 1 km, whereas the other markers are found up to 450 km beyond the chromosomal boundary. Past biogeographical and present environmental considerations suggest that the zone has moved and may still be moving, leaving in its wake a series of neutral markers while retaining its sharp chromosomal profile. Biogeographical analysis also suggests that its current location may be environmentally determined.

The chromosomal rearrangements that highlight the hybrid zone are themselves involved in concerted, clinal patterns of variation within one of the taxa. The position of the centromere gradually moves up the chromosome, transforming the northern metacentric genome at the hybrid zone to an acro/telocentric one at its southern limit. This unique gradual transformation of genomic structure from metacentric to acrocentric within an otherwise genetically homogeneous taxon has allowed us to partition the components of postmating reproductive isolation that exist at the hybrid zone. Almost 50% of the reproductive isolation is attributable to the chromosomal differences that distinguish the hybridizing taxa, irrespective of the genic differentiation that has accompanied chromosome evolution.

Such concerted changes to the entire genome are unprecedented and have no conventional genetic or evolutionary explanations for their origin, establishment, and possible function. Even so, the chromosomal differences that distinguish the Moreton and Torresian taxa point to the fact that chromosome evolution—in the form of centromeric movement—in *Caledia captiva* may fulfill roles in two important evolutionary processes: adaptation and speciation. First, the chromosomal rearrangements act as major isolating mechanisms at the hybrid zone between the two chromosomally differentiated taxa. Despite apparent long-term hybridization and introgression, chromosomal organization involving the entire genome has been maintained. Second, the same chromosomal changes that implement strong isolation between taxa may themselves be involved in an adaptive role within a taxon. Our studies have concentrated on these two unique features, and we are attempting to relate the systemic changes in the genome to phenotypic characters along the cline and at the hybrid zone. We have concentrated on life history characters and found clinal patterns of phenotypic variation in the length of development time that correlate strongly with the patterns of genomic change. Any causal relation between genome structure and life history parameters remain to be resolved, although unconventional systems may require unconventional interpretations.

SPECIES COMPLEX AND ITS DISTRIBUTION

The grasshopper *Caledia captiva* was first collected by Joseph Banks in 1770. It is now known to be distributed along the entire eastern and northern seaboards of the Australian continent and in southwestern Papua New Guinea, where it occurs sympatrically with the only other member of the genus, *Caledia species nova 1.* The distribution covers 35° of latitude and includes both tropical and temperate zones that are occupied by open eucalyptus forest and coastal wallum heath characterized by high average moisture indices during the summer season (Bryan, 1973). The species is common across its entire range and is considered to be a tropical species that has gradually moved south into the temperate regions, which have been subjected to considerable fluctuations in vegetation and climate under the influence of global Ice Age effects (Nix and Kalma, 1972; Kershaw, 1974). Within this broad geographic distribution *C. captiva* can be subdivided into several taxa that display varying levels of divergence in their nuclear and cytoplasmic DNA, karyotypic organization, and reproductive isolation.

The two most widely distributed taxa are the Torresian and the Moreton cytotypes (Fig. 7-1). The Torresian taxon is found in Papua New Guinea, the Northern Territory, and along the entire coast of Queensland. The Moreton taxon replaces the Torresian in southeastern (S.E.) Queensland and extends as far south as coastal Vic-

Figure 7-1. Known distributional range of *Caledia*. There are only two species in the genus: *C. captiva* and *C. species nova 1*. *C. captiva* is made up of a series of taxa characterized by distinctive patterns of chromosomal organization. The Daintree taxon is totally reproductively isolated from the remaining taxa of *C. captiva* and is regarded as a sibling species. The Moreton and Torresian taxa are only partially reproductively isolated and are considered to be subspecies.

toria. These two taxa are parapatric in S. E. Queensland where they form a hybrid zone that is over 250 km long but less than 1 km wide (Fig. 7-1; see also Fig. 7-6, below).

LEVELS OF GENETIC DIFFERENTIATION

We have estimated the levels of genetic differentiation in these two taxa and their close relatives using a variety of molecular (mitochondrial DNA and ribosomal DNA spacer region restriction fragment length polymorphisms (RFLPs); highly repeated DNA sequence variation), enzyme electrophoretic (28 loci), and cytological (pericentric rearrangements and C-band variation) marker systems. To determine the taxonomic status of the Torresian and Moreton taxa in relation to other members of the genus, we have examined variation at 28 enzyme loci using standard enzyme electrophoretic techniques (Daly et al., 1981). The observed patterns of variation produce a phylogenetic reconstruction that recognizes four major taxa, which correlate directly with the levels of reproductive isolation that exist between taxa within the genus (Fig. 7-2a) (Shaw et al., 1980). The Moreton and Torresian taxa are separated by a Nei's D of 0.23 with fixed allelic differences at four loci (MPI, PGI, ICD-1, GOT-2).

(a)

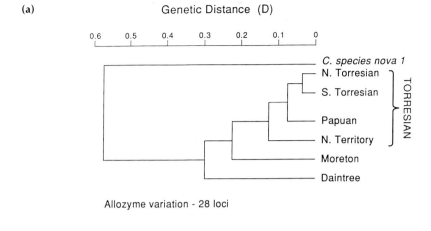

Allozyme variation - 28 loci

(b)

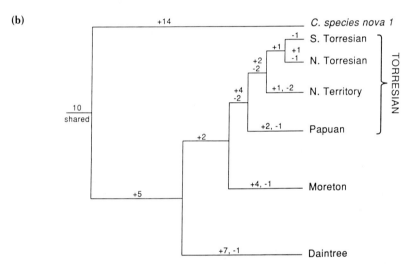

Mitochondrial DNA - 9 restriction enzymes

Figure 7-2. (a) Dendrogam constructed using the mean genetic distance between populations and all known taxa of *Caledia* after analyzing the patterns of variation at 28 allozyme loci. **(b)** Phylogenetic reconstruction of the evolutionary relations between the taxa of *Caledia* using restriction fragment length variation in the mitochondrial genome after digestion with nine restriction enzymes. The entire *C. captiva* mitochondrial genome, cloned as three fragments, was used as a probe. Note the similarity of the reconstructions.

More recently, Marchant and Shaw (1991) have analyzed the RFLPs of their respective mitochondrial genomes using nine restriction enzymes to digest genomic DNA and using the entire *Caledia* mitochondrial genome as a radiolabeled probe. In this case, the phylogenetic reconstruction is similar to that derived from the allozyme analysis with the recognition of four Torresian taxa and a single homogeneous More-ton taxon (Fig. 7-2). Arnold et al. (1987) have also identified a diagnostic difference in the *Cla I* restriction fragment length of the ribosomal DNA spacer region between the Moreton (2.9 kb fragment) and Torresian (2.1 kb fragment) taxa.

The most spectacular difference between the two taxa is seen in the contrasting patterns of chromosome organization within their respective genomes. Both taxa possess the standard acridid karyotype of 22 autosomes and an XX/XO sex chromosome system (Hewitt, 1979), but the structure of these chromosomes is dramatically different. The Torresian genome consists of 24 acrocentric and telocentric chromosomes with only one variant so far recognized. Chromosome 4 of the Northern Torresian is always telocentric, whereas in the remaining Torresian taxa, this chromosome is invariably acrocentric (Fig. 7-3) (Arnold and Shaw, 1985).

In complete contrast, the Moreton genome is highly variable owing to changes in the location of the centromere along each chromosome (Figs. 7-3 and 7-4). Moreover, the observed variation shows a concerted pattern of change along a latitudinal gradi-

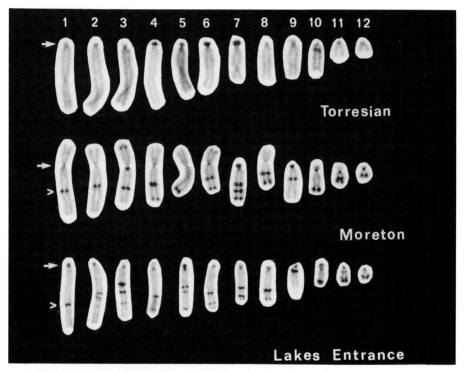

Figure 7-3. The c-banded haploid genomes of the Torresian, Moreton, and Lakes Entrance taxa. The Torresian and Lakes Entrance genomes have their centromeres located at the distal end of each chromosome, whereas in the Morerton chromosomes the centromeres are located more centrally along most of the chromosomes. Despite the major chromosomal differences between them, the Lakes Entrance and Moreton taxa share the same patterns of allozymic, rDNA, mtDNA, and highly repeated DNA profiles, but they differ markedly from those present in the Torresian taxon. Closed arrows = centromere position; open arrows = location of a 168bp highly repeated DNA sequence that is C-band-positive after treating the chromosomes with Ba(OH)$_2$.

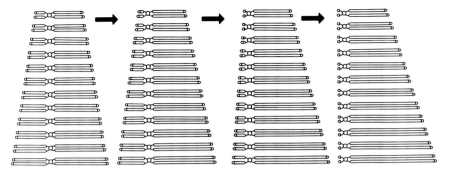

Figure 7-4. Change from a metacentric karyotype seen in northern Moreton populations to be acro/telocentric karyotype of southern Moreton populations. Note that the concerted change in karyotypic structure is gradual and is caused by the movement of the centromere from medial to distal locations along a latitudinal temperature gradient over 1500 km.

ent. Populations of the Moreton taxon in the north and close to the hybrid zone (from Fraser Island, coastal S.E. Queensland and northern New South Wales) (Fig. 7-5) have karyotypes composed predominantly of metacentric chromosomes. As one moves south, the entire karyotype is transformed by the gradual movement of the centromere from medial to distal locations involving all chromosomes (Fig. 7-4) until at its southern limit (Lakes Entrance) the entire genome is now acro/telocentric and its gross morphology closely resembles that of the Torresian taxon (Fig. 7-5). So far, more than 270 different chromosomal morphs have been identified, 21 of which involve the X chromosome (Shaw et al., 1988).

Such clinal patterns of concerted change are unique, and we emphasize the following features. First, the observed changes involve the entire genome. Second, the same kind of directional change is evident among nonhomologous chromosomes. Third, the mechanism for such grand scale transposition may involve the activation/deactivation of latent centromeres distributed along the length of each chromosome. Fourth, there is no evidence of any restriction to gene flow along this genomic cline, with all populations showing homogeneous patterns of allozyme variation, rDNA and mtDNA RFLPs, and the distribution and abundance of highly repeated DNA.

In addition to the spectacular centromeric changes, the Moreton chromosomes carry a series of C-bands that occupy unique positions along each chromosome (Fig. 7-3). Arnold et al. (1986) have shown that these bands are composed of a 168bp repeat sequence of which there are 1.5×10^5 copies per haploid genome. The Torresian genome contains only 3.5×10^3 copies that are restricted to the telomeric regions of autosomes 10, 11, and 12 (Arnold and Shaw, 1985). The characteristic distribution of these C-banded regions and the location of the centromere on each Moreton chromosome not only allows recognition of individual chromosomes during mitosis and meiosis but also permits the chromosomes in hybrid individuals to be classified as recombinant or nonrecombinant (Coates and Shaw, 1982; Shaw et al., 1982). These cytological features have proved to be of considerable value when analyzing the chromosomal structure of the hybrid zone.

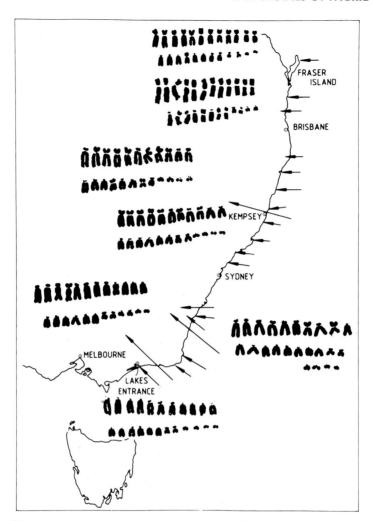

Figure 7-5. Representative karyotypes of the Moreton taxon taken from a variety of sites (arrows) along the southeastern coastal region of Australia. As one moves south, the predominantly metacentric genome at Fraser Island is gradually transformed into an acrocentric one at Lakes Entrance. The transformation involves the gradual movement of the centromere down the chromosome, giving rise to highly complex chromosomal polymorphisms in intermediate populations. To date, more than 270 chromosomal morphs have been identified along the cline.

STRUCTURE OF THE HYBRID ZONE

The Torresian and Moreton taxa are contiguous in S.E. Queensland, where they form a narrow zone of hybridization that is over 250 km long but less than 1 km wide (Fig. 7-6). The region is predominantly open forest that has been systematically cleared during the past 150 years and pasture improved for grazing (Kohlmann-Cuesta, 1988). The grasshopper is abundant and continuously distributed across the region, where it occurs at densities of around 2800 per hectare (Craft, unpublished data). Kohlmann-

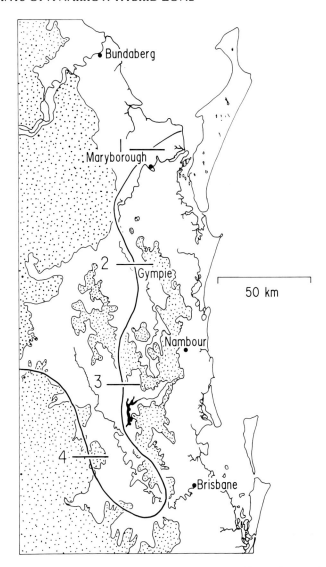

Figure 7-6. Position of the hybrid zone between the Moreton and Torresian taxa in S.E. Queensland. The Torresian taxon occupies the region to the north of the zone, whereas the southern region is Moreton. Transects 1–4 indicate the locations of the four regions used to analyze the 12 climatic variables at the parapatric limit of both taxa. Transect 3 has been studied in detail over several years to determine the detailed structure of the hybrid zone.

Cuesta (1988) has estimated that the average dispersal per generation is 150–200 meters, a figure that accounts for the observed cline widths at the hybrid zone (Barton, 1981). By sampling grasshoppers at 200-meter intervals, Moran (1979) defined the structure of the hybrid zone in terms of its chromosomal characteristics (Fig. 7-7). There is a concordant change in chromosomal type from a Moreton metacentric genome at site TB to an acrocentric Torresian genome at site TA, a distance of 1000

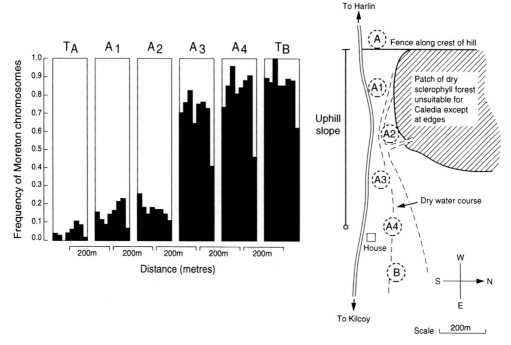

Figure 7-7. Transect across the hybrid zone showing the locations of the sample sites (A, A_1–A_4, and B) taken at 200-meter intervals. The histograms indicate the frequencies of Moreton chromosomes present in these sample sites. Note the abrupt and concerted change of at least 50% between sites A_2 and A_3, a distance of only 200 meters.

meters. A similar steep transition is also seen in the distribution of the 168bp highly repeated DNA sequence that occupies interstitial and terminal locations on all the Moreton chromosomes (Arnold and Shaw, 1985) (Fig. 7-3). Using thees two markers—centromere position and the distribution of 168bp repeat DNA—we can conclude that there is little if any movement of these genomic components across the hybrid zone.

Within the zone itself, there is a steep change in frequency of more than 60% for all the diagnostic chromosomal differences between sites A_2 and A_3, a distance of only 200 meters. Moreover, 85% of all individuals between sites A_1 and A_4 can be classified as hybrids on the basis of their chromosomal characteristics, indicating high levels of intertaxon mating within the hybrid zone. The fact that the hybrid zone involves all members of the genome and each member can be diagnostically distinguished both within and between genomes permits an assessment of linkage disequilibrium to be made between pairs of nonhomologous chromosomes. Moran (1979) identified high levels of disequilibrium between Moreton and Torresian chromosomes, which was asymmetrical and predominated on the Torresian side of the zone at sites A_1 and A_2. Here, highly significant deficits of recombinant chromosomal genotypes implied the operation of strong selection to remove such combinations from the Torresian side of the zone but not on the Moreton side. In complete contrast, a reanalysis of the same populations 6 years later revealed a reversal of this situation, with significant disequilibrium now predominating on the Moreton side of the zone at sites A_3 and A_4 and

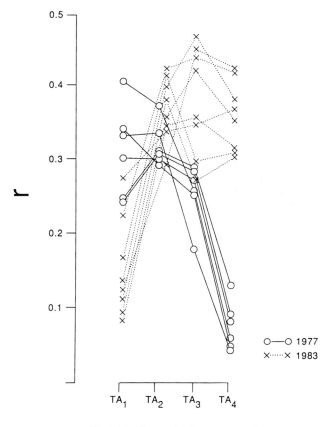

Distribution of "r" averaged
over all chromosomes

Figure 7-8. Distribution of r, the correlation between all pairwise combinations of chromosomes within sample sites across the hybrid zone In 1977 significant disequilibria (Nr^2) were evident in samples A_1 and A_2, whereas in 1983 the situation was reversed, with siginificant values now located at sites A_3 and A_4. This change in the direction of equilibrium coincided with a drought in 1977 and a return to more mesic conditions during 1978–1983. $r = D/[p(1-p)q(1-q)]^{1/2}$.

equilibrium values on the Torresian side (Fig. 7-8) (Shaw et al., 1985). This relatively sudden reversal of disequilibrium values coincided with a major environmental perturbation during 1977 in the form of a drought in S.E. Queensland, which prompted more detailed investigations of the possible role of environmental factors in determining both the structure and location of the hybrid zone.

ENVIRONMENTAL VARIATION ACROSS THE HYBRID ZONE

The bioclimate prediction system (BIOCLIM) has been used to generate bioclimatic envelopes for all of the taxa of *Caledia* (Kohlmann et al., 1988). Climatic estimates for each of 179 collection sites were calculated from mathematical surfaces fitted to data obtained from a network of almost 1000 temperature recording stations and more than 15,000 rainfall recording stations across Australia. Twelve basic bioclimatic indices considered to have biological significance (Nix, 1986) and indicative of mean, seasonal, and extreme values were derived from these estimates. For each of the 12 bioclimatic indices, maximum, minimum, and cumulative frequency distributions at the 5 and 95 percentile levels were calculated (Table 7-1). These profiles constitute the bioclimatic envelopes of the Moreton and Torresian taxa and give a quantitative description of their climatic environment (Kohlmann-Cuesta, 1988). BIOCLIM operates by simple matching of selected thresholds and limits for each of the 12 bioclimatic indices across a grid of points used for predicting potential distribution.

The Torresian and Moreton taxa can be separated into two distinct clusters on the basis of their macroclimatic profiles with a clear, well defined environmental demarcation in the S.E. Queensland region. If a finer scaled analysis is performed on a 0.1° grid using data obtained from four climatic transects that span the Moreton-Torresian contact zone (Fig. 7-6), BIOCLIM predicts a sharply defined environmental boundary. This analysis reveals a marked contrast in rainfall seasonality along a north–south gradient over a distance of approximately 40 km with a strong seasonal component in the north and a more evenly distributed annual rainfall in the south. Two climatic parameters are correlated with the distributional limits of the Moreton and Torresian taxa. The abrupt change to lower annual rainfall associated with high maximum temperatures in the warmest month at the northern and western limits of the zone approaches and sometimes exceeds values that lie at the extreme limit of the climatic envelope of the Moreton taxon. In contrast, the climatic conditions in the region 25–30 km to the south of the hybrid zone still lie within the climatic envelope of the Torresian taxon (Fig. 7-9). The high precipitation values in the driest quarter

Table 7-1. Climate Profile for the Moreton and Torresian Taxa of *C. captiva.*

| Climatic Parameter | Torresian | | | | Moreton | | | |
| | | Percentile | | | | Percentile | | |
	Min	5	95	Max	Min	5	95	Max
Annual mean temperature	17.2	18.9	27.6	27.8	16.9	17.1	21.2	21.2
Minimum temperature coolest month	2.4	4.6	19.4	20.3	3.3	4.1	10.8	10.8
Maximum temperature warmest month	28.6	29.1	37.0	38.0	27.2	27.5	29.8	29.8
Annual temperature range (Max–Min)	11.9	13.2	26.7	29.1	17.9	18.2	23.0	25.3
Mean temperature wettest quarter	22.6	23.5	28.9	30.5	21.6	22.0	24.8	24.8
Mean temperature driest quarter	13.6	14.5	25.4	25.6	12.2	12.6	17.7	19.1
Annual mean precipitation	420	541	2384	3226	732	797	1898	2028
Mean precipitation wettest month	74	88	538	654	104	115	306	329
Mean precipitation driest month	0.6	1.4	37	73	32	32	58	61
Annual precipitation range	52	63	474	580	72	82	252	271
Mean precipitation wettest quarter	218	250	1436	1740	293	326	883	948
Mean precipitation driest quarter	4	7	151	249	108	111	205	215

Source: Data derived from BIOCLIM.
All temperatures in C°.
All precipitations in mm.

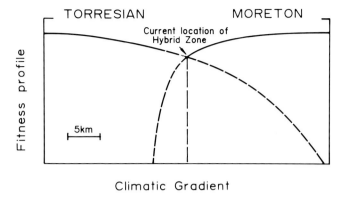

Figure 7-9. Bioclimatic envelopes of the Moreton and Torresian taxa. The northern extreme climatic limit of the Moreton distribution occurs approximately 5 km beyond the present position of the hybrid zone and is determined by the rapid change to low annual rainfall values. The climatic envelope of the Torresian taxon extends 25–30 km into the Moreton region, and its limit is set by the lower temperatures during the wettest quarter (summer).

(winter) and low mean temperatures in the wettest quarter (summer) coincide with the extreme limits of the climatic envelope for this taxon. The fact that the Moreton and Torresian taxa approach extreme values of their known climatic tolerances in a region that includes their parapatric distributional limits suggests that the location of the hybrid zone may represent a dynamic state under the influence of both environmental and genetic factors.

In biological terms, this analysis has revealed abrupt and significant seasonal differences within a narrow region currently occupied by the hybrid zone. The climatic envelopes of the Moreton and Torresian taxa, which reflect ecological tolerances, overlap in this region. Perhaps more significantly, the overlap is asymmetrical (Fig. 7-9), with climatic features capable of sustaining the Torresian taxon located 25–30 km to the south of the current location of the zone. Hence it is reasonable to propose that the zone is not static but may move during periods of climatic change. The asymmetry of the climatic envelopes also raises the possibility that the zone has not yet reached its climatically determined equilibrium and may still be moving in a southeasterly direction. We already know through preliminary work that the phenologies of the two taxa are different in terms of their developmental profiles (see Fig. 7-17, below), diapause requirements, and voltinism, all of which are plausible adaptations to climatic seasonality (Groeters and Shaw, 1991). The evidence to date is mainly inferential, but, as we shall show in the following two sections, such a possibility is supported by two independent lines of evidence.

DISTRIBUTION OF MOLECULAR AND BIOCHEMICAL MARKERS ACROSS THE HYBRID ZONE

In addition to the gross karyotypic differences, the Moreton and Torresian taxa also show diagnostic differences in alleles at four allozyme loci (MPI, GPI, IDH-1, GOT-2) and in restriction fragment length patterns of their ribosomal and mitochondrial

DNA (Arnold et al., 1987; Marchant, 1988). The distribution of these marker systems has been analyzed in samples taken from across the entire range of the Moreton and Torresian taxa and has revealed two very different patterns of distribution across the hybrid zone.

Unlike the steep clines seen for the chromosomal rearrangements and the 168bp repeat sequences, the mtDNA, rDNA, and allozyme markers have been detected in Torresian populations up to 450 km to the north of the hybrid zone (Fig. 7-10). All

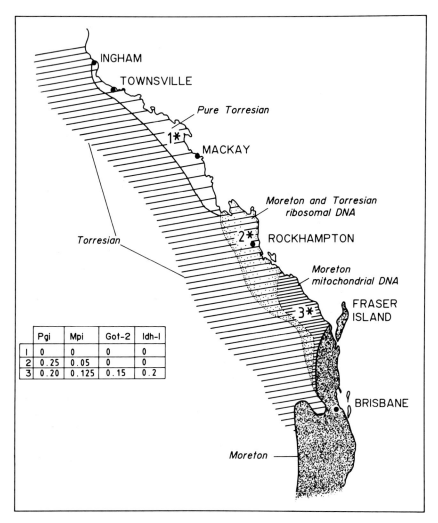

	Pgi	Mpi	Got-2	Idh-I
I	0	0	0	0
2	0.25	0.05	0	0
3	0.20	0.125	0.15	0.2

Figure 7-10. Map of coastal Queensland showing the location of the chromosomal hybrid zone and the Torresian regions that carry Moreton mtDNA and rDNA restriction fragments and the presence of Moreton allozymes (sites 2* and 3*). It is proposed that the hybrid zone may have occupied this region during more mesic conditions in the past and has gradually moved south, leaving behind these essentially neutral markers in Torresian populations while retaining its sharp chromosomal profile.

individuals examined by Marchant et al. (1988) among samples up to 250 km north of the zone were characterized by either of two mitochondrial haplotypes that are of Moreton origin. One of these haplotypes is a feature of all Moreton populations, from the hybrid zone to its southern limit at Lakes Entrance. None of the Torresian mtDNA haplotypes has ever been detected beyond 1 km of the Moreton side of the zone.

The rDNA restriction fragment length analysis reveals a similar pattern to the mtDNA. In southern and central Queensland, up to 450 km to the north of the zone, the Moreton 2.9kb fragment is present in chromosomally Torresian populations at frequencies ranging from 11% to 100% (Fig. 7-10). Similarly, we have detected Moreton alleles at the GPI, MPI, GOT-2, and IDH-1 loci at two Torresian sites located 130 and 350 km to the north of the zone (nos. 2 and 3) (Fig. 7-10). Frequencies of these Moreton alleles range from 5% to 25%; and, like the rDNA and mtDNA, they have never been detected in other, more northern Torresian populations.

There are three possible explanations for the presence of these Moreton markers in Torresian populations. First, they represent shared ancestral polymorphisms that were present prior to the development of the chromosomal differences. Such an explanation seems highly unlikely when one considers that none of these markers is present in any of the more northern Torresian populations in north Queensland, the Northern Territory, or Papua New Guinea, which, on the basis of allozyme and mtDNA analysis, are more closely related to each other than to any other taxon of *Caledia* (Fig. 7-2). Second, the Moreton rDNA, mtDNA, and allozymes have achieved their present distributional limits by selective incorporation into the Torresian taxon following hybridization at the hybrid zone and subsequent introgression up to 450 km into Torresian territory. Such a high incidence of introgression would necessitate the involvement of strong selection processes favoring all three independent marker systems to explain such long range movement. The third possibility is that the current distribution of these Moreton markers reflects a previous occupation of this region by the Moreton taxon. It would then have retreated southward to its current location, leaving behind a series of essentially neutral markers (Fig. 7-11). It is envisaged that the hybrid zone, as defined by its chromosomal features, was previously located in the region close to the current limit of the Moreton rDNA markers, about 450 km to the north of its present position. Implicit in this scenario is the retention of the chromosomal profile of the hybrid zone as it retreated as a narrow chromosomal "front." The current location of the hybrid zone would then represent a finely tuned boundary under environmental control, as was proposed in the previous section. Again, there is evidence from past climatic reconstruction and pollen sequence analysis to show that this region has been subjected to considerable climatic change at least during the past 20,000 years.

PAST CLIMATIC CHANGE IN NORTHEASTERN AUSTRALIA

There are many examples of hybrid zones in temperate regions currently occupying areas that were glaciated in the recent past (Hewitt, 1987; see Ch. 6). Such zones have obviously formed subsequent to climatic amelioration and range expansion. Even though tropical and subtropical regions of the world were not physically influenced by glaciation, they were profoundly modified in terms of their macroclimatic environ-

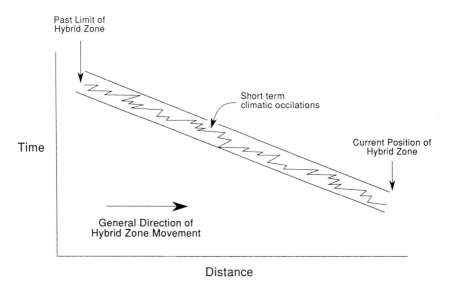

Figure 7-11. Postulated pattern of movement of the hybrid zone that may have occurred during past climatic changes resulting in the transfer of the neutral markers from the Moreton to the Torresian taxon as the zone moved south to its current location.

ment and vegetation. This fact has been clearly revealed in northeastern Australia by pollen sequence analysis (Kershaw, 1974), computerized climatic reconstruction (Nix and Kalma, 1972), and paleoclimatic studies (Chappell et al., 1983).

   If we assume that the current distributional limits of *Caledia* are determined by climatic and environmental factors, as has been proposed by Kohlmann et al. (1988), and that these same factors also defined its distribution in the past, it is evident that the distributional boundaries of *Caledia* would have changed dramatically in eastern Australia at least over the past 20,000 years. *Caledia captiva* is currently restricted in northern and eastern Australia to an open forest habitat that requires a high water table (Fig. 7-12a). The computer reconstructions of climatic change over the past 20,000 years between the equator and the Tropic of Capricorn by Nix and Kalma (1972) clearly imply that the distribution and abundance of this habitat has oscillated dramatically (Fig. 7-12). The open forest habitat showed its maximum distribution about

Figure 7-12. Reconstructed vegetational types associated with paleoclimatic change postulated for northern Australia and Papua, New Guinea, between the equator and the Tropic of Capricorn over the past 20,000 years. (Adapted from Nix and Kalma, 1972). (**a**) Present climate and vegetation, which has changed only slightly over the past 3000 years. Open circles represent the known distribution of *Caledia* in the region. Its distribution is associated with open forest (30–70% foliage cover and high moisture index values for the wettest and driest quarters). (**b**) Distribution of vegetational types 8000 years BP when precipitation was 1.5 times its present value, evaporation remained the same, and air temperatures were 1°C higher than present. The open forest vegetation had a much broader distribution. Under these conditions *Caledia* would have shown a similar expanded distribution.

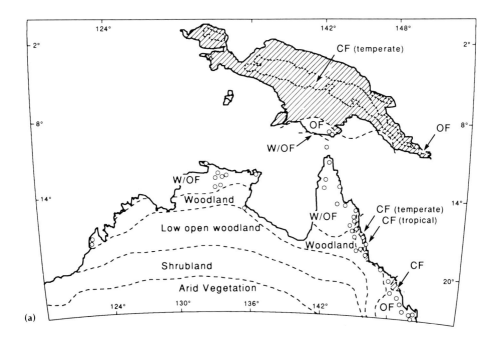

(a)

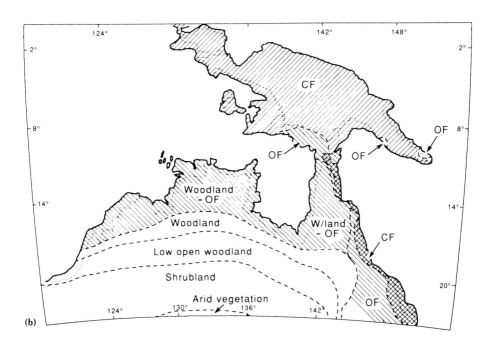

(b)

181

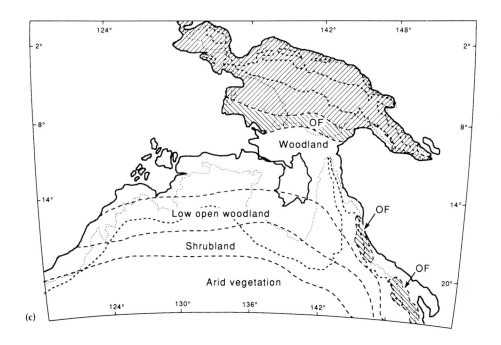

(c)

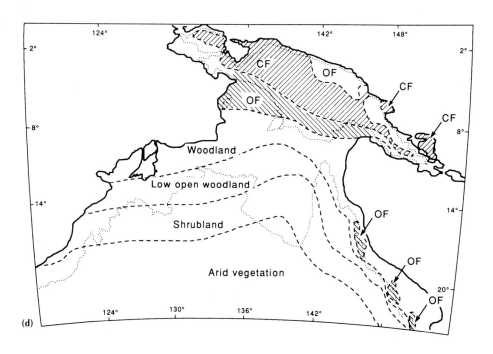

(d)

182

8000 years ago when it is postulated that precipitation levels were 1.5 times higher than present values. Around 17,000–14,000 years BP the climate was dominated by a greatly reduced rainfall pattern with marked seasonal temperature fluctuations giving hot summers and cold winters (Fig. 7-12c). During this period the Moreton and Torresian taxa of *C. captiva* would have been severely restricted to isolated refugia of open forest along the east coast and in southwest Papua.

At around 20,000 years BP, mean air temperatures were greatly reduced and precipitation was reduced to about one-half of its current value (Fig. 7-12d). This situation is thought to have shifted the tropical/temperate floristic disjunction, providing a continuous chain of "thermally suitable islands" for temperate-adapted species, such as the Moreton taxon, along the coastal ranges of northeastern Australia.

Hence it can be seen from this broad overview that our proposal of zone movement along the east coast of Australia is consistent with past climatic perturbations in the region. Although more accurate limits cannot be assigned to these changes, they do reveal that the distribution of *Caledia* must have changed markedly at least in the recent past and probably over much longer time scales. Moreover, as we shall see in a later section, one of the most important factors in determining the distributional limits of the Moreton and Torresian taxa concerns climatic seasonality and developmental profiles.

CHROMOSOMAL VARIATION AND THE EVOLUTIONARY PROCESS

There are two mutually exclusive processes that are generally invoked to explain the persistent existence of chromosomal variation in natural populations. First, homozygous or fixed differences between closely related taxa are assumed to achieve this condition in small demes by stochastic processes (Lande, 1979). They must attain fixation rapidly after the mutation event because of the presumed deleterious effects of the rearrangement as a heterozygote during meiosis (White, 1978). Such a scenario has never attributed any phenotypic effects to the rearrangements either as homo- or heterozygotes. Alternatively, those chromosomal rearrangements that exist as polymorphisms at nontrivial frequencies in natural populations (e.g., the paracentric inversion system in *Drosophila pseudoobscura*) do so because they protect so-called coadapted gene complexes in heterozygotes from breakdown by recombination, an extreme form

---

Figure 7-12. *Continued* (c) At 17,000–14,000 BP, the precipitation rate was only half the present amount and was associated with an increase in evaporation and a slight reduction in temperature. This climatic pattern was associated with increased diurnal and seasonal amplitudes with much lower winter temperatures. The increase in aridity resulted in a greatly reduced open forest vegetation distributed as refugia along the east coast. Similarly, *Caledia* would have been severely restricted in its distribution. (d) Postulated paleoclimate 20,000 years BP was dominated by the global cooling due to glaciation at higher latitudes. Mean air temperatures were 3.5°C lower than present. Open forest habitat was restricted to the edges of the continental shelf. Temperature-adapted species could have expanded north along the coastal region, and it is during climatic changes such as this that the Moreton taxon could have expanded ts range into northern Australia.

of balancing selection (Dobzhansky, 1970). The chromosomal variation we have described in the Moreton taxon is complex, concerted, and genomic. It could not have attained its current status by either stochastic processes or heterozygous advantage for obvious reasons. This point is further endorsed by the presence of the complex chromosomal clines within the Moreton taxon, which transform the northern metacentric genome at the hybrid zone into an acrocentric genome at Lakes Entrance (Fig. 7-5). Such a systemic change to genome structure creates some conceptual problems in terms of the mutational process responsible for this nonrandom pattern of change and the forces responsible for establishing and maintaining these chromosomal clines. The changes to chromosome structure cannot simply be interpreted in terms of pericentric inversions that maintain coadapted genes within inverted segments by balancing selection, which would imply a highly ordered gene arrangement with all the "important" genes localized within the centromeric region of every member of the genome, a highly unlikely and unprecedented phenomenon. The homozygosity for alternative chromosomal forms at each end of the cline also makes the "coadaptation" argument unacceptable. Similarly, in *Caledia* the patterns of the chromosomal rearrangements clearly offer no support to the notion of fixation of the rearrangements by stochastic processes during bottlenecks. Marchant and Shaw (1991) have provided clear evidence of gene flow between all populations along the cline. Moreover, they have shown that, after comparative analyses of mtDNA restriction fragment patterns, the karyotypic restructuring within the Moreton taxon has occurred rapidly relative to the time needed for neutral elements to accumulate mutations.

If we examine in more detail the putative evolutionary role of the chromosomal rearrangements, we must do so from two perspectives: one at the interface between the Moreton and Torresian taxa and their involvement in the long-term structure of the hybrid zone; the other in terms of a possible adaptive role within the Moreton taxon itself. Any adaptive role could, of course, also be operative at the hybrid zone.

Maintenance of the Narrow Hybrid Zone. The most obvious feature of the hybrid zone's structure is the concordant nature of the chromosomal changes across the zone. Most of the genome changes from Torresian to Moreton occurred over a distance of only 200 meters. There is no evidence to indicate the independent assortment and spatial separation of any of the chromosomal rearrangements, giving the impression that each genome is selectively maintained as a coherent unit. We have already presented evidence to indicate significant pairwise interactions between nonhomologous chromosomes in the form of linkage disequilibrium, the direction of which can change with time (Fig. 7-8). We also know from laboratory hybridization experiments that the F_1 generation obtained after crossing the Moreton and Torresian taxa is both fully viable and fertile. When heterozygous, the chromosomal differences are not responsible for any mechanical instability during meiosis, and hence the stasipatric model of chromosomal speciation (White, 1978) does not apply here. However, the F_2 generation is totally inviable, and all backcrosses suffer approximately 50% inviability due to embryonic breakdown (Shaw and Wilkinson, 1980). Shaw et al. (1982) and Coates and Shaw (1982) have analyzed in detail the patterns of recombination in parents and F_1 hybrids and segregation of chromosomes among the backcross survivors. Heterozygosity for pericentric chromosomal rearrangements in *Caledia* is responsible for inducing a major redistribution of recombination along the bivalent with a total inhibition of recombination within the region bounded by the two centromeres (Fig. 7-13), leading to the production of a high frequency of novel recombinant chromosomes among the gametes of F_1 individuals; according to an analysis of backcross progeny, they are generally less fit than nonrecombinant parental chromosomes (Shaw et al., 1982). Furthermore, all of the deficits are caused by intrachromosomal effects with no evidence of the linkage disequilibrium detected in the hybrid zone itself. Thus it is plausible that the redistribution of recombination within chromosomes is the major factor in creating the embryonic breakdown among the F_2 and backcross generations. If true, it implies that some critical internal, *cis*-acting organizational differences exist between the Moreton and Torresian chromosomes that are disrupted by alterations to the normal pattern of recombination, which in turn is responsible for generating a postmating isolating mechanism in the form of embryonic breakdown within the hybrid zone (Shaw et al., 1985). Thus it seems that there are two major effects of the chromosomal rearrangements: (1) an *intrachromosomal effect,* which disrupts the internal organization of the chromosome by novel recombination in heterozygotes; (2) an *interchromosomal effect,* which manifests in the form of linkage disequilibrium within the zone, favoring either a Moreton metacentric or a Torresian acrocentric genome (Fig. 7-8). However, the Moreton and Torresian taxa are also differentiated genically with a Nei's D of 0.23 (Fig. 7-2a). If the mutations at enzyme loci are representative of mutations in the remainder of the genome, our evaluation of the role of the chromosomal rearrangements in generating hybrid breakdown is confounded by the unknown contribution of this genic component. We have already pointed out that at its southern limit (Lakes Entrance) the karyotype of the Moreton taxon is composed

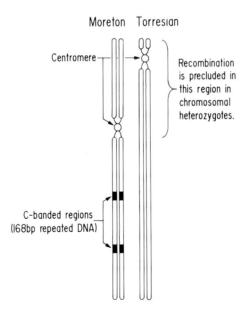

Figure 7-13. Moreton and Torresian chromosomes showing the effects of heterozygosity on recombination during meiosis. Crossing over between the two centromeric regons is precluded, and recombination is pushed to more medial positions along the chromosome, giving rise to novel recombinant chromosomes.

of acro- and telocentric chromosomes and equivalent in its gross morphology to that of the Torresian (Fig. 7-4). Furthermore, Coates and Shaw (1984) have shown that the patterns of recombination in both the Torresian and the Lakes Entrance karyotypes and in their F_1 hybrids are equivalent and do not show the dramatic redistribution seen in metacentric Moreton × Torresian hybrids (Fig. 7-14). Hence by crossing the Torresian to both the northern metacentric and to the southern acro/telocentric karyomorphs, we now have a unique situation that should allow the F_2 and backcross inviability to be partitioned into their genic and chromosomal components (Fig. 7-15). After such an analysis (Table 7-2) it is apparent that approximately 42% of the F_2 and 45–65% of the backcross breakdown can be attributed solely to the recombinational change induced by the chromosomal rearrangements when heterozygous. Thus even a redistribution of recombination among the progeny of chromosomally different but genically equivalent taxa (southern Moreton × northern Moreton) is sufficient to generate high levels of postmating isolation: 42% F_2 inviability. This finding is surprising

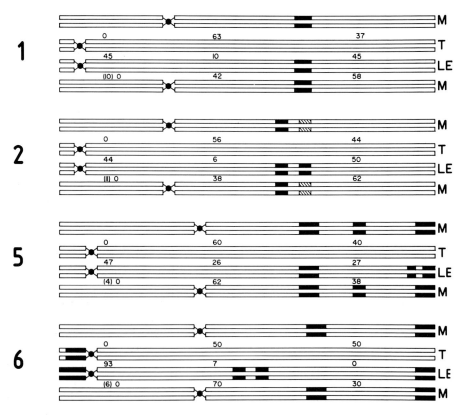

Figure 7-14. Distribution of recombination during meiosis in Moreton × Torresian, Torresian × Lakes Entrance, and Lakes Entrance × Moreton F_1 hybrids. In structurally different chromosome pairs, e.g., T × M and LE × M, recombination is redistributed, whereas the pattern observed in T × LE is the same as that observed in both of the pure parental taxa. The figures indicate the percent of recombination events distributed along the chromosome during diplotene. The figures in brackets represent the frequency of "u" type exchanges occasionally seen in LE × M F_1 hybrids.

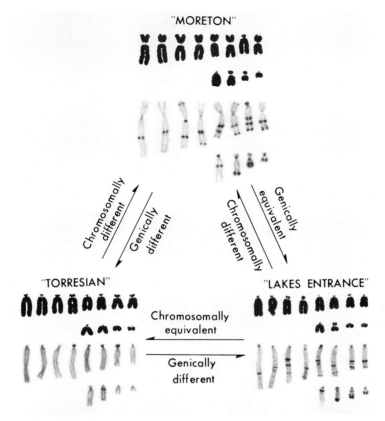

Figure 7-15. Relation between the Torresian, Moreton, and Lakes Entrance taxa that allows postmating reproductive isolation (embryonic inviability) to be separated into its genic and chromosomal components.

because it reveals not only a significant role for the chromosomal rearrangements as isolating mechanisms within the hybrid zone but that the isolation generated is a latent component *within* the Moreton taxon itself.

The evidence for high levels of postmating isolation between the Moreton and Torresian taxa prompted Marchant (1984) to assess mating preference in both allopatric and parapatric populations of these taxa to estimate their levels of sexual isolation. Surprisingly, matings between Moreton and Torresian individuals taken from allopatric sites far from the hybrid zone show significant levels of positive assortative mating, whereas individuals taken 500 meters from either side of the zone are mating at random (Table 7-3). Shaw (unpublished data) obtained a similar result in an independent experiment, and Moran (1979) detected random mating in all samples taken from within the hybrid zone. Thus despite the existence of high levels of sexual isolation between allopatric Moreton and Torresian populations ($I = 0.31$–0.55), there is no evidence to indicate any premating reinforcement of the F_2 and backcross postmating isolation at the hybrid zone. If the zone has moved, however, as we advocate, its movement may have precluded the time needed to refine the reinforcement process.

Table 7-2. The Viabilities of Eggs Laid by Parental, F_1, F_2 and Backcross Generations Involving the Three Chromosomally and Genically Differentiated Taxa. Removal of the Chromosomal Component from the Backcrosses and F_2 Generations Improves Viabilities by 45–65% and 42% Respectively.

Cross	Components	No. of Embryos	Embryonic Viability (%)
Moreton × Moreton (MM)		271	
Torresian × Torresian (TT)		320	>95
Lakes Entrance × Lakes Entrance (LE)		225	
MM × TT		463	
TT × MM		391	
LE × MM		216	>95
LE × TT		198	
TT × LE		217	
(TT × LE) F_1 × TT	Genic	256	69
(TT × MM) F_1 × TT	Genic + chromosomal	169	40
LE × (LE × TT) F_1	Genic	74	70
MM × (MM × TT) F_1	Genic + chromosomal	412	34
(LE × TT) F_1 × TT	Genic	209	69
(MM × TT) F_1 × TT	Genic + chromosomal	502	47
TT × (TT × LE) F_1	Genic	162	61
TT × (TT × MM) F_1	Genic + chromosomal	137	40
(LE × MM) F_1 × (LE × MM) F_1	Chromosomal	113	58
(LE × TT) F_1 × (LE × TT) F_1	Genic	557	46
(MM × TT) F_1 × (MM × TT) F_1	Genic + chromosomal	635	0

We do not know the nature of the mechanism that is directly responsible for creating the postrecombinational breakdown, nor can we use it to explain the concerted nature of the change from one genome to the other across the hybrid zone simply in terms of "within chromosome" recombinational change. There is no *a priori* reason to assume that all chromosomes contribute equally to the observed hybrid breakdown. Hence the observed concordance can be attributed only to either the recent origin of the hybrid zone or to the selective maintenance of alternative genomes on either side of the zone. The long distance penetrance of the zone by such a wide array of other markers (allozymes, mtDNA and rDNA RFLPs) clearly does not support the former. The latter is, however, indirectly supported by the bioclimatic analyses described earlier where it was proposed that the distributions of the Torresian and Moreton taxa are limited by different bioclimatic envelopes and hence may be subjected to different selection regimes. How the structure of the genome could possibly be directly implicated as a selectable character affecting phenotype is unknown. Evidence to warrant the investigation of such an unconventional concept may be found within the Moreton taxon itself.

Concerted Genomic Change and Life History Variation in the Moreton Taxon. The concerted pattern of chromosomal change with latitude in the Moreton taxon is unprecedented and warrants further consideration of both the cause of the genomic changes and the reasons for their subsequent establishment as clines along a latitudinal gradient. The process responsible for altering the location of the centromere could

Table 7-3. The Estimation of Premating Isolation between Allopatric and Parapatric Populations of the Moreton and Torresian Taxa. Equal Numbers of Coded Males and Females of Both Taxa Were Maintained in a Population Cage and the Number of Intra- and Inter-Taxon Matings Was Recorded. I is the Sexual Isolation Coefficient and Ranges in Value from -1.0 (Negative Assortative Mating) to $+1.0$ (Positive Assortative Mating) and 0.0 = Random Mating.

Population Location	Taxonomic Status	Association	No. of Recorded Matings	Sexual Coefficient (I)
Bongmuller $\times$ Miriam Vale	Torresian $\times$ Moreton	Allopatric	58	0.31 (p = .025)
Neara Creek $\times$ M. S. Creek	Torresian $\times$ Moreton	Allopatric	49	0.55 (p = .005)
TA $\times$ $\times$ TA$_4$	Hybrid Zone	Parapatric	56	0.14 (p = ns)
Bongmuller $\times$ Neara Creek	Torresian $\times$ Torresian	Allopatric	41	0.02 (p = ns)

plausibly involve two-break pericentric inversions with the breakpoints asymmetrically disposed on either side of the centromere. Because orthopteran chromosomes cannot be G-banded, we cannot substantiate this possibility. We do know, however, that the expected inversion loops are not evident during meiosis in *Caledia* (Coates and Shaw, 1982). A second possibility is a three-break centric transposition that involves the movement of the centromere with no change to the orientation of the chromosome, as has been found in *Eusimulium aureum* (Rothfels and Mason, 1975). Again, we cannot definitively assess the validity of this process in *Caledia.* The final possibility that deserves consideration concerns the activation and deactivation of latent centromeres distributed along each chromosome. We have acquired evidence to strongly support this possibility. By examining embryonic neuroblast cells in which the chromosomes are exceptionally large, we have discovered that many of the chromosomes carry secondary constrictions at locations that are known to be actively centromeric in homologous chromosomes in other individuals and populations (Fig. 7-16). Thus it is possible that the Moreton chromosomes are multicentric but only one of the centromeres is active during cell division. It is known, in murine and human cultured cell lines that contain stable dicentric chromosomes, that one of the centromeres is preferentially inactivated during cell division (Therman et al., 1986). The mechanism of activation and deactivation is totally unknown but appears to involve mutations that render them nonfunctional. One of the three centromere-associated polypeptides (CENP-C) involved in the attachment of the chromosomes to the spindle during mitosis in humans has been shown to be absent from the inactivated centromeric region (Earnshaw et al., 1989). Such dicentric chromosomes maintain their mitotic stability because they behave as monocentrics during spindle attachment and division. Similar constrictions have not been observed in the other taxa of *Caledia.* The presence of latent centromeres distributed along each of the Moreton chromo-

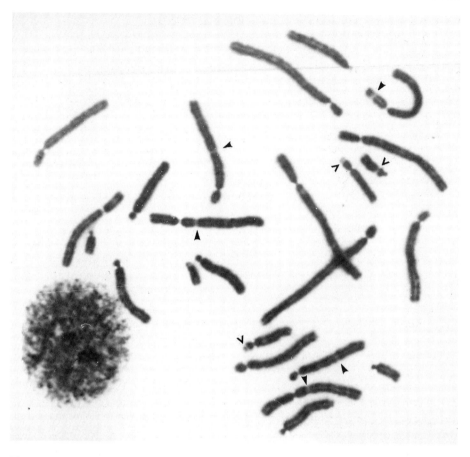

Figure 7-16. Embryonic neuroblast cell from a central New South Wales population showing the presence of at least five secondary constrictions (solid arrowheads). These constrictions may represent inactivated centromeres, implying that the chromosomes in this taxon are multicentric. Hence changes to the centromere position seen along the cline could be caused by changes in centromere activation. It is known that the positions of these inactive centromeres shown in this cell represent active centromeres in other individuals and populations. Open arrowheads indicate the sites of the nucleolar organizing regions as determined by in situ hybridization.

somes is an interesting possibility, but our current knowledge of the molecular structure and function of higher eukaryote centromeres is still minimal. We have as yet no substantive evidence concerning the nature of the centromeric shifts, but the discovery of the secondary constrictions provides us with an avenue for further investigation.

Patterns of clinal variation for inherited characters are generally used to indicate the action of selective forces, although the nature of the selection has rarely been determined (Levitan, 1978; van Delden and Kamping, 1989). Similarly, the chromosomal clines in *Caledia* immediately suggest the involvement of selection acting on the genome. However, the very nature of the patterns of chromosomal variation makes it difficult to establish a relation between chromosomal reorganization and natural selection. What phenotypically variable effect can be caused by gradually changing the shape of the karyotype, and what kind of external selection pressures would operate on such a character? The answers to these questions are not easy to find because of the novel nature of the phenomenon. It is crucial to our thesis, however, that a direct relation between chromosomal architecture, phenotype, and natural selection exists and indeed is a dominant component in the maintenance of both the hybrid zone between chromosomally different taxa and the concerted chromosomal clines within the Moreton taxon. All the evidence so far gathered from several independent sources clearly implicates the chromosomal changes as factors that respond to selection. How and why they respond remains a mystery.

We have commenced an investigation to search for any causal relations between genome structure and phenotypic variation. We already know that seasonality is an important environmental component at the hybrid zone and along the chromosomal cline. We also know that the number of generations per year changes from two in coastal S.E. Queensland to one at both the hybrid zone and at the southern end of the chromosome cline. All coastal populations of the Moreton taxon in Queensland and northern New South Wales regularly undergo two generations a year during which development is highly synchronized. However, as one moves south, both the incidence and abundance of the second, winter generation becomes gradually less frequent until at Lakes Entrance this generation is totally absent. The Lakes Entrance population is thus univoltine, like the Torresian populations in the north with which it shares a similar acrocentric genome. The gradual change from a metacentric to an acro/telocentric along the north-south gradient is accompanied by a gradual decrease in the occurrence and abundance of the winter generation. We have speculated that the metacentric genome of the Moreton taxon provides a selective advantage to its carriers by permitting them to synchronize their development with the winter and summer periods of food abundance. In order to investigate the existence of developmental variation in both the Torresian and Moreton taxa, we have taken samples from across their ranges and estimated the length of development during embryogenesis under controlled laboratory conditions. All Torresian embryos from a range of latitudes showed no differences in their rates of development (Fig. 7-17). In complete contrast, the Moreton samples show a highly significant negative correlation between length of development and latitude with northern populations, taking approximately 10% longer than those at the southern limit. These initial experiments were performed at 33°C over only one-fourth of the entire life history. If this result is extrapolated to the real world situation, it is apparent that the length of the life history could differ by 20–30 days at each end of the cline. This latitudinal relation between genome structure and developmental time is the only correlation we have been able to detect, and it clearly warrants more detailed investigation to reveal the strength of the correlation and if variation in developmental time is a phenotypic expression of chromosome structure.

CONCLUSIONS

The evolutionary status of the hybrid zone between the Torresian and Moreton taxa has been assessed in terms of its broad genetic structure and its relation with environmental and climatic profiles. The comparative genetic analysis of the zone's structure clearly separates the hereditary component into two markedly different patterns: (1) analysis of allozyme variation and ribosomal and mitochondrial DNA restriction frag-

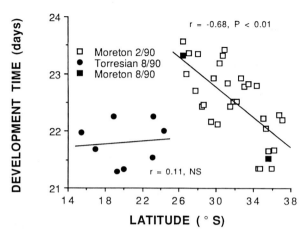

Figure 7-17. Estimation of embryonic development time at 33°C in Torresian and Moreton samples taken from a range of latitudes. Note the similarity in developmental time among Torresian samples from different latitudes, which contrasts markedly with the highly significant negative correlation observed among samples from the chromosomal cline in Southeastern Australia (SEA).

ments reveals an asymmetrical distribution with diagnostic Moreton markers present in Torresian populations up to 450 km north of the hybrid zone; (2) pericentric chromosomal rearrangements and C-band markers do not penetrate beyond 500 meters from the center of the zone, giving rise to the sharp transition from one genome to the other. From these data and analyses of both past and present climatic reconstructions we have argued that the hybrid zone may have moved to its current location, leaving behind a series of essentially neutral markers while retaining its sharp chromosomal profile. Despite the possibility that the zone could exist simply because of conservative dispersal rates coupled with F_2 and backcross hybrid breakdown, there is a growing amount of evidence to indicate that the chromosomal rearrangements that characterize the Moreton taxon are selectively maintained. An examination of the patterns of chromosomal variation within this taxon has revealed an unprecedented and concerted change to the structure of the entire genome from a northern metacentric to a southern acro/telocentric genome over a distance of 1500 km. Moreover, the transition is not simply a consequence of changes in the frequency of two alternative chromosomal morphs but involves the gradual movement of the centromere from medial (northern populations) to distal (southern populations) positions on each of the 24 chromosomes within the genome. Intuitively, such a gradual transformation of the genome from metacentric to acrocentric along a latitudinal cline implies the involvement of selective forces.

It is difficult to apply a conventional genetic explanation to this complex case because the observed changes to chromosome structure do not conform to current concepts of chromosome function, which focus on two processes: (1) the ability to maintain adaptive genic combinations by balancing selection and an extension of the genetic theory of natural selection (Dobzhansky, 1970); and (2) the ability to undergo regular meiosis—the "chromosome-genetic connection" of Darlington (1972). There are many aspects of chromosome structure that were either not considered or were unknown during the development and integration of classical genetic and cytological theory. More recent work on nuclear structure and function clearly indicates that chromosomes are involved in a variety of complex processes and are not simply convenient vehicles for the transmission of encoded biochemical information through evolutionary time. They also possess fundamental biophysical properties that include site-specific attachment to the nuclear envelope (Comings and Okada, 1971), involvement in reassembly of the nucleus after cell division (McKeon et al., 1984), nonrandom disposition during the cell cycle (Korf et al. 1982; Heslop-Harrison and Bennett, 1983), and formation of the chromosomal scaffold and nucleosomal domains that are thought to modify rates of transcription and gene expression (Mirkovitch et al., 1987). More recently, Earnshaw and Cooke (1991) have identified a series of "chromosomal passenger" proteins that are carried by the chromosomes to the equator during mitosis, where they are released to become part of the cytoskeleton during cytokinesis. Although these properties of the chromosomes are still contentious, the concept of nucleotypic, as well as genic, control over nuclear and cellular kinetics is becoming more prominent in cell biology studies (Bennett, 1982; Cavalier-Smith, 1985).

The systemic changes to genome structure that we have observed in *Caledia* may require a change in our perspective of the chromosome as an evolutionary unit. If the Moreton chromosomes are indeed multicentric, as our preliminary studies seem to indicate, with one or more latent centromeres distributed along the length of the chromosome, it would have quite profound consequences on our interpretation of chromosomal evolution in *Caledia*. Although the molecular structure of the centromere is virtually unknown in higher eukaryotes, it is probably composed of several thousand basepairs of DNA organized as tandem repeats. Hence the presence of one or more additional centromeres to the Moreton chromosome and their absence from Torresian chromosomes could generate meiotic pairing problems that lead to the production of duplication/deficiency gametes with subsequent hybrid inviability. Clearly, the verification of these putative additional centromeres on Moreton chromosomes would necessitate rethinking our interpretation of the hybrid zone's maintenance. It does, however, provide us with an opportunity to define its evolutionary implications more specifically.

Similarly, the structure of the genomic cline in S.E. Australia becomes more amenable to analysis with the discovery of multicentric chromosomes, but the origin of such structures on such a large scale as constitutive components of the genome remains enigmatic. Their direct involvement in creating phenotypic variation is, again, highly speculative, but their association with changes in development time warrants the further investigation of genome structure and cell cycle patterns during embryogenesis.

It is clear that our attempts to provide a direct role for chromosomal variation both as an effective isolating mechanism at the hybrid zone and as an adaptive character within the Moreton taxon are still ambiguous and based largely on correlations. We make no apologies for this ambiguity because the current state of our knowledge of chromosome structure and its function in the evolutionary process awaits a paradigm.

ACKNOWLEDGMENTS

We thank Maureen Whittaker, Gary Hanson, and James Whitehead for preparing the illustrations and photographs presented in this chapter.

REFERENCES

Arnold, M. L., Appels, R., and Shaw, D. D. 1986. The heterochromatin of grasshoppers from the *Caledia captiva* species complex. I. Sequence evolution and conservation in a highly repeated DNA family. Mol. Biol. Evol. 3:29–43.

Arnold, M. L., and Shaw, D. D. 1985. The heterochromatin of grasshoppers from the *Caledia captiva* species complex. II. Cytological organisation of tandemly repeated DNA sequences. Chromsoma 93:183–190.

Arnold, M. L., Shaw, D. D., and Contreras, N. 1987. Ribosomal RNA encoding DNA introgression across a narrow hybrid zone between two subspecies of grasshopper. Proc. Natl. Acad. Sci. USA. 84:3946–3950.

Barton, N. H. 1981. The width of the hybrid zone in *C. captiva*. Heredity 47:279–282.

Barton, N. H., and Hewitt, G. M. 1989. Adaptation, speciation and hybrid zones. Nature 341:497–503.

Bennett, M. D. 1982. Nucleotypic basis of spatial ordering of chromosomes in eukaryotes and the implications of order for genome evolution and phenotypic variation. In G. A. Dover and R. B. Flavell, eds. Genome Evolution. London: Academic Press, pp. 239–261.

Bryan, W. W. 1973. Tropical and subtropical forests and heaths. In R. Milton Moore, ed. Australian Grasslands. Canberra: Australian National University Press, pp. 101–111.

Carson, H. L. 1971. Speciation and the founder principle. Stadler Symp. 3:51–70.

Cavalier-Smith, T. 1985. Cell volume and the evolution of eukaryotic genome size. *In* T. Cavalier-Smith, ed. The Evolution of Genome Size. Chichester: Wiley, pp. 105–184.

Chappel, J., Chivers, A., Wallensky, E., Polach, H. A., and Aharon, P. 1983. Holocene paleoenvironmental changes, central to north Great Barrier Reef inner zone. BMR J. Aust. Geol. Geophys. 8:223–235.

Coates, D. J., and Shaw, D. D. 1982. The chromosomal component of reproductive isolation in the grasshopper *Caledia captiva*. I. Meiotic analysis of chiasma distribution patterns in two chromosomal taxa and their F_1 hybrids. Chromosoma 86:509–531.

Coates, D. J., and Shaw, D. D. 1984. The chromosomal component of reproductive isolation in the grasshopper *Caledia captiva*. III. Chiasma distribution patterns in a new chromosomal taxon. Heredity 53:85–100.

Comings, D. E., and Okada, T. 1971. Condensation of chromosomes onto the nuclear envelope during prophase. Exp. Cell Res. 63:471–473.

Daly, J. C., Wilkinson, P., and Shaw, D. D. 1981. Reproductive isolation in relation to allozymic and chromosomal differentiation in the grasshopper *Caledia captiva*. Evolution 35:1164–1179.

Darlington, C. D. 1972. The place of the chromosomes in the genetic system. In J. Wahrman and K. Lewis, eds. Chromosomes Today. Vol. 4. New York: Wiley, pp. 1–13.

Dobzhansky, T. 1970. Genetics of the Evolutionary Process. New York: Columbia University Press.

Earnshaw, W. C., and Cooke, C. A. 1991. Analysis of the distribution of INCEPS throughout mitosis reveals the existence of a pathway of structural changes in the chromosomes during metaphase and early events during cleavage furrow formation. J. Cell Sci. 98:443–461.

Earnshaw, W. C., Rattrie, H., III, and Stetten, G. 1989. Visualisation of centromere proteins CENP-B and CENP-C on a stable dicentric chromosome in cytological spreads. Chromosoma 98:1–12.

Groeters, F. R. and Shaw, D. D. 1992. Association between latitudinal variation for embryonic development time and chromosome structure in the grasshopper *Caledia captiva* (Orthoptera: Acrididae). Evolution 46:245–257.

Heslop-Harrison, J. S., and Bennett, M. D. 1983. The spatial order of chromosomes of root tip metaphases of *Aegilops umbellulata*. Proc. R. Soc. Lond. Biol. 218:225–239.

Hewitt, G. M. 1979. Orthoptera. In B. John, ed. Animal Cytogenetics. Berlin: Gebruder Borntraeger, pp. 1–170.

Hewitt, G. M. 1987. Hybrid zones—natural laboratories for evolutionary studies. Tree 3:158–167.

Hewitt, G. M. 1988. Speciation in the Orthoptera—insights from hybrid zones. In B. Baccetti, ed. Evolutionary Biology of Orthopteroid Insects. Chichester: Halstead Press, pp. 252–259.

Kershaw, A. P. 1974. A long continuous pollen sequence from north-eastern Australia. Nature 251:222–223.

Kohlmann, B., Nix, H., and Shaw, D. D. 1988. Environmental predictions and distributional limits of chromosomal taxa in the Australian grasshopper *Caledia captiva* (F.). Oecologia 75:483–493.

Kohlmann-Cuesta, B. 1988. The ecological genetics of a narrow hybrid zone: the case of *Caledia captiva*. Ph.D. thesis, Australian National University, Canberra.

Korf, B. R., Gershez, E. L., and Diakumakos, E. G. 1982. Centromeres are arranged in clusters throughout the muntjac cell cycle. Exp. Cell Res. 139:393–396.

Lande, R. 1979. Effective deme sizes during long term evolution estimated from rates of chromosomal rearrangement. Evolution 33:234–251.

Levitan, M. 1978. Studies of linkage in populations. IX. The effects of altitude on X-chromosome arrangement combinations in *Drosophila robusta*. Genetics 89:751–763.

Lewontin, R. C. 1974. The Genetic Basis of Evolutionary Change. New York: Columbia University Press.

Marchant, A. D. 1984. Assortative mating in parapatric subspecies which form hybrid zones. Honors thesis, Australian National University, Canberra.

Marchant, A. D. 1988. Apparent introgression of mitochondrial DNA across a narrow hybrid zone in the *Caledia captiva* species complex. Heredity 60:39–46.

Marchant, A. D., and Shaw, D. D. 1992. Contrasting patterns of geographic variation shown by mtDNA and karyotype organisation in two subspecies of *Caledia captiva* (Orthoptera). Mol. Biol. Evol. (in press).

Marchant, A. D., Arnold, M. L., and Wilkinson, P. 1988. Gene flow across a chromosomal tension zone. I. Relicts of ancient hybridisation. Heredity 61:321–328.

Mayr, E. 1963. Animal Species and Evolution. Oxford: Oxford University Press.

McKeon, F. D., Tuffanelli, D. L., Kobyashi, S., and Kirschner, M. W. 1984. The redistribution of a conserved nuclear envelope protein during the cell cycle suggests a pathway for chromosome condensation. Cell 36:83–92.

Mirkovitch, J., Gasser, S. M., and Lemmli, U. K. 1987. Relation of chromosome structure and gene expression. Phil. Trans. R. Soc. Lond. Biol. 317:563–574.

Moran, C. 1979. The structure of the hybrid zone in *Caledia captiva*. Heredity 42:13–32.

Nix, H. A. 1986. A biogeographical analysis of Australian elapid snakes. In R. Longmore, ed. Atlas of Elapid Snakes of Australia. Australia Flora and Fauna Series 7. Canberra: Bureau of Flora and Fauna, pp. 4–15.

Nix, H. A., and Kalma, J. D. 1972. Climate as a dominant control in the biogeography of northern Australia and New Guinea. In D. Walker, ed. Bridge and Barrier: The Natural and Cultural History of Torres Strait. Canberra: Australian National University Press, pp. 61–91.

Rothfels, K. H., and Mason, G. F. 1975. Achiasmate meiosis and centromere shift in *Eusimulium aureum* (Diptera: Simulidae). Chromosoma 51:111–124.

Shaw, D. D., Coates, D. J., and Arnold, M. L. 1988. Complex patterns of chromosomal variation along a latitudinal cline in the grasshopper Caledia captiva. Genome 30:108–117.

Shaw, D. D., Coates, D. J., Arnold, M. L., and Wilkinson, P. 1985. Temporal variation in the chromosomal structure of a narrow hybrid zone and its relationship to karyotypic repatterning. Heredity 55:293–306.

Shaw, D. D., Moran, C., and Wilkinson, P. 1980. Chromosomal reorganisation, geographic differentiation and the mechanism of speciation in the genus *Caledia*. In R. L. Blackman, G. M. Hewitt, and M. Ashburner, eds. Insect Cytogenetics Symposium of the Royal Entomological Society of London. Oxford: Blackwell, pp. 171–194.

Shaw, D. D., and Wilkinson, P. 1980. Chromosome differentiation, hybrid breakdown and the maintenance of a narrow hybrid zone in *Caledia*. Chromosoma 80:1–31.

Shaw, D. D., Wilkinson, P., and Coates, D. J. 1982. The chromosomal component of reproductive isolation in the grasshopper *Caledia captiva*. II. The relative viabilities of recombinant and non-recombinant chromosomes during embryogenesis. Chromosoma 86:533–549.

Templeton, A. R. 1982. Adaptation and the integration of evolutionary forces. In R. Milkman, ed. Perspectives on Evolution. Sunderland, MA: Sinauer, pp. 15–31.

Therman, E., Trunca, C., Kuhn, E. M., and Sarto, G. M. 1986. Dicentric chromosomes and the inactivation of the centromere. Hum. Genet. 72:191–195.

Van Delden, W., and Kamping, A. 1989. The association between the polymorphisms at the Adh and the Gpdh loci and the In (26) t inversion in *Drosophila melanogaster* in relation to temperature. Evolution 43:775–793.

White, M. J. D. 1978. Modes of Speciation. San Francisco: W. H. Freeman.

8

Nature of Selection in
the Northern Flicker Hybrid Zone and
Its Implications for Speciation Theory

WILLIAM S. MOORE AND JEFF T. PRICE

Hybrid zones are regions of steep genotypic transition between populations that have reached a level of divergence comparable to that between "good" species. Remarkably, however, hybrid zones appear to be equilibrium configurations where the parental populations (whether they are called species, subspecies, or races) retain their taxonomic identity and integrity even though reproductive isolation is incomplete or, in many instances, wholly lacking. Although species definitions remain very much in contention (Mayr, 1957, 1987; Ehrlich, 1961; Sokal and Crovello, 1970; Mishler and Donoghue, 1982; Cracraft, 1983; Ghiselin, 1987; McKitrick and Zink, 1988; Endler, 1989; Templeton, 1989; Moore, 1990), we think there would be unanimity that an essential property of a species is that there is uniformity of genotype within the species and a discontinuity between the genotype of one species and that of another. In other words, genotypic variance within a species is low compared to the variance between species. Reproductive isolation, of course, would enforce this essential property of species, but the sustained cohesion (*sensu* Templeton, 1989) of the hybridizing taxa and the abrupt discontinuity of genotype seen across hybrid zones imply that forces more fundamental than reproductive isolation are involved in speciation. Hybrid zones provide *natural* experimental systems where modern genetic analysis can be applied to discern the forces that drive the origin of species.

The period since the mid-1970s has been one of substantial growth in both empirical and theoretical studies of hybrid zones. This proliferation of studies has been fostered by the application of cline theory and related population genetics models to the interpretation of empirical studies of hybrid zones. Two important classes of models have been developed. The first, exemplified by the models of Slatkin (1973, 1975), May et al. (1975), and Endler (1977), imagines a geographical selection gradient that determines the relative fitnesses of genotypes. This kind of model is useful for analyzing the adaptive response of populations to forms of ecological selection that vary spatially. The second is the set of "tension zone" models developed by Barton (1979a, b,

1983), Barton and Hewitt (1981, 1985, 1989), Hewitt and Barton (1980), and Szymura and Barton (1986), building on early work by Bazykin (1969), in which selection is simply against hybrids and does not vary along a spatial gradient. In their simplest forms, both kinds of models consider selection on a single locus with two alleles, but both can be extended to multiple loci (Slatkin, 1975; Barton, 1983).

The geographical-selection-gradient model and tension-zone model are remarkably similar mathematically (Barton and Hewitt, 1985, 1989); however, the biological implications of their assumptions are profoundly different. Specifically, the geographical-selection-gradient model implies that ecological variables determine the position and width of hybrid zones and hence that the coherent nature of species is maintained by ecological selection. In tension zone models, in contrast, selection is *endogenous*[1] in the sense that it is generated purely by interactions within the genome and does not have an *exogenous* or ecological component (Barton, 1979b, 1983). In tension zones models, melding distinctly coadapted gene complexes results in hybrids with reduced fitness. This situation stabilizes the width of the hybrid zone but not its position; more intense selection against hybrids results in a narrower hybrid zone.

A third, unique, hybrid zone type has been analyzed by Mallet (1986), Mallet and Barton (1989a, b), and Mallet et al. (1990). Hybrid zones exist between Müllerian mimetic morphs of several species of *Heliconius* butterflies. The structure of these hybrid zones is determined by frequency-dependent selection; the hybrid zone has the dynamics of a tension zone, but selection is caused by an ecological agent—predatory birds.

The essential features of the exogenous selection model (geographical selection-gradient) and the endogenous selection model (tension zone of Barton and Hewitt, 1985) are summarized in Figure 8-1. [We have illustrated one-dimensional models, which would be applicable, for example, to populations distributed in riparian woodlands along a river or populations along a coast, but the models can be extended to two dimensions (Slatkin, 1975).] In each model the change in allele frequency (p) at locale x is determined by selection at x and migrants dispersing to and from x. The dispersal submodels are identical, but the selection submodels differ, most importantly in that the exogenous selection model has a function f(x) that affects fitness as a function of locale (x). In the endogenous selection model hybrids are equally unfit at all locales, whereas in the exogenous selection model heterozygote (hybrid) fitness varies

---

1. A cautionary note should be made with regard to the application of tension zone models in hybrid zone studies. A disparity may exist between theoretical and empirical studies of tension zones with regard to which genotypes are selected against. In the mathematical formulation, selection is against the heterozygote (Barton, 1979b). Considering multiple loci, the most heterozygous individuals, F_1 hybrids, are the least fit (Barton, 1983). In empirical studies, however, reduced fertility, embryonic mortality, and developmental aberrations are often cited as evidence of a tension zone (e.g., Woodruff, 1979; Barton and Hewitt, 1981, 1985; Kocher and Sage, 1986; Szymura and Barton, 1986). This type of selection is expected when coadapted gene complexes are disrupted, but, presumably, most of this selection operates on backcross progeny and F_2 progeny, rather than on F_1 progeny. Despite this imprecise representation of fitness, tension zone models appear to render a reasonably accurate portrayal of hybrid zones for a diversity of situations where selection operates against hybrids. Mallet and Barton (1989a), for example, found that hybrid unfitness resulting from epistasis produced a tension zone similar to that resulting from heterozygote disadvantage. In any case, a biological implication of the tension zone model is that adaptations for harmonious gene interactions maintain narrow hybrid zones, and these adaptations are fundamental to the coherent nature of species.

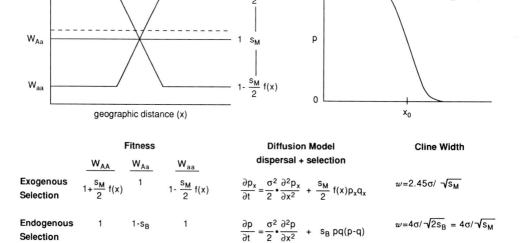

Figure 8-1. Cline models that portray selection in a hybrid zone. (**A**) Selection along a geographical gradient where the AA homozygote has fitness $W_{AA} = 1 + sf(x)/2$ at the left end of the transect and fitness $W_{AA} = 1 - sf(x)/2$ at the right extreme. The solid horizontal line illustrates heterozygote fitness. The dashed line is the special case where hybrids are more fit in the transition area of the selection gradient. (**B**) Equilibrium cline in allele frequency (p) that would result from the selection pattern illustrated in (**A**), balanced by dispersal (σ). A similar sigmoidal cline would result if selection were against hybrids rather than along a geographical gradient. Relative fitnesses, diffusion equations, and expressions for equilibrium cline widths for an endogenous and exogenous selection model are tabulated below the figures. The diffusion equation for exogenous selection is from Eq. 10 in May et al. (1975); the equation for endogenous selection is from Eq. 1 in Barton and Hewitt (1989). For understanding the relation between the two models, it is important to note that the selection coefficients (s) used by May et al. (1975) and Barton and Hewitt (1989) are not identical. Using s_M to denote May et al.'s selection coefficient and s_B to denote that of Barton and Hewitt, the relation appears to be $s_M = 2s_B$.

geographically. In Figure 8-1A, the solid horizontal line represents the case in which heterozygote fitness is intermediate everywhere, whereas the dashed horizontal line is a case where heterozygotes are actually more fit in transitional locales (bounded hybrid superiority) (Moore, 1977; Moore and Buchanan, 1985). An important result is expressed in the equations that give the equilibrium widths (w) of clines maintained by endogenous and exogenous selection. These similar equations are important because they allow one to solve for selection, given the width of the cline and the dispersal parameter (σ, root-mean-square dispersal). The width of a hybrid zone is easy to estimate, the dispersal can usually be estimated, although with greater difficulty; hence this relation provides a method for estimating the strength of selection maintaining a hybrid zone.

A critical point, however, is that these models can be used to determine the intensity of selection, but as a practical matter they *cannot* be used to determine whether the selection that is operating is endogenous or exogenous in nature. This inference must be based on data derived from other sources; for example, evidence of hybrid sterility or increased developmental aberrations would imply endogenous selection. This is apparent from comparison of the two equations in Figure 8-1 that give the equilibrium width (w) of clines maintained by exogenous and endogenous selection. The two differ only in the value of a constant; for comparable amounts of selection, a cline maintained by endogenous selection would be 1.63 times wider than one maintained by a stepped selection gradient ($\Delta X < 1$) (May et al., 1975), but the two forms of selection cannot be distinguished by the application of these models to the data set. If one applied the wrong model, it would simply result in an error in the estimate of selection. (The magnitude of this error is probably not large compared to that resulting from a poor estimate of σ.) The length of the tails of the cline would differ (Barton and Hewitt, 1989), but large sample sizes would be required to detect the differences, and the analysis would be confounded by deviations from the assumptions of the model, e.g., slight temporal fluctuations in the position or width of the hybrid zone, asymmetry in dispersal, and mixtures of endogenous and exogenous selection.

The ideal way to distinguish between endogenous and exogenous selection would be in a manipulative experiment where hybrid and parental phenotypes were transplanted to various parts of the hybrid zone, which would allow relative fitness to be measured directly—and in its entirety. If, for example, selection is endogenous, the fitness of hybrids would be uniformly low everywhere. Unfortunately, this type of experiment is impossible for most animal hybrid zones, but plant hybrid zones hold promise in this respect (see Ch. 5).

An indirect, but nonetheless convincing, inference that exogenous selection is operating in a hybrid zone was exemplified in a study by Rand and Harrison (1989) on a hybrid zone between two species of field crickets, *Gryllus pennsylvanicus* and *G. firmus.* They found that the hybrid zone is so intricately associated with an environmental variable (soil type) that it is incredible to think that there is not a causal relation. Such correlations are useful for detecting natural selection but of course by themselves do not prove natural selection (Endler, 1986). Given the biology of the organisms, however, it seems likely that soil type directly determines the distribution of crickets.

In any case, the low vagility of crickets allows this kind of fine-grained study. The problem when undertaking this kind of study with an avian hybrid zone is one of scale. With crickets, a study over an area of a few hectares is both logistically possible and informative. With birds, however, one would need to work on a regional if not a continental scale, which does not seem logistically possible. In addition to guessing what the causative variables might be, one must find an intricate pattern of variation on a continental scale and then gather the data on the response variables, again on a continental scale, that could be correlated with the putative causative variables.

Our major claims in this Chapter are that the requisite data for this kind of analysis are available for the northern flicker hybrid zone, and that this hybrid zone, like the *Gryllus* hybrid zone, has an intricate association with a suite of ecological variables. Based on this information we conclude that exogenous selection is important in maintaining the northern flicker hybrid zone. The analysis differs from that of Rand and Harrison in that it requires a continental perspective to see the association. The data that make this analysis possible are from the United States Fish and Wildlife Service (USFWS) Breeding Bird Survey and an assortment of data on the distributions of other animal groups, vegetational patterns, and climatic variables.

According to the exogenous selection hypothesis, the divergence seen across a cline (hybrid zone) could have evolved in either allopatry or parapatry (*sensu* Endler, 1977). Historical and biogeographical evidence suggests that the flickers probably diverged in allopatry and the hybrid zone resulted from a secondary contact (discussed below). It is useful to make this assumption because it focuses attention on certain kinds of ecological variables that would likely determine the structure of the hybrid zone. Specifically, the position and width of the hybrid zone ought to be determined by the position and width of the ecotone between the ecological communities within which the parental taxa evolved or with abiotic variables that determine the geography of these communities. We hasten to add that, regardless of whether a "hybrid zone" resulted from parapatric divergence or allopatric divergence and secondary contact, the test for exogenous selection would be the same. For this reason, because there is evidence that the flicker hybrid zone is a secondary contact, and for the sake of brevity, we do not discuss the parapatric divergence hypothesis, although it cannot be completely dismissed.

It should be noted also that the term ecotone is used here in a somewhat specialized sense. More broadly, an ecotone is the interface between any two ecological communities, e.g., between tundra and boreal forest. Here the ecotones of interest would be those that resulted from relatively recent vicariance events that divided an "ancestral community" into refugia, the refuge communities became distinct as incipient species formed, and a subtle ecotone formed when changing climatic conditions permitted the refugia to expand into secondary contact. We coin the term *vicariance ecotone* to refer to an ecotone formed in this manner. The exogenous selection hypothesis then predicts that hybrid zones should be associated with vicariance ecotones and should track their geographic variation: Where the ecotone changes direction, the hybrid zone should change in the corresponding direction; where the ecotone is narrow, the hybrid zone should be correspondingly narrow; where the ecotone is more gradual, the hybrid zone should be relatively broad.

One might expect other correlates as well. For example, climatic variables such as precipitation, temperature (averages and extremes), and evapotranspiration might be correlated with a hybrid zone either because the hybridizing taxa are adapted directly to climatic parameters or indirectly in that they are adapted to distinct vegetational associations that are determined by climatic parameters.

These predictions are clear enough, but how does one identify assemblages of sister taxa that arose in the same vicariance event, and how does one map the position of the ecotone between the constituent ecological communities? Several phylogenetic methods might be applied to the first problem. For example, a common vicariance history should produce similar phylogenetic patterns in numerous unrelated groups (e.g., Zink and Hackett, 1988). Thus one might expect to find the range boundaries of numerous mammalian sister species co-occurring with each other and with the range boundaries of numerous avian species, ant species, and so on. (The choice of indicator species is arbitrary so long as there is an expectation that they would have experienced comparable levels of isolation during the phase of allopatry.) As indicators of more recent vicariance-secondary contact cycles, one might expect a comparable pattern but now involving hybridizing subspecies. This is the essence of Remington's (1968) concept of suture zones. A suture zone is a geographical area where two ecological communities have come into secondary contact and numerous pairs of sister species or subspecies have begun to hybridize. A key prediction of this hypothesis is that hybrid zones should be clustered, geographically, rather than randomly distributed. If one could apply a molecular clock to the analysis, one would find that most of the taxa that define the ecotone diverged at the same time.

Although potentially collectable, precious few of these data are actually available (Zink and Hackett, 1988). The detailed lower level phylogenies for relevant taxa are not often available, and the molecular clock remains controversial and uncalibrated for comparisons between disparate groups (e.g., mammals and birds). It is important to note, however, that *a priori* identification of taxa that have a common biogeographical history is not necessary; it would just greatly reduce statistical "noise" and increase the probability of discerning vicariance ecotones. Presumably, the number of taxa forming a vicariance ecotone should be large, and thus it should be possible to discern the ecotone even though taxa that experienced different histories are included in the analysis.

Once an assemblage of appropriate taxa has been identified, mapping ecotones is straightforward, provided the data on range boundaries are available. All that is required is to map the range boundaries and to determine where there are significant concentrations. These data are available for some groups in the form of distribution maps (e.g., mammals: Hagmeier and Stults, 1964; birds: Zink and Hackett, 1988), but they are limited with regard to precise placement of range boundaries, and there usually is no information on variation in population density. Moreover, extensive, detailed species and subspecies accounts are available for only a few groups (e.g., vertebrates), and then only birds and mammals could be said to be well known. This detailed alpha-level taxonomic work is a prerequisite for proper identification of subspecies, sibling species, and otherwise closely related taxa that may be the products of recent vicariance events. The data for avian species are outstanding in all these regards. The status of most North American avian taxonomic entities is relatively well known. Moreover, an extraordinary database has been constructed over more than a quarter-century in the form of the Breeding Bird Survey (BBS) (M. B. Robbins et al., 1986). Although not without problems and shortcomings, the BBS database is outstanding relative to data available for other taxonomic groups. It is comprehensive; it spans the United States and most of Canada; it is quantitative and provides data on geographic variation in the relative densities of populations.

In the remainder of this chapter we briefly describe the northern flicker hybrid zone and review previous studies. We then analyze associations of position and width of the hybrid zone with three sets of biotic variables and a set of climatic variables.

NORTHERN FLICKER HYBRID ZONE

The northern flicker *(Colaptes auratus),* a woodpecker, is one of the most abundant birds in North America. Flickers excavate nest cavities in tree trunks and large limbs but show no strong preference for particular species of tree; the most important criteria seems to be the availability of wood (usually diseased or dead) sufficiently soft to excavate. They do show a marked preference for forest edges or woodlands with low tree density, approaching a savanna-like character (Conner and Adkisson, 1977). On the Great Plains, where the hybrid zone between red- and yellow-shafted flickers occurs, flickers are abundant in riparian woodlands that traverse the plains, running generally west to east, in sheltered areas that support arboreal vegetation and in isolated conifer forests such as the Pine Ridge area of the Nebraska panhandle and the Black Hills.

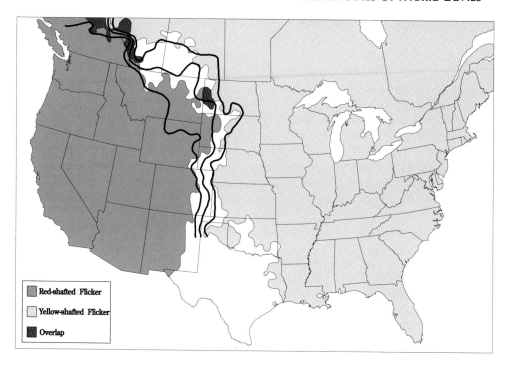

Figure 8-2. Distributions of red- and yellow-shafted flickers and their hybrid zone in North America based on the Breeding Bird Survey. The range boundaries are the contours where fewer than 0.5 birds per year were recorded. The contour lines are the hybrid index isopleths for the 20%, 50%, and 80% transitions from yellow-shafted to red-shafted phenotype.

There is no published comprehensive analysis concerned with the origin of the northern flicker hybrid zone. There is unequivocal evidence that the hybrid zone existed on the Great Plains at approximately its present position when European explorers and settlers first arrived (Short, 1965; Moore and Buchanan, 1985; Moore, unpublished data). Scenarios based on reasonable assumptions about the habitat associations of flickers strongly suggest that red- and yellow-shafted flickers diverged as a result of one or more vicariance events during the cyclical occurrence of glacial and interglacial periods that characterize the Pleistocene and that the hybrid zone is a secondary contact. According to this scenario, the yellow-shafted flicker evolved adaptations to the mesic forests that characterize eastern North America, whereas the red-shafted flicker evolved adaptations to the generally more xeric forests of western North America.

Figure 8-2 maps the breeding ranges of the red- and yellow-shafted flickers and the hybrid zone in the United States and Canada south of 56° N and east of 128° W. [Range boundary and contour maps were made with the computer software Surfer, Version 4, Golden Software, Inc., Golden, CO (1989) using a quadrant search method and an inverse distance gridding method.] Figure 8-3 is a more detailed map where the hybrid zone is represented by contour curves of hybrid index score (Short, 1965; Moore and Buchanan, 1985). The contour map is based on average hybrid index

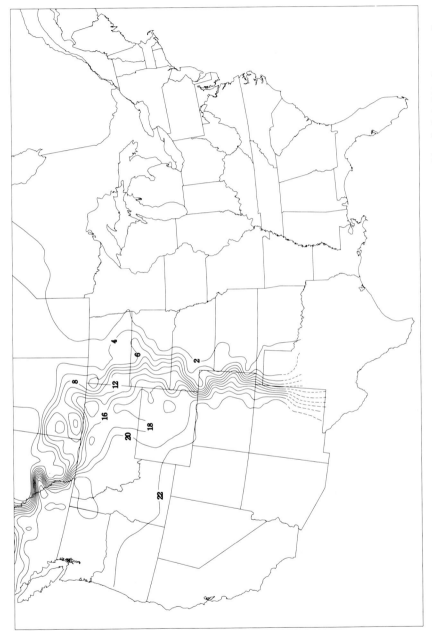

Figure 8-3. Contour map of the northern flicker hybrid zone. The contours represent the transition in hybrid index score from pure yellow-shafted flickers (0) in the east to pure red-shafted Flickers in the west (23). The hybrid index is based on six plumage traits that are diagnostic for the two forms (Short, 1965; Moore and Buchanan, 1985).

scores determined for museum specimens from 112 locales (Short, 1965; Moore and Buchanan, 1985; Moore, unpublished data). Figure 8-3 is the most accurate and detailed map of the hybrid zone published to date, but it is not accurate in western British Columbia because few data points were available for that region.

The hybrid zone is generally narrow but extends in length nearly 4000 km from the Texas panhandle to southern Alaska. Considerable curvature of the hybrid zone and variation in width are apparent. The hybrid zone is narrow at its southern terminus along the Canadian River in the Texas Panhandle, broadens slightly as it is traced northward through western Kansas and eastern Colorado, and broadens more dramatically in the northern plains where the center contour of the hybrid zone (hybrid index = 11.5) (Short, 1965; Moore and Buchanan, 1985) turns toward the northwest and west. Approaching the Rocky Mountains, near the point where the Montana, Alberta, and British Columbia boundaries come together, the hybrid zone narrows radically, turns sharply to the north, and then back to the west where it again broadens. Too few collections are available from Alaska and northern British Columbia to accurately map the hybrid zone in this region, but the red-shafted side of the zone appears to continue west, whereas the center and yellow-shafted side of the hybrid zone bulge to the north. Over its transcontinental length, the hybrid zone exhibits dramatic variation in width and orientation. Correlated variation in ecological variables would provide evidence that exogenous selection is important to the dynamics of this hybrid zone.

It was established in previous studies that the two flickers differ with regard to six plumage traits (five in females) and in several morphometric traits that are related to an overall size difference (Short, 1965). The traits appear to be inherited independently, and step clines for all of the traits are more or less congruent (Short, 1965). The hybrid zone has been remarkably stable in historical time (Moore and Buchanan, 1985), the reproductive success of hybrids is equal to that of the parental subspecies (Moore and Koenig, 1986), and there is no assortative mating in the hybrid zone (Bock, 1971; Moore, 1987). A survey of allozyme allele frequencies indicates that the northern flicker species population has levels of polymorphism typical of birds, but there is little or no geographic variation in allele frequencies (Grudzien and Moore, 1986; Grudzien et al., 1987). There is geographic variation in mtDNA haplotype frequencies, but the pattern involves divergence of southwestern red-shafted flicker populations from northern and eastern populations of both subspecies; that is, the pattern of divergence is not associated with the hybrid zone (Moore et al., 1991). The genetic studies also indicate that gene flow is high in flickers. An estimate of dispersal based on the USFWS Banding Recovery data indicates that root-mean-square dispersal (σ) is near 100 km per generation (Moore and Buchanan, 1985). This estimate is based on a small sample size, but it is consistent with the genetic studies in indicating that gene flow is high. As is discussed below, the inference that selection operates to maintain the flicker hybrid zone is based on the assumption that gene flow is high.

ECOLOGICAL CORRELATES OF THE FLICKER HYBRID ZONE

Mammalian Range Boundaries. As argued above, an ecotone with which a hybrid zone is associated should involve the range boundaries of many disparate taxa, and the choice of taxa used to identify and map the ecotone is arbitrary. Thus one can use

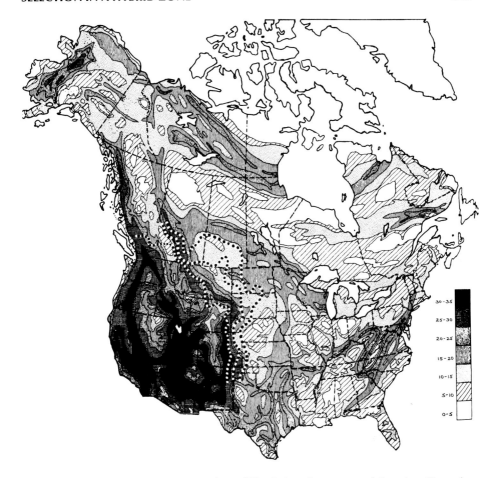

Figure 8-4. Range boundary concentrations of North American mammals based on Hagmeier and Stults' (1964) mammal map overlaid with the northern flicker hybrid zone *(dotted contours).*

whatever data happen to be available. One such data set and analysis was compiled by Hagmeier and Stults (1964) for North American mammal species. Their primary concern was to devise objective, quantitative criteria for identifying mammalian faunal provinces. They overlaid North America (United States and Canada) with a grid, where blocks were proportionally 50 miles to a side. An index of faunal change (IFC) was computed for each block as the percentage of mammalian species occurring in the block with a range limit also occurring within the block. For example, if 50 mammalian species occurred in a block and 6 of those species had a range limit within that block, the IFC for the block would be 12%. A contour map, based on IFCs, was then generated. This map is reproduced in Figure 8-4 with an overlay of the flicker hybrid zone. (In this map, and in succeeding maps, the hybrid zone is represented by the 20%, 50%, and 80% hybrid index contours.)

The central Great Plains is an area of low faunistic change but is bounded on both east and west by areas of relatively great change. Of particular interest is the western

Great Plains, which is bounded by a steep slope of faunistic change that runs generally north from the Texas panhandle to central South Dakota, where it turns to the northwest in much the same position and direction as the flicker hybrid zone. As the "slope" approaches the United States–Canadian border near the Montana, Alberta, and British Columbia political boundaries, it turns more to the north; and the "ridge" capping the slope follows the flicker hybrid zone rather closely along the eastern slope of the Canadian Rockies along the Alberta–British Columbia Provincial line. Although this association is not convincing in itself, it is one of several biotic boundaries and ecological transition zones that appear to be associated with each other and with the flicker hybrid zone.

Avian Range Boundaries. We undertook a similar analysis based on the USFWS BBS data. The BBS has been run on a continental scale annually since 1968. The survey now comprises approximately 2400 routes, of which 1800–1900 are run annually. Each route is run on a day in June with good weather and as close as possible to the date the route was run the previous year. The numbers of individuals for each species seen on each route are entered in a computerized database available from the USFWS (C. S. Robbins et al., 1986; Droege, 1990). Our objective was to use the BBS data to determine whether avian boundaries are concentrated in certain geographical areas; and if they are, whether the flicker hybrid zone is associated with one of the concentrations. Achieving this objective is computationally demanding, and so we devised a sampling procedure that limited the number of species mapped to a manageable number. In addition, we limited the geographical extent of sampled routes to the area between 24° and 56° N latitude and between 66° and 128° W longitude. This area includes the 48 contiguous states and the parts of Canada for which a useful number of routes has been run. To limit the number of species, we drew a random sample of 72 species from a list of 392 species. The complete list of species (and subspecies) recognized in the BBS was reduced from 509 to 392 by *a priori* elimination of species obviously irrelevant to identifying an ecotone between woodland species. Specifically, virtually all species whose habitat includes some woody plant association (woodland or shrub) were *retained* on the list, resulting in the elimination, for example, of most shorebirds, pelagic birds, and purely grassland birds. Species on the residual list were then assigned a random number, and the list was sorted by ascending value of the random numbers. We then plotted the range boundaries for species in the order of the random list until we reached 72 species. This stopping point was somewhat arbitrary, but it seemed at, or beyond, a point of diminishing return in that the emergent pattern of range boundary concentrations was not changing significantly as new species were added, and the process is computationally arduous. The precise determination of a range boundary for a species is also somewhat arbitrary. The data we used were the number of observations of a particular species on a route averaged over the period 1976–1985. We defined the range boundary to be the 0.5 birds per route per year contour; that is, the range of the species is the area that contains routes where the species is seen at least every other year, on average. A contour map of the density of species range limits was then generated from the sample of 72 species. This map, overlaid with the flicker hybrid zone, is illustrated in Figure 8-5.

The contour map of BBS-range-limit densities has a number of interesting features. First, it bears a strong resemblance to Hagmeier and Stults' map of mammalian

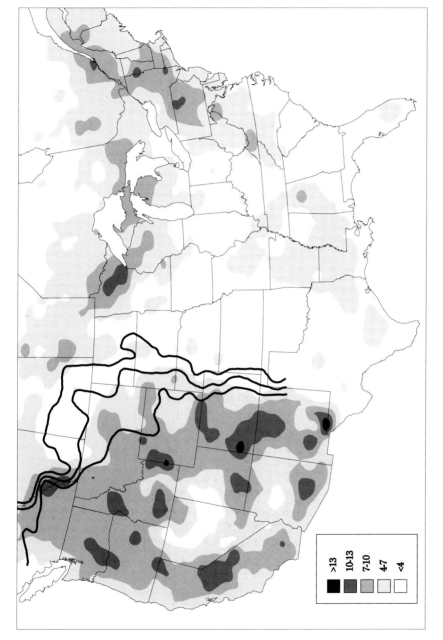

Figure 8-5. Range boundary concentrations of North American birds based on the USFWS BBS overlaid with the northern flicker hybrid zone. See text for a full explanation.

■	>13
▨	10-13
▨	7-10
▨	4-7
□	<4

faunistic changes (Fig. 8-4). In both maps, the mountainous areas of western North America have numerous range limits, whereas the Great Basin and Great Plains have few range limits. Of greater interest is that the bird map, like the mammal map, has a sharp increase in range boundaries along the western Great Plains that traces over essentially the same geography as that previously described for the mammals. Both maps also have a concentration of range boundaries associated with the southern limit of the boreal forest. This area extends toward the northwest across central Michigan, Wisconsin, and Minnesota; and it traces the complex ecotones involving aspen parklands that are transitional between the grasslands of the northern prairies in Manitoba, Saskatchewan, and Alberta and the boreal spruce forest to the north. This "ridge" of boundaries continues to the west where it coalesces, in southwestern Alberta, with the concentration of boundaries extending northwest along the western edge of the Great Plains. Most importantly, it can be seen from Figure 8-5 that the western edge and center of the flicker hybrid zone follow this concentration of range limits north along the western edge of the southern Great Plains and then parallel the turn toward the northwest in northern South Dakota. The concentration of western plains range limits turns north at approximately the United States–Canadian boundary and then back toward the west. The flicker hybrid zone roughly parallels these two curves to the point where the accuracy of the hybrid zone map becomes suspect. This area of concentrated avian range boundaries along the eastern edge of the Canadian Rocky Mountains is associated with a succession of parallel forest communities, succeeding from characteristically western to eastern and boreal tree species (see below).

The random sample of 72 avian range boundaries contains many species (probably most) that arose in biogeographical episodes completely distinct from those that formed the ecotone with which the flicker hybrid zone is hypothetically associated. The data analysis would be more powerful if there were an objective way to identify taxa that likely experienced the same biogeographic history as the flickers without falling into the circularity of simply picking taxa whose range boundaries are correlated with the hybrid zone. One possibility would be to look at the range boundaries of *all* taxa that are involved in hybrid zones. This method seems reasonable because, although the levels of divergence between parental taxa that form hybrid zones is variable, a sample based on the criteria that the taxa from hybrid zones should contain relatively many sister taxa that diverged at roughly the same time as the flickers (which hybridize) and relatively few pairs that diverged in more ancient vicariance events, because more of the older pairs would have diverged beyond the point where they could hybridize. Of course it has long been known that the flicker hybrid zone is one of several avian hybrid zones occurring on the Great Plains (Rising, 1983b), but how many of the North American hybrid zones occur on the Great Plains and how intimate are the associations? A liberal inclusion of reported hybridizing taxa as hybrid zones yields a list of 10 avian hybrid zones in North America (north of Mexico). They are listed in Table 8-1.

Complete maps of these hybrid zones, comparable in quality to the flicker hybrid zone map, are not available, but the approximate position of each hybrid zone can be inferred from maps of the range boundaries of the parental "species" derived from the BBS data. Five of the ten hybrid zones actually occur on the Great Plains (flickers, orioles, grosbeaks, buntings, and towhees). A sixth, the warbler hybrid zone, occurs along the eastern slope of the Canadian Rockies in close proximity to the flicker hybrid

Table 8-1. Avian Hybrid Zones in the United States and Canada

Hybridizing Taxa	References
Red-shafted flicker × yellow-shafted flicker[a,b]	Short (1965), Anderson (1971), Moore and Buchanan (1985), Moore and Koenig (1986), Moore (1987), Rising (1983b)
Bullock's oriole × Baltimore oriole[a,b]	Sibley and Short (1964), Sutton (1968), Rising (1970, 1973, 1983a,b), Anderson (1971), Corbin and Sibley (1977), Misra and Short (1974), Corbin et al. (1979)
Black-headed grosbeak × red-breasted grosbeak[a,b]	West (1962), Anderson and Daugherty (1974), Kroodsma (1974a,b), Rising (1983b)
Audubon warbler × myrtle warbler[b]	Hubbard (1969), Barrowclough (1980)
Indigo bunting × lazuli bunting[a,b]	Sibley and short (1959), Emlen et al. (1975), Kroodsma (1975), Rising (1983b)
Spotted towhee × rufous-sided towhee[a]	Sibley and West (1959), Rising (1983b)
Black-capped chickadee × Carolina chickadee[b]	Braun and Robbins (1986), Robbins et al. (1986), Brewer (1963), Johnston (1971), Mack et al. (1986)
Black-crested titmouse × tufted titmouse	Dixon (1955), Rising (1983b), Braun et al. (1984)
Bronzed grackle × purple grackle	Huntington (1952), Yang and Selander (1968)
Red-breasted sapsucker × red-naped sapsucker	Johnson and Johnson (1985), Johnson and Zink (1983)

[a]Hybrid zone occurs on the western Great Plains.
[b]Recognized as distinct taxa in the Breeding-Bird Survey.

zone in that area. The "species" comprising the hybrid zones are recognized as distinct in the BBS for only 6 of the 10 North American hybrid zone pairs (flickers, orioles, grosbeaks, warblers, buntings, and chickadees). The flicker hybrid zone is the most geographically extensive of these zones and four of the remaining five appear to be associated with the flicker hybrid zone (Figs. 8-6 to 8-9).

The red- and yellow-shafted flicker range map (Fig. 8-2) provides an indication of how well the BBS range maps reflect the position of a hybrid zone. The range boundaries for the two flickers were determined by the 0.5 birds per route criterion applied to the BBS data and the resultant map overlaid by an accurate map of the hybrid zone based on museum collections. It can be seen in Figure 8-2 that the approximate position of the flicker hybrid zone is accurately portrayed, but the extent of hybridization could not be determined from the BBS map alone.[2]

The following are brief summaries of the ranges of the other hybridizing avian taxa.

2. In this analysis and in the analyses of the other avian hybrid zones (Figs. 8-6 to 8-9) we mapped only the distributions of the parental "species" because the distinction between parental "species" and hybrids is made inconsistently in the BBS; and even when a distinction is made, incorrect categorization of hybrids and parental "species" near the hybrid zone is probably common. Thus one would expect to find hybrids in the unshaded areas between the range boundaries, but the abundance of hybrids is not indicated in these maps.

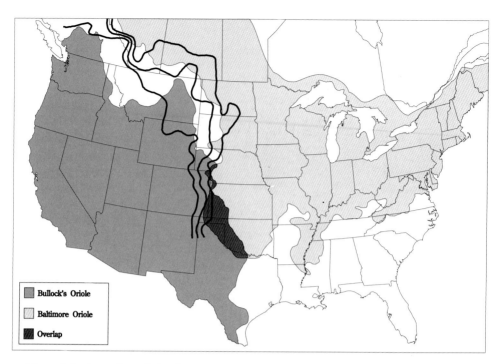

Figure 8-6. Range boundaries for Bullock's oriole and the Baltimore oriole overlaid with the flicker hybrid zone.

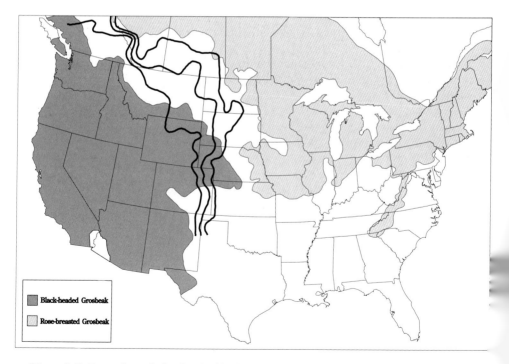

Figure 8-7. Range boundaries for the black-headed grosbeak and the rose-breasted grosbeak overlaid with the flicker hybrid zone.

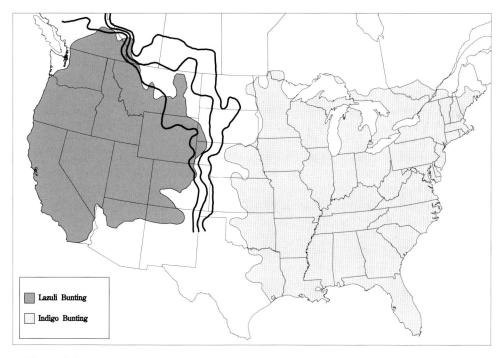

Figure 8-8. Range boundaries for the lazuli bunting and the indigo bunting overlaid with the flicker hybrid zone.

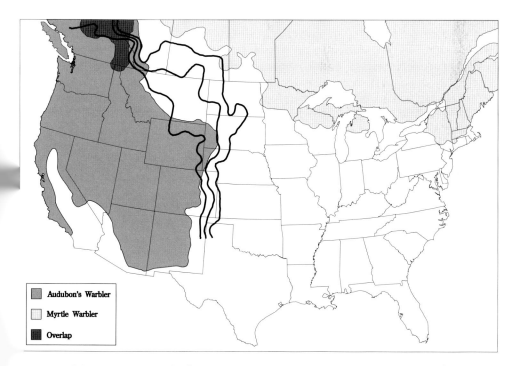

Figure 8-9. Range boundaries for Audubon's warbler and the myrtle warbler overlaid by the flicker hybrid zone.

The Baltimore oriole does not occur in the southeast, but its range contacts that of Bullock's Oriole along the Oklahoma-Texas State line (Fig. 8-6). The BBS range limits of the two subspecies broadly overlap in the southern plains but separate somewhat to the north and northwest. It is apparent from Figure 8-6 that the range boundaries of both Orioles and the implied hybrid zone track the flicker hybrid zone closely.

The eastern rose-breasted grosbeak does not range as far south as the yellow-shafted flicker (Fig. 8-7). The BBS ranges of the rose-breasted and black-headed grosbeaks come into close proximity only on the central Great Plains, particularly Nebraska, and it is only there that hybridization appears to be extensive (Rising, 1983b). Nonetheless, the range boundary of each species roughly parallels the respective eastern or western edge of the flicker hybrid zone.

Hybridization between the lazuli and indigo buntings appears to be limited, being most frequent in Nebraska and South Dakota, but even there samples containing hybrids are dominated by specimens with lazuli or indigo bunting phenotypes (Rising, 1983b). The BBS range boundary map (Fig. 8-8) shows that their ranges are in closest proximity in Nebraska and southern South Dakota. The indigo bunting boundary parallels, to its northern limit, the eastern edge of the flicker hybrid zone. The range boundary for the Lazuli Bunting is more informative because it extends farther north; it not only parallels the western edge of the hybrid zone north across the Great Plains but continues this strong parallel track as the flicker hybrid zone turns toward the northwest and along the Canadian Rockies.

The range boundaries for Audubon's warbler and the myrtle warbler overlaid by the flicker hybrid zone are shown in Figure 8-9. The myrtle warbler is restricted to the mixed conifer-deciduous and boreal forests of the north, and so its range boundary does not extend south to the Great Plains (Hubbard, 1969); however, it probably occurred farther south and southwest during the Wisconsin glacial maximum when boreal spruce forest occurred on the central Great Plains (Hubbard, 1969; Wright, 1970). Audubon's warbler breeds in conifer woodlands of western North America including outlying ponderosa pine woodlands on the western Great Plains and in the Black Hills (Hubbard, 1969). Its range boundary strongly parallels the western edge of the flicker hybrid zone from its southern anchor in New Mexico all the way to western British Columbia (where the accuracy of the flicker hybrid zone map breaks down). The hybrid zone between these two warblers occurs in the overlap area illustrated in Figure 8-9. Although the warbler hybrid zone and the flicker hybrid zone have not been mapped on a single, composite map, they appear to occur in close proximity in the Canadian Rocky Mountains (Hubbard, 1969; Barrowclough, 1980).

Vegetational Ecotones. The relation between the flicker hybrid zone and several vegetational ecotones is illustrated in Figure 8-10.[3] [Vegetational ecotones were digitized from Kuchler (1966) and the National Atlas of Canada (1973), Vegetation Map, pp. 45–46.) The eastern boundary of the hybrid zone parallels the ecotone between tall-

---

3. We selected the major vegetational ecotones in the vicinity of the northern flicker hybrid zone. We ignored numerous other vegetational ecotones in North America that obviously are not associated with the hybrid zone. Our objective here is not to test whether the flicker hybrid zone has more than a chance association with ecotones drawn at random but, rather, to observe how intimate the association is between the hybrid zone and nearby ecotones.

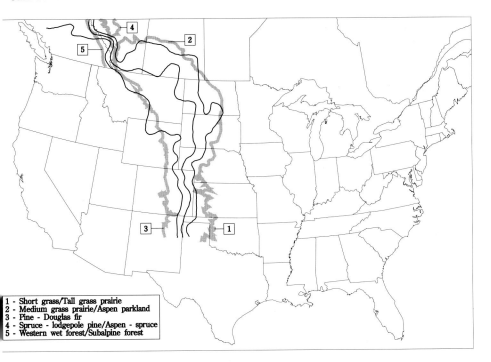

1 - Short grass/Tall grass prairie
2 - Medium grass prairie/Aspen parkland
3 - Pine - Douglas fir
4 - Spruce - lodgepole pine/Aspen - spruce
5 - Western wet forest/Subalpine forest

Figure 8-10. Vegetational ecotones correlated with the northern flicker hybrid zone. The hybrid zone is represented by the 20%, 50%, and 80% transition contours from yellow-shafted to red-shafted phenotype. The broader, shaded boundaries represent vegetational ecotones digitized from Kuchler (1966) for the United States and from the National Atlas of Canada. The numbered ecotones separate the vegetational communities or correspond to the boundaries indicated in the legend (west/east). See text for details.

grass prairie (bluestem–grama prairie in the south, wheatgrass–bluestem–needlegrass in the north) (Kuchler, 1966) and short-grass prairie (grama-buffalo grass in the south; wheatgrass–needlegrass in the north) (Kuchler, 1966) from the southern limit of the hybrid zone north to the Canadian border. Vegetation types as defined in the National Atlas of Canada are not exactly comparable with Kuchler's (1966) classification. Nonetheless, the tall-grass–short-grass prairie ecotone can be extrapolated north to contact an ecotone between medium-grass prairie (wheatgrass–grama grass–needle-grass) and aspen–bur oak parkland, just north of the United States–Canadian boundary. This deciduous parkland forms a "cap" over the northern extent of prairie and is succeeded to the north by boreal forest. The yellow-shafted edge of the flicker hybrid zone closely follows the arc of this prairie-deciduous parkland ecotone through southern Saskatchewan and Alberta to the vicinity of the Alberta–British Columbia Provincial boundary where the hybrid zone narrows dramatically and turns sharply to the north. From this point to the northwest, the entire hybrid zone is largely confined to a narrow belt of forest, described in the National Atlas of Canada as subalpine forest comprising Engelmann spruce, alpine fir, and lodgepole pine, often with an open distribution. This belt is closely paralleled by western hemlock–western red-cedar–douglas fir (western wet forest) or ponderosa pine associations to the southwest and a boreal spruce–lodgepole pine association to the northeast (National Atlas of Canada, 1973).

The red-shafted edge of the flicker hybrid zone strongly parallels the eastern limit of pine–Douglas fir forests of the front range of the Rocky Mountains in the southern and central United States, although the two are not closely associated. At approximately 45°N, where the hybrid zone turns northwest, the red-shafted edge of the hybrid zone comes in close association with this cordilleran forest–grassland ecotone and follows it closely along the Alberta–British Columbia provincial line to the point where the hybrid zone map becomes inaccurate.

Climatic Gradients. Our analysis of climatic correlates of the norther flicker hybrid zone is limited because of difficulties we encountered when obtaining computerized Canadian weather data in a timely manner. We were able to examine correlations with mean annual precipitation data for the United States and with measures of water balance for the United States and Canada. The precipitation data were derived from CLI-MATEDATA (EarthInfo, Inc., Denver, CO.) and are data from the National Climate Data Center. The water balance data were derived from Thornthwaite (1964a,b). A moisture index overlaid by the flicker hybrid zone is mapped in Figure 8-11. We do not present a precipitation map because our analysis was limited to the United States and the moisture index reflects precipitation.

The association between the flicker hybrid zone and a precipitation gradient in the United States is fairly strong. The hybrid zone is confined to a region of 10–20 inches of annual precipitation. The eastern and center contours of the hybrid zone are associated with the 20-inch and 15-inch isopleths, respectively, from the Texas panhandle to southern North Dakota, where the hybrid zone turns toward the northwest, but the precipitation contours continue north and even northeast. The width of the region between 10 and 20 inches of annual precipitation also corresponds with the width of the hybrid zone. The precipitation gradient is steepest in the southern plains and broadens to the north in rough correlation with the broadening hybrid zone.

On a continental scale, water balance is thought to be a primary determinant of the geography of vegetational communities (Stephenson, 1990). Thornthwaite (1964a) quantified water balance with a moisture index (MI) defined as $MI = 100(s - d)/PE$, which he used to classify climates (s = water surplus, d = water deficit, PE = potential evapotranspiration; -40 to -20 = semiarid; -20 to 0 = dry subhumid; 0 to 20 = moist subhumid). We extracted the mean annual PE, s, and d for 1000 sites recorded by Thornthwaite (1964a,b) and used them to study the association of the flicker hybrid zone with MI. Data were extracted for all sites in states and provinces (up to 56°N) where the hybrid zone occurs and in adjacent states and provinces. Data for a smaller number of sites (two to eight) were extracted for states and provinces remote from the hybrid zone. We calculated the MI for the 1000 sites and plotted the data as a contour map overlaid by the flicker hybrid zone (Fig. 8-11).

As was the case with certain mean annual precipitation contours, the MI contours between $+10$ and -40 take on a strong south-to-north orientation on the Great Plains with strong parallelism between successive contours. This area is the rain shadow of the Rocky Mountains, and this map and the annual precipitation map (not illustrated) suggest that the hybrid zone parallels this rain shadow, at least in the United States. In the northern plains, the generally parallel MI contours bend toward the northwest and

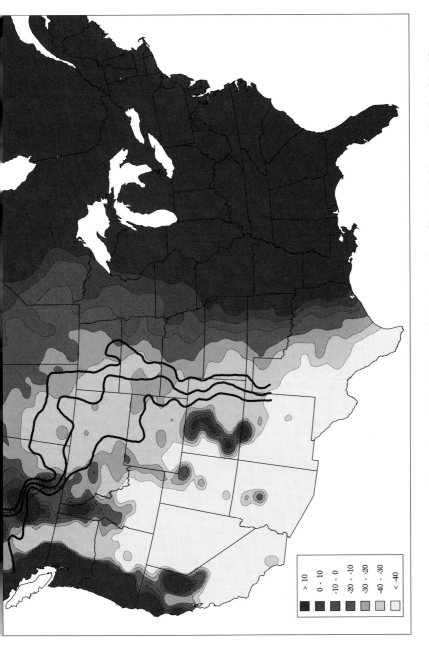

Figure 8-11. Contour map of the moisture index (MI) overlaid by the northern flicker hybrid zone. Shading identifies areas between successive contour intervals as indicated in the legend. The 20%, 50%, and 80% hybrid zone contours are overlaid as heavy curves.

215

make the familiar arcing pattern across the northern limit of the prairie in Saskatch-ewan and Alberta. This pattern is particularly apparent for MI values between -30 and -20. From the point defined by the junction of the Alberta–British Columbia provincial boundary with the Montana state line toward the northwest, the hybrid zone does not closely parallel any MI contour but seems to cross a "ridge" of high MI. At this point, however, the accuracy of the hybrid zone contour map becomes suspect.

ROLE OF SELECTION IN DETERMINING THE STRUCTURE OF THE FLICKER HYBRID ZONE

The apparent association of the northern flicker hybrid zone with ecological gradients is, in itself, evidence that selection is involved in maintaining the hybrid zone. How-ever, even in the absence of this evidence, arguments based on cline theory suggest that selection is a factor and, indeed, is probably strong—although, it could not be deter-mined whether the selection is endogenous or exogenous. Stated verbally, the argu-ment is that selection must be strong in order to maintain the narrow hybrid zone in the face of the apparently high dispersal rate estimated for flickers. Extending cline theory to polygenic characters, Barton (1983) defined a variable termed effective selec-tion (s^*), which is the selection operating on a particular allele plus all of the alleles at other loci with which it is in linkage disequilibrium. Linkage disequilibria appear to be high in hybrid zones (Szymura and Barton, 1986; Bert and Harrison, 1988) and are probably attributable to dispersal of parental genotypes into the hybrid zone, selection favoring parental genotypes (either as endogenous selection against hybrids or exoge-nous selection favoring parental genotypes on either side of the hybrid zone), and in some instances physical linkage between alleles that define the parental taxa. Thus selection operating on an individual marker locus can be weak, and yet the average amount of selection experienced by whole organisms can be high because effective selection is a function of selection operating on all correlated alleles. Szymura and Bar-ton (1986), for example, estimated that 300 loci are selected in the hybrid zone involv-ing fire-bellied toads, and the effective selection is 0.17.

Specific marker genes that differ in allelic frequency across the flicker hybrid zone have not been identified. Nonetheless, this reasoning can be applied to estimate roughly the effective selection maintaining the hybrid zone. Each of the six plumage traits that contribute to the hybrid index is determined by polygenes (Moore, unpub-lished pedigree data), as is the *hybrid index,* which is the sum of the scores for individ-ual traits. The effective selection operating on a typical gene involved in the determi-nation of a polygenic trait, e.g., hybrid index, can be estimated. Obviously it is a crude—but useful—estimate, particularly considered against the alternative of no esti-mate at all. The equation that relates effective selection to hybrid zone width is iden-tical to the equilibrium-width equation for a single gene given in Figure 8-1 but with s^* substituted for s (Szymura and Barton, 1986, p. 1154). Solving for s^* gives $s^* = (4\sigma/w)^2/2$, where σ = the root mean square dispersal (RMS dispersal), and w = the width of the hybrid zone (w is defined as the inverse of the maximum slope, or approximately the distance from the 12%–88% transition in hybrid index) (Barton and Hewitt, 1981). The width of the flicker hybrid zone in the southern plains is approximately 300 km, and Moore and Buchanan (1985) estimated RMS dispersal for the flicker as 100.7 km per generation; thus the estimated effective selection is 0.90. This selection is so intense that the diffusion model of the hybrid zone is not applicable because it assumes that s* is small. Even if the conservative assumption were made that $\sigma = 10$ km, an s* of 0.067 would still be required to explain the narrow hybrid zone (i.e., selection of nearly 7%). Thus if the estimated dispersal rate for the flickers is correct to within an order of magnitude, it is reasonable to conclude that the selection operating on the hybrid zone is strong.

This inference is sensitive to the estimated RMS dispersal, and we acknowledge that the estimate for flickers may be substantially incorrect because it was based on a small number of USFWS banding recoveries (Moore and Buchanan, 1985). However, the low level of genetic structure observed in flicker populations is consistent with the inference that gene flow is high (Grudzien et al., 1987). Moreover, estimates of RMS dispersal for two other migratory North American birds, the red-winged blackbird and the common grackle, are comparably high (Moore and Dolbeer, 1989), and genetic analysis of several additional North American bird species reveal high indirect estimates of gene flow (Zink and Remsen, 1986; Rockwell and Barrowclough, 1987; Johnson and Marten, 1988). Thus gene flow appears to be generally high in North American migratory birds, which is not surprising given that they annually fly thousands of miles between wintering and breeding grounds.

In the case of the northern flicker, the apparent association of the hybrid zone with several ecological variables and indeed the similar geographic patterns of all the variables strongly suggest that much of the selection maintaining this hybrid zone is exogenous and hence ecological in nature. All of the correlates generally parallel the hybrid zone as it runs north from the Texas panhandle to the vicinity of the Black Hills. Near the point where the South Dakota, Wyoming, and Montana state lines form a corner, both the western edge of the hybrid zone and the western correlates turn toward the northwest and continue to the vicinity where the Alberta–British Columbia provincial line joins the Montana state line; there the hybrid zone turns to the north, and at least the mammalian range boundary concentrations, the BBS avian range boundary concentrations, the vegetational boundaries, and the MIs roughly parallel this turn. The eastern edge of the hybrid zone strongly parallels the boundary between short-grass and tall-grass prairie in the United States, which continues as an ecotone between prairie and aspen parkland in Canada. Virtually all of the correlates parallel the arc of this ecotone across the northern limit of the prairies, with the exception of the indigo bunting range boundary, which does not extend that far north.

The width of the hybrid zone also appears to vary with the steepness of ecological gradients. The hybrid zone narrows as it enters Canada at the point where the vegetational ecotones come together. The gradients in avian and mammalian range boundaries are also steep in this area as is the gradient in MI. In the northern United States the hybrid zone becomes widest where the MI gradient is flattest and the vegetational ecotones are farthest apart. The hybrid zone is narrower in the southern plains, where the MI gradient steepens and the vegetational ecotones draw close together.

We have not performed a statistical analysis with tests for significant association between the hybrid zone and the various ecological variables. The appropriate probability model would be akin to scattering pipe cleaners that had been crumpled and then partially restraightened on a map of North America and asking about the probability that they would fall, by chance alone, with the extent of association observed. This kind of analysis would be formidable; but even without a rigorous analysis, it is obvious that this probability is small given that the pipe cleaners would have random crumples and fall with random orientations.

Moreover, our estimates of the tracks the ecological variables follow are probably "noisy" because residual variance is added to the data by the sampling procedures involved in collecting the original data (by various investigators), temporal variation both within and between data sets, and by variance in the spatial distribution of data points (which affects the contour mapping algorithms). In other words, with this much potential residual variance in the data, one would expect to see the associations at all only if they were fairly strong.

Two issues remain to be addressed with regard to the association between the flicker hybrid zone and ecological variables. The first is whether a hypothesis other than exogenous selection could explain the associations, and the second is to infer which variables are directly involved (e.g., vegetation, climate), assuming that exogenous selection is operating.

With regard to the first issue, there are alternative hypotheses: (1) The hybrid zone had an historical origin as part of an ecotone, with which it remains associated, but the ecotone does not exert selection. (2) The hybrid zone is maintained by endogenous selection, and the position of the hybrid zone is determined by a population density trough that, not implausibly, covaries with the ecological variables studied. The first hypothesis posits that hybrid zones have inertia, of sorts, and tend to stay put or move only slowly relative to the life expectancy of an evolutionary biologist. Thus hybrid zones tend to stay where they were formed. Moreover, if the hybrid zone is indeed a secondary contact between two incipient species that diverged along with a host of other species during a refugial epoch, the hybrid zone would be part of an ecotone (or suture zone) formed by the secondary contact of the whole ecological communities expanding from the refugia, and the hybrid zone would be correlated with biotic gradients. This hypothesis also posits that selection is not a factor (the alleles fixed across the hybrid zone are adaptively neutral), and so the hybrid zone would not tend to straighten out, as would be the case with a tension zone (Barton, 1979b; Barton and Hewitt, 1981, 1985, 1989), or track an environmental variable. In other words, the hybrid zone would simply stay where it was formed, although it would broaden as the adaptively neutral marker alleles diffused across the hybrid zone. If RMS dispersal for the species were low, however, the hybrid zone would broaden only gradually. We have discounted this hypothesis because RMS dispersal appears to be too high to account for the observed narrowness of the hybrid zone; but, as mentioned above, it is possible that we have overestimated dispersal. Also the width of the hybrid zone has been stable in historical time (Moore and Buchanan, 1985).

The second hypothesis—that the hybrid zone is maintained purely by endogenous selection but is "trapped" in a population density trough—has some plausibility. Although, there are no troughs in the density of breeding flickers along specific riparian woodlands (Moore and Buchanan, 1985) the Great Plains as a whole can be viewed as a population density trough because of a paucity of arboreal vegetation there. This pattern is apparent in contour maps of flicker population density based on the BBS data (not illustrated). The rain shadow of the Rocky Mountains doubtless causes both the observed gradients in ecological variables and the population density trough. Although we cannot determine with certainty which factor is directly linked in the chain of causation that determines the position of the flicker hybrid zone, it seems unlikely that this population density trough, in context of the endogenous selection model, could explain the observed variation in position and width of the hybrid zone. The tension zone model predicts that the hybrid zone would tend to straighten to the shortest distance between its endpoints (the Texas panhandle and southern Alaska) and would be uniform in width. The flicker hybrid zone bows substantially from the straight line between its termini, makes three major turns in orientation, and varies substantially in width. Thus although the flicker hybrid zone may be associated with a population density trough, its geometry is not consistent with the predictions of a tension zone model. It would be possible to accommodate these anomalies in the tension zone model, but it would require several *ad hoc* suppositions about variation in the magnitude and direction of gene flow along the length of the hybrid zone. Moreover, the hybrid zone is not correlated as strongly with the population density trough (not illustrated) as it is with the ecological variables. Specifically, the track of the hybrid zone appears to be unaffected by areas of high density in southeastern Montana and where it tracks the woodland ecotones along the Alberta–British Columbia provincial boundaries. Thus we conclude that the intense selection apparently operating to maintain the position and width of the northern flicker hybrid zone is primarily exogenous in nature.

The second issue concerns the ecological variables that are directly involved in the inferred exogenous selection. We believe that the rain shadow of the Rocky Mountains determines the position and width of the northern flicker hybrid zone, and the other avian hybrid zones of the Great Plains as well as the correlated vegetational patterns. However, we are unable to make a reasonable inference whether climate per se (particularly aridity) is the direct cause or it is mediated by either the flora or fauna of the region, or both.

It is plausible that climatic variables directly determine the ranges of the hybridizing taxa. For example, Rising (1969) showed that the Baltimore oriole has a higher metabolic rate under heat stress than Bullock's oriole. Rising argued that the higher metabolic rate demands increased evaporative cooling and hence water loss, which would translate into decreased fitness for the Baltimore oriole in hot, arid regions such as those inhabited by Bullock's oriole.

It is plausible that the red- and yellow-shafted flickers have metabolic differences comparable to those seen between the orioles. It is of interest in this regard that there is significant geographic variation in the frequencies of mitochondrial DNA haplotypes in flickers, and the pattern is one of divergence between southwestern populations, on the one hand, and northern and eastern populations on the other (Moore et al., 1991). Moore et al. (1991) hypothesized, as one of two alternatives, that variation in haplotype frequency was determined by selection, and that the southwestern mitochondrial genomes adapted the flickers to hot and arid conditions prevalent in the southwestern United States.

Other plausible explanations are that either biotic variables alone or a combination of abiotic and biotic variables determine the ranges of the hybridizing taxa. For example, Rohwer and Manning (1990) reported a difference in the molting and migratory patterns of the two orioles that are probably adaptations to the mesic and xeric regions of North America inhabited by the two subspecies. The Baltimore oriole migrates in the spring to its breeding grounds in eastern North America from southern Mexico, Central America, and northern South America; it breeds and then undergoes its prebasic molt before returning to its winter grounds in the fall. Rohwer and Manning argued that this molting and migratory pattern are parts of an adaptation for a migratory route across the Gulf of Mexico that demands unworn feathers for the long overwater flight. Bullock's oriole, in contrast, migrates from the southwestern United States and western Mexico to its breeding grounds in the western United States; it breeds and soon begins the first stage of a two-stage southern migration. In the first stage Bullock's oriole moves as far as the southwestern United States before it undergoes its prebasic molt. It then undertakes the second stage of migration to its wintering grounds in Mexico. Rohwer and Manning (1990) hypothesized that this adaptation avoids drought conditions that characterize the northern part of its range in the late summer and takes advantage of the high productivity of the southwest brought on by late summer monsoons. Bullock's oriole is not penalized for migrating in juvenal and worn plumages because migration is strictly over land. The adaptation in Bullock's oriole hypothesized by Rohwer and Manning (1990) would be an adaptation to biotic differences resulting from climatic differences.

Of the several ecological variables examined in this chapter, the association between vegetational ecotones and the hybrid zone appear strongest. It is obvious, however, that an ecotone between two grassland types would not directly determine the position of hybrid zones involving woodland species. Moreover, the hybrid zone does not map onto either of the vegetational ecotones in the United States but only strongly parallels them. There are indications, however, of vegetational ecotones occurring in the riparian woodlands of the rivers that traverse the plains in the area of the avian hybrid zones.

We did an ordination study, based on data from Currier (1982), of riparian vegetation along part of one drainage that traverses the hybrid zone. Currier sampled and characterized the flood plain vegetation along the Platte and North Platte Rivers in Nebraska from longitude 98.2° to 101.7°. The western edge of Currier's study is approximately where the flicker hybrid zone begins, but his sampled region would cover the other avian hybrid zones along the Platte River. The percent ground cover was determined for 84 species from each of several plots on 55 transects across the flood plain. We did a principal component (PC) analysis and correlated the PC scores with longitude of the transect along the River. PC1 is significantly correlated with longitude ($r = -0.43$; $p<0.0009$). Eastern plots tend to have positive PC1 scores, and plot scores become more negative traveling west. Species that load heavily on PC1 in the positive direction all have primarily eastern distributions or are found in relatively moist, rich woodlands (Rydberg, 1971; Stephens, 1973). These species include, in order of loading, rough-leaved dogwood, woodbine, poison ivy, white avens, starry false Solomon's seal, and snakeroot. Species with large negative loadings, and hence that dominate PC1 to the west, are either herbs or grasses that tend to favor less shaded habitat (ragweed, marsh grass, false indigo, *Sphenopholis obtusata,* prairie sunflower, and wheatgrass, in order of decreasing negative magnitude). The obvious interpretation of PC1 is that it is a manifestation of a trend toward more open woodlands, with scattered trees and a less shrubby understory, which correlates with increasing aridity. Thus it is possible that the position of the hybrid zone is determined by vegetational changes in the riparian woodlands themselves.

At this juncture there is no point in belaboring hypotheses about the specific roles of ecological variables in determining the structure of the flicker hybrid zone because some of the data needed to test these hypotheses are unavailable, and none of the requisite analyses have been done. It is worth mentioning, however, that these hypotheses are testable, and the work presented in this chapter suggests that the appropriate studies would be worthwhile.

ECOLOGICAL AND SEXUAL SELECTION

We close by briefly exploring the paradox that ecological selection appears to determine the position and width of the hybrid zone, and yet the traits that differ across the hybrid zone are primarily plumage patterns that probably evolved in context of intrasexual selection, specifically territorial defense (Moore, 1987), and do not have obvious ecological significance. As a tentative explanation, we propose that linkage between sexually selected genes and genes selected along an ecological gradient could produce a hybrid zone such as that observed for the northern flicker (and most other avian hybrid zones), provided the sexually selected phenotypes diverged in allopatry—which implies that the hybrid zone arose as a secondary contact.

The origin and establishment of novel phenotypes that communicate social status (e.g., mimetic patterns) are not easily understood, but understanding the origin of red- and yellow-shafted phenotypes is not critical for the sexual selection hypothesis; we assume they arose in allopatry, somehow. However, frequency-dependent selection is involved in their evolution (Moore, 1979, 1980; Barton and Hewitt, 1985; Mallet and Barton, 1989a,b; see Ch. 9). In terms of the northern flicker, we posit that the bright and conspicuous plumage patterns that define the northern flicker hybrid zone are important in territorial defense, and that flickers that cross the hybrid zone have difficulty defending territories because individuals of the resident subspecies are not intimidated by the status signals communicated by the immigrant flicker.

The population genetics of this hypothesis must be complex because it presumably would involve genetically determined sending and receiving systems, both of which probably would be polygenic. Frequency-dependent selection would occur, however, because minority phenotypes would not send, or correctly receive, social signals preferred by the majority. Moreover, selection would be directional, favoring a single phenotype, regardless of how many traits and genes contributed to the phenotype. The phenotype would be a composite of all the traits involved in communicating social status. For the flicker, it may include all of the plumage traits diverged across the hybrid zone. Thus linkage of a single gene among the polygenes that determine the social signaling system to an ecologically selected gene would result in the cline in the ecological gene determining the position of the linked social-status gene, which in turn would determine the clines for all of the polygenes determining the social-status phenotype.

The cline in social-signaling phenotype would actually have the dynamics of a tension zone (Barton and Hewitt, 1985, 1989; Mallet and Barton, 1989a; see Ch. 9), but the position of the hybrid zone would be determined by an ecological gradient. Presumably, the width of such a hybrid zone would be determined by the effective selection resulting from both the ecological gradient and the frequency-dependent sexual selection. In the absence of an appropriate mathematical model, making an inference about the dynamics of this system is risky. However, it is intuitive that the hybrid zone would be narrower than a hybrid zone determined solely by an ecological gradient because of the effective selection mediated by the social-signaling genes (Lande, 1982).

IMPLICATIONS FOR SPECIATION THEORY

A synthesis of observations on the northern flicker hybrid zone, in context of cline theory, suggests that several factors interact to maintain the cohesion of phenotype in each of the two subspecies involved in the hybrid zone. Gene flow affects many aspects of the hybrid zone, but more fundamental questions concern how selection and the genetic system interact to maintain the integrity of the subspecies. It is clear that eco-logical selection is a major factor, but it is likely that sexual selection and linkage dis-equilibria between genes that determine ecological and sexually selected phenotypes are essential factors as well. To the extent that frequency-dependent sexual selection is involved, the flicker hybrid zone has properties of a tension zone, but it is biologi-cally important to recognize that it does not result from hybrid breakdown.

The cohesion of phenotypes maintained across hybrid zones would involve some but not all genes in the genomes of hybridizing taxa. Specifically, adaptively neutral genes and genes that have positive fitness everywhere would pass through the hybrid zone; and in this sense, hybrid zones are semipermeable barriers to gene flow. None-theless, the cohesion of genes that define the parental taxa would be maintained unless there was some change in the selection regimes that maintain the clines. If our analysis of the northern flicker hybrid zone is approximately correct, such change would occur only if the rain shadow of the Rocky Mountains disappeared. This disappearance will not occur over a time interval that is relevant to this evolutionary process, and so the genes that cohere to define the red- and yellow-shafted flickers are locked onto diver-gent evolutionary pathways. For this reason, the red-shafted and yellow-shafted flick-ers should be considered distinct species.

ACKNOWLEDGMENTS

We thank Theresa M. Bert for her thoughtful discussion and comments during the early development of this manuscript. We also thank Theresa M. Bert, D. Carl Freeman, Daniel J. Howard, James Mallet, and Terry Root for providing critical readings of an early draft of the manuscript. This research was supported in part by a grant from the National Science Foundation (BSR 87-05374).

REFERENCES

Anderson, B. W. 1971. Man's influence on hybridization in two avian species in South Dakota. Condor 73:342–347.

Anderson, B. W., and Daugherty, R. J. Charac-terization and reproductive biology of gros-beaks *(Pheucticus)* in the hybrid zone in South Dakota. Wilson Bull. 86:1–11.

Barrowclough, G. F. 1980. Genetic and pheno-typic differentiation in a wood warbler (genus *Dendroica*) hybrid zone. Auk 97:655–668.

Barton, N. H. 1979a. Gene flow past a cline. *Heredity* 43:333–359.

Barton, N. H. 1979b. The dynamics of hybrid zones. Heredity 43:341–359.

Barton, N. H. 1983. Multilocus clines. Evolution 37:454–471.

Barton, N. H., and Hewitt, G. M. 1981. Hybrid zones and speciation. In W. R. Atchley and D. S. Woodruff, eds. Essays on Evolution and Spe-ciation in Honor of M. J. D. White. Cam-bridge: Cambridge University Press.

Barton, N. H., and Hewitt, G. M. 1985. Analysis of hybrid zones. Ann. Rev. Ecol. Syst. 16:113–148.

Barton, N. H., and Hewitt, G. M. 1989. Adaptation, speciation and hybrid zones. Nature 341:497–503.

Bazykin, A. D. 1969. Hypothetical mechanism of speciation. Evolution 23:685–687.

Bert, T. M., and Harrison, R. G. 1988. Hybridization in western Atlantic stone crabs (genus *Menippe*): evolutionary history and ecological context influence species interactions. Evolution 42:528–544.

Bock, C. E. 1971. Pairing in hybrid flicker populations in eastern Colorado. Auk 88:921–924.

Braun, D., Kitto, G. B., and Braun, M. J. 1984. Molecular population genetics of tufted and black-crested forms of *Parus bicolor.* Auk 101:170–173.

Braun, M. J., and Robbins, M. B. 1986. Extensive protein similarity of the hybridizing chickadees *Parus atricapillus* and *P. carolinensis.* Auk 103:667–675.

Brewer, R. 1963. Ecological and reproductive relationships of black-capped and Carolina chickadees. Auk 80:9–47.

Conner, R. N., and Adkisson, C. S. 1977. Principal component analysis of woodpecker nesting habitat. Wilson Bull. 89:122–129.

Corbin, K. W., and Sibley, C. G. 1977. Rapid evolution in orioles in the genus *Icterus.* Condor 79:335–342.

Corbin, K. W., Sibley, C. G., and Ferguson, A. 1979. Genetic changes associated with the establishment of sympatry in orioles in the genus *Icterus.* Evolution 33:624–633.

Cracraft, J. 1983. Species concepts and speciation analysis. Curr Ornithol. 1:159–187.

Currier, P. J. 1982. The floodplain vegetation of the Platte River: phytosociology, forest development, and seedling establishment. Ph.D. thesis, Iowa State University.

Dixon, K. L. 1955. An ecological analysis of the interbreeding of crested titmice in Texas. Univ. Calif. Publ. Zool. 54:125–206.

Droege, S. 1990. The North American Breeding Bird Survey. In: Survey Designs and Statistical Methods for the Estimation of Avian Population Trends. U.S. Fish Wildl. Serv. Biol. Rep. 90:1–4.

Ehrlich, P. R. 1961. Has the biological species concept outlived its usefulness? Syst. Zool. 10:167–176.

Emlen, S. T., Rising, J. D., and Thompson, W. L. 1975. A behavioral and morphological study of sympatry in the indigo and lazuli buntings of the Great Plains. Wilson Bull. 87:145–179.

Endler, J. A. 1977. Geographic variation, speciation, and clines. Princeton, NJ: Princeton University Press.

Endler, J. A. 1986. Natural selection in the wild. Princeton, NJ: Princeton University Press.

Endler, J. A. 1989. Conceptual and other problems in speciation. In D. Otte and J. A. Endler,

eds. Speciation and Its Consequences. Sunderland MA: Sinauer.

Ghiselin, M. T. 1987. Species concepts, individuality, and objectivity. Biol. Philos. 2:127–143.

Grudzien, T. A., and Moore, W. S. 1986. Genetic differentiation between the yellow-shafted and red-shafted subspecies of the northern flicker. Biochem. Syst. Ecol. 14:451–453.

Grudzien, T. A., Moore, W. S., Cook, J. R., and Tagle, D. 1987. Genetic population structure of the northern flicker *(Colaptes auratus)* hybrid zone. Auk 104:654–664.

Hagmeier, E. M., and Stults, C. D. 1964. A numerical analysis of the distributional patterns of North American mammals. Syst. Zool. 13:125–155.

Hewitt, G. M., and Barton, N. H. 1980. The structure and maintenance of hybrid zones as exemplified by *Podisma pedestris.* In R. L. Blackman, G. M. Hewitt, and M. Ashburner, eds. Insect Cytogenetics. Symp. R. Entomol. Soc. Lond. 10:149–170.

Hubbard, J. P. 1969. The relationships and evolution of the *Dendroica coronata* complex. Auk 86:393–432.

Huntington, C. E. 1952. Hybridization in the purple grackle, *Quiscalus quiscula.* Syst. Zool. 1:149–170.

Johnson, N. K., and Johnson, C. B. 1985. Speciation in sapsuckers *(Sphyrapicus).* II. Sympatry, hybridization, and mate preference in *S. ruber daggetti* and *S. nuchalis.* Auk 102:1–15.

Johnson, N. K., and Zink, R. M. 1983. Speciation in sapsuckers *(Sphyrapicus).* I. Genetic differentiation. Auk 100:871–884.

Johnson, N. K. and Marten, J. A. 1988. Evolutionary genetics of flycatchers. II. Differentiation in the *Empidonax difficilis* complex. Auk 10:177–191.

Johnston, D. W. 1971. Ecological aspects of hybridizing chickadees *(Parus)* in Virginia. Am. Midl. Natur. 85:124–134; 85:124–134.

Kocher, T. D., and Sage, R. D. 1986. Further genetic analyses of a hybrid zone between leopard frogs (*Rana pipiens* complex) in central Texas. Evolution 40(1):21–33.

Kroodsma, R. L. 1974a. Species recognition behavior of territorial male rose-breasted and black-headed grosbeaks *(Pheucticus).* Auk 91:54–64.

Kroodsma, R. L. 1974b. Hybridization in grosbeaks *(Pheucticus)* in North Dakota. Wilson Bull. 86:230–236.

Kroodsma, R. L. 1975. Hybridization in buntings *(Passerina)* in North Dakota and eastern Montana. Auk 92:66–80.

Kuchler, A. W. 1966. Potential vegetation of the United States, University of Kansas, 1966 (Revised, 1985). National Atlas of the United States of America. Reston, VA: Department of the Interior, U.S. Geological Survey.

Lande, R. 1982. Rapid origin of sexual isolation and character divergence in a cline. Evolution 36:213–223.

Mack, A. L., Gill, F. B., Colburn, R., and Spolsky, C. 1986. Mitochondrial DNA: a source of genetic markers for studies of similar passerine bird species. Auk 103:676–681.

Mallet, J. 1986. Hybrid zones of *Heliconius* butterflies in Panama and the stability and movement of warning colour clines. Heredity 56:191–202.

Mallet, J., and Barton, N. 1989a. Inference from clines stabilized by frequency-dependent selection. Genetics 122:967–976.

Mallet, J., and Barton, N. H. 1989b. Strong natural selection in a warning color hybrid zone. Evolution 43:421–431.

Mallet, J., Barton, N. H., Lamas, G. M., Santisteban, J. C., Muedas, M. M., and Eeley, H. 1990. Estimates of selection and gene flow from measures of cline width and linkage disequilibrium in *Heliconius* hybrid zones. Genetics 124:921–936.

May, R. M., Endler, J. A., and McMurtrie, R. E. 1975. Gene frequency clines in the presence of selection opposed by gene flow. Am. Naturalist 109:659–676.

Mayr, E. 1957. Species concepts and definitions. In E. Mayr, ed. The Species Problem. American Association for the Advancement of Science Publication No. 50. Washington, DC: AAAS.

Mayr, E. 1987. The ontological status of species: scientific progress and philosophical terminology. Biol. Philos. 2:145–166.

McKitrick, M. C. and Zink, R. M. 1988. Species concepts in ornithology. Condor 90:1–14.

Mishler, B. D., and Donoghue, M. J. 1982. Species concepts: a case for pluralism. Syst. Zool. 31:491–503.

Misra, R. K., and Short, L. L. 1974. A biometric analysis of oriole hybridization. Condor 76:137–146.

Moore, W. S. 1977. An evaluation of narrow hybrid zones in vertebrates. Q. Rev. Biol. 52:263–277.

Moore, W. S. 1979. A single locus mass-action model of assortative mating with comments on the process of speciation. Heredity 42:173–186.

Moore, W. S. 1980. Assortative mating genes selected along a gradient. Heredity 46:191–195.

Moore, W. S. 1987. Random mating in the northern flicker hybrid zone: implications for the evolution of bright and contrasting plumage patterns in birds. Evolution 41:539–546.

Moore, W. S. 1990. The species paradox. Bioscience 40:313–314.

Moore, W. S., and Buchanan, D. B. 1985. Stability of the northern flicker hybrid zone. Evolution 39:135–151.

Moore, W. S., and Koenig, W. D. 1986. Comparative reproductive success of yellow-shafted, red-shafted and hybrid flickers across a hybrid zone. Auk 103:42–51.

Moore, W. S., and Dolbeer, R. A. 1989. The use of banding recovery data to estimate dispersal rates and gene flow in avian species: case studies in the red-winged blackbird and common grackle. Condor 91:242–253.

Moore, W. S., Graham, J. H., and Price, J. 1991. Geographic variation of mitochondrial DNA in the northern flicker *(Colaptes auratus)*. Mol. Biol. Evol. 8:327–344.

National Atlas of Canada. 1973. Surveys and Mapping Branch, Department of Energy, Mines and Resources, Ottawa, Canada.

Rand, D. M., and Harrison, R. G. 1989. Ecological genetics of a mosaic hybrid zone: mitochondrial, nuclear, and reproductive differentiation of crickets by soil type. Evolution 43:432–439.

Remington, C. L. 1968. Suture-zones of hybrid interaction between recently joined biotas. In T. Dobzhansky, M. K. Hecht, and W. C. Steere, eds. Evolutionary Biology, Vol. 2. New York: Appleton-Century-Crofts, pp. 321–428.

Rising, J. D. 1969. A comparison of metabolism and evaporative water loss of Baltimore and Bullock orioles. Comp. Biochem. Physiol. 31:915–925.

Rising, J. D. 1970. Morphological variation and evolution in some North American orioles. Syst. Zool. 19:315–351.

Risking, J. D. 1973. Morphological variation and status of the orioles, *Icterus galbula, I. bullockii,* and *I. abeillei,* in the northern Great Plains and Durango, Mexico. Can. J. Zool. 51:1267–1273.

Rising, J. D. 1983a. The progress of oriole hybridization in Kansas, USA. Auk 100:885–897.

Rising, J. D. 1983b. The Great Plains hybrid zones. Curr. Ornithol. 1:137–157.

Robbins, M. B., Braun, M. J., and Tobey, E. A. 1986. Morphological and vocal variation across a contact zone between the chickadees *Parus atricapillus* and *P. carolinensis.* Auk 103:655–666.

Robbins, C. S., Bystrak, D., and Geissler, P. H. 1986. The Breeding Bird Survey: Its First Fifteen Years, 1965–1979. United States Fish and Wildlife Service, Resource Publication 157. Washington, DC: U.S. Fish and Wildlife Service.

Rockwell, R. F., and Barrowclough, G. F. 1987. Gene flow and the genetic structure of populations. In F. Cooke and P. A. Buckley eds. Avian Genetics, Orlando, FL: Academic Press, pp. 223–255.

Rohwer, S., and Manning, J. 1990. Differences in timing and number of molts for Baltimore and Bullock's orioles: implications to hybrid fitness and theories of delayed plumage maturation. Condor 92:125–140.

Rydberg, P. A. 1971. Flora of the Prairies and Plains of Central North America. Vol. I and II. New York: Dover Publications.

Short, L. L. 1965. Hybridization in the flickers *(Colaptes)* of North America. Bull. Amr. Mus. Nat. Hist. 129:307–428.

Sibley, C. G., and Short, L. L. 1959. Hybridization in the buntings *(Passerina)* of the Great Plains. Auk 76:443–463.

Sibley, C. G., and Short, L. L. 1964. Hybridization in the orioles of the Great Plains. Condor 66:130–150.

Sibley, C. G., and West, D. A. 1959. Hybridization in the rufous-sided towhees of the Great Plains. Auk 76:326–338.

Slatkin, M. 1973. Gene flow and selection in a cline. Genetics 75:733–756.

Slatkin, M. 1975. Gene flow and selection in a two-locus system. Genetics 81:787–802.

Software, Golden. 1989 SURFER, Version 4 Reference Manual. Golden, CO: Golden Software.

Sokal, R. R., and Crovello, T. J. 1970. The biological species concept: a critical evaluation. Am. Naturalist 104:127–153.

Stephens, H. A. 1973. Woody Plants of the North Central Plains. University Press of Kansas. Lawrence, Kansas.

Stephenson, N. L. 1990. Climatic control of vegetation distribution: the role of water balance. Am. Naturalist 135:649–670.

Sutton, G. M. 1968. Oriole hybridization in Oklahoma. Bull. Okla. Ornithol. Soc. 1:1–7.

Szymura, J. M., and Barton, N. H. 1986. Genetic analysis of a hybrid zone between the fire-bellied toads, *Bombina bombina* and *B. variegata,* near Cracow in southern Poland. Evolution 40:1141–1159.

Templeton, A. R. 1989. The meaning of species and speciation: a genetic perspective. In D. Otte and J. A. Endler, eds. Speciation and Its Consequences. Sunderland, MA: Sinauer.

Thornthwaite, C. W. 1964a. Average climatic water balance data of the continents. Part VI. North America (excluding United States). Lab. Climatol. 17(2).

Thornthwaite, C. W. 1964b. Average climatic water balance data of the continents. Part VII. United States. Lab. Climatol. 17(3).

West, D. A. 1962. Hybridization in grosbeaks *(Pheucticus)* of the Great Plains. Auk 79:399–424.

Woodruff, D. S. 1979. Postmating reproductive isolation in *Pseudophryne* and the evolutionary significance of hybrid zones. Science 203:561–563.

Wright, H. E., Jr. 1970. Vegetational history of the central Plains. In W. Dort Jr. and J. K. Jones Jr. eds. Lawrence: University of Kansas Press, pp. 157–172.

Yang, S. Y., and Selander, R. K. 1968. Hybridization in the grackle *Quiscalus quiscula* in Louisiana. Syst. Zool. 17:107–143.

Zink, R. M., and Remsen, Jr., J. V. 1986. Evolutionary processes and patterns of geographic variation in birds. Curr. Ornithol. 4:1–69.

Zink, R. M., and Hackett, S. J. 1988. Historical biogeographic patterns in the avifauna of North America. In H. Ouellet ed. Acta XIX Congress International Ornithology, Vol. 2 Ottawa: National Museum of Natural Sciences, University Ottawa Press.

9

Speciation, Raciation, and Color Pattern Evolution in *Heliconius* Butterflies: Evidence from Hybrid Zones

JAMES MALLET

Hybrid zones often involve either morphological traits such as color patterns of vertebrates with poorly understood genetics or genetic traits such as chromosomes, allozymes, and mitochondrial DNA, which tell us little about selection. In neotropical *Heliconius* butterflies, hybrid zones for warning, mimetic wing patterns are known in which both genetics and selection can be comprehended. While learning to avoid unpalatable prey, predators cause frequency-dependent selection against rare color-pattern morphs. Good evidence for this evolutionary constraint on color pattern change comes from both sympatric Müllerian mimicry between *Heliconius* butterflies and narrow hybrid zones between color pattern races within *Heliconius* species. Given selection against rare morphs (which occurs even though a new morph might be advantageous if common), it is difficult to explain the rampant geographic variation we see in *Heliconius* color patterns (Fig. 9-1). Once divergence *has* occurred, it is preserved by local selective pressures, even in parapatry, but the explanation of the initial divergence remains elusive.

It has been generally accepted that *Heliconius* races differentiated in allopatric Pleistocene refugia and that differentiation was in response to divergent mimetic pressures within each refugium (Turner, 1965, 1971a; Brown et al., 1974; Sheppard et al., 1985; Brown, 1987a). In this chapter, it is argued that allopatry is not necessary for divergence, and that de novo warning color evolution must have been partially responsible for raciation in *Heliconius*. I develop hypotheses of divergence and test them against data from hybrid zones and from distribution patterns of the races and species of *Heliconius*, which they separate. Little evidence is found for the importance of refugia in divergence once one admits that a null hypothesis of allopatric divergence is invalid on current theoretical grounds. The implications of these studies for those of other systems of hybrid zones and refugia are discussed. Conservation strategies that employ refugium theory as a means of choosing conservation areas should be urgently reexamined; it would be better to conserve areas that house particularly endangered species or areas with high levels of species diversity than to trust a poorly supported evolutionary theory as a guide.

HELICONIINE BUTTERFLIES AND THEIR EVOLUTION

Heliconius butterflies are a characteristic element of the neotropical biota from Texas and Florida to Argentina. The color patterns of *Heliconius* are famous for their warning colors (aposematism) and mimicry, indeed the "Heliconiidae" formed part of the basis for Bates' (1862) original theory of mimicry[1]. Bates might have dismissed as a coincidence similarities of a few butterflies in any one area, but his extensive collections across many regions of the Amazon basin revealed a striking geographic pattern. First, there were often 10 or more species in each "mimicry ring" within any area. Second, these color patterns changed regionally (see, for example, Fig. 9-1): "In tropical South America a numerous series of gaily-colored butterflies and moths, of very different families, which occur in abundance in almost every locality a naturalist may visit, are found all to change their hues and markings together, as if by the touch of an enchanter's wand, at every few hundred miles" (Bates, 1879).

Bates argued that rare unprotected butterflies such as dismorphiines came to imitate protected Ithomiinae because of selection by predators against unprotected, rare species. Bates recognized that *Heliconius* were protected because they were themselves often the objects of mimicry (Bates, 1862, p. 510), but he also argued that rare protected heliconiines such as the "silvaniform" *Heliconius* (Eltringham, 1916; Brown, 1976a) and ithomiines such as *Napeogenes* might gain an advantage by mimicking commoner protected ithomiines such as *Melinaea* (Bates, 1862, pp. 507, 549–550). Essentially, Bates proposed an early version of "Müllerian" mimicry 17 years before Müller; Müller's (1879) major advance was to clarify and generalize the argument to any pair of protected species with arbitrary relative abundance. Bates (1879) was unimpressed by Müller's explanation of sympatric mimicry between protected species, perhaps in part because Müller's theory did not explain how mimicry rings diverged geographically. Few people now read Bates' original papers, and it is not generally realized that Darwin (1863), Wallace (1865), and Bates himself thought that the evolution of mimicry was a highly convincing case of geographic speciation by natural selection, rather than just another example of trivial microevolution. Following Wilson and Brown (1953) and Mayr (1963, 1970), many biologists view species as qualitatively different taxa from races and subspecies, and they dismiss racial variation such as that in *Heliconius* as unimportant for speciation (e.g., Grimaldi, 1984). In this chapter I use theory and data from *Heliconius* to revive the idea that mimetic races have much to say about the nature and biogeography of speciation.

Heliconius and allies (e.g., *Eueides, Dryas,* and *Agraulis*—hereafter heliconiines) are closely related to the holarctic fritillaries *Argynnis, Speyeria,* and *Boloria* according to detailed morphological work by Harvey (1991). The caterpillars of heliconiines are almost completely restricted to Passifloraceae (Benson et al., 1976; Brown, 1981). These plants have many secondary chemicals, but is is still not clear to what extent the

---

1. Bates (1862) included the Ithomiinae as "danaiform Heliconiidae" within his "Heliconiidae" and referred to the butterflies we now term Heliconiini (Brown, 1981) or Heliconiiti (Harvey, 1991) as "acraeiform Heliconiidae," although he clearly understood that these groups were unrelated. Bates also used the name "Leptalidae" for what we now term Dismorphiinae (Pieridae).

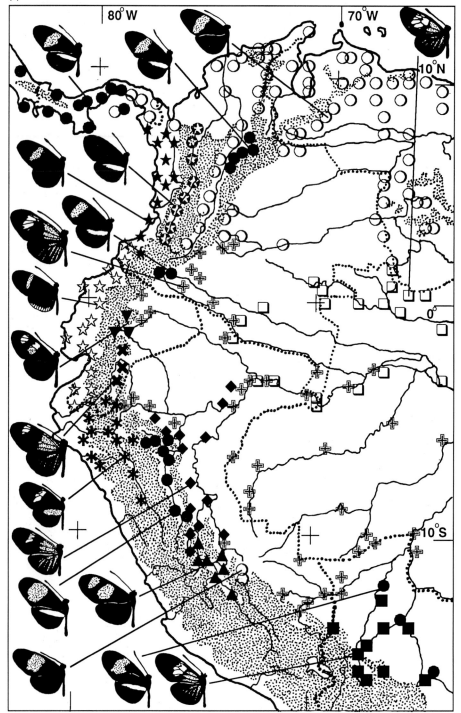

Figure 9-1. Distribution of mimetic color pattern races of *Heliconius* in Andean South America. (**A**) *Heliconius himera* (* symbols) and *H. erato* (all other symbols). (**B**) *Heliconius melpomene.* Land higher than about 1500 meters above sea level is stippled. Colors of butterflies: stippled = red; white = yellow; black = dark melanin. In *H. erato venus, H. melpomene vulcanus, H. e. cyrbia,* and *H. m. cythera* (filled and empty stars on west coast of Ecuador and Colombia) the

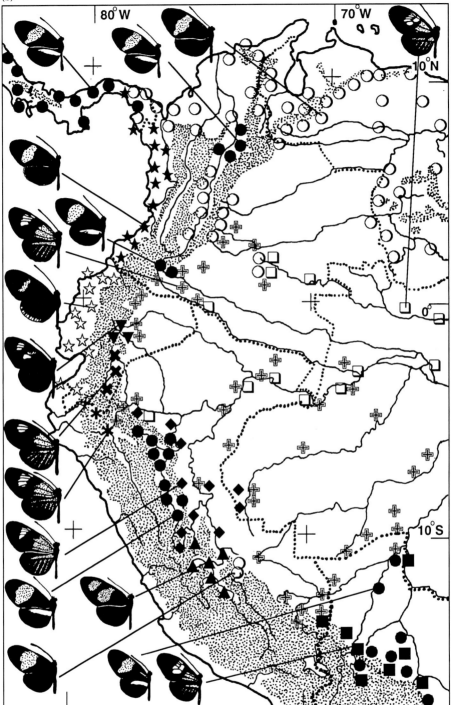

(b)

dark area of the forewing has a strong blue iridescence, there is a yellow bar on the underside of the hindwing, and the pale fringe on the hindwing is white. Hybrid zones are known between most parapatric races, although hybrids are rare between *H. himera* and adjacent races of *H. erato* (see also Tables 9-1 and 9-3). (Redrawn from Brown, 1979, and Sheppard et al., 1985, with some modifications based on the author's work.)

heliconiines obtain their unpalatability from their food plants, adult feeding, or chemical synthesis (Brower, 1984; Gilbert, 1991). Regardless of the origins of their unpleasant taste, tests with caged birds and field studies have confirmed unpalatability and efficacy of mimicry within *Heliconius* (Brower et al., 1963; Brower and Brower, 1964; Benson, 1972; Boyden, 1976; Chai, 1986; Mallet and Barton, 1989b), and these butterflies conform to a classic unpalatable body and wing shape (Chai and Srygley, 1990; Srygley and Chai, 1990). *Heliconius* species are able to hover and fly slowly because of their protection against predators, and they have become adept at precise oviposition on young tendrils and shoot tips of *Passiflora*. The genus has also developed an unusual coevolved pollinator mutualism with *Psiguria* and *Gurania* (Cucurbitaceae). The plants provide amino-acid-rich pollen used in the nutrition of adults (Gilbert, 1975), as well as nectar, a more usual food for butterflies. These novel trophic adaptations of *Heliconius* are also associated with memorized home ranges (including gregarious roosting sites), low levels of dispersal and migration, and highly derived aposematic and mimetic color patterns (Benson, 1971; Turner, 1971a,b; Gilbert, 1975, 1984; Mallet, 1986b,c) in comparison to other heliconiine genera such as *Dryas* or *Agraulis*.

Although the systematics on which Bates based his theories was correct in broad outline, more recent work (Eltringham, 1916; Emsley, 1964, 1965; Turner, 1971a,c; Brown, 1979) has clarified the taxonomy of heliconiines to the point where names have stabilized. Little is known on the phylogeny of the heliconiine species, but named entities that were confused because of mimicry and geographic divergence have been identified and assigned mostly as geographic subspecies to polytypic species (Turner, 1971a; Brown, 1976b). This work culminated with the list published by Brown (1979), whose taxonomy is currently the basis for most workers. Species affiliations of the racial forms have been deduced chiefly from collections made in hybrid zones, which show connections between divergent races (Kaye, 1916; Emsley, 1964, 1965; Turner, 1971c; Brown, 1976b, 1979). Table 9-1 summarizes some of the better-studied hybrid zones and clinal polymorphisms in *Heliconius*. An example of the geographic diversity and mimicry of a pair of mimetic heliconiine species is shown in Figure 9-1.

Species, Races, Color Patterns, and Hybrid Zones: Definitions. Mayr (1963, pp. 369, 381; 1970, pp. 214–227) proposed that "allopatric hybridization" in narrow hybrid zones was always caused by "secondary intergradation" and could be easily distinguished from clinal variation that evolved in parapatry by "primary intergradation." These connotations make it impossible to discuss the possible origins of hybrid zones, so I here attempt to use theory-free definitions. In this chapter the term species refers to reproductively isolated "biological species" (Wallace, 1865; Kaye, 1916; Mayr, 1963, 1970). Though this definition makes the nature of species somewhat arbitrary in allopatry (Wallace, 1865), species can always be recognized in sympatry and parapatry. Speciation is therefore the acquisition of reproductive isolation. Geographic races are divergent forms (including formally named subspecies) that are not reproductively isolated but that are relatively constant over wide areas; they are connected to other races by hybrid zones or clines that are narrow relative to the distributions of the races they connect. Clines are more or less continuous geographic changes in the frequency of alternative forms of a single gene, chromosome, or char-

Table 9-1. Large-sample Field Studies of Hybrid Zones and Polymorphisms in *Heliconius* Known to the Author

Species[a]	Subspecies[a]	Place	Reference
doris	various (usually polymorphic)	Neotropics	Sheppard (1963); Turner (1971a)
ethilla	*ethilla* (polymorphism)	Trinidad	Sheppard (1963); Turner (1968)
numata	all (often polymorphic)	Amazon basin	Brown and Benson (1974); Brown (1976a)
cydno	*galanthus* × *pachinus*	Costa Rica	L. Gilbert (unpublished)
cydno	*weymeri* × *gustavi* (× others)	Colombia (Cauca valley)	M. Linares (unpublished)
telesiphe	*telesiphe* × *sotericus* × *cretacea*	Andes of Peru and Ecuador	Vane-Wright et al. (1975); J. Mallet (unpublished)
erato	*erato* × *hydara*	Surinam, French Guiana	Sheppard (1963); Benson (1982)
erato	*hydara* × *amalfreda*	Brazil (Amazon), S. Guyana	Benson (1982); P. & J. Mallet (unpublished)
melpomene	*melpomene* × *meriana* × *thelxiopeia*	Surinam, French Guiana, Guyana	Sheppard (1963); Turner (1971c); P. & J. Mallet (unpublished)
erato	various	Neotropics	Emsley (1964)
melpomene	various	Neotropics	Emsley (1964)
erato	*amazona* × *venustus*	Brazil (Riozinho/ Mato Grosso)	Brown and Mielke (1972)
erato	*notabilis* × *lativitta*	E. Ecuador	P. Brakefield (unpublished); K. Brown (unpublished); J. Mallet (unpublished)
melpomene	*plesseni* × *aglaope*	E. Ecuador	
erato	*petiverana* × *hydara* × *venus*	Panama	Mallet (1986a)
melpomene	*rosina* × *melpomene* × *vulcanus*	Panama	Mallet (1986a)
erato	*favorinus* × *emma*	N. Peru (Huallaga)	Lamas (1976); Mallet (1989); Mallet et al. (1990)
melpomene	*amaryllis* × *cognata*	N. Peru (Huallaga)	Lamas (1976); Mallet (1989); Mallet et al. (1990)
erato	*dignus* × *lativitta*	S. Colombia (Putumayo)	Mallet (this chapter)
melpomene	*bellula* × *aglaope*	S. Colombia (Putumayo)	Mallet (this chapter)
erato	*microclea* × *emma*	C. Peru (Río Palcazu)	J. Mallet (unpublished)
melpomene	*xenoclea* × *cognata*	C. Peru (Río Palcazu)	J. Mallet (unpublished)
erato	*phyllis* (polymorphism)	S.E. Brazil	Pansera and Araújo (1983)
erato	*phyllis* × various	S.E. Brazil	K. Brown (unpublished)

Table 9-1. Large-sample Field Studies of Hybrid Zones and Polymorphisms in *Heliconius* Known to the Author (*Continued*)

Species[a]	Subspecies[a]	Place	Reference
himera × *erato*	*cyrbia*	S. Ecuador	Descimon and Mast de Maeght (1984); J. Mallet (unpublished); Brown (unpublished)
himera × *erato*	*favorinus*	N. Peru (Mayo valley)	König (1986); J. Mallet (unpublished)
himera × *erato*	*lativitta*	N. Peru (Río Marañon)	Mallet (this chapter)
hecalesia	*hecalesia* × *ernestus* × *longarena*	W. Colombia	Brown and Benson (1975)
sapho	*sapho* × *chocoensis*	W. Colombia	Brown and Benson (1975)

[a]Nomenclature used is that of Brown (1979) rather than necessarily that of original authors.

acter; and hybrid zones are then narrow clines, or, more usually, clusters of narrow clines that connect races or incipient species together.

"Warning color," "aposematism," and "mimicry" also include evolutionary hypotheses within terms for phenomena, but in this case the hypotheses seem generally accepted. By warning color or aposematism, I mean colors that seem adapted for reminding predators of previous unpleasant experiences. By mimicry I mean a color pattern that is too similar to the pattern of an unrelated species to be explained by chance and that is not due to common ancestry; the pattern of the mimetic species must have converged on a warning pattern in the model species.

POPULATION STRUCTURE AND COLOR PATTERN EVOLUTION

Theory of Warning Colors in a Single Deme. Novel warningly colored morphs in cryptically colored unpalatable species are selected against for two reasons: (1) They are more visible to predators; and (2) predators lack experience with these new morphs (reviewed by Mallet and Singer, 1987). The first effect places the more conspicuous morph at a constantly greater risk of predation. The second effect is more interesting, as predator learning causes frequency-dependent selection against rarity, as shown below. The argument is framed in terms of discrete morphs because it is applied to *Heliconius* butterflies in which major genes control color pattern. Similar arguments can be made for continuous variation.

Any unpalatable species risks damage or death during an encounter with a naive predator because it always takes one or more encounters for the predator to learn the morphology of that species. A novel morph is not recognized even by predators experienced with the typical form, and the learning process begins again. (The two morphs can of course be confused by predators, but to the extent it occurs it is ignored here because it does not lead to selection.) Because the first individual of a novel morph is rare, a high rate of damage or death is expected on a per-individual basis relative to that of the wild type (approximately in the ratio $r_a : r_c / N_c$, where r_a, r_c = total risk or fitness reduction expressed as numbers killed by local predators during learning of the novel aposematic and the cryptic morphs respectively; N_c = local population size of the cryptic insect morph). It is difficult to imagine that warning color evolves by individual selection because the advantages of a new pattern (e.g., enhanced memorability) rarely reduce the risk of death by as much as $1/N_c$ compared with a cryptic morph: For aposematic coloration to evolve, we must have $N_c r_a < r_c$, with N_c assumed large in a panmictic population. It is even more difficult for a rare cryptic morph to invade an unpalatable, warningly colored species because we must have $N_a r_c < r_a$, where we expect from the definition of aposematism that $r_a < r_c$. The selection is thus density-dependent against the rarer morph (Harvey et al., 1982). In a population of approximately constant size (perhaps maintained by density-dependent regulation acting on larval stages), this density-dependent selection is also frequency-dependent because frequency is proportional to numbers.

Most discussions of the evolution of warning color consider warningly colored morphs in cryptic but unpalatable species. However, in *Heliconius* we are more concerned with novel warningly colored morphs in already warningly colored species. This problem has similar dynamics, differing only in parameter values from the situation in a cryptic species (Mallet and Singer, 1987). In the simplest case both patterns are equally conspicuous and do not differ in memorability or palatability ($r_a \approx r_c$), so only frequency-dependent selection is involved. Within a species, frequency-dependent selection favors commoner aposematic morphs, impedes the evolution of novel patterns, selects against rare immigrants from populations with different colors, and, in doing so, stabilizes hybrid zones between different races (Brown et al., 1974; Benson, 1982; Harvey et al., 1982; Turner, 1982; Sheppard et al., 1985; Mallet, 1986a; Mallet and Singer, 1987; Endler, 1988). Frequency-dependent selection is also the engine that drives Müllerian mimicry between two or more unpalatable species. Mimicry, which would not evolve unless there were selection against rare aposematic forms, in fact provides some of the best evidence for a frequency-dependent advantage of common warning patterns (Mallet and Singer, 1987; Mallet, 1990). Purifying selection therefore explains (1) stasis of color patterns and Müllerian mimicry within regions and (2) maintenance of an interregional diversity of warningly colored races within species (Brown et al., 1974; Turner, 1982; Mallet, 1986a; Mallet and Barton, 1989a). In contrast, the origin of new aposematic patterns is seemingly prohibited by this purifying selection, which makes the rampant color pattern diversity within genera and even species of butterflies such as *Heliconius* especially puzzling.

One model for the evolution of novel warning colors is usually known as "kin selection" (Harvey and Greenwood, 1978). However, important differences were recognized between warning color and other possibly kin-selected altruisms such as unpalatability and eusociality, so the process for the evolution of warning color has also been called a "family model" (Harvey et al., 1982), kin-based "green beard selection" (Guilford, 1985), and "indirect selection" (Wiklund and Sillén-Tullberg, 1985). To my mind, these alternative names and the terms "altruism" and "kin selection" in this context confuse the process that establishes the novel form. The underlying idea is that warning colors can evolve if there are only a few families of an unpalatable species in an area containing few predators. A novel morph can then increase suddenly in the progeny of one of these families, thereby exceeding a threshold frequency such that the novel pattern becomes locally favored by frequency-dependent selection (Harvey and Greenwood, 1978; Harvey et al., 1982). This model is similar to the process of establishment of a novel underdominant chromosomal morph by genetic drift (Wright, 1941); essentially, N_c is assumed small, so a reasonable memorability advantage can cause local evolution; $N_c r_a < r_c$. Although Harvey et al. (1982, p. 712) proposed that novel warning colors increase "deterministically," small numbers of families per predator territory are required; and thus the model involves genetic drift or something closely resembling it (Mallet and Singer, 1987). "Kin-founding," used by Hedrick and Levin (1984) to describe a possible mode of chromosomal evolution in plants, is perhaps a better name than "kin selection."

Kin-founding and drift, which require small local populations, can be contrasted with standard "mass" or "individual" selection. There are at least three possible major ways in which warning colors can evolve by mass selection despite the frequency dependence: preadaptation, evolutionary enhancement, and mimicry. First, many palatable butterflies have bright colors that are used for signaling to predators and potential mates. If such species became unpalatable, their colors would act as preadaptations for warning color. Second, the patterns could then become enhanced by mass selection if modified patterns acted as "supernormal stimuli" of the original pattern memorized by predators (Mallet and Singer, 1987).

Mimicry is a third way in which warning colors can evolve by mass selection; it includes the evolution of frightening stimuli such as eyespots and color flashes, which can be interpreted as generalized mimicry of dangerous attacks by enemies of the insect predator (Mallet and Singer, 1987). In *Heliconius,* most races of most species have clearly evolved novel warning colors by means of Müllerian mimicry because there are usually more than two species within any mimicry ring, although this finding still does not explain the initial geographic divergence of the patterns. An interesting feature of mimicry is that genes with a major effect on the color pattern may be needed to bridge the phenotypic gap between the original nonmimetic pattern and the new mimetic pattern, which might explain why *Heliconius* races have only a small number of major Mendelian factors involved in pattern expression (Turner, 1976, 1984; Sheppard et al., 1985).

Warning Colors in Multidemic or Continuous Populations: Shifting Balance.

The evolution of warning color, unlike typical altruisms between relatives, does not require continued proximity of relatives. Therefore occasional kin-founding could cause the evolution of novel warning colors, which would then be conserved by ordinary natural selection within a deme. Because this occasional population structure depends on random factors influencing local population size, it is a form of genetic drift. The kin-founding process can lead in multidemic or continuous populations to Wright's "shifting balance" (Wright, 1977, pp. 454–455). In general, models of kin selection *(sensu lato),* originally suggested by Fisher (1958) in connection with aposematic insects, are closely related to Wright's ideas, despite the strong disagreements between these two authors about the shifting balance. Both theories depend on interactions between small numbers of individuals within groups.

The shifting balance consists of three phases (using warning color as an example): (1) genetic drift (or kin-founding, or both) within a deme or group of demes leading to local establishment of a superior color pattern at a critical frequency; (2) selection within the deme or demes to fix or stabilize the new pattern; and (3) spread of the new pattern by "interdemic selection" to other demes. The term "interdemic selection" was used because deterministic or at least biased competition between demes can quickly spread advantageous adaptive peaks to new areas, even though there is initial selection against the new pattern during its introduction to any deme. In the absence of interdemic selection, each deme would have to evolve the new peak separately by genetic drift, and the process would be greatly slowed.

The shifting balance envisioned by Wright is only one of a series of models that seem to encapsulate his original idea. Extensions to the model are as follows: (1) extension from multiple epistatic loci to any genetic systems that exhibit multiple equilibria; (2) extension from demic to continuous populations; (3) extension of interdemic selection (the third phase) to any asymmetry between adaptive peaks that move the peaks from one area to another; (4) extension of the concept of interdemic selection to include movements of clines or hybrid zones within continuous populations.

Wright usually referred to multiple epistatic genes, but the first extension to the shifting balance includes the simplest example of a genetic system with more than one equilibrium: a single gene or chromosome with heterozygous disadvantage (Wright, 1978; Barton, 1979; Lande, 1985; Barton and Rouhani, 1991). Frequency-dependent selection on a single warning color gene is similar to selection on an underdominant trait, such as a chromosomal rearrangement (Mallet and Barton, 1989a), and should also be included. It is of course true that warning colors themselves are usually determined by two or more epistatic genes (Sheppard et al., 1985; Mallet, 1989).

Second, Wright and some of his followers (Wright, 1977; Crow et al., 1990; Wade and Goodnight, 1991) usually phrased the shifting balance in terms of discrete demes, but as migration between demes increases populations become effectively continuous (Slatkin and Barton, 1989). For shifting balance in a continuous population, substitute "neighborhoods" for "demes" in the above description (see also Barton and Rouhani, 1991).

Spread of Warning Colors to New Areas by Habitat-Independent Moving Clines.
The two other extensions to the shifting balance involve the third phase of the shifting balance: interdemic selection. Wright proposed that interdemic selection would occur because demes that had reached the new, more adaptive peak would produce more emigrants than demes stabilized at the less adaptive peak. When selection affects population numbers, as in this case, it is known as "hard selection" (Wallace, 1968). This migration asymmetry due to adaptation would preserve and spread the new adaptive peak. In the third extension to the shifting balance, selection asymmetries on their own might also aid the spread of a new peak, even if selection is "soft" (i.e., does not affect local population size) (Wallace, 1968). For example if a new warning color is more memorable (i.e., less selected against when rare) than the old pattern, it can spread even if parasites and predators of larvae determine population size, so there is no migration asymmetry (Mallet, 1986a; Mallet and Barton, 1989a).

In the final extension of the shifting balance, the preservation and spread of a new morph behind a moving cline or hybrid zone in a continuous population is considered equivalent to interdemic selection in a demic population (Barton, 1979; Mallet, 1986a; Barton and Hewitt, 1989). Contact between two areas differing at a single locus or chromosome, with underdominant or purifying frequency-dependent selection, causes a cline or "tension zone" (Key, 1981; Hewitt, 1988) to form. Migrants crossing the cline are selected against, and this selection stabilizes the cline to a width that is a multiple of $\sigma/\sqrt{s}$, where σ = the migration distance, and s = the selection pressure (Bazykin, 1969). This result can also be generalized to multilocus hybrid zones (Barton, 1983; Mallet and Barton, 1989a) or to any other kind of genetic system where more than one equilibrium occurs (e.g., Turelli and Hoffmann, 1991).

Hybrid zones or narrow clines between forms differing in habitat-independent traits can form anywhere and are free to move from place to place (Bazykin, 1969; Barton, 1979; Mallet, 1986a; Turelli and Hoffmann, 1991) because selection is independent of habitat, or "endogenous" *sensu* Moore and Price (see Ch. 8). By "habitat-independent" traits, I mean only those "general" adaptations (W. L. Brown, 1958) with more than one stability peak and with unstable intermediate equilibria. Such traits include warning color patterns (Mallet, 1986a), chromosomal rearrangements or other types of heterozygous disadvantage (Bazykin, 1969; Barton, 1979), some kinds of genes affecting assortative mating (e.g., chirality in snails of the genus *Partula*) (Johnson et al., 1987, 1990), and traits affected by multiple epistatic loci (Wright, 1977). In the case of warning color, cline movement could occur if one form warned predators more effectively, but anything causing asymmetry of selection or migration across the zone, such as dominance of a color pattern gene or differences in population density, can also cause cline movement (Barton, 1979; Mallet, 1986a; Mallet and Barton, 1989a). Moving clines become trapped at density troughs because a density gradient causes asymmetric migration, and clines move to the bottom of the trough where migration is minimized (Bazykin, 1969; Barton, 1979; Hewitt, 1988; Barton and Hewitt, 1989). Because of this effect, habitat-independent clines are often associated with habitat discontinuities. Intermediate stages of evolution by the shifting balance in a continuous population then consist of a patchwork of different races, each connected to others by stationary or moving tension zones, similar to those shown by Wright (1977, p. 459) or those in Figure 9-1.

A good criticism of my view that moving clines can explain geographic patterns is that clines move slowly and often become trapped by slight barriers to dispersal or density troughs (Barton, 1979; Barton and Hewitt, 1989). Because cline movement is part of the third phase of the shifting balance, this amounts to a critique of Wright's model; if cline movement rarely occurs except by accidental effects of the environment, it is difficult to justify the shifting balance as having much importance in adaptive evolution. Clines should move at a speed proportional to (and have units of) $\sigma\sqrt{s}$ (Barton, 1979; Mallet and Barton, 1989a; Turelli and Hoffmann, 1991). In the three-locus color pattern hybrid zone in *Heliconius erato* studied by Mallet et al. (1990), dominance of two of the color pattern genes is theoretically expected to move the hybrid zone at about $0.07\sigma\sqrt{s}$ (Mallet and Barton, 1989a). Because $\sigma \approx 2.6$ km and $s \approx 0.23$ per locus (Mallet et al., 1990), we expect movement of about 0.42 km per generation. At four generations per year, this movement is about 1700 km in 10,000 years—sufficient to blur biogeographic patterns created during the Pleistocene. This analysis assumes that the pure patterns are equally able to warn away predators; if one pattern is more memorable, cline speeds may be considerable increased or reversed. With selective asymmetry (S) and heterozygote disadvantage ($\approx$ frequency-dependent selection) maintaining the cline (s), cline speed is proportional to $\sigma S/\sqrt{s}$ (Barton, 1979); thus any increase in selective asymmetry causes the speed of a moving cline to increase in direct proportion.

Although hybrid zone movement seems potentially rapid for *Heliconius,* the criticism that slight density gradients or weak dispersal barriers commonly trap such zones remains serious. The trapping effect depends on whether spatial variation in population density in the field is large relative to movement tendency ($\sigma S/\sqrt{s}$), but there is a problem with the models that predict local hybrid zone trapping: a constant value for σ may be unrealistic. On encountering an area of low resource density, individuals of many mobile species are likely to disperse further, thereby increasing σ locally. This dispersal can at least partially compensate for small density troughs that would trap clines if the species had the purely diffusive dispersal assumed in the models.

Neutral and Habitat-Dependent Clines. Wright (1977, 1978) did not claim that races such as those in Figure 9-1 were completely neutral, as some have assumed (Provine, 1986, pp. 287–291). Wright thought only that neutral genetic drift might cause the initial divergence between two races, thereby leading to subsequent divergent selected evolution. The problem is with the definition of the term "neutral" on a local and global scale. Racial differences could be neutral in the sense that they are not specially adapted for the environment in which they are found; any *Heliconius* pattern could probably teach many types of predator in a wide variety of habitats. On the other hand, such "neutral" races may be strongly selected toward the local adaptive peak that exists within their distribution range, as is clear from *Heliconius* warning color, which involves mimicry. There are of course other genetic differences, such as allozymes or third basepairs in codons, that might indeed be effectively neutral, both locally and globally. Clines of such elements could be caused by neutral secondary contact or local genetic drift, in which case the clines will slowly decay until their widths are much greater than the dispersal distance (Endler, 1977).

Biogeographic patterns such as those shown in Figure 9-1 might also be caused by habitat-dependent selection ("special" adaptations *sensu* W. L. Brown, 1958, or "exogenous" selection *sensu* Moore and Price, Ch. 8); that is, selection depends on the environment of the taxon under consideration, including geography, climate, vegetation, food sources, competitors, parasites, predators, and symbionts. Habitat-dependent traits include adaptation to sooty bark by melanics in the peppered moth *Biston betularia* (Bishop, 1972), plant adaptations to local soil or wind conditions (Jain and Bradshaw, 1966), physiological adaptations to climate (e.g., Hagen, 1990), and mimicry (Turner, 1971a, 1982). Clines between traits that are under habitat-dependent selection quickly move to areas of environmental or biotic change and thereafter remain stationary, unless the environment itself changes.

In the case of Müllerian mimicry rings, the whole ring evolves relatively independently of habitat, given the existence of predators that can learn, whereas rare species within the mimicry ring evolve as though their environment were determining the local selection. In addition, a particular color pattern might have a habitat-dependent advantage (e.g., melanism in a cool climate) but may be unable to evolve because of a habitat-independent disadvantage (e.g., frequency-dependent selection against rare morphs in an unpalatable species). Thus habitat-independent and habitat-dependent selection are not mutually exclusive. We can also imagine that either neutral or habitat-dependent divergence between parapatric or completely allopatric populations might eventually result in habitat-independent reproductive isolation. Pleiotropic habitat-independent incompatibilities in the mating system or in postmating barriers seem likely to cause hybrid breakdown, such as that in "Haldane's rule" (Charlesworth et al., 1987; Barton and Hewitt, 1989). Geographic variation leading to speciation is often a mix of neutral, habitat-dependent, and habitat-independent components (Hewitt, 1989).

Evidence from* Heliconius *Population Ecology and Hybrid Zones. Although mass selection for mimicry can occur in any population structure, the probability of kin-founding or the shifting balance requires small local population sizes (Wright, 1941, 1977; Harvey et al., 1982; Lande, 1985; Barton and Rouhani, 1991). Current population structure may be inferred from ecological or genetic studies, but it might not help in understanding the past population structures leading to divergence. However, because species of *Heliconius* have similar population biology (Gilbert, 1984), and most *Heliconius* species have evolved multiple novel color patterns (Brown, 1979), it seems likely that population studies will be useful for understanding raciation in the genus. The method is not likely to be so useful for elucidating ancient evolution, such as that of the first warning colors in the progenitor of the genus.

Heliconius individuals repeatedly visit feeding, mating, and nocturnal gregarious roosting sites within small home ranges about 100 meters wide (Turner, 1971a,b; Gilbert, 1975; Mallet, 1986b) and survive as adults for up to 6 months, more than six times the egg to adult development time, so that altruistic acts and kin selection between and within generations would be possible. These observations together suggested that *Heliconius* was a good candidate for kin selection and kin-founding in the evolution of unpalatability, warning colors, and knowledge-sharing within gregarious roosts (Benson, 1971; Turner, 1971a; Gilbert, 1975, 1977; Wilson, 1975; Harvey and Greenwood, 1978; Harvey et al., 1982).

Any genetic markers, such as allozymes, could give some information about population structure and the likelihood of kin-founding. In hybrid zones between heliconiine color pattern races, however, we can observe polymorphisms similar to those that must have existed when novel color patterns first arose. Sheppard et al. (1985) reviewed inheritance of warning color patterns in *Heliconius.* Color patterns in hybrid zones are therefore especially useful as genetic markers for testing hypotheses of kin selection and the shifting balance, as they are the very genes in whose evolution we are interested.

If population structure in *Heliconius* were suitable for "deterministic" kin-founding, we might expect that as we pass from the home range or territory of one predator to another within a hybrid zone there will be sharp changes of color pattern gene frequencies. It might occur in two ways. Dispersal might be as limited as measured in mark-recapture studies, and hybrid zones would be little wider than a single butterfly home range ($\sim$ 100 meters) because butterflies would rarely disperse, and the interface would be formed by adjacent predator home ranges containing divergent butterfly color patterns. Alternatively, dispersal might be greater, leading to a broad hybrid zone. Under these conditions, because most populations are founded by few individuals, a broad mosaic of mostly fixed populations with alternative color patterns should be present in the transition area. In this case the mosaic would consist of locally selected color patterns, based on the alleles present in initial founder populations; in contrast to other mosaic hybrid zones (Harrison and Rand, 1989), local genetic rather than environmental conditions would determine the selection.

Table 9-2. Allelic and Genotypic Frequencies for *H. erato* near Villa Garzón, Putumayo, Colombia

	No. of Alleles		No. of Genotypes		No. of Genotypes	
	D^{Ry}	d^{rY}	$Cr-$	$crcr$	$Sd-$	$sdsd$
N. of Río Mocoa and W. of Villa Garzón	15	25	11	9	15	5
N. of Río Mocoa and E. of Villa Garzón	13	29	13	8	15	6
S. of Río Mocoa and E. of Villa Garzón	22	36	12	17	25	4
S. of Río Mocoa and W. of Villa Garzón, (mark-recap. site);						
Total	68	132	52	48	63	37
Subsite A	4	4	2	2	2	2
Subsite B	10	26	9	9	8	10
Subsite C	10	26	12	6	9	9
Subsite D	10	20	8	7	13	2
Mobile individuals	34	56	21	24	31	14

Villa Garzón (formerly Villa Amazónica) is situated on the southern bank of the Río Mocoa, at 420 m above sea level; *H. e. dignus* from the upper Putumayo (Fig. 9-1A; ● symbols at about 2°N) here hybridizes with the Amazonian race *H. e. lativitta* (Fig. 9-1A; + symbols). Color pattern genetics are assumed similar to that near Tarapoto, Peru (Mallet, 1989; Mallet et al., 1990) on the basis of the appearances of hybrids, although the yellow bar of *Sd- crcr* genotypes is apparently more strongly expressed in Colombia than in Peru. Allelic frequencies are shown for the codominant locus D^{Ry}/d^{rY}, and phenotypic frequencies are shown for the two dominant loci *Sd/sd* and *Cr/cr*. All sites were within easy walking distance of Villa Garzón, about 5 km, though exact distances are not known. A site immediately to the west of Villa Garzón on the road to Mocoa was studied intensively using mark-recapture; individuals could be classified into those that had restricted home ranges (about 100–300 m across) and those that moved between subsites A, B, C, and D. Groups of gregarious roosts were discovered for subsites B, C, and D.

There is now good evidence from *Heliconius erato* and its Müllerian mimic *H. melpomene* that hybrid zones between races are relatively broad and do not consist of mosaics, suggesting that dispersal is greater than measured in mark-recapture experiments. For example, populations of *Heliconius erato* of the order of 1–5 km apart near Villa Garzón in Colombia are remarkably similar in gene frequency (Table 9-2): Homogeneity tests for D^{Ry}/d^{rY} (allelic frequencies), G = 0.71, 3df; for Cr/cr (genotypic frequencies), G = 2.22, 3 df; and for Sd/sd (genotypic frequencies), G = 6.68, 3 df—giving a sum of G = 9.61, 9 df overall ($p > 0.05$). This gene frequency heterogeneity is equivalent to an average (but nonsignificant) F_{st} of 0.02. Each population can be subdivided again and again, even down to individual gregarious roosting sites, and still only small gene frequency differences result. For instance, in the intensively studied S./W. site in Table 9-2, the overall heterogeneity was again insignificant (lumping adjacent subsites A and B, and including the mobile individuals as a separate group, summed G = 12.40, 9 df, $F_{st} \approx 0.06$). The Fisher-Ford estimates of population size for the S./W. sites A, B, and C summed (D was observed as roosts only) varied between 26 and 40, with average life expectancy of 41 days. The low population sizes, together with the low turnover rate and high recapture rate (86%) shows that virtually all individuals were captured in this site, so these estimates of F_{st} must be close to actual. Therefore it is possible that the measured F_{st} is entirely due to the small local population size, rather than because there is underlying gene frequency variation from population to population. More extensive data showing similarly small and insignificant gene frequency differences between adjacent sites has been collected from *H. erato* and *H. melpomene* hybrid zones in Panama and Peru (Mallet, 1986a; Mallet et al., 1990). The transitions from one color pattern to another in Panama and Peru are smooth gene frequency clines about w ≈ 10–100 km wide with little "noise" attributable to strong gene frequency differences between adjacent populations (Mallet, 1986a; Mallet et al., 1990).

In addition, local hybrid zone populations have no significant deviations from Hardy-Weinberg at color pattern loci, suggesting that the populations are not subdivided. Near Villa Garzón, the lack of a heterozygote deficit at D^{Ry}/d^{rY}, F ≈ 0.14, G = 3.45, 1 df ($p > 0.05$, using all individuals in Table 9-2) shows that there is no significant substructuring even after lumping the populations studied; again, this lack of heterozygote deficit is also true for Panama and Peru (Mallet, 1986a; Mallet et al., 1990) and for other hybrid zones (Turner, 1971c). Methods based on linkage disequilibria (see Ch. 2) and direct field experiments have indicated that selection coefficients required to stabilize some hybrid zones of *H. erato* and *H. melpomene* must be strong: s ≈ 0.23–0.26 per locus against the foreign alleles in the narrowest (10 km) hybrid zones (Mallet and Barton, 1989b; Mallet et al., 1990). Indirect estimates of dispersal made from the *H. erato* hybrid zones in Peru show that gene flow, measured as the standard deviation in parent-offspring distances, is $\sigma \approx 2.6$ km gen$^{-1/2}$ (Mallet et al., 1990). Linkage disequilibria were not significant at Villa Garzón, perhaps because of the small total sample size, and so could not be used to estimate dispersal. This work has mostly been done with *H. erato,* but preliminary evidence from disequilibria in *H. melpomene* suggests an even greater tendency to movement ($\sigma \approx 3.7$ km gen$^{-1/2}$) (Mallet et al., 1990; see also Turner, 1971c).

Detailed mark-recapture studies in Colombia and Costa Rica show how this population structure is realized. Although some individuals have home ranges centered on faithfully attended gregarious roosts, other individuals are much more prone to change roosts. In the Villa Garzón mark-recapture site (Table 9-2), 45% of individuals moved between roosting groups or disappeared after one capture; the number moving between roosts within subsites was higher still. Further data showing movement between roosting sites in Costa Rica is given by Mallet (1986b). Home ranges of individuals attending different roosts overlap markedly, and there is no evidence for communal feeding behavior exclusively involving roostmates (Mallet, 1986b). Most dispersal of *H. erato* occurs before learned home ranges are set up, as is normal for mammals and birds; this point would not have been detectable in earlier mark-recapture studies. Movements of recently eclosed adults of *H. erato* suggest that gene flow, $\sigma \geq 0.3$ km gen.$^{-1/2}$ (Mallet, 1986c). Because the sampling area for this mark-recapture study was small, it seems likely that the indirect estimate of gene flow obtained from hybrid zones (above), $\sigma \approx 2.6$ km gen$^{-1/2}$, is more accurate.

The population structures of these two species are therefore difficult to reconcile with a simple-minded "deterministic" kin-founding model for the evolution of aposematic patterns in gregarious roosting families or with any of the other population-level kin selection models for the evolution of altruisms proposed for *Heliconius.* To sum up the evidence: individuals roost and fly together with unrelated individuals most of the time; home ranges overlap, allowing for gene flow between roosting groups; individuals on the same roost have different home ranges, suggesting a lack of communication about feeding, mating and oviposition sites, and further possibility for gene flow; newly eclosed individuals often leave their parental home range; gene flow is of the order of $\sigma \approx 2\text{--}4$ km $\text{gen}^{-1/2}$, in contrast to the 0.05 km $\text{gen}^{-1/2}$ observed in earlier mark-recapture studies; F_{st} is low and insignificant for warning color genes in hybrid zones, suggesting that the effective population size is larger than local population sizes centered on gregarious roosts; hybrid zones between color pattern races are broad in comparison with the butterflies' home ranges and form smooth clines, suggesting that local populations do not usually exhibit sudden frequency changes because of founder events. These results for *Heliconius* seem to reflect a general problem for kin selection in unpalatable butterflies, as *Heliconius* were thought to be one of the best examples of kin grouping. Distasteful ithomiines, danaines, and troidines appear to lack home ranges and are much more migratory than *Heliconius* (Gilbert, 1969; Brown and Benson, 1974; Brown and Neto, 1976; Eanes and Koehn, 1978; Brown et al., 1981) and so would be even less likely to evolve unpalatability or warning colors by kin selection or deterministic kin-founding.

Instead, population structural conditions for kin-founding and kin-selection must occur only periodically (if at all), as in classical genetic drift. Genetic drift and kin-founding could lead to habitat-independent establishment of a new warning color pattern in a local area or subpopulation. Although this initial change is likely to be very rare, once an area of reasonable size is fixed for an advantageous new pattern it is resistant to invasion by the old pattern because stable clines, maintained by frequency-dependent selection, form around it (see above; also Benson, 1982, p. 633). The new form is preserved and spread if it is selectively superior or has some other advantages, such as a high rate of emigration, or is dominant; clearly this mode of divergence would be an example of Wright's "shifting balance."

Some calculations (Rouhani and Barton, 1987; Barton and Rouhani, 1991) indicate that local shifting balances are fairly likely in continuous populations with restricted movement such as those in *Heliconius. Heliconius* population structure, however, is probably still too poorly known for a convincing estimate of the probability of this mode of divergence. In the rest of the chapter, I instead concentrate on whether the biogeography of *Heliconius* and allies gives evidence for either habitat-dependent selection or the shifting balance in color pattern evolution.

EVOLUTION OF NOVEL COLOR PATTERN RACES AS A MODEL FOR SPECIATION

As we have seen, geographic variation may be neutral, selected and habitat-dependent, or selected and habitat-independent. We normally think of the evolution of reproductive isolation, or "biological" speciation, as habitat-independent; we expect mating incompatibilities or hybrid breakdown to be maintained under most field or laboratory circumstances. How might habitat-independent reproductive isolation evolve? *Heliconius* color pattern races are partially reproductively incompatible because there is selection against hybrids (heterozygous color patterns are "fuzzier" and presumably less effective as warning signals), because of frequency-dependent selection against rare introgressing patterns, and because the pattern of any given race consists of a self-adapted set of alleles at epistatically interacting loci (Mallet, 1989; Mallet et al., 1990).

With *Heliconius,* both habitat-dependent and habitat-independent routes for the evolution of this kind of incompatibility seem likely.

Warningly colored species that are unpalatable and common cause strong habitat-dependent selection for Müllerian mimicry of local species with similar color patterns (Brown et al., 1974; Turner, 1982). Other traits related to habitat, such as host-plant survival or choice, or physiological adaptations to temperature or altitude, are also likely to diverge geographically under selection in continuous populations (Benson, 1982; Endler, 1982). There can be no doubt that mimicry is the major reason for color pattern divergence in most heliconiines and ithomiines with geographically differentiated color patterns (Turner, 1982): There are almost always more than two species in a local mimicry ring, implying that more species are mimics than models. However, it is unlikely that mimetic convergence alone can cause all this divergence: some independent divergence of color patterns in the major "model" species seems necessary to trigger novel color patterns in these butterflies. This initial diversification could be provided by the shifting balance. Indeed, the shifting balance is suggested by the existence of the alternative stable states demonstrated by Müllerian mimicry and strongly selected narrow hybrid zones.

An important feature of both mass selection and the shifting balance is that neither requires allopatry. It has been clear since the 1940s that habitat-dependent adaptation can take place in the absence of geographic barriers, provided the patches under selection for divergence are greater than about $\sqrt{(3\sigma^2/s)}$ across (Haldane, 1948; Fisher, 1950; Slatkin, 1973; Endler, 1977). Genetic drift can also cause divergence for habitat-independent or neutral traits in viscous but continuous populations; once again, the scale of divergence is some critical small multiple of σ (see Barton and Charlesworth, 1984; Barton and Hewitt, 1989; Slatkin and Barton, 1989; Barton and Rouhani, 1991).

Because it causes partial reproductive isolation, color pattern change in warningly colored species has some of the characteristics of speciation, but further change is required to complete the process. Any mass selection or habitat-dependent adaptation may produce pleiotropic effects causing pre- or postmating incompatibilities. Pleiotropic reproductive isolation may seem questionable if color pattern alone is involved. Nonetheless, mimicry is often limited to females, which suggests that mimicry often interferes with mating behavior and sexual selection, at least in Batesian mimics (Turner, 1978; Silberglied, 1984). This effect of color pattern evolution does not seem to be important in *Heliconius,* as there is no evidence for heterozygote deficits in hybrid zones (see above), and sexual dimorphism is rare. Color pattern divergence due to a shifting balance could potentially have similar pleiotropic effects on reproductive traits. However, it is perhaps more likely that habitat-dependent evolution or the shifting balance at noncolor traits would cause the pleiotropic effects that lead to speciation. In addition, population structures suitable for color pattern changes via the shifting balance are also likely to produce shifting balances for other habitat-independent traits such as chromosomal morphs and other underdominant loci, epistatic gene complexes, and sexually selected traits. Similarly, chromosomal evolution may be associated with rapid speciation, rather than itself being the prime cause of speciation (Bush et al., 1977; Barton and Charlesworth, 1984).

The heliconiines display a rich spectrum of geographic divergence: weakly differentiated forms separated by broad clines, e.g. within *Heliconius doris* or *H. hecale*

Table 9-3. Captures of *Heliconius himera, H. erato lativitta,* and *H. melpomene aglaope* in the Marañon Valley of Peru: July 1984 and June 1986

Species	Captures of Heliconius at Certain Distances (km) North of Muyo													
	−10.2	−9.7	−8.0	−6.0	−1.5	1.0	3.0	9.0	9.5	20	30	60	130	160
H. himera	6	26	12	21	11	0	0	1	0	0	0	0	0	0
H. erato	0	0	0	0	0	0	0	1	5	1	5	1	2	2
H. melpomene	0	0	0	0	1	1	5	1	4	0	2	4	2	0

Distances are expressed as kilometers north of Muyo (near Aramango), Amazonas. Muyo is about 30 km NNE of Bagua and is the site of the abrupt change from thorn scrub upriver toward Bagua and moist forest downriver toward Chiriaco and Sarameriza. Aramango is at +1.0 km, Chiriaco is at about +20 km, and Sarameniza is at +160 km on the Rio Maranon/Amazonas. One of the specimens of *H. melpomene* from 3 km north of Muyo has reduced rays, presumably mimicking *H. himera,* and therefore corresponds to the form designated "ssp. nov." (Fig. 9-1B; • symbols) by Brown (1979).

(Turner, 1971a; Brown and Benson, 1974; Brown, 1976a) or among rayed races of *H. erato* and *H. melpomene* (Fig. 9-1) (Brown and Mielke, 1972), strongly differentiated forms separated by narrow hybrid zones with extensive hybridization, e.g., rayed and unrayed races of *H. erato, H. melpomene,* and *H. hecalesia* (Fig. 9-1, Table 9-1) (Emsley, 1964; Turner, 1971c; Brown and Benson, 1975; Brown, 1976b; Benson, 1982; Mallet, 1986a; Mallet et al., 1990), parapatric "good species" abutting at narrow contact zones with few or no hybrids produced, e.g., *H. erato* and *H. himera, H. erato* and *H. clysonymus* (Fig. 9-1A; Tables 9-1 and 9-3) (Benson, 1978, 1982; Descimon and Mast de Maeght, 1984; König, 1986; G. Lamas, unpublished), and finally sympatric closely related species that rarely if ever hybridize, e.g., *H. melpomene* and *H. cydno; H. eleuchia, H. antiochus,* and *H. sapho; H. sara* and *H. leucadia* (Kaye, 1916; Emsley, 1964, 1965; Brown and Mielke, 1972; Brown and Benson, 1975; Brown, 1976b). Even if we do not know exactly how speciation occurs, this spectrum of reproductive isolation implies that speciation is gradual (but not necessarily slow), and that usually many genetic changes are involved in the completion of speciation. Single genes or chromosomes rarely cause complete reproductive isolation, as suggested by White (1978) in his "stasipatric" speculation model.

BIOGEOGRAPHIC EVIDENCE FOR ALLOPATRIC AND PARAPATRIC DIVERGENCE IN *HELICONIUS*

The remainder of this chapter uses the preceding models of evolution as a basis for reviewing evidence for and against hypotheses of divergence in *Heliconius.* The argument focuses on the possible role that Pleistocene refugia have played in the evolution of *Heliconius* color pattern races and their hybrid zones. I am here interested less in whether Pleistocene refugia existed than in whether they contributed to morphological divergence and current geographic distributions. On the basis of the distribution of neotropical taxa and the hybrid zones between them, I argue not only that parapatric hypotheses for divergence in neotropical species have not been falsified (see also Connor, 1986) but also that parapatric hypotheses seem at least as likely (based on current evidence) as allopatric ones. Two sorts of allopatric hypotheses might be proposed: First, raciation or speciation could have occurred initially in allopatry (e.g., Mayr,

1963); and second, allopatric spread from refugia could be the most important factor that has determined current distributions of taxa and the positions of contact zones, even though the initial divergence might have been in parapatry (Barton and Hewitt, 1985; Hewitt, 1988). I shall argue for the plausibility of extreme parapatry—parapatric divergence as well subsequent parapatric maintenance—because this model must be effectively rejected before we can think about accepting one or both allopatric hypotheses. At present, I believe that none of these models can be accepted; we do not know the relative likelihoods of different modes of speciation, and so it does not seem sensible to accept either model as a null hypothesis (Turner, 1982).

Refugium Hypothesis. Ideas about Pleistocene refugia were developed to explain an apparent paradox. Evolutionary biologists were at one time largely convinced by the arguments of Mayr (1963) that speciation required allopatry, and many continue to hold this conviction (e.g., Futuyma and Mayer, 1980; Haffer, 1985). However, the rain forest on the neotropical mainland is one of the most speciose areas in the world for birds, insects, and flowering plants despite a pronounced lack of geographic barriers, especially in Amazonia. Jürgen Haffer (1967, 1969, 1987a) summarized geological evidence for previous drier neotropical climates associated with glaciation in the temperate zone, which may have limited the extent of the forest during the Pleistocene. Haffer (1967, 1969, 1974, 1985, 1987b) suggested that forest birds were able to persist and speciate in the remnant forest refugia, giving rise to forms that subsequently spread out and formed zones of secondary contact during the more humid interglacials. The huge species diversity observed today in many neotropical groups might have been produced via several such cycles (Haffer, 1969).

Similar arguments have been used to explain divergence in aposematic tropical Heliconiini and Ithomiinae (Turner 1965, 1971a, 1976; Brown, 1976b, 1979, 1981, 1982, 1987a,b,c; Brown et al., 1974; Lamas, 1973, 1982); but see also Turner's (1982) "faunal drift" refugium hypothesis, discussed under Mimicry in Refugia below.

Geological Evidence and the Supposed Necessity for Allopatry. Although some have emphasized geological evidence for drier Pleistocene climates, others argue that much of the Amazon remained wet throughout the glacial periods (Colinvaux, 1989). As to whether allopatry was necessary, I have already pointed out that parapatric divergence now seems much more likely on the basis of population genetic theory than seemed credible during the 1960s and 1970s (Endler, 1982; Connor, 1986). Mechanisms for divergence in allopatric refugia are rarely discussed, but the implication of Mayr's (1963) founder effect speciation and similar models is that genetic drift and habitat-dependent selection in small populations can combine to initiate speciation. It is difficult to imagine, however, how refugia would help the operation of genetic drift unless the refugia were extremely small (e.g., less than a few kilometers across). Such refugia are more likely to cause extinction than speciation (Barton and Charlesworth, 1984; Barton, 1989).

Importance of Habitat for Speciation and Raciation in Heliconius. Some species of *Heliconius* are characteristic of particular ecological zones, rather than being centered on regions of high rainfall. *Heliconius himera* is restricted to the dry, scrubby vegetation of the upper Marañon drainage and western slopes of the Andes Mountains in

Peru and Ecuador (Fig. 9-1A, * symbols). This species has been treated on the basis of larval and adult morphology as a race of *H. erato* (Brown, 1979). Hybrids of *H. himera* with *H. erato* have been found where the two meet (Descimon and Mast de Maeght, 1984; König, 1986); however, the hybrids are rare in comparison with the parental species, so it is more sensible to regard *H. himera* as a good species. In my own collections at the contact zone between *H. himera* and *H. erato lativitta* of the Marañon, Peru, I found no hybrids (Table 9-3), though König has found a single *erato* × *emma* hybrid in this area (G. Lamas, pers. comm.). The position of the contact zone correlates perfectly with the change of vegetation from the acacia scrub and dry forest habitat of *H. himera* to moist tropical forest near Muyo and Aramango, Amazonas, Peru; the latter is typical habitat for *H. erato* (Table 9-3). The other contact zones between *H. himera* and *H. erato* are on similar ecotones (pers. obs.; Keith Brown, pers. comm.). The himera–erato contact zones are not by themselves proof of a biogeographic pattern, but essentially the same distributions are known for many of the other aposematic and nonaposematic butterflies of the Marañon (Brown, 1979; Lamas, 1982). In the case of the Marañon fauna, endemism is centered on a valley habitat that is dry owing to the rain shadow effect of surrounding mountains, rather than being a likely area for a moist forest Pleistocene refugium.

Another species that is closely related to *H. erato* is *H. clysonymus*. This species has a color pattern similar to that of *H. himera* (Fig. 9-1A, * symbols), although it is unclear whether the patterns in the two species are homologous. *Heliconius clysonymus* is found only above about 800 meters in the Andean chain, as well as at similar altitudes in Central America. *Heliconius clysonymus* overlaps but does not hybridize with *H. erato* at intermediate elevations. There is some evidence for competitive exclusion between the two species (Benson, 1978). *H. himera* and *H. clysonymus*, both close relatives of *H. erato*, provide good evidence for the importance of habitat-dependent divergence in speciation: Both are found in habitats different from those usually occupied by *H. erato*, and both are less interfertile with *H. erato* than any of the races within *H. erato*. Other close relatives of *H. erato*—*H. charitonia*, *H. hermathena*, *H. hortense*, *H. telesiphe*, and *H. hecalesia*—are also found either in drier or more montane or marginal habitats than typical *H. erato*. Moreover, there is good evidence among the ithomiines for restriction of certain species with lowland rain forest relatives to montane or semiarid forest types (Brown, 1979; Lamas, 1982; Mallet and Lamas, in prep).

Some races of polytypic species, as well as sibling species of *Heliconius*, appear to be adapted to particular environments, but here the correlation is less clear. Andean races of species in *Melinaea*, *Mechanitis*, and *Hypothyris* (Ithomiinae), as well as *Heliconius*, among other genera, are often more melanic than their conspecifics in the Amazon basin (Brown 1977). In *Heliconius erato* and *H. melpomene* the large basal patch on the forewing and the greater degree of yellow scaling on the thorax of the Amazonian rayed races contrast with the blacker bodies and blacker distal portion of wings of Andean races (Fig. 9-1). Although experimental work has never been performed on *Heliconius*, this type of butterfly melanism is often associated with an ability to attain daytime operating temperature rapidly in cooler climates (Watt, 1968; Douglas and Grula, 1978; Roland, 1982), as might be expected in the higher altitudes of the Andes. For humid areas near the equator, an elevation increase of 100 meters corresponds approximately to a temperature decrease of 0.6°–1°C (MacArthur, 1972).

Although this general correlation between climate and color pattern exists and is undoubtedly important, it does not explain all of the differentiation patterns. For example, *H. erato emma* is an Amazonian rayed race that also invades the Andean valley of the Pozuzo river in Peru together with its *H. melpomene* mimic (Fig. 9-1; ♦ symbols at about 10°S). These co-mimics reach an altitude of 1200–1500 meters in this valley; similar neighboring valleys contain the melanic Andean races *H. erato microclea* (Fig. 9-1; ▲ symbols to the south) and *H. erato favorinus* (Fig. 9-1; ● symbols to the north, at about 6°–9°S) as well as their *H. melpomene* mimics, which reach altitudes as low as 200 meters, and exclude *H. e. emma* and similar rayed *H. melpomene*. Even within valleys, racial distributions are not clearly altitude-limited. *H. erato emma* hybridizes with the Andean *H. e. favorinus* at Pongo de Aguirre at 210 meters altitude in the Huallaga River Valley, whereas the same hybrid zone is at about 300–320 meters near Pongo de Cainarache on the eastern slopes of the Andes. There are large areas of the Mayo and Huallaga River Valleys upriver from Pongo de Aguirre that are less than 300 meters in elevation, but that have pure Andean *H. e. favorinus.* *Heliconius erato hydara* appears adapted to coastal and open areas, as well as to low-lying areas along the lower Amazon River, in Brazil, Suriname, and Cayenne; rayed races *H. e. erato* and *H. e. amalfreda* are found in denser inland forest (Benson, 1982). However, this correlation breaks down in Guyana, where *H. e. hydara* is found in tall inland rain forest, as well as savannah and more open areas. This race occurs from the Atlantic coast inland to Marudi Mountain, where there is a hybrid zone between *H. e. hydara* and *H. e. amalfreda*. Andean forms which are good species, such as *H. congener* and numerous ithomiines, are found only above about 500 meters in mountain chains ringing river valleys (Lamas, 1982; Mallet and Lamas, in prep.), but *Heliconius* races within species rarely follow climatic or habitat factors so exactly. Poor correlations between Amazonian hybrid zones and habitat change have been used as evidence for refugia (Haffer, 1982, 1985; Brown, 1987c) as a cause of divergence. An alternative explanation is parapatric, habitat-independent divergence, such as by the shifting balance.

The habitat-related biogeographic patterns that do exist might be explained by direct climatic selection on color patterns (above), by an influence of habitat type on the effectiveness of warning colors or mimicry (Benson, 1982; Endler, 1982) or, alternatively, by a buildup of associations between genes for mimicry and genes for host-plant or climatic adaptation and in which partial or complete reproductive isolation might maintain the association (Mallet and Lamas, in prep.). Whatever the causes, strong habitat adaptation to arid or montane habitats seems often to have led to speciation, as exemplified by *H. clysonymus* and *H. himera* versus *H. erato,* whereas the weaker differences between habitats occupied by the various lowland rain forest races of *H. erato* seem to have caused less reproductive isolation.

Correlations of Hybrid Zones with Rainfall Patterns. Some apparent centers of endemism correlate approximately with regions of high rainfall, especially on the slopes of the Andes. This evidence has been used to support the idea that such areas, which are purportedly more likely to have retained forest during a dry period, were indeed refugia during the Pleistocene (Haffer, 1969; Brown, 1979, 1987b). However, many of the proposed correlations may not stand up under close scrutiny. There is a need to map hybrid zones and races accurately in order to investigate this proposed correlation. The patterns mentioned above for Marañon Valley endemics show that, rather than always involving high rainfall, centers of endemism often occur in areas with unusually low levels of rainfall, as well as in areas of higher altitude, with hybrid zones being often found in ecotones between habitats (see also Benson, 1982).

There are also other patterns of correlation with rainfall within species. Rather than being centered in drier zones that are often considered to be interrefugial (Brown, 1979, 1987c), some hybrid zones in neotropical butterflies occur almost exactly at orogenic rainfall peaks at the base of the eastern slopes of the Andes. A number of hybrid zones for *Heliconius erato* and *H. melpomene* between Andean Valley forms and forms from the Amazonian lowlands are arranged in this way; at Pongo de Cainarache, in the centers of hybrid zones between races of *H. erato* and of *H. melpomene,* the rainfall peaks (3637 mm per year), compared with nonhybrid populations of the Andean Valley *H. e. favorinus* and *H. m. amaryllis* at Tarapoto (1004 mm) and Rioja (1728 mm) and the Amazonian *H. e. emma* and *H. m. aglaope* at Yurimaguas (2279 mm) on either side of the hybrid zone (Fig. 9-1; ● and ◆ symbols at about 6°–9° S in the eastern Andes) (Mallet and Barton, 1989a,b; Mallet et al., 1990; rainfall data were obtained from SENAMHI, Peru). Andean valley races of *H. erato* (and *H. melpomene* equivalents) hybridize with Amazonian races in other high-rainfall lowland areas east of the easternmost Andes. They include the Peruvian races *H. erato favorinus* near Tingo María, *microclea* near Iscozacín, and *euryades* on the Río Urubamba below Quillabamba, all of which hybridize with rayed races from the Amazon basin. All three of these races (and their *melpomene* equivalents) are centered on drier areas in the rain shadows of the easternmost Andes chain (Fig. 9-1; ●, ◆, and ○ symbols in the Andes between 6° and 14° S) (Benson, 1982). These distribution patterns, which are repeated many times in other heliconiines and ithomiines from the same areas (Mallet and Lamas, in prep.), again point to a lack of correlation of centers of endemism with regions that currently have high rainfall.

Why should hybrid zones be found in areas with high rainfall? A possible explanation of the anomaly is that moving clines might become trapped by rainfall peaks if the latter act as migration troughs. Although aposematic butterflies probably fly more than palatable butterflies during wet weather (Mallet and Singer, 1987; Chai and Srygley, 1990), cloudy weather and rainfall do considerably reduce the activity of heliconiines and ithomiines, which in turn could reduce their reproductive rate and survival. Butterflies must mate and find adult and larval host plants in order to reproduce; and these activities are reduced during rainy weather. In 1977 I observed at the Río Negro on the eastern slopes of the Andes near Caqueza, Colombia, *Heliconius* that had decomposing wing margins owing to long-term exposure to rain and humidity; local people reported rain had fallen every day for a month before my visit. Such anecdotes suggest the possibility that orogenic rainfall peaks might act as barriers to the free movement of butterfly clines, even though populations of the same species are distributed throughout the zone. In contrast, we have seen that the refugium hypothesis predicts the reverse: Centers of endemism, rather than hybrid zones, should correlate with rainfall peaks.

Correlations of Hybrid Zones with Rivers, Mountains, and Other Features. Under the allopatric model, it is rather puzzling that hybrid zones for mammals, birds, and butterflies are often associated with major rivers in Amazonia (Hershkovitz, 1968; Haffer, 1974, 1985, 1987a,b; Brown, 1979; Lamas, 1982; Beven et al., 1984), as the rich alluvial soils and the high probabilities for gallery forests remaining near rivers during dry periods would tend to cause refugia to be associated with the rivers themselves, rather than with higher ground between major rivers. Many probable "model" species for Müllerian mimicry rings can be currently found year-round in gallery forests, even near smaller, seasonally dry rivers, in the savannah and Llanos regions of Guyana and Venezuela (Brown and Fernandez, 1985; pers. obs.). They include *Lycorea* (Danainae), *Melinaea, Tithorea, Mechanitis,* and *Hypothyris* (Ithomiinae), as well as *Heliconius erato, H. melpomene,* and other heliconiini. It would probably take complete desertification to remove this "weedy" fauna (many species of which have among the highest rates of geographic divergence) from near major rivers. Partial barriers such as rivers should not in any case much delay advancing population fronts emerging from refugia: These fronts should spread until they contact another race in a location that is independent of partial barriers (*contra* Turner, 1982; Sheppard et al., 1985; Hewitt, 1989).

Butterfly hybrid zones in the neotropics often seem to occur at rather major discontinuities: along major rivers, associated with low mountain ridges (though not usually on the tops of these ridges; see above, under Correlations of Hybrid Zones with Rainfall Patterns), at passes through high mountain ranges, and at or near the Isthmus of Panama (Fig. 9-1) (Brown, 1979; Lamas, 1982). Because we do not especially expect secondary contact along partial barriers to dispersal, these distributions imply that hybrid zones can move to areas that act as major partial barriers and are not trapped as easily by very minor local population density troughs as theory leads us to expect.

Correlations Between Clines Within Taxa. Multiple genetic changes within a single pair of hybridizing taxa have been used as evidence for secondary contact at hybrid zones (e.g., Hewitt, 1988, 1989). This within-species pattern can also be understood under a parapatric hypothesis because linkage disequilibria can build up and cause clines to coalesce (Slatkin, 1974; Key, 1981; Barton, 1983; Mallet and Barton, 1989a). Clines are not attracted to each other from any great distance; but if two clines are moving at different rates, a collision results; and the clines tend subsequently to stay together. Both habitat-dependent and habitat-independent clines can accumulate other clines, so that moving habitat-independent clines can become trapped at ecotones where stationary habitat-dependent clines are found. Similarly, clines trapped at a partial dispersal barrier cause other clines to accumulate, in addition to the effect of the density trough itself. Any epistatic interactions that have built up further enhance the effect of multiple clines. Once started, this process of accumulation results in multilocus hybrid zones and causes an increasingly strong barrier to gene flow and further cline movement, even without reinforcement by the evolution of mating barriers (Barton and Hewitt, 1989). Concordance of gene clines within a taxon therefore gives only weak evidence for secondary contact.

Habitat-dependent clines are also likely to occur together if environmental changes are steep (Fig. 9-2). For example, *Papilio glaucus* hybridizes with *P. canadensis* in the northeastern United States and Canada; clines for host plant adaptation, voltinism, and mimetic color pattern all exist in this hybrid zone (Hagen, 1990).

Reality of Subspecies. As well as being important for the interpretation of multiple-locus hybrid zones, theory and data on clines can justify the category of race or sub-

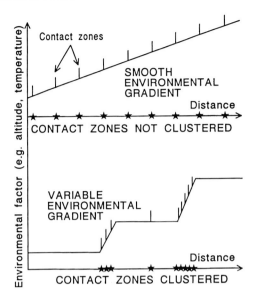

Figure 9-2. Reasons for independent contact zones becoming clumped. If differentiation is habitat-dependent, maps of contact zones in different taxa do not appear clumped, provided the habitat gradient is linear (upper portion of figure). More generally, there are geographically abrupt changes in habitat gradients, leading to an apparent clustering of contact zones on a map (lower portion of figure), even though each contact zone is associated with the same level of environmental parameter as before. For a similar reason, independent clines within a species also appear clumped.

species. Hybrid zones that have accumulated many clines cause severe barriers to gene flow, even though production of heterozygotes may be near Hardy-Weinberg. At neutral or habitat-dependent loci, or at loci with multiple stable equilibria, such hybrid zones separate races, which can evolve virtually independently (Barton, 1983). Only alleles that are advantageous everywhere easily penetrate the hybrid zones. Even less "congealed" hybrid zones, such as those in *Heliconius,* often separate races differing at more than one character or gene (Mallet, 1986a; Mallet et al., 1990). Wilson and Brown (1953) argued that subspecies could not be defined because characters involved in subspecies definitions were uncorrelated. We now have theory predicting the buildup of multilocus hybrid zones, as well as theory predicting that such hybrid zones may become major barriers to gene exchange (Barton, 1983). We also have data showing multiple character or multilocus changes across hybrid zones (Barton and Hewitt, 1983, 1989). When defined properly, subspecies are "real."

Correlations of Hybrid Zones Between Taxa. Groups of hybrid zones forming a "suture zone" between two endemic regions have also been used as evidence for multiple secondary contact of distinct, allopatrically derived faunas (Remington, 1968). In some cases, evidence for suture zones is poor: For example, statistical tests failed to reveal concordant patterns of distribution in the birds studied by Haffer (1969), except that hybrid zones were often found along major rivers (Beven et al., 1984). Plant "centers of endemism" may also be artifacts due to uneven collecting coverage in the Amazon basin (Nelson et al., 1990).

In other cases, concordant distributions definitely do exist: Lamas and I have studied a suture zone manifested by heliconiines and ithomiines in the Huallaga region of Peru (Mallet and Lamas, in prep.). Suture zones can be explained by a common response of species to a habitat gradient as well as by secondary contact, and we have good evidence that a number of contact zones are responding to habitat gradients (for example, see evidence from *Heliconius* and ithomiines above). The steepness of habitat gradients is liable to vary (for example, there are relatively flat plains, valleys, and plateaus with little altitudinal gradient, contrasting with steep-sided mountains). Clines and contact zones for habitat-dependent traits tend to accumulate at the steep parts of an ecotone, even if each pair of hybridizing taxa has a different point on the gradient at which the contact zone is liable to occur, independently of all the others (Fig. 9-2). In addition, hybrid zones of all species with habitat-independent variation will accumulate on the same partial barriers to gene flow (see above), which can again cause concordance between species.

Patterns of Disjunction. One prediction of the shifting balance hypothesis for color pattern evolution is that new taxa should frequently originate in the center of the range of a species—a centrifugal pattern of speciation (W. L. Brown, 1957)—rather than a pattern that might be explained by peripheral isolation, with more derived taxa invading from the periphery (Mayr, 1963). K. S. Brown (1979, 1987b) has compiled examples of disjunct or nearly disjunct distributions that suggest "inundação" (inundation) of one race of ithomiine or heliconiine by another: He has interpreted them as movements of races occurring since refugia expanded and united. If racial redistributions were slightly more rapid than envisaged by Brown, however, distributions might become too blurred to make inferences about refugia. "Leapfrog" (i.e., disjunct) distribution patterns in parapatrically distributed neotropical birds (Remsen, 1984a) seem likely to be caused by derived taxa arising in the center of a distribution and displacing ancestral taxa to the peripheries. *Heliconius erato* (as well as its mimic *H. melpomene*) provides good examples of disjunct apparently primitive color patterns (Fig. 9-1) (see discussion by Sheppard et al., 1985). Phenotypes with red forewing bands and yellow hindwing bars are found in five disjunct locations: Central America and Mexico, the northern Magdalena Valley of Colombia, the upper Putumayo in Southern Colombia, the Mayo and upper Huallaga Valleys of Peru, and Bolivia to southeastern Brazil (Fig. 9-1, ● symbols). Western Colombia and western Ecuador also have red-banded races but with yellow hindwing bars expressed on the underside only (Fig. 9-1, filled and hollow ★ symbols). Forms with red bands but no yellow bar occur in northern South America, as well as in the Urubamba Valley of southern Peru (Fig. 9-1, ○ symbols). The races with yellow hindwing bars are separated from similar forms by other races, particularly by rayed phenotypes with yellow forewing bands. Hybrid zone phenotypes and some crosses suggest that the genetic basis of the color patterns in all races with red forewing bands and yellow hindwing bars is the same within each species (though different between species) (Sheppard et al., 1985; Mallet, 1989). The most parsimonious explanation for these disjunctions is that rayed races evolved in Amazonia and then expanded, isolating older unrayed patterns in the periphery of the distribution of *H. erato* (and *H. melpomene*). Although the initial evolution could

have occurred in allopatry, the disjunct patterns in Figure 9-1 do at least suggest extensive past hybrid zone movements that enlarged the distributions of the rayed races at
the expense of unrayed forms.

It seems possible to use cladistics to test between parapatric evolution involving
moving clines and peripheral isolation. Under peripheral isolation, the peripheral
morphologies would be evolutionarily advanced, and the central taxa would have
more plesiomorphic morphologies. Unless parallel evolution is common, there should
be few disjunctions between peripheral isolates. Under centrifugal evolution brought
about by the shifting balance, the innovations would come from the center of the
range, and the isolated valleys should have plesiomorphic patterns, which might very
well show disjunction patterns. Unfortunately, mimetic color patterns are not ideal
for testing between these hypotheses because of the tendency for color patterns to converge on those of other species, but the frequent disjunctions in the peripheries of *Heliconius* distributions tend to support the centrifugal pattern rather than peripheral isolation. Disjunctions in Andean birds (Remsen, 1984a) seem to provide similar
evidence of centrifugal evolution.

Mimicry in Refugia. Turner (1982) is unusual among refugium biogeographers in
agreeing that differentiation among taxa might easily occur in parapatry. Turner suggested instead that allopatric refugia would be important because of the effect of "faunal drift," where random extinctions affect biotic interactions in the remaining species. Different species of butterflies are expected to have gone extinct in each refugium,
which would have caused frequencies of aposematic patterns to vary between refugia.
Subsequently, mimicry rings in each refugium would diverge under the pressure of the
differing model species.

One problem for this hypothesis is that there are now more racial mimetic patterns of butterflies than there are species mimicking each other (Benson, 1982). Some
species, such as *H. erato* and *H. melpomene,* have more than 25 races, although they
belong to local mimicry rings of only 2–10 heliconiine species (Brown, 1979). To
explain the divergence of the current races of 10 species in the Amazon basin (*Eueides
tales, Neruda aoede, H. burneyi, H. astraea, H. egeria, H. demeter, H. xanthocles, H.
elevatus, H. erato,* and *H. melpomene*) by means of mimicry alone, we require about
10 refugia. In each of these refugia, one of the available 10 species must have become
established; and, simultaneously, it must have gone extinct or become rare in the other
refugia. Only with this unlikely scenario could about 10 patterns be produced with
only 10 available species. This hypothesis would also entail that eight of these rayed
species (all except *H. astraea* and *H. egeria*) secondarily recolonized and adapted to
virtually every local mimicry ring formed in these 10 refugia.

Outside the Amazon basin, there are few extant model species to explain divergence in *H. erato* and *H. melpomene.* The model species that have been postulated for
these extra-Amazonian races, such as *Pereute* (Pieridae) or *Altinote* (Acraeinae) (Sheppard et al., 1985) are unconvincing to me because the analogous mimetic similarity of
the two *Heliconius* is so much greater than the resemblance of either *Heliconius* to any
of the purported model species. In other cases, there are simply no available models.

How could so much divergence occur when there are so many fewer species than
there are color patterns? One possibility is that many model species have gone extinct
(Sheppard et al., 1985); although possible, this *ad hoc* hypothesis has no evidence.

Another possibility was suggested by Sheppard et al. (1985). New mimicry rings could form that combine components of older mimicry rings. However, this second explanation would still require separate mimicry rings to become more and more alike with every cycle of refugia. With no independent tendency to diverge, all mimicry rings would ultimately converge to a single pattern. The third possibility that must be considered is that an intraspecific process, such as the shifting balance, leading to truly novel color patterns, has caused geographic divergence of aposematic patterns in a common and unpalatable model. This divergence in a model species would then enforce divergence in the rest of the mimicry ring.

Mimicry has clearly been important in causing color pattern change of some mimic species, but it would be difficult to explain such rampant divergence on the basis of mimetic convergence alone. It seems likely that at least some model species must have diverged despite, rather than because of, Müllerian mimicry. As we have demonstrated, this kind of divergence would not require refugia at all.

One way to evaluate the effect of refugia on mimicry is to investiate the fauna of current refugia. Obervations in a tiny present-day wet forest refugium isolated by the Guajira desert in Colombia show that the butterfly fauna is indeed depauperate, but that *H. erato hydara* and the ithomiine species *Greta andromica* are little altered in their color patterns from populations in the forests hundreds of kilometers away (Knappett et al., 1976) despite the apparent extinction of all other heliconiines and ithomiines. Similar observations have been made by K. S. Brown on *H. erato phyllis* in the Serra Negra modern forest refuge in NE Brazil (Sheppard et al., 1985, p. 588). *H. erato* is a common species found in wet forest habitats as well as in scrubby vegetation such as that in the Guajira Mountains and gallery forests and other woodlands in currently dry savannahs (see above under Correlations Between Clines Within Taxa). The abundance of *H. erato* makes it a likely candidate to drive evolution in a noncoevolutionary mimicry ring (Bates, 1862; Müller, 1879, Eltringham, 1916; Gilbert, 1983), and the species apparently survives better than its co-mimics under poor conditions. Thus extinction is typically nonrandom (Benson, 1982, p. 632), and Turner's faunal drift does not occur. Once again, there seems to be evidence that divergence involved forces other than simple mimicry or the "faunal drift" refugium hypothesis. We need more observations from small refugia of this kind or from other areas of reduced species diversity such as the Guyana savannahs mentioned above. Under the faunal drift hypothesis, we would expect to see reduced numbers of species having divergent but mimetic color patterns in refugia; under parapatric, centrifugal divergence, there would be no reason to expect greater geographic divergence in modern refuges than in modern continuous rain forest.

Correlations Between Cladograms of Different Taxa. One proposed method for distinguishing parapatric from allopatric divergence is to compare the cladistics of groups of overlapping taxa in an area. Cladograms that are geographically concordant across groups would demand a common explanation (Cracraft and Prum, 1988); in Amazonian birds, it might indicate a common set of origins in Pleistocene refugia. Two criticisms of this viewpoint can be raised. First, we expect related forms to inhabit ecologically similar sites, provided some of the divergence is habitat-dependent: concordant cladograms might indicate similar environments rather than common patterns of vicariance. Second, testing for concordance between cladograms is still in its infancy. What few data we have seem to indicate rather little concordance. Cracraft and Prum (1988) chose the Amazonian bird taxa they analyzed because these taxa appeared to give similar cladograms, but they admitted that cladograms of excluded groups appeared to differ, suggesting that a variety of histories may have caused divergence in the Amazon basin. This fallacy of proposing a common historical explanation on the basis of selected concordant data has been pointed out by Page (1988).

The use of phylogenies of molecular markers within a species is rather more convincing evidence for secondary contact because the markers are likely to be nearly neutral (Rand and Harrison, 1989). However, once a hybrid zone containing a number of loci has formed, a partial barrier to gene flow is erected that excludes neutral, habitat-dependent, or habitat-independent genes much more effectively than genes advantageous everywhere (Barton, 1983). Thus these loci can evolve independently after hybrid zone formation. On the other hand, there seems to be strong evidence for some sort of secondary contact between patches when there is concordance across molecular markers in "mosaic hybrid zones" (Rand and Harrison, 1989), provided the markers are entirely neutral. However, these mosaics do not prove that the race originated in allopatry, only that isolated patches in the mosaic originated via dispersal to the appropriate area. For *Heliconius* or other neotropical organisms, molecular work remains to be done.

Cladistic Evidence for Secondary Contact. Hybrid zones between unrelated taxa can also be used to indicate "secondary contact" (Thorpe, 1984; Nixon and Wheeler, 1990). However, this type of secondary contact does not necessarily indicate that differences evolved in allopatry. Suppose a race of *Heliconius* in the western Amazon basin adapts to montane conditions. Its novel adaptations, evolved in parapatry, may then enable it to spread up and down the Andean chain, bringing it into "secondary contact" with other races of its species, even though there could have been continual parapatric contact via other races in such a "ring" species.

Spatial Scale of Differentiation. Different groups of forest organisms show different scales of divergence. For example, *Heliconius* races are more finely divided than the birds studied by Haffer, and races of small ithomiine butterflies much more finely divided than those of *Heliconius* (Brown, 1979). Presumed warning color patterns in some *Dendrobates* frogs have even narrower ranges than those of the ithomiines; Rainer Schulte (pers. comm.), a resident of Tarapoto, Peru, has found that *Dendrobates bassleri* patterns diverge on either side of the Río Mayo, and that there are strong local differences even within the mountains to the northeast of Tarapoto. *Heliconius erato* and *H. melpomene* are completely monomorphic throughout this area. Silverstone (1975) reported similar localized color pattern variation in *Dendrobates histrionicus* and *D. quinquevittatus*. The differences in the scales of divergence between major groups suggest that the characteristic dispersal distances of a species, rather than gross forest changes, are chiefly responsible for divergence between its forms or races.

Direct Historical Evidence. Some differentiation seems to have occurred more recently than the last glaciation. The Ilha de Marajó on the mouth of the Amazon was probably completely inundated about 5000 years ago, and yet unusual forms of *Eueides, Heliconius, Tithorea,* and *Napeogenes* now appear there (Brown, 1979). This phenomenon is especially surprising because the Marajó has a small land area compared with other lowland Amazonian areas containing endemic races of these butterflies.

There are one or two reports of rapid historically documented movement by butterfly contact zones or geographic races, suggesting that neotropical butterfly biogeography is dynamic even on a 100-year time scale. For example, during the building of the Panama Canal during the early twentieth century, *Heliconius hecale* in the canal zone was polymorphic and included mimetic forms with white-spotted forewings. This polymorphism has been interpreted as a hybrid zone between the Colombian race *H. hecale melicerta* and the Central American *H. h. zuleika* (Brown, 1976a). The canal zone is today monomorphic for *H. h. melicerta,* and Gordon B. Small (pers. comm.) has found a hybrid zone between *H. h. zuleika* and *H. h. melicerta,* including white-spotted forms, about 100 km to the west of the canal zone. There is a tantalizing possible example of the swamping of a race of *H. erato* in eastern Peru. Red-banded *H. erato amphitrite* and *H. melpomene euryades* (Fig. 9-1; ○ symbols) have been collected on the Ríos Urubamba, Cosñipata, and Inambari (Lamas, 1976) around the turn of the century. Today, only yellow-banded rayed *erato* and *melpomene* are found in the latter two valleys, and red-banded forms are present only in the Río Urubamba. Unfortunately, we cannot be sure of the accuracy of the label data of the earlier Peruvian specimens. It is well worth carefully mapping as many neotropical hybrid zones as possible, so that future generations can detect potential movement 50–100 years from now. Similar studies performed in birds seem to indicate similarly rapid evolution of races and species (e.g., Remsen, 1984b).

CONCLUSIONS AND IMPLICATIONS FOR CONSERVATION

I have attempted to show how knowledge of the traits under selection in hybrid zones enables one to make more detailed models of divergence, and how an understanding of divergence can provide information about the geographical circumstances, whether allopatric or parapatric, in which divergence occurs. Warning color is a particularly useful tool for this work because it provides a visual model of partial reproductive isolation due to the strong selection in hybrid zones. For neotropical Lepidoptera, warning color changes seem to be associated strongly with various levels of speciation, and it is not easy to exclude parapatric differentiation. For color patterns, habitat-dependent associations such as mimicry and climatic adaptation seem important; but there seems a clear possible role for the shifting balance in generating utterly new patterns. Neither of these modes of divergence requires allopatry.

Although I have dealt only with neotropical butterflies, similar arguments could be applied to other taxa, including taxa in the temperate zone. Some work in Europe seems definitely to implicate secondary contact as a reason for the current positions of hybrid zones. In *Bombina* toads genetic differences are great, implying divergence so old that known glaciations must have repeatedly separated currently hybridizing taxa (Szymura and Barton, 1986, 1991; see Ch. 10). For *Podisma,* a flightless grasshopper, low levels of dispersal ($\sigma \approx 20$ meters) imply slow potential cline movement and a great tendency for clines to become stuck on minor density troughs. Population restructuring caused by glaciation is thus likely to be faster than inherent tendencies of *Posidma* clines to move. The hypothesis of secondary contact is further enhanced by the position of the *Podisma* hybrid zone on the ridge of the Alpes Maritimes, where retreating ice would be expected to leave it (Nichols and Hewitt, 1988; Hewitt, 1989). It is unclear, however, whether these arguments apply to less well studied systems, such as bird hybrid zones in the Great Plains of North America (Rising, 1983; see Ch. 8). As with the neotropics, I suggest that much, perhaps most, of the divergence ascribed to refugia in the temperate zone should more be carefully tested against alternative hypotheses before a causative role of refugia can be contemplated, however likely the existence of actual refugia during Pleistocene glaciation.

Currently, refugium theory is being used by conservation authorities in Brazil, Peru, Venezuela, and other countries in the neotropics (Brown, 1979; Myers, 1982; Brown and Fernandez, 1985; Rylands, 1990). One of the aims of this chapter is to cast doubt on the role of refugia in causing divergence because I believe the hypothesis is weakly supported at best. In our current state of ignorance, a naive approach to conservation might be optimal: Conserve areas with interesting species or high species diversity, regardless of the possibility that these areas were or were not in Pleistocene refugia. Species are a far less renewable resource than theories about their evolution.

ACKNOWLEDGMENTS

I am grateful to Nick Barton for introducing me to the theory of hybrid zones and linkage disequilibria, and for discussions on the relation between kin selection and genetic drift. However, Barton does not believe that moving clines often produce adaptive change (Barton and Hewitt, 1989). Discussions and correspondence with Keith Brown, Guy Bush, John Endler, Larry Gilbert, Godfrey Hewitt, John R. G. Turner, and Marcy Uyenoyama have also strongly influenced my ideas. Dorothy Jackson helped with the observations in Villa Garzón, Colombia in 1978. This work was financed by the Natural Environment Research Council, the Royal Society, and the Nuffield Foundation. Additional support was given by Sandra Knapp, my grandmother Mrs. G. W. Borlase, and the long-suffering parents, P. L. V. and M. M. G. Mallet, of a "perpetual student."

REFERENCES

Barton, N. H. 1979. The dynamics of hybrid zones. Heredity 43:341–359.

Barton, N. H. 1983. Multilocus clines. Evolution 37:454–471.

Barton, N. H. 1989. Founder effect speciation, In D. Otte and J. A. Endler, eds. Speciation and Its Consequences. Sunderland, MA: Sinauer, pp. 229–256.

Barton, N. H., and Charlesworth, B. 1984. Genetic revolutions, founder effects, and speciation. Ann. Rev. Ecol. Syst. 15:133–164.

Barton, N. H., and Hewitt, G. M. 1983. Hybrid zones as barriers to gene flow. In G. S. Oxford and D. Rollinson, eds. Protein Polymorphism: Adaptive and Taxonomic Significance. London: Academic Press, pp. 341–359.

Barton, N. H., and Hewitt, G. M. 1985. Analysis of hybrid zones. Ann. Rev. Ecol. Syst. 16:113–148.

Barton, N. H. and Hewitt, G. M. Adaptation, speciation and hybrid zones. Nature 341:497–503.

Barton, N. H., and Rouhani, S. 1991. The prob-

ability of fixation of a new karyotype in a continuous population. Evolution 45:499–517.

Bates, H. W. 1862. Contributions to an insect fauna of the Amazon valley. Trans. Linn. Soc. Lond. 23:495–566.

Bates, H. W. 1879. Reservations on Müllerian mimicry. Trans. Entomol. Soc. Lond. xxviii–xxix.

Bazykin, A. D. 1969. Hypothetical method of speciation. Evolution 23:685–687.

Benson, W. W. 1971. Evidence for the evolution of unpalatibility through kin selection in the Heliconiinae (Lepidoptera). Am. Naturalist 105:213–226.

Benson, W. W. 1972. Natural selection for müllerian mimicry in *Heliconius erato* in Costa Rica. Science 176:936–939.

Benson, W. W. 1978. Resource partitioning in passion vine butterflies. Evolution 32:493–518.

Benson, W. W. 1982. Alternative models for infrageneric diversification in the humid tropics: tests with passion vine butterflies. In G. T.

Prance, ed. Biological Diversification in the Tropics. New York: Columbia University Press, pp. 608–640.

Benson, W. W., Brown, K. S., and Gilbert, L. E. 1976. Coevolution of plants and herbivores: passion flower butterflies. Evolution 29:659–680.

Beven, S., Connor, E. F., and Beven, K. 1984. Avian biogeography in the Amazon basin and the biological model of diversification. J. Biogeogr. 11:383–399.

Bishop, J. A. 1972. An experimental study of the cline of industrial melanism in *Biston betularia* (L.) (Lepidoptera) between urban Liverpool and rural North Wales. J. Anim. Ecol. 41:209–243.

Boyden, T. C. 1976. Butterfly palatability and mimicry: experiments with Ameiva lizards. Evolution 30:73–81.

Brower, L. P. 1984. Chemical defence in butterflies. In R. I. Vane-Wright and P. R. Ackery, eds. The Biology of Butterflies (Symposia of the Royal Entomological Society of London No. 11). London: Academic Press, pp. 109–134.

Brower, L. P., and Brower, J. V. Z. 1964. Birds, butterflies and plant poisons: A study in ecological chemistry. Zoologica 49:137–159.

Brower, L. P., Brower, J. V. Z., and Collins, C. T. 1963. Experimental studies of mimicry. 7. Relative palatability and Müllerian mimicry among neotropical butterflies of the subfamily Heliconiinae. Zoologica 48:65–84.

Brown, K. S. 1976a. An illustrated key to the silvaniform Heliconius (Lepidoptera: Nymphalidae) with descriptions of new subspecies. Trans. Am. Entomol. Soc. 102:373–484.

Brown, K. S. 1976b. Geographical patterns of evolution in neotropical Lepidoptera: systematics and derivation of known and new Heliconiini (Nymphalidae: Nymphalinae). J. Entomol. (B) 44:201–242.

Brown, K. S. 1977. Geographical patterns of evolution in neotropical Lepidoptera: differentiation of the species of *Melinaea* and *Mechanitis* (Nymphalidae, Ithomiinae). Syst. Entomol. 2:161–197.

Brown, K. S. 1979. Ecologia Geográfica e Evolução nas Florestas Neotropicais. Vols. 1 and 2. Campinas, Brazil: Universidade Estadual de Campinas.

Brown, K. S. 1981. The biology of *Heliconius* and related genera. Annu. Rev. Entomol. 26:427–456.

Brown, K. S. 1982. Paleoecology and regional patterns of evolution in neotropical forest butterflies. In G. T. Prance, ed. Biological Diversification in the Tropics. New York: Columbia University Press, pp. 255–308.

Brown, K. S. 1987a. Biogeography and evolution of neotropical butterflies. In T. C. Whitmore and G. T. Prance, eds. Biogeography and Quaternary History in Tropical America (Oxford Monographs on Biogeography, No. 3.). Oxford: Oxford University Press, p. 66–104.

Brown, K. S. 1987b. Areas where humid tropical forest probably persisted. In T. C. Whitmore and G. T. Prance, eds. Biogeography and Quaternary History in Tropical America (Oxford Monographs on Biogeography, No. 3.). Oxford: Oxford University Press, p. 45.

Brown, K. S. 1987c. Conclusions, synthesis, and alternative hypotheses. In T. C. Whitmore and G. T. Prance, eds. Biogeography and Quaternary History in Tropical America (Oxford Monographs on Biogeography, No. 3.). Oxford: Oxford University Press, pp. 175–196.

Brown, K. S., and Benson, W. W. 1974. Adaptive polymorphism associated with multiple Müllerian mimicry in *Heliconius numata* (Lepid.: Nymph.). Biotropica 6:205–228.

Brown, K. S. and Benson, W. W. 1975. West Colombian biogeography: notes on *Heliconius hecalesia* and *H. sapho* (Nymphalidae). J. Lepid. Soc. 29:199–212.

Brown, K. S., and Fernandez Yepez, F. 1985. Los Heliconiini (Lepidoptera, Nymphalidae) de Venezuela. Bol. Entomol. Venez. N. S. 3:29–76.

Brown, K. S., and Mielke, O. H. H. 1972. The heliconians of Brazil (Lepidoptera: Nymphalidae). Part II. Introduction and general comments, with a supplementary revision of the tribe. Zoologica 57:1–40.

Brown, K. S., and Neto, J. V. 1976. Predation on aposematic ithomiine butterflies by tanagers (Pipraeida melanonota). Biotropica 8:136–141.

Brown, K. S., Damman, A. J., and Feeny, P. 1981. Troidine swallowtails (Lepidoptera: Papilionidae) in southeastern Brazil: natural history and foodplant relationships. J. Res. Lepid. 19:199–226.

Brown, K. S., Sheppard, P. M., and Turner, J. R. G. 1974. Quaternary refugia in tropical America: evidence from race formation in Heliconius butterflies. Proc. R. Soc. Lond. Biol. 187:369–378.

Brown, W. L. 1957. Centrifugal speciation. Q. Rev. Biol. 32:247–277.

Brown, W. L. 1958. General adaptation and evolution. Syst. Zool. 7:157–168.

Bush, G. L., Case, S. M., Wilson, A. C., and Patton, J. L. 1977. Rapid speciation and chromosomal evolution in mammals. Proc. Natl. Acad. Sci. USA 74:3942–3946.

Chai, P. 1986. Field observations and feeding experiments on the responses of rufous-tailed jacamars (*Galbula ruficauda*) to free-flying butterflies in a tropical rainforest. Biol. J. Linn. Soc. 29:166–189.

Chai, P., and Srygley, R. B. 1990. Predation and

the flight, morphology, and temperature of neotropical rain-forest butterflies. Am. Naturalist 135:748–765.

Charlesworth, B., Coyne, J. A., and Barton, N. H. 1987. The relative rates of evolution of sex chromosomes and autosomes. Am. Naturalist 130:113–146.

Colinvaux, P. A. 1989. Ice-age Amazon revisited. Nature 340:188–189.

Connor, E. F. 1986. The role of Pleistocene forest refugia in the evolution and biogeography of tropical biotas. Trends Ecol. Evol. 1:165–168.

Cracraft, J., and Prum, R. O. 1988. Patterns and processes of diversification: speciation and historical congruence in some neotropical birds. Evolution 42:603–620.

Crow, J. F., Engels, W. R., and Denniston, C. 1990. Phase three of Wright's shifting-balance theory. Evolution 44:233–247.

Darwin, C. 1863. A review of H. W. Bates' paper on "mimetic butterflies." In P. H. Barrett, ed. The Collected Papers of Charles Darwin. Vol. 2. Chicago: University of Chicago Press, pp. 87–92.

Descimon, H., and Mast de Maeght, J. 1984. Semispecies relationships between *Heliconius erato cyrbia* Godt. and *H. himera* Hew. in south western Ecuador. J. Res. Lepid. 22:229–239.

Douglas, M. M., and Grula, J. W. 1978. Thermoregulatory adaptations allowing ecological range expansion by the Pierid butterfly *Nathalis iole* Boisduval. Evolution 32:776–783.

Eanes, W. F., and Koehn, R. K. 1978. An analysis of genetic structure in the monarch butterfly, *Danaus plexippus* L. Evolution 32:784–797.

Eltringham, H. 1916. On specific and mimetic relationships in the genus *Heliconius.* Trans. Entomol. Soc. Lond. 1916:101–148.

Emsley, M. G. 1964. The geographical distribution of the color-pattern components of *Heliconius erato* and *Heliconius melpomene* with genetical evidence for the systematic relationship between the two species. Zoologica 49:245–286.

Emsley, M. G. 1965. Speciation in *Heliconius* (Lep., Nymphalidae): morphology and geographic distribution. Zoologica 50:191–254.

Endler, J. A. 1977. Geographic Variation, Speciation, and Clines. Princeton, NJ: Princeton University Press.

Endler, J. A. 1982. Pleistocene forest refuges: fact or fancy? In G. T. Prance, ed., Biological Diversification in the Tropics. New York: Columbia University Press, pp. 641–657.

Endler, J. A. 1988. Frequency-dependent predation, crypsis and aposematic coloration. Phil. Trans. Roy. Soc. London B 319:459–472.

Fisher, R. A. 1950. Gene frequencies in a cline determined by selection and diffusion. Biometrics 6:353–361.

Fisher, R. A. 1958. The Genetical Theory of Natural Selection. 2nd ed. New York: Dover.

Futuyma, D. J., and Mayer, G. C. 1980. Nonallopatric speciation in animals. Syst. Zool. 29:254–271.

Gilbert, L. E. 1969. Some aspects of the ecology and community structure of ithomid butterflies in Costa Rica. O.T.S. Report 1969:68–90.

Gilbert, L. E. 1975. Ecological consequences of a coevolved mutualism between butterflies and plants. In L. E. Gilbert and P. R. Raven, eds. Coevolution of Animals and Plants. Austin: University of Texas Press, pp. 210–240.

Gilbert, L. E. 1977. The role of insect-plant coevolution in the organization of ecosystems. In V. Labeyrie, ed. Le Comportement des Insectes et les Signaux Issus du Milieu Trophique (Colloques Internationaux du C.N.R.S, 265.). Paris: CNRS, pp. 399–413.

Gilbert, L. E. 1983. Coevolution and mimicry. In D. J. Futuyma and M. Slatkin, eds. Coevolution. Sunderland, MA: Sinauer, pp. 263–281.

Gilbert, L. E. 1984. The biology of butterfly communities. In R. I. Vane-Wright and P. R. Ackery, eds. The Biology of Butterflies (Symposia of the Royal Entomological Society of London, No. 11.). London: Academic Press, pp. 41–54.

Gilbert, L. E. 1991. Biodiversity of a Central American *Heliconius* community: pattern, process, and problems. In P. W. Price, T. M. Lewinsohn, T. W. Fernandes, and W. W. Benson, eds. Plant-Animal Interactions: Evolutionary Ecology in Tropical and Temperate Regions. New York: Wiley, pp. 403–427.

Grimaldi, D. 1984. (Review of) Biological Diversification in the Tropics. G. T. Prance, ed. Proceedings of the Fifth International Symposium for the Association of Tropical Biology, 1982. New York: Columbia University Press. J. N.Y. Entomol. Soc. 92:94–96.

Guilford, T. 1985. Is kin selection involved in the evolution of warning coloration? Oikos 45:31–36.

Haffer, J. 1967. Speciation in Colombian forest birds west of the Andes. Am. Mus. Novitat. 2294:1–57.

Haffer, J. 1969. Speciation in Amazonian forest birds. Science 165:131–137.

Haffer, J. 1974. Avian speciation in tropical South America. Publ. Nuttall Ornithol. Club 14:1–390.

Haffer, J. 1982. General aspects of the refuge theory. In G. T. Prance, ed. Biological Diversification in the Tropics. New York: Columbia University Press, pp. 6–24.

Haffer, J. 1985. Avian zoogeography of the neotropical lowlands. Ornithol. Monogr. 36:113–145.

Haffer, J. 1987a. Quaternary history of tropical America. In T. C. Whitmore and G. T. Prance, eds. Biogeogrpahy and Quaternary History in Tropical America (Oxford Monographs on Biogeography No. 3.). Oxford: Oxford University Press, pp. 1–18.

Haffer, J. 1987b. Biogeography of neotropical birds. In T. C. Whitmore and G. T. Prance, eds. Biogeography and Quaternary History in Tropical America (Oxford Monographs on Biogeography No. 3.). Oxford: Oxford University Press, pp. 105–150.

Hagen, R. H. 1990. Population structure and host use in hybridizing subspecies of *Papilio glaucus* (Lepidoptera: Papilionidae). Evolution 44:1914–1930.

Haldane, J. B. S. 1948. The theory of a cline. J. Genet. 48:277–284.

Harrison, R. G., and Rand, D. M. 1989. Mosaic hybrid zones and the nature of species boundaries. In D. Otte and J. A. Endler, eds. Speciation and its Consequences. Sunderland, MA: Sinauer, pp. 111–133.

Harvey, D. J. 1991. Appendix B: higher classification of the Nymphalidae. In H. F. Nijhout, ed. The Development and Evolution of Butterfly Wing Patterns. Washington, DC: Smithsonian Institution Press, pp. 255–273.

Harvey, P. H., Bull, J. J., Pemberton, M., and Paxton, R. J. 1982. The evolution of aposematic coloration in distasteful prey: a family model. Am. Naturalist 119:710–719.

Harvey, P. H., and Greenwood, P. J. 1978. Antipredator defence strategies: some evolutionary problems. In J. R. Krebs and N. B. Davies, eds. Behavioural Ecology. Oxford: Blackwell Scientific, pp. 129–151.

Hedrick, P. W., and Levin, D. A. 1984. Kin-founding and the fixation of chromosomal variants. Am. Anturalist 124:789–797.

Hershkowitz, P. 1968. Metachromism or the principle of evolutionary change in mammalian tegumentary colors. Evolution 22:556–575.

Hewitt, G. M. 1988. Hybrid zones—natural laboratories for evolutionary studies. Trends Ecol. Evol. 3:158–167.

Hewitt, G. M. 1989. The subdivision of species by hybrid zones. In D. Otte and J. A. Endler, eds. Speciation and its Consequences. Sunderland, MA: Sinauer, pp. 85–110.

Jain, S. K., and Bradshaw, A. D. 1966. Evolutionary divergence among adjacent plant populations I. The evidence and its theoretical analysis. Heredity 21:407–441.

Johnson, M. S., Clarke, B., and Murray, J. 1990. The coil polymorphism in *Partula suturalis* does not favor sympatric speciation. Evolution 44:459–464.

Johnson, M. S., Murray, J., and Clarke, B. 1987. Independence of genetic subdivision and variation for coil in *Partula suturalis.* Heredity 58:307–313.

Kaye, W. J. 1916. A reply to Dr. Eltringham's paper on the genus *Heliconius.* Trans. Entomol. Soc. Lond. 1916:149–155.

Key, K. H. L. 1981. Species, parapatry, and the morabine grasshoppers. Syst. Zool. 30:425–458.

Knappett, C. P., Mallet, J., and Sugden, A. M. 1976. Oxford expedition to the Serranía de la Macuira, Colombia, 1975. Bull. Oxford Univ. Explor. Club New Series 2:7–19.

König, F. 1986. Ein *Heliconius erato himera*—Hybrid aus Nord-Peru (Lepidoptera, Heliconiinae). Z. Arbeitsgem. Osterr. Entomol. 38:49–50.

Lamas, G. 1973. Taxonomia e evolucão dos gêneros Ituna Doubleday (Danainae) e Paititia, gen. n., Thyridia Hübner e Methona Doubleday (Ithomiinae) (Lepidoptera, Nymphalidae). Ph.D. thesis, Universidade de São Paulo.

Lamas, G. 1976. Notes on Peruvian butterflies (Lepidoptera). II. New Heliconius from Cusco and Madre de Dios. Rev. Peruana Entomol. 19:1–7.

Lamas, G. 1982. A preliminary zoogeographical division of Peru based on butterfly distributions (Lepidoptera, Papilionoidea). In G. T. Prance, ed. Biological Diversification in the Tropics. New York: Columbia University Press, pp. 336–357.

Lande, R. 1985. The fixation of chromosomal rearrangements in a subdivided population with local extinction and colonization. Heredity 54:323–332.

MacArthur, R. H. 1972. Geographical Ecology. New York: Harper & Row.

Mallet, J. 1986a. Hybrid zones in *Heliconius* butterflies in Panama, and the stability and movement of warning colour clines. Heredity 56:191–202.

Mallet, J. 1986b. Gregarious roosting and home range in *Heliconius* butterflies. Natl. Geogr. Res. 2:198–215.

Mallet, J. 1986c. Dispersal and gene flow in a butterfly with home range behaviour: *Heliconius erato* (Lepidoptera: Nymphalidae). Oecologia (Berl.) 68:210–217.

Mallet, J. 1989. The genetics of warning colour in Peruvian hybrid zones of *Heliconius erato* and *H. melpomene.* Proc. R. Soc. Lond. Biol. 236:163–185.

Mallet, J. 1990. Is mimicry theory unpalatable? Trends Ecol. Evol. 5:344–345.

Mallet, J., and Barton, N. 1989a. Inference from clines stablized by frequency-dependent selection. Genetics 122:967–976.

Mallet, J., and Barton, N. 1989b. Strong natural selection in a warning color hybrid zone. Evolution 43:421–431.

Mallet, J., Barton, N., Lamas, G., Santisteban, J.,

Muedas, M., and Eeley, H. 1990. Estimates of selection and gene flow from measures of cline width and linkage disequilibrium in *Heliconius* hybrid zones. Genetics 124:921–936.

Mallet, J., and Singer, M. C. 1987. Individual selection, kin selection, and the shifting balance in the evolution of warning colours: the evidence from butterflies. Biol. J. Linn. Soc. 32:337–350.

Mayr, E. 1963. Animal Species and Evolution. Cambridge: Harvard University Press.

Mayr, E. 1970. Populations, Species, and Evolution. Cambridge: Harvard University Press.

Müller, F. 1879. *Ituna* and *Thyridia;* a remarkable case of mimicry in butterflies. Trans. Entomol. Soc. Lond. 1879:xx–xxix.

Myers, N. 1982. Forest refuges and conservation in Africa, with some appraisal of survival prospects for tropical moist forests throughout the biome. In G. T. Prance, ed., Biological Diversification in the Tropics. NY: Columbia University Press, pp. 658–672.

Nelson, B. W., Ferreira, C. A. C., Da Silva, M. F., and Kawasaki, M. L. 1990. Endemism centres, refugia and botanical collection density in Brazilian Amazonia. Nature 345:714–716.

Nichols, R. A., and Hewitt, G. M. 1988. Genetical and ecological differentiation across a hybrid zone. Ecol. Entomol. 13:39–49.

Nixon, K. C., and Wheeler, Q. D. 1990. An amplification of the phylogenetic species concept. Cladistics 6:211–223.

Page, R. D. M. 1988. Quantitative cladistic biogeography: constructing and comparing area cladograms. Syst. Zool. 37:254–270.

Pansera, M. C. G., and Araújo, A. M. 1983. Distribution and heritability of the red raylets in *Heliconius erato phyllis* (Lepid.: Nymph.). Heredity 51:643–652.

Provine W. 1986. Sewall Wright and Evolutionary Biology. Chicago: University of Chicago Press.

Rand, D. M., and Harrison, R. G. 1989. Ecological genetics of a mosaic hybrid zone: mitochondrial, nuclear, and reproductive differentiation of crickets by soil type. Evolution 43:432–449.

Remington, C. L. 1968. Suture-zones of hybrid interaction between recently joined biotas. Evol. Biol. 1:321–428.

Remsen, J. V. 1984a. High incidence of "leapfrog" pattern of geographic variation in Andean birds: implications for the speciation process. Science 224:171–173.

Remsen, J. V. 1984b. Geographic variation, zoogeography, and possible rapid evolution in some Cranioleuca spinetails (Furnariidae) of the Andes. Wilson Bull. 96:515–523.

Rising, J. D. 1983. The Great Plains hybrid zones. Curr. Ornithol. 1:131–157.

Roland, J. 1982. Melanism and diel activity of alpine *Colias* (Lepidoptera: Pieridae). Oecologia (Berl.) 53:214–221.

Rouhani, S., and Barton, N. 1987. Speciation and the "shifting balance" in a continuous population. Theoret. Pop. Biol. 31:465–492.

Rylands, A. B. 1990. Priority areas for conservation in the Amazon. Trends Ecol. Evol. 5:240–241.

Sheppard, P. M. 1963. Some genetic studies of Müllerian mimics in butterflies of the genus *Heliconius.* Zoologica 48:145–154.

Sheppard, P. M., Turner, J. R. G., Brown, K. S., Benson, W. W., and Singer, M. C. 1985. Genetics and the evolution of Mullerian mimicry in *Heliconius* butterflies. Philos. Trans. R. Soc. Lond. Biol. 308:433–613.

Silberglied, R. E. 1984. Visual communication and sexual selection among butterflies. In R. I. Vane-Wright and P. R. Ackery, eds. The Biology of Butterflies (Symposia of the Royal Entomological Society of London, No. 11.). London: Academic Press, pp. 207–223.

Silverstone, P. A. 1975. A Revision of the Poison-Arrow Frogs of the Genus *Dendrobates* Wagler. Scientific Bulletin, No. 21. Los Angeles: Natural History Museum of Los Angeles.

Slatkin, M. 1973. Gene flow and selection in a cline. Genetics 75:733–756.

Slatkin, M. 1974. Gene flow and selection in a two-locus system. Genetics 81:209–222.

Slatkin, M., and Barton, N. H. 1989. A comparison of three indirect methods for estimating average levels of gene flow. Evolution 43:1349–1368.

Srygley, R. B., and Chai, P. 1990. Flight morphology of neotropical butterflies: palatability and the distribution of mass to the thorax and abdomen. Oecologia (Berl.) 84:491–499.

Szymura, J. M., and Barton, N. H. 1986. Genetic analysis of hybrid zone between the fire-bellied toads, *Bombina bombina* and *B. variegata* near Cracow in southern Poland. Evolution 40:1141–1159.

Szymura, J. M., and Barton, N. H. 1991. The genetic structure of the hybrid zone between the fire-bellied toads *Bombina bombina* and *B. variegata:* comparisons between transects and between loci. Evolution 45:237–261.

Thorpe, R. S. 1984. Primary and secondary transition zones in speciation and population differentiation: a phylogenetic analysis of range expansion. Evolution 38:233–243.

Turelli, M., and Hoffmann, A. A. 1991. Rapid spread of an inherited incompatibility factor in California Drosophila. Nature 353:440–442.

Turner, J. R. G. 1965. Evolution of complex polymorphism and mimicry in distasteful South American butterflies. Proc. XII Int. Cong. Entomol. London 1964:267.

Turner, J. R. G. 1968. Natural selection for and

against a polymorphism which interacts with sex. Evolution 22:481–495.

Turner, J. R. G. 1971a. Studies of Müllerian mimicry and its evolution in burnet moths and heliconid butterflies. In E. R. Creed, ed. Ecological Genetics and Evolution. Oxford: Blackwell, pp. 224–260.

Turner, J. R. G. 1971b. Experiments on the demography of tropical butterflies. II. Longevity and home-range behaviour in *Heliconius erato*. Biotropica 3:21–31.

Turner, J. R. G. 1971c. Two thousand generations of hybridization in a *Heliconius* butterfly. Evolution 25:471–482.

Turner, J. R. G. 1976. Müllerian mimicry: classical "beanbag" evolution and the role of ecological islands in adaptive race formation. In S. Karlin and E. Nevo, eds. Population Genetics and Ecology. Orlando, FL: Academic Press, pp. 185–218.

Turner, J. R. G. 1978. Why male butterflies are non-mimetic: natural selection, sexual selection, group selection, modification and sieving. Biol. J. Linn. Soc. 10:385–432.

Turner, J. R. G. 1982. How do refuges produce tropical diversity? Allopatry and parapatry, extinction and gene flow in mimetic butterflies. In G. T. Prance, ed. Biological Diversification in the Tropics. New York: Columbia University Press, pp. 309–335.

Turner, J. R. G. 1984. Mimicry: the palatability spectrum and its consequences. In R. I. Vane-Wright and P. R. Ackery, eds. The Biology of Butterflies (Symposia of the Royal Entomological Society of London, No. 11.). London: Academic Press, pp. 141–161.

Vane-Wright, R. I., Ackery, P. R., and Smiles, R. L. 1975. The distribution, polymorphism and mimicry of *Heliconius telesiphe* (Doubleday)

and the species of *Podotricha* Michener (Lepidoptera: Heliconiinae). Trans. R. Entomol. Soc. Lond. 126:611–636.

Wade, M. J., and Goodnight, C. J. 1991. Wright's shifting balance theory: an experimental study. Science 253:1015–1018.

Wallace, A. R. 1865. On the phenomena of variation and geographical distribution as illustrated by the Papilionidae of the Malayan region. Trans. Linn. Soc. Lond. 25:1–71.

Wallace, B. 1968. Topics in Population Genetics. New York: Norton.

Watt, W. B. 1968. Adaptive significance of pigment polymorphisms in *Colias* butterflies. I. Variation of melanin pigment in relation to thermoregulation. Evolution 22:437–458.

White, M. J. D. 1978. Modes of Speciation. San Francisco: W. H. Freeman.

Wiklund, C., and Sillén-Tullberg, B. 1985. Why distasteful butterflies have aposematic larvae and adults, but cryptic pupae: evidence from predation experiments on the monarch and European swallowtail. Evolution 39:1155–1158.

Wilson, E. O. 1975. Sociobiology: The New Synthesis. Cambridge, MA: Belknap.

Wilson, E. O., and Brown, W. L. 1953. The subspecies concept and its taxonomic application. Syst. Zool. 2:97–111.

Wright, S. 1941. On the probability of fixation of reciprocal translocations. Am. Naturalist 75:513–522.

Wright, S. 1977. Evolution and the Genetics of Populations. Vol. 3. Experimental Results and Evolutionary Deductions. Chicago: University of Chicago Press.

Wright, S. 1978. Review: Modes of Speciation, by M. J. D. White. Paleobiology 4:373–379.

10

Analysis of Hybrid Zones with *Bombina*

JACEK M. SZYMURA

The study of fire-bellied toads in Europe has a long history, and for much of it debate has centered around whether the two species, *Bombina bombina* and *B. variegata,* hybridize in nature. The existence of individuals with intermediate morphology in areas where the ranges of *B. bombina* and *B. variegata* meet (overlap) was explained by invoking either hybridization (Méhely, 1892; Karaman, 1922; Mertens, 1928) or response to local environmental conditions. Artificial hybrids could be obtained in the laboratory, and heterospecific pairs in amplexus were observed in the field (Michał-owski, 1958), but fertility of hybrids was unknown and morphology of F_1 hybrids did not resemble most intermediate forms found in nature (Michałowski and Madej, 1969). Spotting pattern and coloration of the belly, although characteristic for each of the forms, are variable and difficult to describe, making simple morphological analysis of presumed hybridization inconclusive. Furthermore, the lowland and the mountain form of *Bombina* have distinct mating calls (Lörcher, 1969), which traditionally have been regarded as effective barriers to gene exchange in many amphibian species. Even more confusing, intermediate morphologies are also found far from the area of contact, and some association between habitat type and morphology has been observed within the species' overlap. Therefore until recently evidence for hybridization in the fire-bellied toads was ambiguous.

   With the advent of molecular techniques, which allowed classification of individuals into clearly defined genotypic classes, the question of hybridization could be answered. Application of protein electrophoresis demonstrated that the fire-bellied toads *B. bombina* and *B. variegata* do interbreed west of Cracow (Szymura, 1976) in a morphological transition described earlier by Michałowski (1958). The same study revealed that there is a genuine transition from *bombina* to *variegata* alleles over just 6 km near Cracow rather than patchy, localized hybridization. All enzymes showed a similar pattern. Central populations consisted almost exclusively of individuals of mixed ancestry, but F_1's were rare, if present at all. This finding is not surprising if pure parental types have little opportunity to meet. Hybrids appeared to be fertile because a whole array of backcross and recombinant genotypes was revealed. Similarly, mating appeared to be random because the genotypic composition of each population could be derived from allelic frequencies. Location of the zone did not seem to be associated with any apparent environmental discontinuity but corresponded to the position determined earlier by morphological studies (Michałowski, 1958). The hybrid zone had therefore all the characteristics of a zone stable for at least the last 20 years.

Although many hybrid zones have been studied, the clinal patterns of variation have rarely been seen as clearly as in *Bombina*. Detailed examination of these clines has provided a rare opportunity to quantify predictions of the strength of selection against hybrids, the number of genes under selection, and the extent of gene flow between the two taxa. The hybrid zone for the fire-bellied toads is almost ideal for studies of hybrid zone dynamics. The two taxa differ in a wide variety of characters, including traits whose adaptive significance can be understood and presumably neutral, diagnostic biochemical markers. The toads are common, so large samples can be collected. For example, analysis of the Cracow transect was based on 3014 individuals and that near Przemysl on 1091 toads. Sampling does not need to be disruptive, as at least six marker loci can be scored from amputated toes. These enzymes show Mendelian inheritance and assort independently (Szymura and Farana, 1978). Moreover, individuals can be recognized from photographs of their belly pattern, which is highly variable, and thus can be followed over several years. If needed, fire-bellied toads can also be bred in the laboratory, the only disadvantage being their long maturation time.

Another great advantage of the fire-bellied toads is the vast amount of information that has been gathered over the past century, including surveys of morphological variation and distribution in several areas of contact. The history of *Bombina* can be reconstructed from the pattern of allozyme and mitochondrial DNA (mtDNA) variation in Europe (which also allow estimates of divergence time) and from paleontological and geological data. In this way the hybrid zone can be viewed in the broader context of historical biogeography and taxonomy.

DISTRIBUTION OF *B. BOMBINA* AND *B. VARIEGATA*

The parapatric distribution of *B. bombina* and *B. variegata* (Fig. 10-1) is well explained by the ecology of the species and by postglacial range expansion from southern refuges. The fire-bellied toad, *B. bombina,* is found in lowlands of central and eastern Europe, from the Ural Mountains in the east to the Weser River in the west. It occurs on the Danish isles and recently disappeared from southern Sweden. In the south, *B. bombina* inhabits lowlands along the Danube and its tributaries as well as the Bohemian-Moravian plateau. These southern populations are separated from the northern ones by intervening *B. variegata* in the Carpathian Mountains or by areas where no *Bombina* is present. Isolated populations of *B. bombina* are found in northeastern Greece and northwestern Turkey (Andrén et al., 1984).

The yellow-bellied toad, *B. variegata,* occupies higher altitudes of southern and western Europe from France through the Alpine countries, most of Germany, the Balkans, and the Carpathian Mountains. There are enclaves of *B. variegata* on isolated mountain ranges in the Hungarian plains (e.g., Mecsek Mountains, Matra Mountains, Papuk Mountains, and Fruška Gora), which are completely surrounded by *B. bombina.* In Italy, *B. variegata* is separated from its main range by the Po River Valley. However, the Italian populations, known as *B. v. pachypus,* may not be conspecific with *B. variegata.* The distribution of the two fire-bellied toad species, although complicated, clearly follows the topography of central Europe.

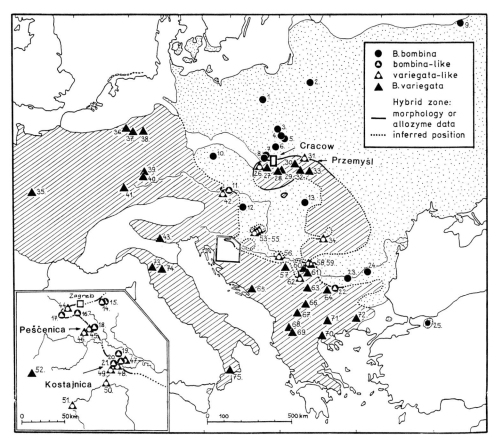

Figure 10-1. Ranges of the fire-bellied toads in Europe and sampling sites for *Bombina bombina* (circles) and *B. variegata* (triangles). A circle with a triangle and a triangle with a circle indicate *B. bombina*-like and *B. variegata*-like hybrid populations, respectively. A dotted line shows the inferred position of hybrid zones in central Europe. A solid line indicates regions where hybridization was detected either by electrophoretic or morphological studies. Distribution of samples in western Yugoslavia is shown in the inset. Position of the transects across the hybrid zones in Poland and western Yugoslavia are indicated by arrows.

CHARACTERISTICS OF FIRE-BELLIED TOADS

Hybridization between *B. bombina* and *B. variegata* may create an erroneous impression of great similarity of these forms and even call into question their specific taxonomic status. In fact, the European fire-bellied toads are distinct forms with contrasting ecologies. Three separate lines of evidence indicate substantial genetic divergence (Table 10-1).

Morphology, Ecology, Behavior. *B. bombina* and *B. variegata* differ in numerous characteristics of their morphology, anatomy, ecology, and behavior. Many of these differences can be regarded as adaptations to the predominant habitat in their respective environments. The coloration of the belly is aposematic. Both species display a

Table 10-1. Differences Between *B. bombina* and *B. variegata*

Measurement	*B. bombina*	*B. variegata*	Reference
Distribution	Lowlands of Eastern and Central Europe	Mountainous and hilly regions of western and southern Europe and the Carpathians	Andrin et al. (1984); Arntzen (1978)
Breeding sites	Large permanent waters	Temporary pools, small ponds	Madej (1973)
Habits	Largely aquatic	More terrestrial	Madej (1973)
Skin thickness (epidermis/ dermis), μm	134.5 (22.8/111.7)	296.6 (65.2/231.4)	Czopkowa & Czopek (1955)
Breeding behavior	Prolonged breeder, territorial	Explosive breeder, non-territorial	Lörcher (1969); Szymura (unpublished)
Rate of calls (min^{-1})	22	95	Lörcher (1969)
Call duration (ms)	210	160	Lörcher (1969)
Fundamental frequency (Hz)	530	580	Lörcher (1969)
Sound pressure at 20 cm, 18°C (dyn cm^{-2})	10.5	1.8	Lörcher (1969)
Vocal sacs	Present	Absent	Boulenger (1886)
Lung volume in 4.5-cm toad (cm^3)	3.0–3.5	2.0–2.2	Lörcher (1969)
Mean fecundity (largest clutches observed)	363 eggs (509, 547, 689)	116 eggs (204, 233, 294)	Rafińska (1991) Rafińska (1991)
No. of eggs per clump (range)	32 (9–76)	17 (4–58)	Rafińska (1991)
Egg size (mm)	1.4	1.9	Rafińska (1991)
Development time at 20°C (egg to toadlet), days	73–75	61–63	Rafińska (1991)
DNA content per nucleus (pg)	18.8	21.1	Olmo et al. (1982)
Chromosome no. (identical karyotypes)	24	24	Morescalchi (1965); Wickbom (1949)
Nei's D, 29 loci		0.37–0.59	Szymura (1983, 1988)
Albumin distance (IDU)		2–4	Maxson & Szymura (1984)
mtDNA divergence (%)		5.6–7.	Szymura et al. (1985, unpublished)
No. of genes under selection		55 (26–88)	Szymura & Barton (1991)

warning Unken reflex when disturbed on land, and their skin possesses numerous poison glands, concentrated on the back and especially on the hind feet (Bajger, 1980). *B. bombina* has smaller orange-red or red, mostly isolated patches that cover less than 50% of the black or bluish-gray underside with white spots. The colored patches are yellow or orange in *B. variegata* and have a tendency to fuse; usually they occupy more than 50% of the surface. Dorsal warts in *B. bombina* are flat and singular, whereas in *B. variegata* they are pointed and surrounded by numerous smaller ones (Méhely, 1891) There are also differences in body proportions: The thigh and shin of *B. variegata* are relatively longer than corresponding parts of *B. bombina* (Michałowski, 1961). These differences also apply to the proportions of internal organs (Herre and Rawiel, 1939). Tadpoles of *B. bombina* have a higher fin, two longitudinal dark stripes on the dorsal side, and a blue belly. Tadpoles of *B. variegata* have a lower fin and are uniformly gray pigmented (Michałowski, 1966). Development of aposematic coloration after metamorphosis proceeds at an unequal pace in the fire-bellied toads; the lifelong individual pattern develops in *B. bombina* earlier than in *B. variegata.* In the latter species, it is still difficult to detect 1 month after metamorphosis (Boulenger, 1886).

Bombina bombina inhabits large permanent ponds and is more aquatic. Males defend their territories, have vocal pouches, and inflate their lungs when calling (augmenting the calls, which are loud and can be heard from hundreds of meters). Males also form large choruses. *B. bombina* is a prolonged breeder, with mating taking place throughout spring and early summer (Juszczyk, 1974). There are fewer permanent breeding sites at higher altitudes, and so *B. variegata* is usually found in temporary pools and puddles. *B. variegata* are more terrestrial, have thicker skin, and tend to breed only after heavy rain from spring to late summer. Males are nonterritorial and have a quieter mating call, as they lack vocal sacs. However, they do possess additional nuptial pads on their hind toes, which are used to dislodge other males. Female *B. variegata* produce smaller clutches of larger eggs: 2.5 times larger in volume than those of *B. bombina.* Larger eggs result in bigger hatchlings that complete metamorphosis sooner (Rafińska, 1991). The developmental time for *B. variegata* tadpoles can be significantly shorter in nature than shown in Table 10-1, as the water temperature in temporary pools often exceeds 30°C.

Molecular Data. The large number of differences observed between the two species for almost any important biological character may nevertheless overestimate divergence of *Bombina* at a genic level. Many characters are correlated, and therefore the pleiotropic effects of a few loci could, at least in principle, account for the apparent differences. If so, divergence of *B. bombina* and *B. variegata* might be relatively recent.

Molecular data can provide a test of that hypothesis. Moreover, they give a less biased estimate of divergence, as nucleotide substitutions at different loci or positions are independent of each other, are less likely to be influenced by selection, and are based on a much larger number of characters encoded either in the nuclear or cytoplasmic DNA.

Estimates of divergence based on 29 protein-coding loci gave a Nei's (Nei, 1972) genetic distance value of $D_N = 0.37–0.59$ between *B. bombina* and populations representative of major subgroups within *B. variegata* (Szymura, 1988). Similarly, divergence in mtDNA sequence between *B. bombina* and *B. variegata* is also high— between 5.6% and 7.0% (Szymura et al., unpublished data). These independent estimates speak against great similarity of the toads. They suggest a pre-Pleistocene divergence of *B. bombina* and *B. variegata* 2–7 Myr ago (Szymura, 1988).

Number of Genes Under Selection. The third line of evidence comes, surprisingly, from analysis of hybrid zones. It is possible to calculate the number of genes under selection from knowledge of the width of a hybrid zone and dispersal rate of the organism. The number of genes responsible for hybrid unfitness—and involved in maintenance of the zone in *Bombina*—was estimated to be 55 (range 26–88) (Szymura and Barton, 1991). This method is based on a model in which all loci have the same effect. The number of genes is thus an effective number, which is weighted toward loci with stronger effects.

In sum, three separate assessments of divergence between *B. bombina* and *B. variegata* that probed different parts of their genomes showed a consistent pattern pointing to highly significant differences between these species. These differences have persisted despite widespread and seemingly unlimited hybridization.

PATTERNS OF GENETIC VARIATION

With few exceptions, European *Bombina* populations can be assigned to either *B. bombina* or *B. variegata* on the basis of their morphology. Upon electrophoretic examination, however, hybrid genotypes become apparent in many samples from the contact area. Therefore when describing patterns of geographic variation, introgressed populations having more than 5% of introgressed alleles are generally disregarded. Such populations are usually located 20–30 km from the zone center in either direction.

More than 70 populations of *B. bombina* and *B. variegata* have been sampled from throughout the distribution of these species, including areas where their ranges meet (Fig. 10-1). Variation of seven enzymes encoded by nine loci has been assayed: adenylate kinase *(Ak)*, creatine kinase *(Ck)*, aspartic aminotransferase *(Aat-1)*, glucose phosphate dehydrogenase *(Gpi)*, isocitrate dehydrogenase *(Idh-1)*, lactate dehydrogenase *(Ldh-1, Ldh-2)*, and malate dehydrogenase *(Mdh-1, Mdh-2)*. These enzymes were chosen for three reasons. They had already been used to study details of hybridization in Poland. The inheritance of these markers was known, and all the enzymes could be analyzed from tissue obtained by amputation of a single hind toe. More extensive comparisons were also made employing 29 gene loci of samples of *B. bombina*, representative populations of major subgroups of *B. variegata*, and two East Asian species, *B. orientalis* and *B. maxima* (Szymura, 1988).

Similarity Among Populations Within Species. The genetic structure of the two *Bombina* species can be evaluated in two ways: (1) using the F-statistics of Wright (1965), which quantify what proportion of total genetic variance is a result of subdivision into different populations; or (2) expressing the degree of genetic divergence among populations with Nei's genetic distance measure (Nei, 1972). Nei's measure can be used to construct dendograms. A UPGMA dendrogram of genetic similarities is shown in Figure 10-2. The two species are well differentiated from each other; *B. bombina* and *B. variegata* form two separate groups. Populations of *B. variegata* are more differentiated than populations of *B. bombina*. I_N range from 1.00 to 0.60 in *B. variegata* and from 1.00 to 0.87 in *B. bombina*. Mean F_{ST} values are higher for *B. variegata* (0.77 ± 0.054) than for *B. bombina* (0.363 ± 0.160), suggesting greater differentiation in the former.

The two *Bombina* species can be further divided into smaller units. A southern group and a northern group can be distinguished within *B. bombina* and four groups (Carpathian, western, Balkan, and Italian) in *B. variegata*. Subdivision of *B. variegata*

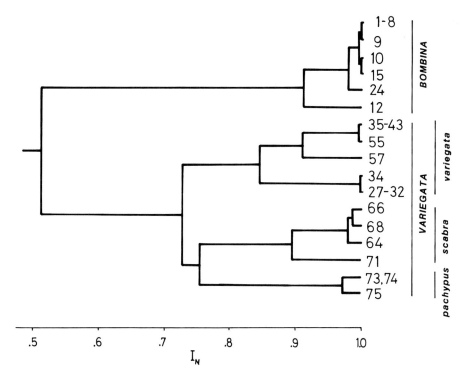

Figure 10-2. UPGMA phenogram constructed from pairwise genetic identities (I_N) among 38 samples of *Bombina*. I_N is based on nine enzyme loci (*Ldh-1, Ldh-2, Mdh-1, Mdh-2, Ck, Ak, Gpi, Idh-1,* and *Aat-1*). Numbers refer to samples in Figure 10-1.

corresponds in part to subspecific categories of Mertens and Wermuth (1960). The Carpathian and western groups are known as *B. v. variegata,* and the Balkan and Italian groups are known as *B. v. scabra* and *B. v. pachypus,* respectively.

Northern populations of *B. variegata,* localities 29–57 and 62, are separated from southern types, i.e., the Italian and the Balkan groups (populations 63–72). The two northern groups are more similar to each other than are the two southern groups. The Carpathian and the western groups of *B. variegata* are connected by intermediate populations inhabiting the northern parts of the Balkan peninsula. The latter come into contact with the Balkan group, *B. v. scabra.*

Genetic distances among populations of *B. variegata* from central Europe are not related in any simple way to their geographic distance. Northern populations inhabit extensive areas and yet seem less differentiated than the two southern groups of *B. variegata.* Groups of populations that show significant differences from other groups are separated either by unpopulated areas or transition zones (clines) that are relatively narrow. Thus the Po River Valley, uninhabited by *Bombina,* separates Italian *B. v. pachypus* from western *B. variegata.* Similarly, the southern branch of *B. bombina* that occupies lowlands along the Danube separates the western group of *B. variegata* from the Carpathian groups. These two groups mix in the Balkans but are separated in turn from the Balkan group *(B. v. scabra)* by a relatively narrow transition zone. The zone is not located on an environmental discontinuity but is a result of coinciding

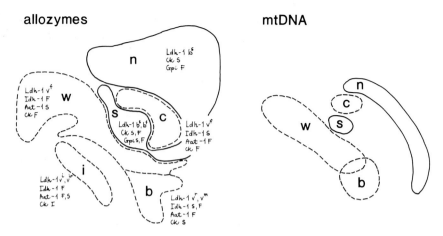

Figure 10-3. Subdivisions of *Bombina*. **n** and **s** = northern and southern groups of *B.bombina*. *B. variegata:* **i** = Italian group *(B. v. pachypus)*; **b** = Balkan group *(B. v. scabra)*; **c** and **w** = Carpathian and western groups, respectively *(B. v. variegata)*. Electromorphs characteristic of each group are also shown. MtDNA in the Italian group of *B. variegata* has not been studied.

clines at three enzyme loci *(Ldh-1, Ck, Idh-1)*, which change over 150–200 km. Parallel change of two mtDNA types characteristic of the Balkan and western groups also occurs in this region.

Patterns of Enzyme Variation. Clear geographic patterns in enzyme variation were discerned in both species. Three of the loci *(Ldh-1, Mdh-1, Ak)* are diagnostic throughout the species ranges. There are two *Ldh-1* alleles in *B. bombina* and five distinct alleles in *B. variegata*. *Mdh-1* and *Ak* are monomorphic but are fixed for alternate alleles in *B. bombina* and *B. variegata*. Other loci *(Idh-1, Aat-1, Ck,* and *Gpi)* are regionally polymorphic for either *Bombina* species and encode two to five electromorphs. These loci become fixed for unique alleles in some parts of the species' range. The loci can therefore differentiate between the species or between groups of populations in more restricted areas (Figure 10-3). Two remaining loci, *Mdh-1* and *Ldh-2*, are monomorphic and show only minor variants in some hybrid zones (Szymura and Święs, 1979).

Amounts of Variation. Though estimates of heterozygosity for *B. bombina* (13 populations) and *B. variegata* (30 populations) vary greatly in both species—0–0.205 in *B. bombina* (mean 0.058) and 0–0.237 in *B. variegata* (mean 0.105)—they show a striking geographic pattern. Northern populations of both species show little or no variation at the seven loci studied. Southern populations, on the other hand, exhibit numerous polymorphisms (Szymura, 1988).

HISTORICAL BIOGEOGRAPHY

Geographic variation in the European fire-bellied toads is not only substantial but complex. It involves divergent populations (i.e., the Italian *B. v. pachypus)*, sharp transition zones (the Balkan transition zone), gradual clinal variation over vast distances (e.g., *Ldh-1* variation in *B. bombina*), and greater divergence per kilometer in the south than in the north.

Mertens (1928) and Arntzen (1978), among others, have suggested that *B. bombina* and *B. variegata* diverged during the Pleistocene as a result of splitting of ancestral populations by the advancing glaciers. Biochemical and paleontological data, however, seem to favor Pliocene divergence of European *Bombina*. At this time ancestors of *B. bombina* colonized large permanent waters in the lowlands, splitting from mountain-adapted *B. variegata* (Szymura, 1983). Several Pleistocene glaciations have nevertheless had a profound impact on the fire-bellied toads. Fossils of *Bombina* from the Middle Pleistocene interglacial were discovered in Poland (Młynarski, 1977) and the former Soviet Union (Chkhikvadze and Sukhov, 1977). Together with findings from the Pliocene (Sanchíz and Młynarski, 1979; Hodrová, 1981), they indicate that the ranges of the fire-bellied toads contracted and expanded periodically following ice-sheet movements. Southeastern Europe, where current populations exhibit high levels of genetic variation, and the Appenine Peninsula probably provided refuges for both taxa.

The species differ in the magnitude of interpopulation differentiation. The explanation should be sought in ecology and history. *B. bombina* inhabits a more uniform, lowland environment than does *B. variegata*, which occupies highlands and separate mountain ridges of Europe. In addition, *B. bombina* has a more northern distribution than *B. variegata*, so more of its present range may have been occupied by founder populations. Pleistocene glaciation could have exerted a stronger homogenizing effect on *B. bombina*.

The complex pattern of variation, especially pronounced for *B. variegata*, can be interpreted as a consequence of Pleistocene history and postglacial range expansion. The order of divergence of *B. variegata* populations is not known, but genetic distances among groups of populations suggest that splitting episodes took place during consecutive glacial maxima. The Italian group perhaps diverged first, followed by a Balkan–northern disjunction. Finally, the northern groups separated into western and Carpathian populations, perhaps only after the last glacial retreat.

A division of *B. bombina* into southern and northern groups is more arbitrary than subdivision of *B. variegata* because differentiation of this species involves clinal variation and loss of variation in peripheral populations. A large portion of *B. bombina*'s range in the former Soviet Union has not yet been sampled.

Climatological data for the Pleistocene together with studies of current patterns of genetic variation can give indications of the location of past refugia (Fig. 10-4). Southern Italy was beyond doubt a refuge for *B. v. pachypus*, especially during the last Würm glaciation, whereas the southern Balkans were a refuge for *B. v. scabra*. A transition zone in the central Balkans, observed for allozymes and mtDNA, suggests a secondary transition in this region between two groups of *B. variegata*: the northern group *(B. v. variegata)* and the Balkan group *(B. v. scabra)*. It is interesting to note that similar transition zones were also observed for the grass snake *(Natrix natrix natrix)* by Thorpe (1984) and the smooth newt *(Triturus vulgaris)* by Kalezić (1984). For these reasons I suggest a refuge in the northwestern part of the Balkan peninsula for *B. v. variegata*.

Variable populations of *B. bombina* occur today along the lower Danube and pre-

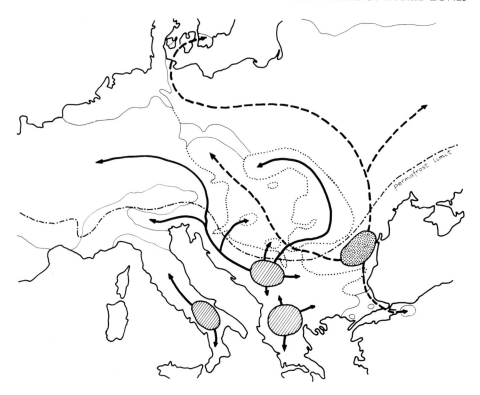

Figure 10-4. Routes of postglacial migrations of *B. bombina* and *B. variegata* from the presumed refugia in southern Europe.

sumably on the plains bordering the Black Sea east of the Carpathians. These populations were probably centers of dispersal for *B. bombina*. The Hungarian plains were proposed as a refuge for *B. bombina* by Lác (1961), but Arntzen (1978) considered this proposition unlikely because enclaves of *B. variegata* exist in this region.

Bombina bombina spread over the plains of eastern and central Europe from refugia after the Würm glaciation. One branch invaded lowlands north of the Carpathian Mountains and reached the Weser River in the west and Sweden via the Danish Islands during the warm postglacial period when a land connection still existed. The southern branch colonized lowlands along the Danube and the Bohemian-Moravian plateau in the west. *B. bombina* apparently displaced *B. variegata* to isolated mountain ridges on the Hungarian plain. *B. bombina* also reached the Ural Mountains in the east and Anatolian Turkey (northwestern Asia Minor) before the Bosphorus appeared 10,000 years BP.

Postglacial migrations of *Bombina* have been discussed at length by Madej (1964), Arntzen (1978), and Gollmann (1984). Electrophoretic data allow us to differentiate between hypotheses. Thus given electrophoretic and mtDNA results, *B. v. variegata* apparently spread from its Balkan refuge, localized perhaps in the northwestern part of the region, into more northern areas in two directions. One group migrated eastward to the southern Carpathians and moved along them toward the Moravian gate. The other group, a western type, occupied western Europe and Italy north of the Po River Valley. Populations that occupy ridges scattered over the Danubian plain are either related to the western type (the Mecsek Mountains populations) or to a type intermediate between the western and Carpathian groups, like those from the Matra Mountains (Gollmann, 1987) or from Fruška Gora (Szymura, 1988). These enclaves are of special interest, as they bear witness to a formerly wider range of *B. variegata.*

After the retreat of the ice sheet, when the fire-bellied toads expanded from southern refuges, their ranges came into contact. How contacts were established is a matter of conjecture. How many times the ranges of *B. bombina* and *B. variegata* met in the past is uncertain, but if we accept a Pliocene origin of these species, numerous opportunities existed. The taxa may have been parapatric in southeastern Europe toward the end of Pleistocene. Thus the contact along the southern edge of the Danubian valley would be the oldest, perhaps several thousand years older than that on the northern slopes of the Carpathians. Parapatry in Poland was probably established 8000–9000 years BP (Szymura and Barton, 1986).

HYBRIDIZATION AND HYBRID ZONES

At present, the ranges of *B. bombina* and *B. variegata* meet in a wide front extending from Austria along the southern edge of the Danube Valley to the Black Sea and surrounding the Carpathian Mountains along their foothills (Fig. 10-1). The contact is complex because *B. bombina* penetrates *B. variegata* territory along large rivers, e.g., the Sava River or the Morava River (northwestern and northeastern Yugoslavia, respectively), forming deep pockets, and because there are disjunct populations of *B. variegata* cut off by the continuous range of *B. bombina* in central Europe. As electrophoretic and mtDNA data have demonstrated, there is considerable regional differentiation within each species, and we may assume that interactions of *B. bombina* and *B. variegata* are subject to local variations according to the environmental conditions and genetic differences between divergent populations now in contact. The outcome of interactions is also influenced by the time elapsed since the *B. bombina* and *B. variegata* populations first met.

Electrophoretic surveys have demonstrated that hybrids occur in widely separated localities in Poland, Romania, Bulgaria, Austria, Hungary, Czechoslovakia, and several places in Yugoslavia, i.e., everywhere expected (Gollmann, 1984; Szymura, 1988). Additionally, morphological studies suggested hybridization in other regions (Méhely, 1905; Karaman, 1922; Horbulewicz, 1927, 1933; Michałowski, 1958; Lác, 1961; Madej, 1964) (Fig. 10-1). Therefore we may safely assume that *B. bombina* and *B. variegata* interbreed in every place where their ranges meet; they are not reproductively isolated from each other. The length of the hybrid zone is remarkable, as it extends over several thousand kilometers.

Several regions have been investigated in considerable detail, i.e., near Cracow (Szymura, 1976; Szymura and Barton, 1986), Przemyśl in southeastern Poland (Szymura and Barton, 1991), the Kostajnica and Peščenica regions in western Yugoslavia (Szymura, 1988), lower Austria (Gollmann, 1984), the Matra Mountains (Gollmann,

1987), and eastern Slovakia (Gollmann et al., 1988). Some generalizations may therefore be attempted.

Hybrid zones in all of the regions investigated share certain common features. The zones are situated at transition from lowlands to mountains in a position expected from *Bombina* ecology. *B. bombina* breeds in large permanent waters, which are found in lowlands, whereas *B. variegata* prefers small ponds and temporary pools that are common in mountainous regions (Madej, 1973). Because *B. bombina* and *B. variegata* differ in many characteristics, it is not surprising that the hybrid zones involve transitions at many characters: morphology, electrophoretic markers, mitochondrial DNA (Szymura and Barton, 1986, 1991), and mating calls (Hödl and Gollmann, 1985; Sanderson et al., 1992). It is significant that many of these characters, such as unlinked enzyme markers (Szymura and Arana, 1978), mitochondrial DNA, and morphology change in parallel and in the same place. Recombinant genotypes are probably at a disadvantage and are removed by selection (see Ch. 2). Another possibility is that *bombina* and *variegata* genotypes are adapted to lowland and mountain environments, respectively. In environmentally complex regions, selection may act over a small scale, favoring *bombina* alleles in *bombina*-like sites and *variegata* alleles in *variegata*-like sites.

The hybrid zones are narrow. Most of the transition takes place within a distance of a few kilometers. The centers of the hybrid zones are bordered on both sides by long tails of introgressing alleles, detected tens of kilometers from the midpoint.

Apart from the above similarities, hybrid zones show different genetic structure in various geographic locations. Three basic types can be distinguished (Fig. 10-5). In some cases, *B. bombina* and *B. variegata* are separated geographically by intermediate populations composed almost exclusively of individuals of mixed ancestry. F₁ hybrids,

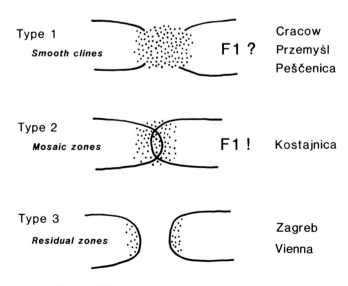

Figure 10-5. Types of hybrid zones for *Bombina.*

if present at all, are rare; and populations are close to Hardy-Weinberg equilibrium, indicating random mating. Hybrid zones of this type have been found near Cracow (Szymura, 1976), Przemyśl (Szymura and Barton, 1991), and Peščenica (Szymura, 1988). In a second type of hybrid zone, discovered in the Kostajnica transect, populations within the zone contain individuals of both species; there is a significant deficit of hybrid genotypes, pointing to a certain degree of isolation between *B. bombina* and *B. variegata*. F_1 individuals seem to be more common here than in the Cracow or the Peščenica areas (about 3% of individuals in the Kostajnica transect, in contrast to 1% near Cracow and 0.3% near Peščenica). The presence of pure individuals of both species as well as a diverse hybrid population on the Slovak Karst Plateau (Gollmann, 1986), an area with a patchy habitat structure that may facilitate breeding site segregation among genotypically different subpopulations, resembles the situation observed in the central part of the Kostajnica hybrid zone. The third category ("residual hybrid zones") includes situations that can be derived from either of the two basic types. They can be interpreted as remnants of previous hybrid zones from which central populations have disappeared because of habitat destruction. Two sets of populations investigated by Gollmann (1984) belong to this type. In the Waldviertel area of lower Austria, hybrid populations at either side of the former hybrid zone are close to Hardy-Weinberg equilibrium; in no place do both parental species occur together, and no presumptive F_1 was found. A similar situation can be observed north of Vienna. Populations of *B. bombina* are separated from those of *B. variegata* by farmed fields where no *Bombina* is present, but marginal populations of either species bear traces of former hybridization. The population from St. Andrä-Wördren (No. 11, Fig. 10-1), more similar to *B. bombina* and located on lowlands close to the Danube, is about 10.5 km from the nearest population of *B. variegata* in Mauerbach studied by Gollmann (1984). These two populations represent opposite sides of a former hybrid zone. The contact area west of Zagreb (Karaman, 1922; Szymura, 1988) also belongs to this category.

TRANSECTS ACROSS HYBRID ZONES

Multilocus genotypic data provided by electrophoresis can be used to describe the structure of a hybrid zone, as well as to estimate the strength of evolutionary forces involved in its maintenance (Szymura, 1988; Szymura and Barton, 1986, 1991; see Ch. 2).

Polish Transects. The area west of Cracow in southern Poland is the best studied section of hybrid zone in which a variety of characters have been investigated (Figs. 10-6 to 10-9). Moreover, the present distribution of *Bombina* can be related to earlier morphological studies by Michałowski (1958). The use of many characters not only helps to understand how selection acts on a large part of the *Bombina* genome, but in this particular case it also gives us confidence that inferences drawn from past morphological studies are not biased and adequately describe the situation in the past.

Studies have revealed many examples in which morphological and molecular evolution do not necessarily proceed at the same rate, apparently related to differences in response to selection or evolutionary constraints. Similar arguments can be applied

(a)   (b)

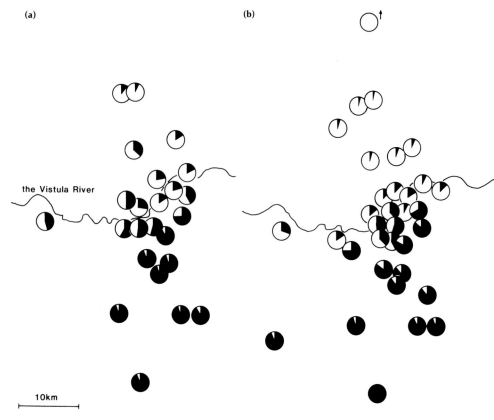

the Vistula River

10km

Figure 10-6. Transect across a hybrid zone near Cracow in southern Poland. (**a**) Average frequency of the *B. variegata* morph averaged over seven diagnostic morphological characters (spots 1, 2, 4, 5, 6, and 7 and toe color) (black sections of pie diagrams). (**b**) Average frequency of the *B. variegata* alleles (black sections of pie diagrams) for five diagnostic enzyme loci (*Ldh-1, Mdh-1, Ck, Ak,* and *Gpi*).

to hybrid zones, where differences in observed pattern relate to varying selection pressures or underlying habitat heterogeneity. There is no *a priori* reason why separate characters should behave in the same way.

Six allozyme loci assort independently in *Bombina,* as shown in laboratory crosses (Szymura and Farana, 1978). Yet, surprisingly, they show a pattern of concordant and coincident clines across the zone in every transect (compare Figs. 10-7 and 10-13). When enzymes are analyzed within populations, locus by locus, genotypic frequencies are close to Hardy-Weinberg expectations, suggesting random mating within hybrid populations. Therefore coincident patterns cannot be explained by either chromosome linkage or assortative mating within populations.

However, within individual toads, alleles at diagnostic loci are not distributed at random, and *bombina* alleles at one locus tend to be associated with *bombina* alleles at other loci; *variegata* alleles behave in the same way. There are therefore strong linkage disequilibria among loci.

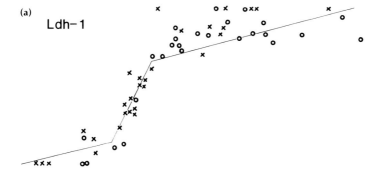

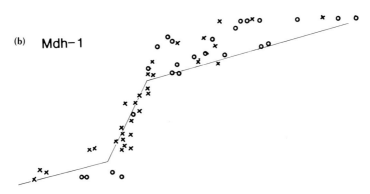

Figure 10-7. Frequencies of the *variegata* allele at each enzyme locus. (a) *Ldh-1*. (b), *Mdh-1*. (c) *Ck*. (d) *Ak*. (e) *Gpi*. (f) *Np*. Frequencies are plotted on a logit scale, with fixed sites being set arbitrarily at $\log_e (p/q) = +5$. Crosses = Cracow transect; circles = Prezemyśl transect. The straight lines shown on each graph show the best fit to the whole data set averaged across loci. See Figure 10-8a for scale.

The most likely explanation for this pattern is a constant influx of individuals into the zone from adjacent populations at the edge. Indeed, capture-recapture studies of fire-bellied toads at the center of the zone near Cracow in Kopanka and elsewhere showed that migration does take place and therefore can produce significant disequilibria (Szymura and Barton, 1986). A direct estimate of dispersal (δ) of 0.43 km gen^{-1} was lower than the indirect estimate of approximately 1 km gen^{-1} (Szymura and Barton, 1986). However, dispersal measured in adults when population density is rather high probably underestimates gene flow.

The pattern of change across *Bombina* hybrid zones is a sharp step at the center, where allelic frequencies change abruptly within about 6 km, bordered by long tails of

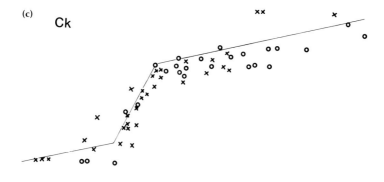

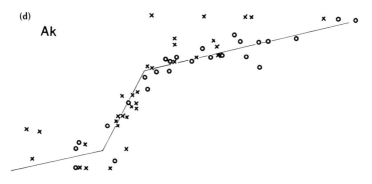

Figure 10-7. *Continued*

introgression (Fig. 10-8). Such a pattern indicates the presence of a barrier to gene flow. There is no environmental obstacle to dispersal in the field, and the cline corresponds to a broad environmental transition from lowlands to mountains. The barrier to gene flow must be of a different kind; for example, hybrids could be at a disadvantage relative to parental genotypes. It is unlikely that strong selection acts directly on enzyme loci (especially as no significant reduction in heterozygote frequencies was observed) but, rather, on loci that are linked to the marker loci. Disequilibria between enzyme loci increase markedly toward the center of the zone, so the effective selection acting against recombinant genomes also increases.

There is direct evidence of selection against hybrids. Early embryonic mortality is increased in the hybrid zone (Koteja, 1984; Szymura and Barton, 1986), and there is a marked increase in developmental abnormalities as well. Morphological abnormalities—ribs fused to vertebrae, reduced number of vertebrae, and asymmetric sacral region in adults, or abnormal mouthparts in tadpoles (Madej, 1965; Czaja, 1980; Szy-

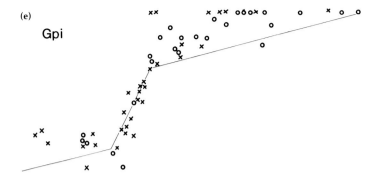

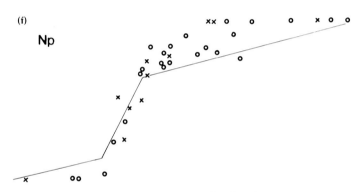

Figure 10-7. *Continued*

mura, unpublished data)—indicate disruption of developmental control and reduced homeostasis in hybrids. Hybrids have substantially lowered fitness relative to parental types.

So far, I have discussed allozyme markers only. Do other characters behave in the same way? Mitochondrial DNA is believed by many authors to introgress more readily than other "neutral" markers because it not linked to any nuclear genes and therefore can recombine freely. No differential introgression was observed in *Bombina*, at least not near Cracow, where the mtDNA cline closely parallels the allozyme clines. In central populations, however, the *variegata* mtDNA type predominates. Because mtDNA is maternally inherited and *B. variegata* males more ready clasp heterospecific females than *B. bombina* males, and because *B. bombina* females exhibit higher fecundity, an excess of the *bombina* mtDNA type is not unexpected. An alternative explanation is that hybridization began when a few *B. bombina* females moved into an otherwise pure *B. variegata* population. No association between hybrid index and mtDNA type

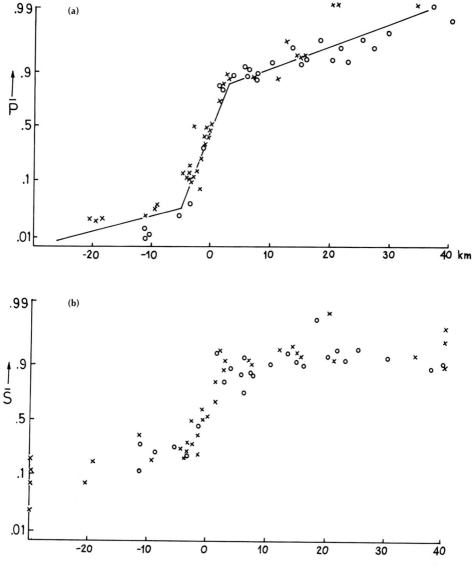

Figure 10-8. Patterns across the Cracow and Przemyśl transects, averaged over all loci. Crosses = Cracow transect; circles = Przemyśl transect. (**a**) Frequency of *variegata* enzyme alleles, plotted in the same way as in Figure 10-7. (**b**) Average frequency of seven morphological characters (spots 1,2,4,5,6, and 7 and toe color), plotted on the same scale as the allozyme frequencies.

was observed in central populations (Szymura et al., 1985). Thus seemingly neutral molecular markers behave in the same way irrespective of their nuclear or cytoplasmic origin.

One might expect that other characters, more directly involved in general adaptations or specific adaptations to particular types of habitat, should show different patterns of change, with clines being steeper and narrower and perhaps displaced with respect to those seen for allozymes.

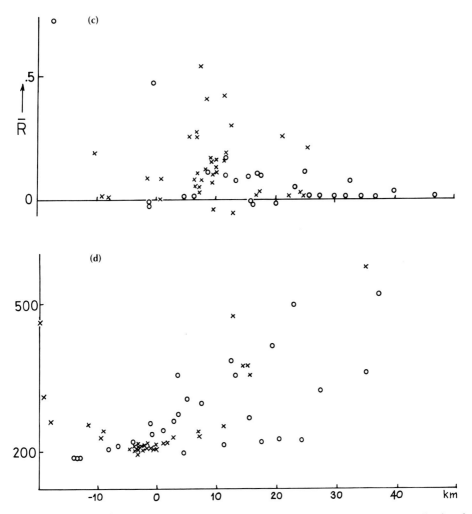

Figure 10-8. *Continued* (**c**). Standardized linkage disequilibrium (R) averaged across all pairs of loci. (**d**) Altitude of each sample, in meters above sea level.

The spotting pattern in *B. bombina* and *B. variegata* is characteristic of the species and can be described in a way analogous to a hybrid index by summing the number of diagnostic spots that fuse together. There are seven such diagnostic spots, which are separate in *B. bombina* and join into larger patches in *B. variegata*. The spots, either yellow or orange, are involved in aposematic coloration. They are exposed during the "Unken reflex," a distinctive warning display for which geographic variation in response is observed (Bajger, 1980). Moreover, there are other characteristics related to spotting pattern (e.g., skin thickness and its vascularization) with obvious adaptive significance.

Analysis of spotting pattern near Cracow shows that the position and width of the morphological cline is the same as for allozymes (Figs. 10-6 and 10-8). Individual spots are nonrandomly associated, as are allozyme alleles. There is also nonrandom asso-

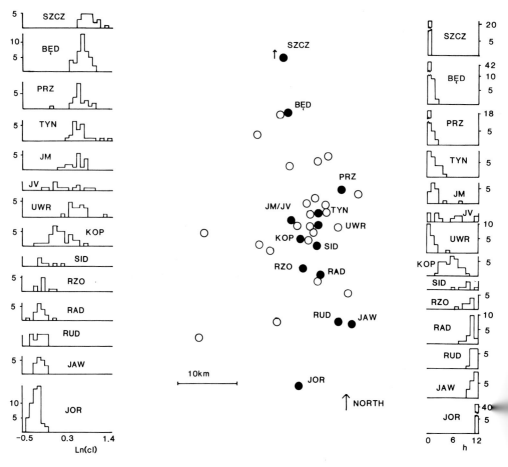

Figure 10-9. Variation in mating call in the Cracow transect. Histograms show call component (Ln[cl]), hybrid index (h), and map location of samples in the transect. Hybrid indexes were constructed by summing numbers of the *variegata* alleles at six diagnostic enzyme loci *(Ldh-1, Mdh-1, Ck, Ak, Gpi, Np)* and therefore vary between 0 *(bombina)* and 12 *(variegata)*. Open circles are samples from an electrophoretic survey (see Fig. 10-6).

ciation of spots with allozymes. Interestingly, the degree of association between allozymes and between spots is roughly the same.

Mating calls of *B. bombina* and *B. variegata* differ in many parameters as well (Table 10-1). The calls pay a role in both mate attraction and maintenance of male territories. Mating calls are also candidates for evolution of reinforcement of preexisting differences. The pattern of change in cycle length (i.e., length of call plus length of a silent interval between two consecutive calls) follows belly pattern and enzyme clines (Fig. 10-9) (Sanderson et al., 1992).

Thus associations between unrelated characters indicate not only that they behave in the same manner but also that a large number of genes are involved in maintaining strong linkage disequilibria. Selection seems to be spread throughout the genome and not concentrated on just a few loci. The estimated number of genes that cause hybrid unfitness and are therefore responsible for zone maintenance is 55 (range 26–88), which is consistent with the large number of differences observed between *B. bombina* and *B. variegata.*

Correlations among enzymes, among morphological characters, and between enzymes and morphology are not peculiar to the Cracow transect but can also be observed in other transects. For example, the Cracow and the Przemyśl transect located 200 km to the east are strikingly similar (compare Fig. 10-6 with Fig. 10-10; see also Figs. 10-7 and 10-8). The two transects were studied 35 and 50 years ago by Michałowski (1958) and Horbulewicz (1933), respectively. Although no detailed comparisons on a population basis can be carried out between the past and present studies, a rather crude comparison shows that patterns of morphological variation have remained stable in time, and the zones have neither moved nor widened (Szymura and Barton, 1991). Hybrid zones for *Bombina* have therefore all of the characteristics of stable hybrid zones maintained by a balance between selection and dispersal.

Selection against hybrids and environmental selection are not easy to distinguish (Moore, 1977; see also Ch. 2 and Ch. 8). Most likely both types of selection contribute to the observed pattern. Evidence for stability comes from repeated sampling of hybrid populations over a relatively short period of time (Szymura and Barton, 1991) as well as from comparisons with earlier morphological studies.

The position of a hybrid zone can change, however. The most compelling evidence for movement of *Bombina* hybrid zones is the presence of *B. variegata* isolates surrounded by *B. bombina* on the Danubian plain (Fig. 10-1). Movement of this zone probably accompanied *B. bombina* range extension (Fig. 10-4) and was most likely fast. A new equilibrium was established at the border of the plains and mountains.

Kostajnica Transect. Two transects in Yugoslavia, although sharing the basic pattern of correlated clines, show significant differences in the structure of central populations, suggesting that habitat choice (or habitat selection) may also play a role in hybrid zone maintenance (Figs. 10-11 and 10-12).

In the Kostajnica transect, five central populations have a bimodal hybrid index distribution (Fig. 10-11). (Note that in Yugoslavia there are four diagnostic loci—*Ldh-1, Ak, Mdh-1, Aat-1*—and therefore hybrid index scores vary between 0 and 8, rather than between 0 and 12, as in Poland). It is unlikely that heterozygote deficiency at individual loci is caused by low viability of hybrids. No such effect was observed in laboratory crosses, and F_1 and backross generations show surprisingly little mortality (Szymura and Farana, 1978; Szymura unpublished data). Nonrandom mating resulting from habitat association is the probable cause of heterozygote deficit and a bimodal hybrid index distribution. Strong association between habitat type and population composition was observed. *Variegata*-like individuals inhabit small ponds, temporary pools with little vegetation, and often murky water, whereas *bombina*-like individuals are found in larger ponds with clear water and more vegetation. This association was especially apparent at the center of the zone, where populations Kostajnica 1 and Kostajnica 2 were only 30 meters apart (Fig. 10-11). Another interesting observation is that both parental types were found within the same ponds or in close proximity, thereby facilitating formation of F_1 hybrids. In Poland, parental populations are separated by hybrid populations, and so F_1 hybrids are rare.

Given sufficient time, a bimodal distribution of genotypes should eventually dis-

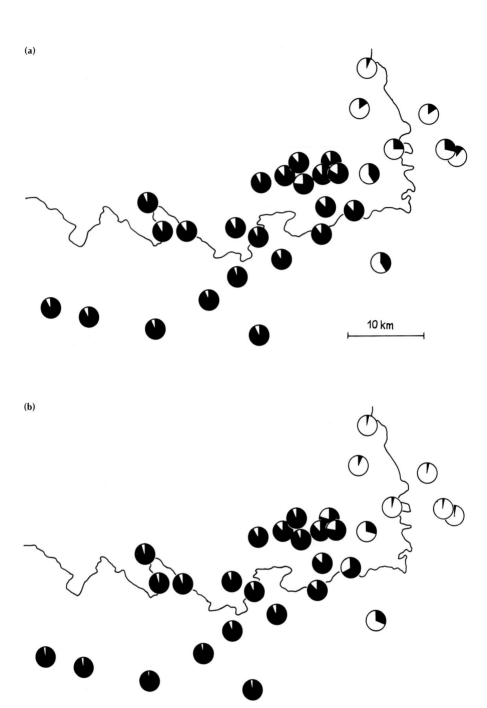

Figure 10-10. Transect near Przemyśl. (**a**) Average frequency of the *variegata* morph, over seven diagnostic characters, which describe belly pattern (spots 1, 2, 4, 5, 6, 7) and toe color. (**b**) Average frequency of alleles characteristic of *variegata* for six diagnostic enzyme loci *(Ldh-1, Mdh-1, Ck, Ak, Gpi, Np).*

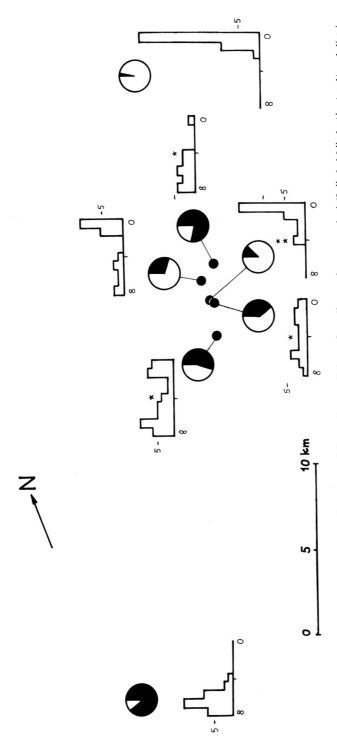

Figure 10-11. Kostajnica transect: average frequency of the *variegata* alleles at four diagnostic enzyme loci *(Ldh-1, Mdh-1, Ak, Aat-1)* and distribution of hybrid indexes. Hybrid indexes were constructed by summing numbers of the *variegata* alleles at four diagnostic enzyme loci *(Ldh-1, Mdh-1, Ak, Aat-1)* and therefore vary between 0 *(bombina)* and 8 *(variegata)*. Asterisks indicate F₁ hybrids, i.e., individuals heterozygous at all four diagnostic loci.

(a)

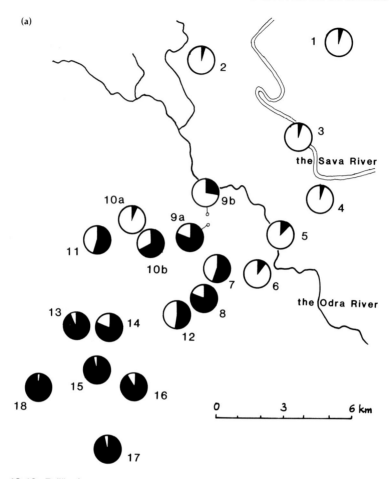

Figure 10-12. Peščenica transect. (a) Average frequency of the *variegata* alleles at four diagnostic enzyme loci *(Ldh-1, Mdh-1, Ak, Aat-1)*, (b) Distribution of hybrid indexes. Hybrid indexes were constructed by summing numbers of the *variegata* alleles at four diagnostic enzyme loci *(Ldh-1, Mdh-1, Ak, Aat-1)* and therefore vary between 0 *(bombina)* and 8 *(variegata)*. Asterisks indicate F_1 hybrids, i.e., individuals heterozygous at all four diagnostic loci. Samples 5, 12, and 14 correspond to samples 18, 45, and 46, respectively, shown in Figure 10-1.

appear, giving rise to more familiar unimodal distributions (Fig. 10-9). Several explanations for the bimodal distribution are possible. First, the Kostajnica contact zone may be of a relatively recent origin with populations that have not yet reached equilibrium. This explanation is unlikely because allozyme markers have introgressed in both directions along this transect 15–25 km from the center. Also, with random mating, heterozygote deficiency would disappear after one generation. Because the Kostajnica transect is only 50 km south of the Peščenica transect, which is more similar to the Polish transects, genetic differences between these transects seems unlikely. A second possibility is recent colonization of the central part of the transect by *bombina*-like and *variegata*-like populations. Third, the situation could be stable, with habitat

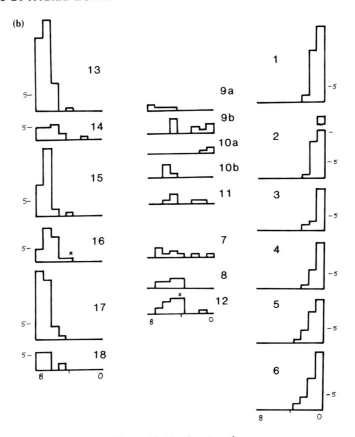

Figure 10-12. *Continued*

preferences and mating preferences, operating in the context of a heterogeneous environment, producing the observed pattern. Indeed, the Kostajnica transect represents a complex environment in which the central populations are located on a saddle between two river systems in a hilly landscape with patchwork fields, meadows, and forests. There is also evidence from other transects that *B. bombina* and *B. variegata* show consistent habitat preferences (i.e., populations 9a/9b or 10a/10b in the Peščenica transect) (Fig. 10-12). How strong these preferences are is unknown, but they could be accentuated in southern Europe for two reasons.

In warmer climates desiccation is a serious risk for amphibians, especially juveniles. *B. bombina* has a thinner skin, and therefore selection may favor either closer association with water or a large size at metamorphosis (i.e., a longer developmental period), both of which would restrict *B. bombina* to larger ponds. These constraints may not be as important for *B. variegata,* a species with thicker skin, which spreads its reproductive effort in time and space, laying small batches of eggs in different pools over a period of several months (Rafińska, 1991).

In natural populations there is probably variation within each *Bombina* species for both habitat preferences and propensity to hybridize (mate choice). In hybrid zones these traits cannot be selected for because of the zone's dynamics. One could argue

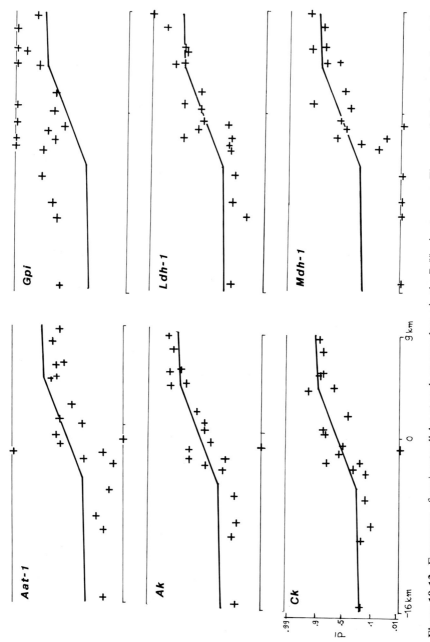

Figure 10-13. Frequency of *variegata* alleles at each enzyme locus in the Peščenica transect. They are plotted on a logit scale in the same way as in Figures 10-7 and 10-8. The line shown on each graph shows the best fit to the whole data set, averaged across loci.

that with *B. bombina* range expansion into the Danubian plains along the Danube through the Iron Gate, only the individuals that possessed traits preventing them from hybridizing with *B. variegata* could get through a narrow corridor. Therefore such genes could be more common in southern *B. bombina* populations.

Peščenica Transect. The Peščenica transect is in many ways similar to both the Polish and Konstajnica transects. The cline width is significantly greater here than in Poland: 9.5 km compared with 6.05 km at Cracow and 6.15 km at Przemyśl. There is also much more scatter around the cline than in Poland (Figs. 10-7, 10-8, and 10-13). The scatter is not due to independent differences at separate loci but, rather, to systematic differences in the fraction of *variegata* alleles. Such scatter is expected if both species show habitat preferences. Unfortunately, samples from the central part of the transect are too small to demonstrate this tendency clearly (compare, however, samples 9a/9b and 10a/10b in Fig. 10-12). A more recent resampling of this area by Nürnberger, MacCallum, and Barton (pers. comm.) showed significant association between spotting patterns and habitat type. The example of Peščenica demonstrates that even topographically simple areas may possess hidden complexity and that great care should be taken when sampling on a sufficiently fine scale to detect it.

Future studies of this and similar zones should help to establish the role of habitat choice, differential selection pressure, and mating preferences. The complexity of hybridization calls also for a better insight into *Bombina* ecology and behavior.

ACKNOWLEDGMENTS

The comments of N. Barton, R. Harrison, A. Marchant, and C. MacCallum regarding this chapter are greatly appreciated. This work was supported by grants from the Polish Academy of Sciences (project MR. II. 6). This chapter is dedicated to Professor Henryk Szarski on the occasion of his 80th birthday.

REFERENCES

Andrén, C., Nilson, G., and Podloucky, P. 1984. The Fire-Bellied Toad *Bombina bombina* (L.). SEH. Göteborg-Hannover: Conservation Committee, Council of Europe.

Arntzen, J. W. 1978. Some hypotheses on post glacial migrations of the fire-bellied toads, Bombina bombina (L.) and the yellow-bellied toad, *Bombina variegata*. J. Biogeogr. 5:339–345.

Bajger, J. 1980. Diversity of defensive responses in populations of fire-toads (*Bombina bombina* and *Bombina variegata*). Herpetologica 36:133–137.

Boulenger, G. A. 1886. On two European species of bombinator. Proc. Zool Soc. Lond. 1886:449–501.

Chkhikvadze, V. M., and Sukhov, V. P. 1977. Amphibians and reptiles from quaternary deposits of Krasnij Bor (River Kama). Abstr. IV Allsoviet Herpetol. Conference. Vopr. Gerpetol. 1977:277–282.

Czaja, A. 1980. Studies on the variability of the oral apparatus of tadpoles of *Bombina bom-*

bina, Bombina variegata, and their natural hybrid populations [in Polish]. M.S. thesis, Jagiellonian University, Cracow, Poland.

Czopkowa, G., and Czopek, J. 1955. The vascularization of the respiratory surfaces in *Bombina variegata.* Bull. Polon. Acad. Sci. Cl.II., 3:313–318.

Gollmann, G. 1984. Allozymic and morphological variation in the hybrid zone between *Bombina bombina* and *Bombina variegata* (Anura: Discoglossidae) in northeastern Austria. Z. Zool. Syst. Evol. Forsch. 22:51–64.

Gollmann, G. 1986. Genetic analysis of *Bombina* hybrids from Eastern Slovakia, Stud. Herpetol. (Prague) 1986:121–126.

Gollmann, G. 1987. *Bombina bombina* and *Bombina variegata* in the Matra mountains (Hungary): new data on distribution and hybridization (Amphibia, Anura, Discoglossidae). Amphibia Reptilia 8:213–224.

Gollmann, G., Roth, P., and Hödl, W. 1988. Hybridization between the fire-bellied toads *Bombina bombina* and *B. variegata* in the karst regions of Slovakia and Hungary. J. Evol. Biol. 1:3–14.

Herre, W., and Rawiel, F. 1939. Vergleichende Untersuchungen an Unken. Zool. Anz. 125:290–299.

Hödl, W., and Gollmann, G. 1985. Bioakustiche Freilanduntersuchungen an hybriden Unkenpopulationen. Verh. Dtsch. Zool. Ges. 78:340.

Hodrová, M. 1981. Plio-Pleistocene frog fauna from Hajnacka and Ivanovce, Czechoslovakia, Vst. Ustř. Úst. Geol. Praha. 54:215–224.

Horbulewicz, L. 1977. Die Verbreitung der Bombinator- und Triton-Arten im Bereiche der Bezirke Sambor, Drohobycz, Stryj (Kleinpolen). Bull. Int. Acad. Polon. Sci. Lett. Cracovie B:87–112.

Horbulewicz, L. 1933. Die geographische Verbreitung der Bombinator Arten im Bereiche der Bezirke Dobromil-Przemyśl-Jarosław [in Polish, with German summary]. Kosmos (Lwów) A58:209–223.

Juszczyk, W. 1974. Plazy i gady krajowe. Warsaw: PWN.

Kalezić, M. L. 1984. Evolutionary divergence in the smooth newt, Triturus vulgaris (Urodela, Salamandridae): electrophoretic evidence. Amphibia Reptilia 5:221–230.

Karaman, S. 1922. Über unsere Bombinatorarten. Glasnik Kroat. Naturwiss. Ges. Zagreb 34:63–70.

Koteja, E. 1984. Embryonic mortality of the fire-bellied toad (Bombina bombina), the yellow-bellied toad (Bombina variegata), and their hybrids [in Polish]. M.S. thesis, Jagiellonian University, Cracow, Poland.

Lác, J. 1961. Verbreitung der Unken (Tiefland-Unke Bombina bombina L. und Berg-Unke B. variegata L.) in der Slowakei und Problematik deren gegenseitigen Kreuzung [in Slovak with German summary]. Biol. Práce SAV Bratislava 7(3):5–32.

Lörcher, K. 1969. Vergleichende bio-akustische Untersuchungen and der Rot- und Gelbbauchunke, Bombina bombina (L.) und B. v. variegata (L.). Oecologia (Berl.) 3:84–124.

Madej, Z. 1964. Studies on the fire-bellied toad (Bombina bombina, Linnaeus, 1761) and the yellow-bellied toad (Bombina variegata, Linnaeus, 1758) of upper Silesia and Moravian gate. Acta Zool. Cracow 9:291–336.

Madej, Z. 1965. Variations in the sacral region of the spina in Bombina bombina (Linneaus, 1761) and Bombina variegata (Linnaeus, 1758) (Salientia, Discoglossidae). Acta Biol. (Cracow Ser. Zool. Crakow 8:185–197.

Madej, Z. 1973. Ecology of European fire-bellied toads (Bombina, Oken 1816) [in Polish with English summary]. Przegl. Zool. Wrocław 17:200–204.

Maxson, L. R., and Szymura, J. M. 1984. Relationships among discoglossid frogs: an albu-min perspective. Amphibia Reptilia 5:245–252.

Méhely, L. 1892. Beitrage zur Kenntnis der Bombinator-Arten, sowie deren Standorte und Verbreitung in Ungarn. Math. Nat. Ber. Ungarn. 10:55–79.

Méhely, L. 1905. Die herpetologischen Verhältnisse des Mecsek Gebirges und der Kapela [in German]. Ann. Hist. Mus. Nat. Hung. (Budapest) 3:256–316.

Mertens, R. 1928. Zur Naturgeschichte der Europäischen Unken (Bombina). Z. Morphol. Ökol. Tiere 11:613–623.

Mertens, R., and Wermuth, H. 1960. Die Amphibien und Reptilien Europas. Frankfurt am Main: W. Kramer.

Michałowski, J. 1958. Geographische Verbreitung der Unken (Bombina Oken) zwischen den Flussen Weichsel, Skawa und Raba (Woiwodschaft Krakow) [in Polish, with Russian and German summaries]. Acta Zool. Cracow 3:247–283.

Michałowski, J. 1961. Studies on species characters in Bombina variegata (L.) and Bombina bombina (L.). I. Applying the L:T indicator to the classifying purposes. Acta Zool. Cracow 6:51–59.

Michałowski, J. 1966. Studies on the relationship of Bombina bombina (Linnaeus) and Bombina variegata (Linnaeus). II. Some taxonomic characters of tadpoles of both species and of tadpoles obtained from crosses under laboratory conditions. Acta Zool. Cracow 6:181–208.

Michałowski, J., and Madej, Z. 1969. Studies on the relationship of Bombina bombina (Linnaeus) and Bombina variegata (Linneaus). III. Taxonomic characters of both species from laboratory and in interspecific hybrids. Acta Zool. Cracow 14:173–200.

Młynarski, M. 1977. New notes on the amphibian and reptilian fauna of the Polish Pliocene and Pleistocene. Acta Zool. Cracow 22:13–36.

Moore, W. S. 1977. An evaluation of narrow hybrid zones in vertebrates. Q. Rev. Biol. 52:263–278.

Morescalchi, A. 1965. Osservatzioni sulla cariologia di Bombina. Boll. Zool. 32:207–218.

Nei, M. 1972. Genetic distance between populations. Am. Naturalist 106:238–292.

Olmo, E., Morescalchi, A., Stingo, V., and Odierna, G. 1982. Genome characteristics and the systematics of Discoglossidae (Amphibia, Salientia). Monitore Zool. Ital. (N.S.) 16:283–299.

Rafińska, A. 1991. Reproductive biology of the fire-bellied toads, Bombina bombina and B. variegata (Anura: Discoglossidae): egg size, clutch size and larval period length differences. Biol. J. Linn. Soc. Lond. 43:197–210.

Sanchíz, F. B., and Młynarski, M. 1979. Remarks

on the fossil anurans from the Polish Neogene. Acta Zool. Cracow 24:153–174.

Sanderson, N., Szymura, J. M., and Barton, N. H. 1992. Variation in mating call across the hybrid zone between the fire-bellied toads *Bombina bombina* and *B. variegata*. Evolution 46:595–607.

Szymura, J. M. 1976. Hybridisation between discoglossid toads *Bombina bombina* and *B. variegata* in southern Poland as revealed by the electrophoretic technique. Z. Zool. Syst. Evol. Forsch. 14:227–236.

Szymura, J. M. 1983. Genetic differentiation between hybridizing species *Bombina bombina* and *Bombina variegata* (Salientia, Discoglossidae) in Poland. Amphibia Reptilia 4:137–145.

Szymura, J. M. 1988. Regional differentation and hybrid zones between fire-bellied toads *Bombina bombina* (L.) and *B. variegata* (L.) in Europe [in Polish]. Rozprawy Habilitacyjne No. 147. Cracow: Uniwersytet Jagielloński.

Szymura, J. M., and Barton, N. H. 1986. Genetic analysis of a hybrid zone between the fire-bellied toads. *Bombina bombina* and *Bombina variegata*, near Cracow in southern Poland. Evolution 40:1141–1159.

Szymura, J. M., and Barton, N. H. 1991. The genetic structure of the hybrid zone between the fire-bellied toads *Bombina bombina* and *B. variegata:* comparisons between transects and between loci. Evolution 45:237–261.

Szymura, J. M., and Farana, I. 1978. Inheritance and linkage analysis of five enzyme loci in interspecific hybrids of toadlets, genus *Bombina.* Biochem. Genet. 16:307–319.

Szymura, J. M., Spolsky, C., and Uzzell, T. 1985. Concordant change in mitochondrial and nuclear genes in a hybrid zone between two frog species (genus *Bombina*). Experientia 41:1469–1470.

Szymura, J. M., and Święs, 1979. Inheritance and geographical distribution of a rare lactate dehydrogenase (E.C 1.1.1.27) polymorphism (Ldh-2 locus) in the fire bellied toads, *Bombina bombina* and *Bombina variegata.* Genet. Polon. 20:415–424.

Thorpe, R. S. 1984. Primary and secondary transition zones in speciation and population differentiation: a phylogenetic analysis of range expansion. Evolution 38:233–234.

Wickbom, T. 1949. Further cytological studies on Anura and Urodela. Hereditas 35:33–48.

Wright, S. 1965. The interpretation of populations structure by F-statistics with special regard to systems of mating. Evolution 19:395–420.

11

Hybridization and Hybrid Zones in Pocket Gophers (Rodentia, Geomyidae)

JAMES L. PATTON

Studies of hybrid zones of pocket gophers of the genera *Thomomys* and *Geomys* are among the earliest of such endeavors with mammals. It is particularly true for analyses employing relatively sophisticated methodologies to identify individuals of mixed ancestry and to examine the nature of hybridization from genetic and ecological perspectives. Thaeler's (1968a) analysis of three zones of hybridization between taxa of *Thomomys* in California based on multivariate morphometry illustrated that hybrids could be identified by their morphological intermediacy, and that qualitative differences existed in the extent of hybridization in different contact populations despite similarities in local ecology and zone widths. Patton was the first to use both chromosomal (Patton and Dingman, 1968; Patton, 1973) and allozymic markers (Patton et al., 1972) as genetic tools to define the nature and extent of hybridization in another *Thomomys* zone in southern Arizona. He and his coworkers showed that F_1 hybrids could be differentiated from subsequent filial or backcross individuals, and that the genetic composition of hybrid populations gave direct evidence about hybrid fitness.

This early focus on hybridization in pocket gophers stemmed primarily from a systematic interest in defining species boundaries, a task made difficult for these subterranean mammals by their allopatric or parapatric distributions (for reviews see Nevo, 1979; and Patton, 1981, 1990). Indeed, from the early days to the present, most studies of pocket gopher hybrid zones have been oriented to answering taxonomic questions, rather than to the analyses of hybrid zones as dynamic evolutionary theaters of their own (as exceptions see Hafner et al., 1983; Baker et al., 1989). It is therefore one purpose of this review to ask what pocket gopher hybrid zones can tell one about the nature of hybridization as an element of evolutionary interest beyond the purely proximal utility of their use for determining species status. In doing so, however, the important issue of hybridization in relation to species boundaries is not ignored.

POPULATION BIOLOGY OF POCKET GOPHERS

As is developed in the pages that follow, hybrid zones between pocket gopher taxa share a common set of attributes regardless of the nature of the genetic interactions that characterize them. It is my belief that their zone dynamics are constrained by an underlying set of populational, demographic, and distributional characteristics common to all gopher taxa. Hence, I take the opportunity early in this chapter to set forth these constraints, and I emphasize how hybridization is influenced by them.

Pocket gophers are small-bodied herbivorous rodents that spend most of their lives in self-dug burrow systems; that is, they are subterranean, with their local and geographic distributions limited to soils of sufficient depth and friability and with sufficient food resources (for reviews see Chase et al., 1982; Patton, 1990). Populations of single species can occur in a wide range of habitats, from desert scrub to alpine communities. Exclusive-use territories characterize both sexes and all ages subsequent to weaning (see, for example, Reichman et al., 1982). Multiple burrow occupancy by adults is present only during the breeding season, with bonding only for the momentary period of mating. Although turnover rates within populations are high (between 50% and 80% annually), individuals with established territories are sedentary, as those surviving from one year to the next retain the same territory. As a result, local abundance is determined by the patch size of available soils and the number of territories that can be supported by the food resources.

Data on dispersal are limited. Maximal dispersal distances recorded have been about 800 meters, but there is great variance in distances moved within and among both the sexes and different age classes. Dispersal success (as measured by the establishment of a territory in the new area) also depends on local density; successful movements into vacant habitats are, not unsurprisingly, considerably higher than those into saturated patches (Daly and Patton, 1990). As far as is known, successful female dispersers have a high probability of breeding. However, not all males, even those that are successful dispersers, breed, as there is a high variance in male reproductive success in high density populations (see below). Estimates of the number of migrants per generation (Nm) between local populations of *T. bottae* in adjacent fields range from one to six in saturated patches to 20–40 in vacant patches based on direct observations of movement, and between 2.5 and 18.0 based on indirect genetic measures, such as F_{ST} or Slatkin's (1981) conditional allele frequency (Lidicker and Patton, 1987; Daly and Patton, 1990). Estimates over larger geographic areas can be more than an order of magnitude less (Patton and Smith, 1990).

The breeding system of pocket gophers is generally characterized by (1) skewed adult sex ratios in favor of females, often by as much as 4:1; (2) extreme sexual dimorphism wherein males may be nearly 50% larger in body mass than females; (3) extreme aggression between males, as judged by the amount of scarring around the mouth and cheeks during the breeding season; and (4) a polygynous mating system with a high variance in male reproductive success but with relatively even success among females (Howard and Childs, 1959; Patton and Feder, 1981; Patton and Smith, 1990).

The "parapatry" of local territories within a given breeding population of pocket gophers is also descriptive of the geographic distribution of species within all genera, and even the genera themselves; true sympatry between adjacent taxa is unusual. Most cases of local sympatry involve horizontal distributional overlap of a few hundred meters at best (see, for example, Vaughan, 1967; Thaeler, 1968b; Reichman and Baker, 1972; Patton, 1973). Parapatric patterns of distribution in pocket gophers are often assumed to result from an inability to subdivide the subterranean niche (Pearson, 1959; Nevo, 1979). In areas where two or more species occur, both competitive exclusion (Miller, 1964) and historical pre-emptive occupancy (Thaeler, 1968b; Patton, 1990) have been offered as explanations for the observed spatial matrix of species' distributions.

SYSTEMATICS AND GENETIC VARIATION

This chapter is not intended to be a review of geographic variation in pocket gophers, but aspects of that variation are important to any discussion of hybrid zones between taxa. Pocket gophers are among the world's most variable mammals. Racial differentiation can be extreme; for example, there are 215 subspecies formally described in the *Thomomys bottae* complex based on morphological characters (Hall, 1981). Equally extensive levels of among-population differentiation are also exhibited at the chromosomal and allozyme levels for these and other pocket gophers (Nevo et al., 1974; Patton and Yang, 1977; Hart, 1978; Thaeler, 1980, 1985; Patton, 1981, 1990).

When coupled with the typically allopatric and patchy distribution, the extensive degree of geographic variation exhibited by pocket gophers renders the determination of species boundaries particularly difficult. It is for this reason that much emphasis has been placed on hybrid zone studies (as noted above). However, even with data on the nature of hybridization, be it limited to F_1 production or evidence of introgression extending into the parental populations, interpretations of species status differ among authors. For example, Hall (1981) argued that any evidence of hybrid occurrence is sufficient to support conspecificity. Patton (1981; Patton and Smith, 1989), on the other hand, recognized that when hybridization is limited to F_1 production the interacting populations are genetically isolated, if not reproductively so, and as such are distinct species. Even when introgression is clearly evident, however, the parental forms interacting in any narrow hybrid zone might be considered by some as distinct species. Because the genetic effects of gene flow dissipate rapidly with distance among pocket gopher populations (Patton and Smith, 1990), widely distributed parental forms that hybridize only locally may still be effectively separate evolutionary entities despite introgression.

The problem of defining species boundaries in pocket gophers is exacerbated by the relatively few points of contact between what are often exceedingly strongly differentiated geographic entities. Average allozyme genetic distances (Rogers, 1972) among local populations of pocket gophers range from 0.05 to 0.08, and those between geographic units can be in excess of 0.30 [see Hafner et al. (1987) and Patton and Smith (1990) for summaries for species of *Thomomys*]. To place these numbers in perspective, conspecific populations of small mammals usually exhibit less than 5% divergence, even over extensive geographic distances. Similarly, pocket gopher populations exhibit more extensive levels of karyotypic differentiation than most other mammals (Hart, 1978; Patton, 1981; Thaeler, 1985). Much of this variation lies in chromosome arm number shifts due to added or deleted blocks of heterochromatin (Patton and Sherwood, 1982; Sherwood and Patton, 1982), but in other cases structural rearrangements are involved (Patton and Sherwood, 1982; Dowler, 1989). The fact of extensive genetic differentiation (allozyme or chromosomal) between hybridizing forms, however does make the elucidation of hybrid zone dynamics in a genetic sense that much easier.

Hybridization does characterize most points of parapatric contact between geographically differentiated units of pocket gophers, but certainly not all. For example, with *Thomomys* no hybridization occurs between any species of the *bottae* group (subgenus *Megascapheus*) and of the *talpoides* group (subgenus *Thomomys*) where they come together, and several species of the latter are found in local sympatry in northern California without evidence of hybridization (Thaeler, 1968b, 1985). On the other hand, narrow hybrid zones characterize the points of contact between all geographically differentiated units (be these recognized as species or subspecies) within both the *bottae* complex and many of those of the *talpoides* group of *Thomomys,* as well as all of those of the *Geomys bursarius* complex (see Table 11-1 for references), whether the interacting forms are considered full species or only geographic races by the separate investigators.

TYPES OF POCKET GOPHER HYBRID ZONES

The characteristics of pocket gopher populations described above exert direct influences on the descriptive features of contact zones between geographically adjacent taxa, regardless of whether hybridization is a feature of these zones. For one, zones are likely to be narrow, with the actual dimensions a function, minimally, of three factors: (1) the limited number of individuals that can actually be in contact due to the combination of exclusive-use territories and high degree of sedentariness; (2) a reduced realized dispersal capability because the general saturation of available habitat means that most dispersal is unsuccessful (Daly and Patton, 1990); and (3) the fact that contacts between geographically adjacent and divergent populations are likely to be at ecotones between different habitats (as between oak woodland and mesquite grassland) (Patton, 1973) or piñon-juniper woodland and desert grassland (Patton et al., 1979).

Second, because of the combination of a limited number of individuals that can interact in any given zone, the narrowness of zones, and short life-spans of interacting animals, hybrid zones in pocket gophers almost invariably are merely collections of individuals whose membership changes substantially on an annual basis. For example, in the hybrid zone between *T. bottae* and *T. umbrinus,* the two species overlap over about a 1-mile section of a narrow canyon that is maximally 200 meters wide (Patton, 1973) (Fig. 11-1a). The total inhabitable area can house the territories of perhaps no more than 70 individuals. As a consequence, neither this nor other zones can be easily characterized mathematically, as separate population samples independently taken from a transitional area between parental forms cannot be obtained.

Finally, even if evidence of hybridization extends beyond the narrow confines of habitat ecotones, the characteristics of populations so affected are greatly influenced by the patchiness of habitat, the size of the population inhabiting each patch, differential gene flow between patches, and the temporal pattern of extinction and recolonization of available patches. Local extirpation of pocket gopher populations, particularly in suboptimal habitat, is likely to be high, and recolonization rates vary depending on the densities of and distances to surrounding source populations (Patton and Feder, 1981; Daly and Patton, 1990). The combination of a generally low density resulting from exclusive-use territories, even in high quality habitat, and high turnover rates within populations means that there is a natural sampling bias for each gen-

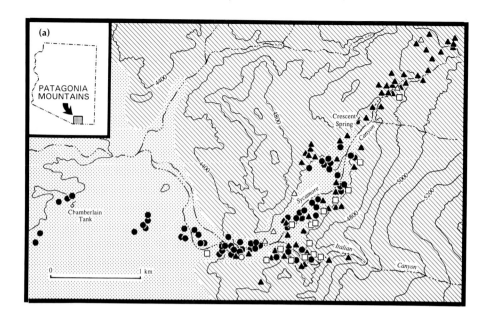

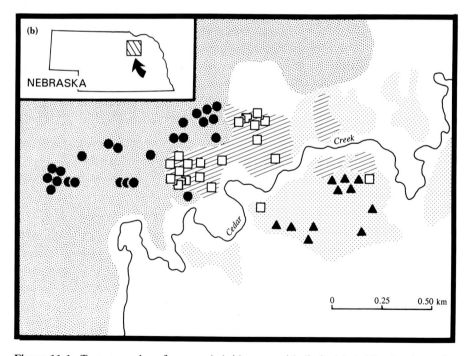

Figure 11-1. Two examples of narrow hybrid zones with limited hybridization in pocket gophers. (**a**) Map of localities of capture of individuals of *Thomomys bottae* (solid circles), *Thomomys umbrinus* (solid triangles), and their hybrids (F₁, open squares; backcross to *bottae*, open circle; backcross to *umbrinus*, open triangles) in Sycamore Canyon, Patagonia Mountains, Arizona. Elevational contours (in feet) are indicated. Hybrids are identified by karyotype supple-

eration that can obscure otherwise uniform patterns and their potential explanations (Patton, 1990).

As emphasized above, contact zones of pocket gophers are typically narrow and often conform to habitat ecotones. With *Thomomys,* for example, most contact zones occur within canyons coursing along elevational gradients in western mountain ranges. In each of these zones, pocket gophers are limited largely to the canyon floor where soils are adequate. Zones of contact in these situations are thus constricted laterally by the inhospitable shallow soils on steep slopes, resulting in a physiographic and ecologic limitation for the extent to which individuals of separate taxa can interact.

Nevertheless, these conditions do not characterize all contact zones between differentiated populations of pocket gophers (including those that can be argued to result from either primary or secondary contact); nor do they identify the extent of hybridization, which can vary greatly even in narrow zones. In addition to narrow zones that occur at habitat ecotones, others, equally narrow, occur through areas of inhospitable terrain where populations are discontinuous. Finally, there are areas of broad clinal change that characterize some pocket gophers. Each of these classes of hybrid zones tells us something about the nature of genetic interactions due to history, to local ecology, and to genomic compatibility.

Narrow Hybrid Zones at Habitat Ecotones. One type of zone is characterized by differentiated taxa that meet at habitat ecotones, where the zone width may range from the diameters of several individual territories to overlap of about 1 km. Populations found in such narrow contacts are characterized by similar levels of allozymic or chromosomal divergence; they meet in similar ecological situations; and the populations exist in roughly the same densities. Examples include most zones of contact examined for the genera *Thomomys* and *Geomys* (Table 11-1); two specific examples are illustrated in Figure 11-1.

Although each of these zones is geographically narrow, those that have been examined with the use of genetic markers fall readily into two rather distinct classes (Fig. 11-2). In the first (Fig. 11-2a), few individuals with pure parental genetic characteristics are present in the zone; rather, most individuals are of hybrid origin (>75% of all individuals in a zone), and the array of hybrid phenotypes (chromosomal and

←————————————————————————————————————

mented by allozyme data. The stippled pattern represents the characteristic mesquite grassland habitat of *T. bottae* and the riparian habitat on the canyon floor where the two species overlap; the parallel lines identify the oak woodland habitat of *T. umbrinus.* Pocket gophers are not found in the shallow soils on the canyon slopes. (Redrawn from Patton, 1973.) (**b**) Map of localities of capture of individuals of *Geomys lutescens* (solid circles), *Geomys bursarius* (solid triangles), and their hybrids (open squares) at a zone of contact west of Oakdale, Antelope County, Nebraska. Hybrids are identified by cranial and external morphology supplemented by some allozyme data. The contact zone is at the interface between pastureland with soils derived from stabilized sand dunes (dense stippling) on the west and northwest that characterize *lutescens* and cultivated areas with loess-derived silt loam soils (light stippling) to the east where *bursarius* is typically found. Hybrids are largely limited to cultivated loam soils (hatching) between these two major soil horizons. Open areas are clay soils of steep slopes with dense trees, habitats unsuitable for pocket gophers. (Redrawn from Heaney and Timm, 1985.)

Table 11-1. Characteristics of Hybrid Zones Described for Taxa of *Thomomys* and *Geomys* Pocket Gophers

Taxa and Location	Type[a]	Characters[b]	Genetic Distance[c]	Width (km)[d]	Shape[e]	% Hybrids	References
Thomomys							
bottae × umbrinus; S.AZ	I	G, C, M	$D_R = 0.155$	5	Asymmet.	11–16	Patton (1973); Patton et al. (1972)
bottae × townsendii; NE, CA	I	G, C, M	$D_R = 0.180$	2	Asymmet.	ca. 10	Patton et al. (1984); Thaeler (1968a)
bottae							
Sacramento Mts., NM	I	G, C, M	$D_R = 0.193$	3	Asymmet.	81	Patton et al. (1979)
Sangre de Cristo Mts., NM, CO	III	G, C, M	$D_R = 0.050$	200	Symmet.	N.A.	Hafner et al. (1983); Nadler et al. (1990)
Rio Grande, NM	II	G, C, M	$D_R = 0.313$	ca. 5	Symmet.	N.A.	Smith et al. (1983)
Colorado River, CA, AZ	II	G, M	$D_R = 0.330$	N.A.[f]	N.A.	N.A.	Smith and Patton (1980)
Owens Valley, CA	II	G, M	$D_R = 0.250$	N.A.	N.A.	ca. 70	Smith and Patton (1988)
talpoides, CO							
2n = 48 × 52	I?	C	N.A.	N.A.	N.A.	15	Thaeler (1974)
2n = 46 × 48	I?	C	N.A.	N.A.	N.A.	24	Thaeler (1974)
2n = 48 × 54	I?	C	N.A.	N.A.	N.A.	37	Thaeler (1974)
Geomys bursarius complex							
breviceps × attwateri; E.TX	I	C, M	N.A.	1	N.A.	4	Dowler (1989); Tucker & Schmidly (1981)
breviceps × major; OK	II	G, M	$D_R = 0.315$	2	N.A.	1	Bohlin and Zimmerman (1982)
bursarius × lutescens; NB	I	G, C, M	N.A.	2	Asymmet.	25	Heaney and Timm (1985)
knoxjonesi × major; NM	I[g]	G, C, M	N.A.	3	Asymmet.	56	Baker et al. (1989); Pembleton and Baker (1978)

[a] I = narrow zones at habitat ecotones; II = discontinuous zones due to habitat and/or topographic barriers; III = broad clines.

[b] G = genetic (allozymes, rDNA, mtDNA); C = chromosomal; M = morphologic.

[c] Rogers' allozyme genetic distance (D_R).

[d] Defined as the maximal observed distributional limits of hybrid individuals.

[e] Asymmetrical = hybrid production and/or individual distribution biased toward one of two parental forms; symmetrical = hybrid production and/or individual distribution relatively balanced between two parental forms in contact.

[f] N.A. = not available.

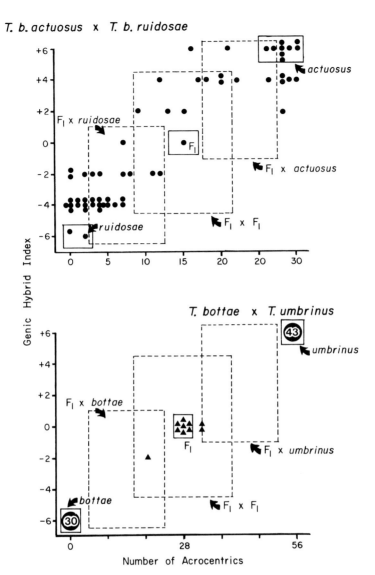

Figure 11-2. Two types of genetic interaction in pocket gopher hybrid zones, as illustrated by examples from the genus *Thomomys*. In both cases, the distribution of individuals with respect to their chromosomal characteristics (number of acrocentric autosomes in the complement) and allozyme genic hybrid index is plotted against the expected distribution of parental, F_1, second filial, and backcross characteristics. (Redrawn from Patton, 1990.) (**a**) Contact zone in the Sacramento and White Mountains of east-central New Mexico between the low-elevation, juniper-grassland subspecies *T. bottae actuosus* and the high-elevation, conifer forest *T. bottae ruidosae* (for details see Patton et al., 1979). These two taxa exhibit fixed differences at three allozyme loci and up to 32 arm number differences in karyotype. Note the paucity of "pure" parental and F_1 individuals, the extensive array of backcross genotypes, and the hybrid bias toward parental *ruidosae*. (**b**) Contact zone in the Patagonia Mountains of southern Arizona (Fig. 11-1) between *T. bottae* and *T. umbrinus*. These taxa also exhibit fixed allele differences at three loci and 56 arm number changes in their respective karyotypes. In contrast to the example in (**a**), note the high proportion of "pure" parental forms, few F_1, and even fewer backcross individuals. (Data from Patton et al., 1972; Patton, 1973.)

allozymic) indicates that most of them resulted from multiple generations of back-crossing. In such cases, there is no evidence of reproductive impairment in either male or female hybrid individuals; and, importantly, evidence of genic introgression usually extends beyond the narrowly defined hybrid zone into the parental populations of both interacting forms (see, for example, Patton et al., 1979). In the second type of zone (Fig. 11-2b), most individuals have pure parental phenotypes, and the small number of hybrids (usually no more than 10–15% of the total number of individuals present within the zone) are largely limited to the F_1 class. In at least one such example (*T. bottae* × *T. umbrinus* in southern Arizona) (Patton, 1973), hybrid males are sterile and hybrid females exhibit reduced fertility, which partially explains the paucity of hybrids. In this zone, there is also no evidence of introgression extending into the parental populations.

In the first type of narrow zone, the parental forms are neither reproductively nor genetically isolated, whereas in the second they are genetically (and hence evolution-arily, if not reproductively) separate. The distinction between these two zone types has therefore been used as an operational definition of species' boundaries for some pocket gophers (see discussion in Patton and Smith, 1989, 1990). It is important to reiterate here that in both outcomes observed in these continuous, narrow zones (Fig. 11-2) the extent of genetic (allozymic and chromosomal) divergence between the parents is the same, as is the degree of ecological transition between them. Genetic and ecological differentiation are thus not satisfactory predictors of the nature of hybridization that will occur.

Narrow but Geographically Discontinuous Zones. The degree of local contact between geographically differentiated forms of pocket gophers is a function of the patchiness of available habitat through the area of contiguity. In cases where popula-tions of two differentiated forms are not continuous through a contact area but are, instead, fragmented into patches defined by local soil mosaics or other geophysical parameters, the ecotonal type of narrow zone described above can be only ephemeral, if ever present at all. Nevertheless, evidence of the genetic interactions between differ-entiated populations that come together under this set of circumstances is still present in transect populations taken through an area of potential contact. In these cases, there may be no difference in the habitat occupied by the parental forms; rather, the points of contact conform to narrowed areas where physiographic, often only temporary, bar-riers exist.

Several such zones have been described in the *Thomomys bottae* and *Geomys bursarius* complexes of taxa (Table 11-1). In each of these examples, interactions between the contact populations are limited owing to constraints on local population distribution combined with specific physiographic barriers, such as inhospitable soil, (e.g., Bohlin and Zimmerman, 1982) or vegetation types (Smith and Patton, 1988) (Fig. 11-3a). In two specific examples within *Thomomys,* geographically differentiated forms meet along major river systems: the lower Colorado River of Arizona and Cal-ifornia (Smith and Patton, 1980) and central Rio Grande in New Mexico (Smith et al., 1983) (Fig. 11-3b). Interactions between the contact forms in both cases are con-strained because local populations are patchily distributed along the river terrace, which itself is (or at least was prior to flood control) subject to seasonal innundation and resulting local population extirpation. In the case of the lower Colorado River

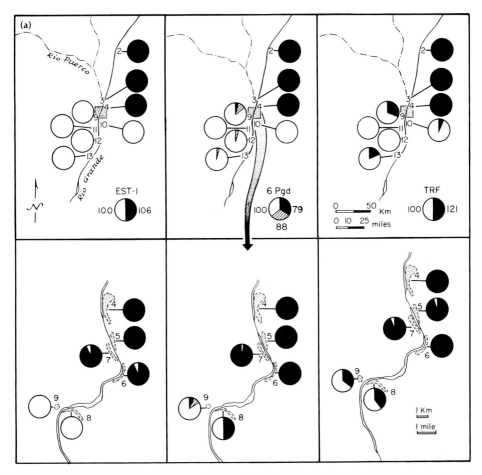

Figure 11-3. Two examples of geographically discontinuous contact zones between taxa of *Thomomys bottae* pocket gophers. (**a**) Electromorphic differentiation at the zone of contact along the central Rio Grande valley south of Albuquerque, New Mexico between the subspecies *T. b. connectens* (northern taxon) and *T. b. opulentus* (southern taxon). These two taxa are fixed for different alleles, or nearly so, at three allozyme loci at localities outside the contact zone. Pocket gophers are distributed along the first terrace of the river but are present in narrow patches of suitable soils (sampling points 4 through 9 in the lower diagrams). The combination of a localized patchy distribution and the likelihood of seasonal disruptions of those patches by fluctuating river levels probably prevents the type of distributional contiguity and overlap as depicted in Figure 11-1. (From Smith et al., 1983. With permission.) (**b**) Contact zone near Lone Pine, Inyo County, California, in the southern end of the Owens Valley. Allozyme genic hybrid index scores are provided for each sample from within the general area of contact as well as from three populations of the parental subspecies (*T. b. perpes* from Butterworth to the south, *T. b. melanotis* from Independence to the north, and *T. b. operarius* from Keeler to the southeast). The subspecies *melanotis* and *operarius* are genetically identical but are either fixed for different alleles or have markedly different frequencies at five loci in comparison to *perpes*. Pocket gophers are absent from the desert scrub habitat, but occasional contact and hence hybridization between *melanotis-operarius* found in the desert riparian habitat of the Owens River and *perpes* in the pasturelands probably occurs periodically as individuals disperse along the drainages that cut across the desert scrub.

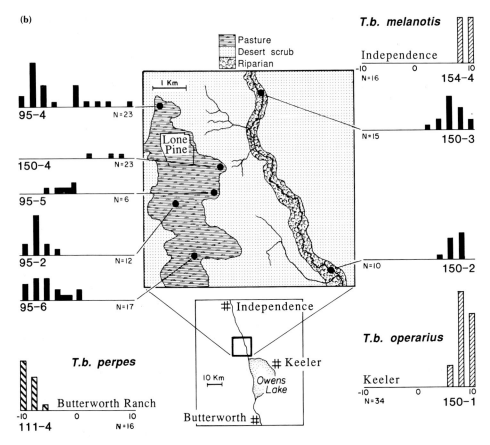

Figure 11-13. *Continued*

contact, however, evidence of past hybridization is present in populations many kilo-meters distant from those that are geographically adjacent along the river, even though the latter are not now in immediate contact due to habitat barriers (Smith and Patton, 1980).

Regardless of whether contact zones are composed of individuals distributed con-tinuously throughout or only in local patches defined by vagaries in habitat, there are several features shared by both ecotonal and discontinuous zones that need to be emphasized. For one, in cases where both chromosomal and allozyme data are avail-able, similar magnitudes of differentiation between the parental forms can be associ-ated with radically different levels of hybridization (see Fig. 11-2, for example). Thus there is no general correspondence between measured levels of genetic divergence and genetic isolation, as is often assumed or expected (e.g., Thorpe, 1982; for further dis-cussion of this point see Hafner et al., 1983).

Second, the results of hybridization are almost always biased toward the more common parental form adjacent to the hybrid zone; a greater proportion of non-F_1 hybrid-class individuals exhibit genetic characteristics of the more abundant parental type, and hybrid individuals of all classes are geographically clustered in the zone closest to that parent. For example, in both cases illustrated in Figure 11-2, the geographic distribution of hybrids in the zone is skewed toward the parental form at the higher elevation. In the case of *bottae actuosus* and *bottae ruidosae* (Fig. 11-2a) (Patton et al, 1979), most individuals of mixed ancestry are also genetically closer to the high elevation parent *(ruidosae)* in terms of both allozyme phenotype and karyotype. Although hybridization is largely limited to the F_1 class in the second case (that of *T. bottae* × *T. umbrinus*) (Patton, 1973), non-F_1 hybrids are skewed toward parental *umbrinus*. These observations suggest that gradients of habitat quality through a contact zone, which determine the local density of individuals, influence the direction and extent of hybridization in cases where backcrossing is possible. In such circumstances, therefore, introgression into one parental population is more likely than into the other merely by virtue of differential density and thus mating opportunities. The same skew in introgression, however, could result from other factors, such as differential viability and fertility of individuals resulting from reciprocal crosses or mating asymmetry in favor of one particular reciprocal cross (see, for example, Baker et al., 1989; see also below).

Finally, all of the hybrid zones that fit either the ecotonal or discontinuous type are most likely the result of secondary contact between geographic elements previously isolated in space and possibly also in time. It is not easy to distinguish between primary and secondary divergence (Endler, 1977, 1982), particularly in cases such as these where the parental forms often meet along habitat gradients. Nevertheless, these hybrid zones represent localized contacts between what are otherwise geographically widespread taxa, each of which is distributed across an extensive habitat diversity and over which populations show similar genetic attributes. For example, a pine-oak population of *T. umbrinus* hybridizes with the desert grassland *T. bottae* in the Patagonia Mountains of southern Arizona (Patton, 1973). Both taxa, however, have extensive geographic and ecological ranges beyond their limited point of contact over which each maintains a relatively uniform set of chromosomal and allozyme characteristics [for *T. umbrinus* see Patton and Feder (1978) and Hafner et al. (1987); for *T. bottae,* see Patton (1972), Patton and Yang (1977), and Patton and Smith (1990)]. Given the limited contact yet extensive extralimital geographic and ecological distributions and the relatively uniform genetic attributes of the hybridizing populations, differentiation in situ due to an ecological gradient (parapatric divergence) seems unlikely.

Broad Geographic Clines. The two classes of hybrid zones identified immediately above are clearly zones of local hybridization, and most probably represent points of secondary contact between taxa that originally diverged in allopatry. A third type of "zone" exists, however, but it is of a distinctly different nature. This one is more appropriately described as a broad cline, with characters shifting across a substantial distance. As a consequence of the geographic extent of these zones, independent samples can be obtained from one end to the other and both temporal and spatial dynamics can be examined by application of mathematical models, such as the static cline theory (Barton and Hewitt, 1989; Hafner et al., 1983).

Only two studies of pocket gophers have revealed this type of cline, and both focus primarily on chromosomal divergence. The first case is that of *Thomomys bottae fulvus,* which is distributed across the Mogollon Plateau from Flagstaff, Arizona to the Rio Grande Valley in New Mexico. Over this distance of about 400 km, the number of uniarmed autosomes in populations varies clinally from a mean near 0 in the west to one averaging 36 in the east (the diploid number is 76) (Patton, 1972; Patton and Yang, 1977). It is not clear if this cline represents primary divergence or secondary contact. Because clinal differentiation is confined to populations distributed across relatively uniform pine-fir forests above 7000 feet, it is difficult to argue for a strong ecological gradient as the basis for primary differentiation. It is equally difficult, however, to develop a convincing scenario of past geographical isolation followed by secondary contact, as the populations that differ clinally in karyotype are rather uniform in allelic variation and are strongly divergent as a group from those to the north or west (Patton and Yang, 1977; Patton and Smith, 1990).

The second example occurs in a narrow, peninsular distribution along the eastern slopes of the Sangre de Cristo Mountains of Colorado and New Mexico. Here three concordant character gradients cut through the distribution of four successive subspecies of *Thomomys bottae* over approximately 200 km. Clines described in this situation are mean diploid number, pelage brightness, and mean morphometric coefficient of variation (Hafner et al., 1983; see also Nadler et al., 1990). These authors applied static cline models to test alternatives to cline development and maintenance. They concluded that, although the operation of natural selection in maintaining these clines could not be excluded, the most parsimonious explanation for the establishment of the present dimensions of the zone involved near neutral diffusion following secondary contact between taxa that differentiated in allopatry. The observed zone width (using the 80% criterion) (Endler, 1977) fit that predicted using reasonable estimates of dispersal (0.1 to 0.5 km per generation; see above) and time since probable contact, based on knowledge of the post-Pleistocene history of the region. Estimated selection gradients were so minimal as to be unimportant in cline mainenance (the value of s ranged from 5×10^{-5} to 6×10^{-6} over different ranges of dispersal distances) (Hafner et al., 1983).

DYNAMIC ATTRIBUTES OF POCKET GOPHER HYBRID ZONES

Virtually all studies of hybrid zones between various taxa of pocket gophers are phenomenological descriptions of character variation, hybrid number and type, and/or ecological associations. With few exceptions, we know little about important aspects such as hybrid fitness, mating asymmetry among parents and hybrid class individuals alike, zone permanence, opportunity for reinforcement, and the variety of other issues of importance to the understanding of hybrid zones as windows through which to view and understand evolutionary processes. In the following sections, I provide what we do know about these and other issues and provide suggestions as to what fruitful avenues of research are open for continued work on pocket gopher hybrid zones.

Mating Asymmetry. Patton (1973) argued that in the zone of contact in southern Arizona between *T. bottae* and *T. umbrinus* the two species do not mate at random. This conclusion was based on the observed distribution of individuals trapped in the hybrid zone and the average expected territory size. These two features combined to suggest that 33–39% of the population in the hybrid zone should result from interspecific matings if mating were simply a function of geographic placement of territories of the two sexes. However, the percentage of hybrids actually observed ranged between 11% and

16% at different sampling periods. Hybrids in this case are largely limited to the F_1 generation, as males are apparently sterile and females exhibit 50% of the fertility of either parent. Nevertheless, some backcrossing must occur based on karyotypic characteristics of hybrid individuals. Interestingly, five of six backcross individuals are closer to parental *T. umbrinus*. Thus not only is mating nonrandom among the parental forms, but there appears to be an asymmetry in the reciprocity of potential crosses.

Direct evidence for asymmetry in reciprocal crosses between parental taxa comes from two hybrid zone studies. One of these studies is in the *Geomys bursarius* complex and involves the subspecies *major* and *knoxjonesi* in eastern New Mexico and adjacent Texas (Baker et al., 1989). This study employed a combination of nuclear (allozyme and ribosonal DNA) and cytoplasmic (mitochrondrial) markers, as well as chromosomal differences, to identify hybrid class individuals. The mtDNA markers showed that hybrid production is asymmetrical; that is, all individuals satisfying the authors' criteria for an F_1 genotype had the *knoxjonesi* mtDNA genome. These data suggest that most if not all crosses were between *knoxjonesi* females and *major* males. From these observations, the authors concluded that premating isolation was partially developed between the interacting parents, specifically between *knoxjonesi* males and *major* females.

A second example of mating asymmetry is known in the hybrid zone between the species *Thomomys bottae* and *T. townsendii* in northeastern California (Patton and Smith, unpubl. data). Here hybridization is limited (about 10% of the total sample at any one time) (Thaeler, 1968a; Patton et al., 1984), and most hybrids are F_1 individuals based on their allozyme scores, although a few apparent backcross individuals are found. A mtDNA marker (a length variant in the control region, or D-loop), however, indicates that all hybrid individuals were the products of *bottae* female and *townsendii* male matings, and that backcross production was sufficient across generations for the *bottae* mitochondrial genome to be present in a few individuals that were otherwise "pure" *townsendii* based on nuclear markers. The asymmetry in the distribution of mtDNA markers relative to nuclear ones in this case is identical to that in the *Geomys* example. However, additional data on body and burrow size suggest an alternative explanation to partial premating isolation.

As noted earlier, pocket gopher populations are often characterized by biased adult sex ratios in favor of females, high variance in male reproductive success, and extensive sexual dimorphism, with males considerably larger than females (Patton, 1990). These features suggest strong male–male competition for mates with reproductive success potentially related to body size, both of which are aspects of great significance in understanding mating asymmetry in pocket gopher hybrid zones. For example, males of *T. townsendii* are approximately three times the body mass of female *T. bottae* (303 grams versus 103 grams, respectively) where they are in contact in northeastern California. Moreover, because of the energetic costs of burrowing, the diameter of burrows of individual pocket gophers closely matches their body diameter. In the *bottae-townsendii* zone, burrows of female *T. bottae* are simply too small to permit entrance of the much larger *T. townsendii* males. This fact, coupled with the asymmetry in the distribution of the mtDNA marker, suggests that females of both species are actively choosing mates and that male body size is a major component on which this choice is based. This finding, in turn, suggests that the mating system of pocket gophers in general is one of female choice.

If, in fact, female choice characterizes the pocket gopher mating system, where the choice is directed to the largest males, most hybrid zones must necessarily exhibit asymmetry as evidenced by nonnuclear genetic markers. The asymmetry should be in favor of the parental form with the smaller female; and backcrossing, if it occurs at all (i.e., if either, or both, F_1 sex is even partially fertile), is in the direction of the parental form with the larger male. This hypothesis means that mitochrondrial markers, for example, penetrate differentially into one parental form (the one with the larger average male) than the other, even though nuclear markers may not indicate such an asymmetry. Interestingly, in the *Geomys* case cited above, this situation is exactly what is found: The mtDNA genome of the smaller *knoxjonesi* characterizes all hybrid class individuals and is found to have differentially penetrated the geographic distribution of parental *major* away from the actual contact zone (Baker et al., 1989). The asymmetry in the distribution of the mtDNA marker may be due therefore only to the normal mating pattern of pocket gophers, where females choose among males on the basis of their body size. If it is the case, such asymmetry does not indicate the development of premating isolation at all. As noted by Baker et al. (1989), and as illustrated by these two examples, analyses of hybrid zones using a combination of nuclear and nonnuclear markers can provide windows of understanding not available by the singular use of either type of marker.

Factors Maintaining Hybrid Zones. There are no direct data relating to the longevity of any hybrid zone in pocket gophers. Circumstantial evidence in some cases, however, suggests that contact and hybridization has occurred between adjacent but differentiated forms potentially for the last several millenia. For example, the geographic and elevational position of the ecotones that often characterize pocket gopher contacts, particularly in the intermontane west, may be as old as the early Holocene, some 8000 years ago (see reviews in Betancourt et al., 1990). Hafner et al. (1983) thus argued that the clines characterizing populations of *T. bottae* along the front range of the Sangre de Cristo Mountains in New Mexico and Colorado could stem from this time period. On the other hand, significant change in vegetation zones that might affect local pocket gopher distribution, and thus prolonged contact, have occurred over the past century (Hastings and Turner, 1965). Thus, even if contact has been old, actual contact in any given circumstance may have been repeatedly punctuated by episodes of noninteraction owing to the combined effects of habitat patchiness in ecotones and the resultant likely ephemeral nature of the populations occupying available patches. Recognition of this fact has considerable consequences for the application of any model to distinguish between competing hypotheses that relate to the maintenance of zones.

Static cline models have been applied only to the clinal situation of *T. bottae* in New Mexico and Colorado (Hafner et al., 1983). As summarized above, the most parsimonious explanation for the patterns observed here is one of near neural diffusion. However, there are key assumptions on which this conclusion is based and that may not actually be met. These assumptions include, minimally, that the clines result from secondary contact, that gene flow rates are in the neighborhood of 0.1–0.5 km per year, that initial contact was several thousand years ago, and that genetic interactions have been rather constant since initial contact. While the gene flow assumption is supported by available evidence, its effect diminishes quickly with distance (see above; and Daly and Patton, 1990). The other assumptions, however plausible, have no real basis in fact.

Even if neutral diffusion following secondary contact underlies the Sangre de Cristo clines, this explanation is not likely to apply to the narrow types of hybrid zone for pocket gophers described above. Most of these zones occur at habitat ecotones, often along rather sharp ecological gradients, and most are very narrow, being less than 2 km in width. Neutral diffusion would be likely in these circumstances only if contact was very, very recent, literally almost yesterday. Given the reasonable dispersal capabilities of pocket gophers on a local scale and the associated habitat gradient, these zones are most likely the product an an equilibrium between dispersal and selection, a conclusion reached by Barton and Hewitt (1985) for most hybrid zones, including those of pocket gophers. In cases where hybridization is limited to the F_1 generation, selection is clearly operating directly on the reproductive fitness of the hybrid individuals (as in the *Thomomys bottae* × *umbrinus* case) (Patton, 1973) (Fig. 11-2b), which in turn acts to maintain the position and width of the hybrid zone. Where hybrids are completely fertile, however, selection would presumably act through the breakdown of local parental adaptations to habitat features such as temperature or moisture tolerances. The White-Sacramento Mountains zone between *T. b. actuosus* and *T. b. ruidosae* (see Patton et al., 1979) (Fig. 11-2a) is a good candidate for this suggestion. Regardless of the basis for selection, however, the narrowness of these zones, particularly in relation to the average home range size, precludes one from obtaining independent samples that are necessary to test any specific hypothesis for their maintenance.

Whatever might be the absolute age of particular pocket gopher hybrid zones, none provide evidence for the development of reinforcement. Moreover, it seems unlikely that reinforcement would ever be an expected outcome in gopher hybrid zones, assuming that such an outcome is even a possibility [see arguments by Butlin (1989) and Howard (Ch. 3). To reiterate from above, all pocket gopher hybrid zones are characterized by a limited area for contact due to habitat patchiness. Moreover, only a small absolute number of individuals of any one parental form can be in contact with their opposite members at each generation. Coupling this fact with the high annual turnover characteristic of pocket gopher populations and the exclusivity of individual territories means that most individuals have but one season during which to mate and leave offspring, and that they can mate only with members of the opposite sex whose territory is adjacent, regardless of the parental form to which those individuals belong. This structure is unlikely to provide the opportunity for reinforcement to operate.

CONCLUSIONS AND PROSPECTUS

Although hybrid zone studies of pocket gophers have provided historically important examples for mammals, the studies have been undertaken primarily as aids in the determination of species boundaries rather than as means to examine hybrid zones as evolutionary phenomena themselves. This bias is perhaps appropriate because pocket gopher hybrid zones are limited by virtue of their typically narrow dimensions, especially in relation to the average size of individual gopher territories and thus to the total number of interacting invidivuals. The constraints of general pocket gopher biology and demography (i.e., low densities, limited movements, exclusive-use territories, high turnover rates, and so forth) are not optimal for using these animals to examine general models of hybrid zone dynamics.

Nevertheless, there are some exciting directions for future studies wherein pocket gophers might provide exceptional opportunities for research. One is the explicit examination of how mating systems determine both patterns of hybridization and the dynamics of introgression using a combination of nuclear and cytoplasmic genetic probes. Here, the exclusivity of individual territories permits an exact knowledge of the geographic positions of *all* individuals in contact areas and thus the total enumeration of the entire population of interacting individuals. There are few organisms for which these critical aspects can be understood so easily.

ACKNOWLEDGMENTS

My work on hybrid zones of pocket gophers has been collaborative, especially with Drs. M. F. Smith, J. C. Hafner, and M. S. Hafner; to each of these individuals, and others, I express my gratitude for their companionship in the field, data analysis, and intellectual insight. The final manuscript was greatly improved by the detailed suggestions of R. G. Harrison, to whom I am also grateful for the invitation to participate in this volume, and an anonymous reviewer. The research reported herein has been supported by grants from the National Science Foundation, for which I am most appreciative.

REFERENCES

Baker, R. J., Davis, S. K., Bradley, R. D., Hamilton, M. J., and Van Den Bussche, R. A. 1989. Ribosomal-DNA, mitochrondrial-DNA, chromosomal, and allozymic studies on a contact zone in the pocket gopher, *Geomys.* Evolution 43:63–75.

Barton, N. H., and Hewitt, G. M. 1985. Analysis of hybrid zones. Annu. Rev. Ecol. Syst. 16:113–148.

Barton, N. H., and Hewitt, G. M. 1989. Adaptation, speciation and hybrid zones. Nature 341:497–503.

Betancourt, J. L., Van Devender, T. R., and Martin, P. S. eds. 1990. The Last 40,000 Years of Biotic Change. Tucson: University of Arizona Press.

Bohlin, R. G., and Zimmerman, E. G. 1982. Genic differentiation of two chromosomal races of the *Geomys bursarius* complex. J. Mammal. 63:218–228.

Butlin, R. 1989. Reinforcement of premating isolation. In D. Otte and J. A. Endler, eds. Speciation and Its Consequences. Sunderland, MA: Sinauer, pp. 158–179.

Chase, J. D., Howard, W. E., and Roseberry, J. J. 1982. Pocket gophers (Geomyidae). In J. A. Chapman and G. A. Feldhammer, eds. Wild Mammals of North America. Baltimore: Johns Hopkins University Press, pp. 239–255.

Daly, J.C., and Patton, J. L. 1990. Dispersal, gene flow and allelic diversity between local populations of *Thomomys bottae* pocket gophers in the coastal ranges of California. Evolution 44:1283–1294.

Dowler, R. C. 1989. Cytogenetic studies of three chromosomal races of pocket gophers (*Geomys bursarius* complex) at hybrid zones. J. Mammal. 70:253–266.

Endler, J. A. 1977. Geographic Variation, Speciation, and Clines. Princeton, NJ: Princeton University Press.

Endler, J. A. 1982. Problems in distinguishing historical from ecological factors in biogeography. Am. Zool. 22:441–452.

Hafner, J. C., Hafner, D. J., Patton, J. L., and Smith, M. F. 1983. Contact zones and the genetics of differentiation in the pocket gopher *Thomomys bottae* (Rodentia: Geomyidae). Syst. Zool. 32:1–20.

Hafner, M. S., Hafner, J. C., Patton, J. L., and Smith, M. F. 1987. Macrogeographic patterns of genetic differentiation in the pocket gopher *Thomomys umbrinus.* Syst. Zool. 36:18–34.

Hall, E. R. 1981. The Mammals of North America. Vol. 1. New York: Wiley.

Hart, E. B. 1978. Karyology and evolution of the plains pocket gopher, *Geomys bursarius.* Occas. Pap. Mus. Nat. Hist. Univ. Kansas 71:1–20.

Hastings, J. R., and Turner, R. M. 1965. The Changing Mile. Tucson: University of Arizona Press.

Heaney, L. R., and Timm, R. M. 1985. Morphology, genetics, and ecology of pocket gophers (genus *Geomys*) in a narrow hybrid zone. Biol. J. Linn. Soc. 23:301–317.

Howard, W. E., and Childs, H. E., Jr. 1959. Ecology of pocket gophers with emphasis on *Thomomys bottae mewa.* Hilgardia 29:277–358.

Lidicker, W. Z., Jr., and Patton, J. L. 1987. Patterns of dispersal and genetic structure in populations of small rodents. In B. D. Chepko-Sade and Z. Halpin, eds. Mammalian Dispersal Patterns. Chicago: University of Chicago Press, pp. 144–161.

Miller, R. S. 1964. Ecology and distribution of pocket gophers (Geomyidae) in Colorado. Ecology 45:256–272.

Nadler, S. A., Hafner, M. S., Hafner, J. C., and Hafner, D. J. 1990. Genetic differentiation among chewing louse populations (Mallophaga: Trichodectidae) in a pocket gopher contact zone (Rodentia: Geomyidae). Evolution 44:942–951.

Nevo, E. 1979. Adaptive convergence and divergence of subterranean mammals. Ann. Rev. Ecol. Syst. 10:269–308.

Nevo, E., Kim, Y. J., Shaw, C. R., and Thaeler, C. S., Jr. 1974. Genetic variation, selection and speciation in *Thomomys talpoides* pocket gophers. Evolution 28:1–23.

Patton, J. L. 1972. Patterns of geographic variation in karyotype in the pocket gopher, *Thomomys bottae* (Eydoux and Gervais). Evolution 26:575–586.

Patton, J. L. 1973. An analysis of natural hybridization between the pocket gophers, *Thomomys bottae* and *Thomomys umbrinus,* in Arizona. J. Mammal. 54:561–584.

Patton, J. L. 1981. Chromosomal and genic divergence, population structure, and speciation potential in *Thomomys bottae* pocket gophers. In O. A. Reig, ed. Ecologia y Genética de la Especiación Animal. Caracas: Equinoccio, pp. 255–295.

Patton, J. L. 1990. Geomyid evolution: the historical, selective, and random basis for divergence patterns within and among species. In E. Nevo and O. A. Reig, eds. Evolution of Subterranean Mammals at the Organismal and Molecular Levels. New York: Alan R. Liss, pp. 49–69.

Patton, J. L., and Dingman, R. E. 1968. Chromosomal studies of pocket gophers, genus *Thomomys.* I. The specific status of *Thomomys umbrinus* (Richardson) in Arizona. J. Mammal. 49:1–13.

Patton, J. L., and Feder, J. H. 1978. Genetic divergence between populations of the pocket gopher, *Thomomys umbrinus* (Richardson). Z. Sauget. 43:17–30.

Patton, J. L., and Feder, J. H. 1981. Microspatial genetic heterogeneity in pocket gophers: non-random breeding and drift. Evolution 35:912–920.

Patton, J. L., Hafner, J. C., Hafner, M. S., and Smith, M. F. 1979. Hybrid zones in *Thomomys bottae* pocket gophers: genetic, phenetic, and ecologic concordance patterns. Evolution 33:860–876.

Patton, J. L., Selander, R. K., and Smith, M H. 1972. Genic variation in hybridizing populations of gophers (genus *Thomomys*). Syst. Zool. 21:263–270.

Patton, J. L., and Sherwood, S. W. 1982. Genome evolution in pocket gophers (genus *Thomomys*). I. Heterochromatin variation and speciation potential. Chromosoma 85:149–162.

Patton, J. L., and Smith, M. F. 1989. Population structure and the genetic and morphologic divergence among pocket gopher species (genus *Thomomys*). In D. Otte and J. A. Endler, eds. Speciation and Its Consequences. Sunderland, MA: Sinauer, pp. 284–304.

Patton, J. L., and Smith, M. F. 1990. The evolutionary dynamics of the pocket gopher *Thomomys bottae,* with emphasis on California populations. Univ. Calif. Publ. Zool. 123:vxiii, 1–161.

Patton, J. L., Smith, M. F., Price, R. D., and Hellenthal, R. A. 1984. Genetics of hybridization between the pocket gophers *Thomomys bottae* and *Thomomys townsendii* in northeastern California. Great Basin Nat. 44:431–440.

Patton, J. L., and Yang, S. Y. 1977. Genetic variation in *Thomomys bottae* pocket gophers: macro-geographic patterns. Evolution 31:697–720.

Pearson, O. P. 1959. Biology of the subterranean rodent, *Ctenomys,* in Peru. Mem. Mus. Hist. Nat. Javier Prado 9:1–55.

Pembleton, E. F., and Baker, R. J. 1978. Studies of a contact zone between chromosomally characterized populations of *Geomys bursarius.* J. Mammal. 59:233–242.

Reichman, O. J., and Baker, R. J. 1972. Distribution and movements of two species of pocket gophers (Geomyidae) in an area of sympatry in the Davis Mountains, Texas. J. Mammal. 53:21–33.

Reichman, O. J., Whittam, T. G., and Ruffner, G. A. 1982. Adaptive geometry of burrow spacing in two pocket gopher populations. Ecology 63:687–695.

Rogers, J. S. 1972. Measures of genetic similarity and genetic distance. University of Texas Publ. Stud. Genet. 8(7213):145–153.

Sherwood, S. W., and Patton, J. L. 1982. Chromosome evolution in pocket gophers (genus *Thomomys*). II. Variation in cellular DNA content. Chromosoma 85:163–179.

Slatkin, M. 1981. Estimating levels of gene flow in natural populations. Genetics 99:323–335.

Smith, M. F., and Patton, J. L. 1980. Relationships of pocket gopher *(Thomomys bottae)* populations of the lower Colorado River. J. Mammal. 61:681–696.

Smith, M. F., and Patton, J. L. 1988. Subspecies of pocket gophers: causal bases for geographic differentiation in *Thomomys bottae.* Syst. Zool. 37:163–178.

Smith, M. F., Patton, J. I.., Hafner, J. C., and Hafner, D. J. 1983. *Thomomys bottae* pocket gophers of the central Rio Grande Valley, New Mexico: local differentiation, gene flow, and historical biogeography. Occasional Papers of the Museum of Southwestern Biology, University of New Mexico, No. 2.

Thaeler, C. S., Jr. 1968a. An analysis of three hybrid zones of pocket gophers (genus *Thomomys*). Evolution 22:543–555.

Thaeler, C. S., Jr. 1968b. An analysis of the distribution of pocket gopher species in northeastern California. Univ. Calif. Publ. Zool. 86:1–46.

Thaeler, C. S., Jr. 1974. Four contacts between ranges of different chromosome forms of the *Thomomys talpoides* complex (Rodentia: Geomyidae). Syst. Zool. 23:343–354.

Thaeler, C. S., Jr. 1980. Chromosome numbers and systematic relations in the genus *Thomomys* (Rodentia: Geomyidae). J. Mammal. 61:414–422.

Thacler, C. S., Jr. 1985. Chromosome variation in the *Thomomys talpoides* complex. Acta Zool. Fenn. 170:15–18.

Thorpe, J. P. 1982. The molecular clock hypothesis: biochemical evolution, genetic differentiation and systematics. Ann. Rev. Ecol. Syst. 13:139–168.

Tucker, P. K., and Schmidly, D. J. 1981. Studies of a contact zone among three chromosomal races of *Geomys bursarius* in east Texas. J. Mammal. 62:258–272.

Vaughan, T. A. 1967. Two parapatric species of pocket gophers. Evolution 21:148–158.

12

Chromosomal Hybrid Zones in Eutherian Mammals

JEREMY B. SEARLE

It has long been realized that many closely related species of eutherian mammals differ in karyotype, sometimes to an extreme degree (e.g., Bush et al., 1977; White, 1978a; Bengtsson, 1980). Furthermore, these species may themselves be subdivided into two or more "karyotypic races," each of which typically occupies a geographically continuous area and is characterized by particular fixed chromosomal variants. This review examines the "hybrid zones" that occur where such races overlap and hybridize. Several well-studied eutherian chromosomal hybrid zones are described, but to aid interpretation various properties of chromosomal variants must first be considered.

CYTOGENETIC OVERVIEW

Chromosomal Variants and Their Description. To minimize terminology, I simplify as follows. First, in terms of chromosome "shape," all chromosomes with a terminal or near-terminal centromere (i.e., uniarmed chromosomes) are called "acrocentrics," and all chromosomes with an internal centromere (i.e., biarmed chromosomes) are known as "metacentrics" (see Figs. 12-1 and 12-2). Second, only two subdivisions of chromosome "structure" are recognized: heterochromatin (highly repetitive, "satellite" DNA) and euchromatin (other DNA, including functional genes). In eutherians, variation in heterochromatin is best studied by the C-banding technique, which causes all blocks of highly repetitive DNA to stain darkly, whereas rearrangements affecting euchromatin are best revealed by G-banding, which causes differential staining within euchromatin (e.g., Therman, 1986) (see Figs. 12-1 and 12-2). Third, consideration is limited to the autosomal complement of chromosomes, as sex chromosome and B-chromosome variation are not important in the best-studied eutherian hybrid zones.

The chromosomal mutations referred to in this chapter are shown in Figures 12-1 and 12-2. In general, the importance of chromosomal rearrangements is the manner in which they affect the disposition of euchromatin (as illustrated), although it is also of interest whether the chromosome breaks occur within euchromatin or heterochromatin (not illustrated). Variation in size of heterochromatic blocks (Fig. 12-1) is

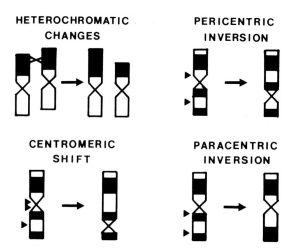

Figure 12-1. Common "within-chromosome" mutations. Heterochromatic changes arising by somatic unequal crossing-over, illustrated by C-banded chromosomes. Euchromatic rearrangements: arrowheads indicate breakpoints, illustrated by G-banded chromosomes. Heterozygous individuals have one copy of the ancestral-type chromosome and one copy of the rearranged chromosome in their karyotypes. Note that for small rearrangements it is difficult to distinguish pericentric inversions and centromeric shifts, even with G-banding (e.g., Wójcik and Searle, 1988), and it is possible that there have been mistakes in the literature.

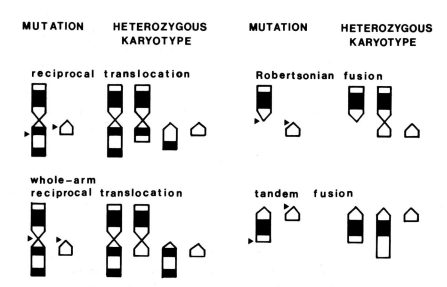

Figure 12-2. Common "between-chromosome" mutations illustrated by G-banded chromosomes (arrowheads indicate breakpoints). Reciprocal translocations involve interstitial breakage within one or both chromosomes. Whole-arm reciprocal translocations involve breaks at centromeres of both chromosomes (one or both chromosomes metacentrics). Robertsonian (Rb) fusions involve breaks at both centromeres (both chromosomes acrocentrics). Tandem fusions include both telomere–telomere and telomere–centromere rearrangements. Rb and tandem fission are the reverse processes from the fusions described above.

310

thought to occur primarily by *somatic* unequal crossing-over (Kurnit, 1979) (crossing-over is reduced or absent in heterochromatin during meiosis) (Stack, 1984). Although the quantity of heterochromatin often varies markedly between apparently healthy and fecund individuals of one species—e.g., the hedgehog *Erinaceus europaeus* (Mandahl, 1978) and humans (Brown et al., 1980)—within-species euchromatic size variation has only occasionally been reported in eutherians, e.g., in the marsh rat *Holochilus* (Nachman, 1992a), and has not been of importance for studies of hybrid zones. Studies of complete or partial aneuploidy suggest that even small additions or deletions of blocks of euchromatin can lead to inviability (Epstein, 1986, pp. 190–195; Kirk and Searle, 1988; Searle, 1989).

The mutation rates for chromosomal rearrangements can be estimated for humans (Lande, 1979). Reciprocal translocations, Robertsonian (Rb) fusions, and inversions arise at rates of 1.3×10^{-4}, 1.0×10^{-4}, and 0.4×10^{-4} per gamete per generation, respectively. Other rearrangements occur rarely in humans (Dutrillaux, 1979). There are few additional data for eutherians, but the mutation rate for reciprocal translocation in house mice is apparently similar to that in humans (Lande, 1979). Data for both humans and mice indicate that heterochromatic additions and deletions are stable genetic markers over many generations (Forejt, 1973; Carothers et al., 1982; Bobrow, 1985), but there is no accurate information on mutation rate. Differences in the frequencies of different types of mutations are to be expected between species: for example, the frequency of Rb fusions and fissions vary according to the numbers of acrocentrics and metacentrics in the karyotype, respectively.

Heterozygotes for chromosomal variants occur at hybrid zones between karyotypic races and in other areas of chromosomal polymorphism. In the simplest case, the heterozygotes merely have both the ancestral-type and derived chromosomes in their karyotypes (Figs. 12-1 and 12-2). However, two hybridizing races may have differently derived karyotypes involving, to an extent, the same chromosomes. In eutherians, this occurs for races that differ by Rb fusions, and therefore a detailed nomenclature is required to describe the various types of hybrid (Fig. 12-3). It should be noted that chain IV-forming and ring IV-forming "complex" Rb heterozygotes (Fig. 12-3) could be produced on hybridization of races that differ either by several Rb fusions or a single whole-arm reciprocal translocation (Fig. 12-2). Other complex Rb heterozygotes could come about as a result of whole-arm reciprocal translocations as well.

Chromosomal Mutations and Phenotype. It may be thought that changes in the disposition of euchromatin or amount of heterochromatin probably do not influence phenotype; and if any effects are observed, they are likely to be of a detrimental nature (e.g., Rutledge et al., 1986; Searle, 1988a; Wu et al., 1989). However, there have been several suggestions as to how chromosomal mutations, through their intrinsic properties, may result in an advantageous phenotype. All between-chromosome rearrangements (Fig. 12-2) and inversions (Fig. 12-1) alter the relative positions of genes; and in the extreme case components of unlinked genes could become spliced together to form an advantageous new gene or gene complex (Wilson et al., 1974). Whether changes in looser linkage relations is of significance in eutherians is unclear; functionally related genes are scattered around the chromosomes (Lyon and Searle, 1989). Chromosomal mutations also change karyotypic properties (chromosomal number, shape, heterochromatin location) that influence the pattern of recombination (Miklos and John,

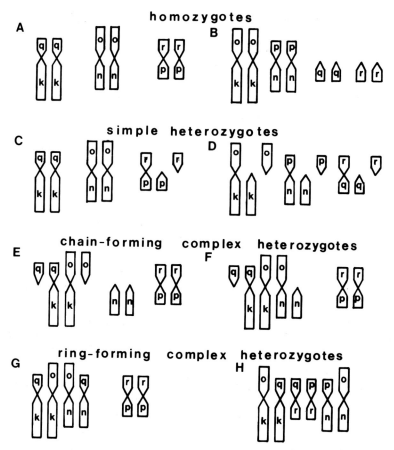

Figure 12-3. Types of Robertsonian (Rb) heterozygotes, as illustrated by the variable chromosomes *k, n, o, p, q,* and *r* of the common shrew *(Sorex araneus).* (a) Individual homozygous metacentric for arm combinations *kq, no,* and *pr.* (b) Individual homozygous metacentric for arm combinations *ko* and *np* and homozygous acrocentric for chromosomes *q* and *r.* (c) Single "simple" Rb heterozygote (arm combination *pr*). (d) Triple "simple" Rb heterozygote. (e) Chain IV-forming "complex" Rb heterozygote. (f) Chain V-forming "complex" Rb heterozygote. (g) Ring IV-forming "complex" Rb heterozygote. (h) Ring VI-forming "complex" Rb heterozygote. For "complex" Rb heterozygotes, the name indicates the type of configuration formed at prophase I of meiosis (see Fig. 12-4).

1979; Dutrillaux and Rumpler, 1987) and hence the genetic variability of gametes. This point may be of little relevance in eutherians, however, as there can be extreme variation in these properties among members of a single genus or species, e.g., *Muntiacus* (Neitzel, 1987) and *Uromys caudimaculatus* (Baverstock et al., 1982).

Empirical data, particularly on Rb fusions, suggest that it is worth considering that chromosomal rearrangements may have an intrinsic phenotypic effect at haplophase as well as at diplophase, as contemplated above. Data on segregation for laboratory mice and humans (but not as yet for wild eutherians) convincingly show that particular chromosomal rearrangements may be preferentially transmitted over the ancestral-type chromosomes or vice versa (for reviews see Martin, 1988; Searle, 1988a; Stene and Stengel-Rutkowski, 1988; but see also Mercer and Searle, in press).

Karyotypic Heterozygosity, Fertility, and Gene Flow. Heterozygote disadvantage has been particularly emphasized as a possible selective force in chromosomal hybrid zones because of the expectation of reduced fertility in karyotypic heterozygotes, arising from meiotic aberrations (e.g., White, 1978a; Lande, 1979; Sites and Moritz, 1987). For viable, chromosomally balanced gametes to be produced by karyotypic heterozygotes, homologous segments within structurally different chromsomes must pair and form chiasmata at prophase I, and such chromosomes must disjoin and segregate appropriately at anaphase I. However, in eutherians, the complexity of this pairing and segregation process in heterozygotes compared with homozygotes may cause a greater proportion of germ cells to die or to contain an unbalanced set of chromosomes, as best elucidated in humans and laboratory mice (Chandley, 1984; de Boer, 1986; Daniel, 1988; de Boer and de Jong, 1989; Speed, 1989).

Considering the male, a slight increase in germ cell death probably causes little or no reduction in fertility, but in some karyotypic heterozygotes there may be oligospermia or azoospermia and hence sterility (de Boer and Searle, 1980). In the female, increased oocyte death in heterozygotes is likely to shorten reproductive life but probably not reduce the number of oocytes per ovulation (Burgoyne and Baker, 1984). Production of chromosomally unbalanced gametes may also significantly affect the fertility of karyotypic heterozygotes. In eutherians, such gametes are as likely to survive and fertilize as those with a normal haploid complement (Ford, 1972; but see Redi et al., 1982, 1984; Stewart-Scott and Bruère, 1987). Therefore chromosomally unbalanced gametes result in embryos with either complete or partial aneuploidy. For blocks of heterochromatin, this aneuploidy may have no effect on viability; but for all except the smallest euchromatic segments, it leads to death at around the time of implantation in the case of monosomies and later during gestation for trisomies (Epstein, 1986; Kirk and Searle, 1988). Therefore in karyotypic heterozygotes there may be primary infertility (as a consequence of germ cell death) or secondary infertility (resulting from early death of their progeny) (Gropp et al., 1982a).

GERM CELL DEATH. The enhanced germ cell death specifically associated with karyotypic heterozygosity is thought to result from aberrations of chromosome pairing at the pachytene stage of meiosis. Chromosome pairing can be studied by visualization of the proteinaceous "axial elements" (Loidl, 1991), which are attached to each chromosome during zygotene and pachytene. Where homologous chromosomes are paired, their axial elements run side by side, forming a "synaptonemal complex" (Fig. 12-4). It is thought that normal transcriptional activity of a meiocyte depends on all autosomal segments becoming fully paired (homologously or even nonhomologously) by the end of pachytene (Burgoyne and Baker, 1984; de Boer and de Jong, 1989; Speed, 1989). In meiocytes of karyotypic heterozygotes, chromosomes may consistently fail to pair in the vicinity of the breakpoints of the heterozygous configuration. Such unpairing seems particularly likely to lead to aberrant gene expression (and hence physiological perturbation and meiocyte death) when it occurs at the ends of chromosomes. Externally unpaired axial elements (Fig. 12-4) often develop a thickened appearance that is perhaps associated with chromosomal condensation and reduced

PACHYTENE   DIAKINESIS   PACHYTENE   DIAKINESIS

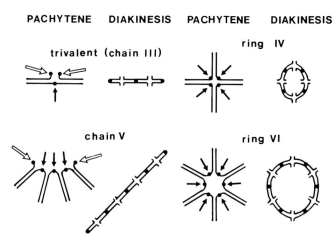

Figure 12-4. Meiotic configurations of "simple" and "complex" Robertsonian (Rb) heterozygotes at prophase I of meiosis. The circles represent centromeres, and the lines indicate axial elements (pachytene) or chromatin (diakinesis); that is, each chromosome forms a single axial element at pachytene but can be seen to be composed of two chromatids at diakinesis. The karyotype of simple Rb heterozygotes includes a metacentric and two homologous acrocentrics for each variable arm combination, such that chain III configurations are formed at prophase I. The karyotype of complex heterozygotes includes metacentrics with monobrachial homology, such that chains (of longer than three elements) or rings are formed at prophase I. For all these heterozygous configurations, the axial elements tend to be paired along most of the length of the chromosome arms at pachytene (these paired structures are known as synaptonemal complexes). However, in the vicinity of the centromeres of the heterozygous configurations, axial elements may be unpaired; potentially internal and external unpaired regions are indicated by closed and open arrows, respectively. The diakinesis configurations illustrated here display one chiasma per chromosome arm.

transcription (de Boer and Searle, 1980; Forejt, 1984; de Boer and de Jong, 1989). In spermatocytes these externally unpaired regions may tend to associated with the unpaired regions of the XY bivalent, and there have been several suggestions as to how this association may enhance the likelihood of cell death (Forejt, 1982, 1984; Luciani et al., 1987; Richler et al., 1989). Perhaps the high frequency of XY–autosome interactions observed in many karyotypic heterozygotes explains why sterility due to germ cell death has been frequently recorded in male karyotypic heterozygotes but not females (Chandley, 1984; de Boer, 1986). This disparity could also result, however, from other differences in the gametogenic process between the sexes.

ANEUPLOID GAMETES. One cause of chromosomally unbalanced gametes in karyotypic heterozygotes is suboptimal localization of chiasmata. Crossing-over between the breakpoints (inversions, centromeric shifts) or between centromere and breakpoints (reciprocal translocations, tandem rearrangements) may lead to aneuploid gametes. The regular occurrence of a single crossover in this critical region normally results in 50% unbalanced gametes in karyotypic heterozygotes (Lewis and John, 1963; White, 1973; Lande, 1979; de Boer, 1986; Daniel, 1988). Clearly, the frequency of such disadvantageous crossovers depends on the size of the critical region. They do not occur in heterozygotes for whole-arm reciprocal translocations or Rb rearrangments but are bound to occur in heterozygotes for reciprocal translocations or tandem fusions, when they involve acrocentric chromosomes.

Unbalanced gametes may also result from malsegregation of heterozygous configurations or univalents, this process being known as "(anaphase I) nondisjunction." [For tandem rearrangements and reciprocal translocations, this term is restricted to numerical malsegregation; see de Boer (1986) for other types of segregation.] The spindle at anaphase I seems best adapted to symmetrical bivalents, and errors are more likely to arise with the segregation of multivalents (Eichenlaub-Ritter and Winking, 1990) and to a lesser degree with heteromorphic bivalents (human pericentric inversions) (Kaiser, 1988, p. 188). Nondisjunction frequency would be expected to increase further in physiologically perturbed meiocytes, e.g., as a sublethal effect of chromosomal unpairing at pachytene (in males) or due to the hormonal irregularity that occurs in old females (Gropp and Winking, 1981, p. 159; Bond and Chandley, 1983).

CROSSOVER SUPPRESSION AND GENE FLOW. Meiotic recombination is expected only between homologous chromosomal regions (Speed, 1989). With regard to heterozygous configurations, there tends to be unpairing or nonhomologous pairing in the vicinity of breakpoints, so these regions would be expected to show a degree of "crossover suppression" (Hale, 1986). Genetic studies in karyotypic heterozygotes of laboratory mice have demonstrated almost complete crossover suppression over the region of two paracentric inversions that characterize the t complex (Lyon, 1991) and significantly reduced crossing-over in the centrometric regions of several Rb fusions (Cattanach, 1978).

In a hybrid zone where karyotypic heterozygotes have reduced fitness, there is a reduction in gene flow that acts most strongly on genes along those chromosomes that differ between the races (Futuyma and Mayer, 1980). Crossover suppression could influence gene flow across such a zone in two ways: (1) indirectly by the effect on hybrid fitness (either reducing errors from crossovers in the critical region or increasing errors due to malsegregation of univalents: see above); and (2) by directly reducing gene flow in the suppressed region. The latter property means that hybridizing karyotypic races may maintain or develop genic differences even when hybrids have near-normal fertility.

KARYOTYPIC HETEROZYGOSITY AND FERTILITY: "NATURAL" HETERO-ZYGOTES. Most of the fertility data on karyotypic heterozygotes with a natural (or near-natural) genome refers to Rb rearrangements (Tables 12-1 and 12-2). The data for non-Rb rearrangements (Figs. 12-1 and 12-2) may be summarized as follows. For heterochromatic size variants, analysis of a substantial data set in humans suggests heterozygotes have normal fertility (Carothers et al., 1982; Bobrow, 1985), whereas pachytene studies of the deer mice *Peromyscus maniculatus, P. sitkensis,* and *P. beatae* indicate that full synapsis in unequal bivalents can be achieved by extensive nonhomologous pairing (Hale and Greenbaum, 1988a; Sudman et al., 1989). Even in hybrids between the northern and southern races of the Australian rat *Uromys caudimaculatus,* which differ by the presence or absence of large telomeric blocks of heterochromatin on up to 15 pairs of chromosomes, there appears to be no fertility reduc-

Table 12-1. Fertility of Eutherian "Simple" (Trivalent-Forming) Robertsonian Heterozygotes with a Natural or Near-Natural Genetic Background

Species (Race)[a]	Sex	No. of CIII Configs[b]	Hets Found in Nature?	Breeding History[c]	AI ND Freq.[d] (%) Hets	AI ND Freq.[d] (%) Homs	GCD[e]	Fecundity[f]	Refs[g]
House mouse (Peru wild stock)	F	1	No	Lab	12–16	0	—	Normal	1
House mouse (undefined)	M	1	No	Lab	8	—	—	Normal	1
House mouse (Zadar)	M	1–3	Yes	Wild	0.5, 2.2[h]	—	—	Normal	2
House mouse (JOG)	M	1	Yes	Wild	—	—	—	Normal	3
House mouse (JOG × stand)	M	1	Yes	Wild	2.7	1.5	Low	—	4
House mouse (JOG × stand)	M	4	No	Wild/wild	—	—	Low	—	5
House mouse (Belgium)	M,F	1	Yes	Wild	—	—	—	Normal	6
House mouse (Pavia)	M,F	1	Yes	Wild	—	—	—	Normal	7
House mouse (Pav × Mi2)	F	7	No	Lab	—	—	High	—	8
House mouse (Mon × stand)	M,F	9	Yes	?	—	—	—	Low	9
Black rat (O × A)	M,F	2	Yes	Lab?	—	—	—	Low	10
Black rat (O × C)	M	2	Yes	Lab?	0	—	—	Low	11
Black rat (O × M)	M,F	1	Yes	Lab?	—	—	—	Mod.	10
	M,F	2	No	Lab?	—	—	—	Normal	12
"Field mouse" (Apodemus speciosus)	M	1	Yes	Wild	22	9	None	—	13
"Field mouse" (Akodon molinae)	M	1	Yes	Lab	20	8[i]	—	Low[i]	14
	M,F	1	Yes	Lab	—	—	—	Low[i]	14
Marsh rat (Holochilus brasiliensis)	M	1–2	Yes	Wild	3.2	2.0	—	Normal	15,16
	F	1–2	Yes	Wild	—	—	—	Normal	15,16
Mole-vole (Ellobius tancrei)	M,F	10	no	?	—	—	—	Low	17
	M,F	9	No	?	—	—	—	Mod.	17
Common shrew (Ox, Herm)	M	1–2	Yes	Wild	1.4	2.3	None	—	18,19

Lemur fulvus	F	1–3	Yes	Wild	4.5–8.3	0–5.4	None	Normal	20,21
L. fulvus × *L. macaco*	M	6	No	Wild/wild	—	—	None/low?	Low	22,23 24
Human (13;14,14;21)	M	8	No	Wild/wild	<1	—	High	None– normal	22,23,24,25
							None– comp.		26,27
Human (13;14)	F	1	—	—	<1	—	—	Normal	26
Human (14;21)	F	1	—	—	~30	—	—	Mod.	26

See Figure 12-3.

[a] The heterozygotes (hets) in most studies were either (1) produced by crosses between races with and without a Rb rearrangement (denoted race 1 × race 2); or (2) collected from hybrid zones at the contact between such races [the race(s) with the rearrangement are given]. JOG = John o'Groats, Scotland; stand = standard $2n = 40$ karyotype; Pav = Pavia, Italy; Mi2 = Milano II, Italy; Mon = Monastir, Tunisia; O = "Oceanian" race; A = "Asian" race; C = "Ceylonese" race; M = "Mauritius" race; Ox = Oxford, England; Herm = Hermitage, England. For the mole-vole, individuals from within the Tadzhikstan hybrid zone (see Table 12-3) were crossed. For the "Peru" mice, the Rb fusion arose within a laboratory wild stock. For *Akodon molinae*, *Holochilus brasiliensis*, and humans, the rearrangements are polymorphic within races. For women, the Rb fusions between chromosomes 13 and 14 [rob(13;14)] and rob(14;21) are considered separately. Homs = homozygotes.

[b] CIII = chain of three elements (trivalent) formed at prophase I of meiosis. In addition, the "Asian" rats could differ by up to three pericentric inversions from the "Oceanian" rats and the marsh rats may be heterozygous for pericentric inversions and/or euchromatic additions and have variable numbers of B chromosomes.

[c] Wild, wild/wild, lab = heterozygotes collected from nature and minimally maintained and bred in captivity, homozygotes collected in nature and crossed after little or no breeding in captivity, or heterozygotes generated from wild (usually homozygous) stocks maintained for several generations in captivity.

[d] Anaphase I nondisjunction frequencies of the Rb heterozygotes in comparison with either form of homozygote, determined from metaphase II counts (double hyperploid count) or fetal karyotypes (various methods). In the data for humans, the entry indicates the increased nondisjunction frequency relative to the standard karyotype. More than one estimate may be given.

[e] None, low, mod., high, comp. = no greater germ cell death in karyotypic heterozygotes than homozygotes of either type, low, moderate, or high germ cell death in heterozygotes, or complete arrest of gametogenesis in heterozygotes.

[f] None, low, mod., normal = no offspring produced by karyotypic heterozygotes, or low, moderate, or apparently normal fecundity compared with homozygotes of either type. Litter size is usually determined from fetal counts or at birth. For humans, reduced fecundity is expressed by a higher frequency of unsuccessful pregnancies.

[g] Key to references: (1) Harris et al., 1986; (2) Winking, 1986; (3) Winking et al., 1988; (4) Wallace et al., 1988; (5) Ferrier et al., 1990; (6) Viroux and Bauchau, 1992; (7) Britton-Davidian et al., 1990; (8) Garagna et al., 1990; (9) Said et al., in press; (10) Yosida, 1980a; (11) Yosida, 1976; (12) Yosida, 1980b; (13) Saitoh and Obara, 1988; (14) Merani et al., 1980; (15) Nachman and Myers, 1989; (16) Nachman, 1992b; (17) Lyapunova and Yakimenko, 1985; (18) Garagna et al., 1989; (19) Mercer et al., 1991; (20) Searle, 1990; (21) Wallace et al., 1991; (22) Dutrillaux and Rumpler, 1977; (23) Ratomponirina et al., 1982; (24) Y. Rumpler, pers. obs.; (25) Rumpler et al., 1985; (26) Stene and Stengel-Rutkowski, 1988; (27) Chandley, 1988.

[h] Anaphase I nondisjunction frequency per heterozygous arm combination.

[i] Data for homozygous metacentric individuals; nondisjunction frequencies of 38% and a lower litter size than heterozygotes is recorded for homozygous acrocentric individuals (Merani et al., 1980).

Table 12-2. Fertility of Eutherian "Complex" (Ring- and Chain-Forming) Robertsonian Heterozygotes with a Natural or Near-Natural Genetic Background

Species (Race)[a]	Sex	No. and Type of Configs[b]	Hets Found in Nature?	Breeding History[c]	AI ND Freq.[d] (%) Hets	AI ND Freq.[d] (%) Homs	GCD[e]	Fecundity[f]	Refs[g]
House mouse (Pos × UV)	M,F	1 CV	Yes	Wild/wild	—	—	—	Low	1
House mouse (Cre × Mi2)	M	1 CV	No	Lab	—	—	Comp.	None	2
House mouse (Cre × Mi2)	F	1 CV	No	Lab	—	—	Mod.	—	3
House mouse (Acr × Lip)	M	1 CXI	No	Lab	—	—	Comp.	—	4
House mouse (Cre × Mi1)	F	1 CIX	No?	Lab	—	—	Mod.	—	3
House mouse (Pos × CD)	M	1 CXVII + 1 CIII	No	Lab	—	—	Comp.	None	5
House mouse (Pos × CD)	F	1 CXVII + 1 CIII	No	Lab	>90	—	—	Low	5
House mouse (Pos × CB)	M	2 CIX	No	Lab	—	—	Comp.	None	5
House mouse (Acr × CD)	M	1 CXIV + 1 CV	No	Lab	—	—	Comp.	—	4
House mouse (Lip × CD)	M	1 CXV + 1 CV	No	Lab	—	—	Comp.	—	4
House mouse (Pos × Zad)	M	1 CVII + 1 CVIII	No	Lab	—	—	Comp.	—	6
House mouse (Mi1 × Mi2)	F	1 RVI	Yes	Lab	—	—	Low	—	3
House mouse (CB × CD)	M	1 RXVI	No?	Lab	~98	—	Mod.	Low	2
House mouse (Cre × CD)	M	1 RXVIII	No	Lab	~100	—	Mod.?	None	2
	F	1 RXVIII	No	Lab	>53	0	—	—	7
Rattus villosissimus × *R. colletti*	M	1 CV + 3 CIII	No	Wild/wild	—	—	—	Low	8
Rattus villosissimus × *R. colletti*	F	1 CV + 3 CIII	No	Wild/wild	—	—	—	Low	8

Marsh rat (Holochilus)	M	1 CIV + 1–2 CIII	Yes	Wild	4.5	3.0	—	—	9
Lemur fulvus collaris × L.f. albocollaris	M	1 CVI + 2 CIII	Yes?	Wild/wild	—	—	High/comp.	—	10,11
Lemur macaco × L. coronatus	M	1 CXI + 1 RVI	No	Wild/wild	—	—	Comp.	None	11,12
Lemur fulvus albocollaris × L. macaco	M	1 CXIV + 2 CIII	No	Wild/wild	—	—	Comp.	None	11,13,14
Lemur fulvus collaris × L. macaco	M	1 CXI + 3 CIII	No	Wild/wild	—	—	High	None	11,14,15
Common shrew, (Ox, Herm)	M	1 CIV or 1 CIV + 1 CIII	Yes	Wild	—	—	Low	—	16
Common shrew (Ox × Ab)	M	1 CVII	No	Wild/wild	12.5	2.3	Low	—	17
Wild pig (Sus s. scrofa × S. s. nigripes)	M,F	1 CIV	No?	?	—	—	—	Normal	18

See Figure 12-3 and Table 12-1.

[a] As for Table 12-1. For the marsh rat, the rearrangements are polymorphic within a race. Pos = Poschiavo, Switzerland (and Italy); UV = Upper Valtellina, Italy; Cre = Cremona, Italy; Mi2 = Milano II, Italy; Acr = Ancarano, Italy; Lip = Lipari, Italy; CD = Cittaducale, Italy; Zad = Zadar, Yugoslavia; Mil = Milano I, Italy; CB = Campobasso, Italy; Ox = Oxford, England; Herm = Hermitage, England; Ab = Aberdeen, Scotland.

[b] CIII = chain of three elements formed at prophase I of meiosis; RIV = ring of four elements, etc. Note that the difference between races could be due to whole-arm reciprocal translocations as well as Robertsonian fusions.

[c, d, e, f] As for Table 12-1.

[g] Key to references: (1) Gropp and Winking, 1981; (2) Gropp et al., 1982a; (3) Garagna et al., 1990; (4) Malorni et al., 1982; (5) Capanna et al., 1976; (6) Winking et al., 1988; (7) Eichenlaub-Ritter and Winking, 1990; (8) Baverstock et al., 1983; (9) Nachman, 1992b; (10) Rumpler and Warter, 1990; (11) Y. Rumpler, pers. obs; (12) Rumpler et al., 1990; (13) Rumpler et al., 1985; (14) Dutrillaux and Rumpler, 1977; (15) Ratomponirina et al., 1982; (16) Garagna et al., 1989; (17) Mercer et al., 1992; (18) Troshina et al., 1985.

tion relative to "pure"-race individuals (Baverstock et al., 1982). For pericentric inversions, meiotic data in *P. sitkensis* heterozygotes suggest that as a result of crossover suppression in the inverted region all gametes are balanced (Greenbaum and Reed, 1984; Hale, 1986; Hale and Greenbaum, 1988b). Among humans, inversion heterozygotes also usually have normal fertility, although a few individuals may produce a high frequency of unbalanced gametes due to crossing-over in the inverted region (Kaiser, 1988; Madan, 1988; Martin, 1988). Reciprocal translocations tend to cause a serious reduction in fertility in both men and women (more detailed data are available for men: Chandley, 1984, 1988; Martin, 1988; Stene and Stengel-Rutkowski, 1988). Crossovers in the critical region, suboptimal segregation (including nondisjunction), or both, result in numerous unbalanced gametes (20–80% of spermatozoa screened, data from several heterozygotes) (Martin, 1988). Some heterozygous males are completely devoid of spermatozoa, generally because of failure of the rearranged chromosomes to synapse properly with their normal homologues, leading to a "chain" configuration (and sometimes univalence) instead of the expected "ring" of four chromosomes at prophase I (Chandley, 1984). Germ cell death is presumably a consequence of the externally unpaired regions present at pachytene in the chain configurations but not in the rings (Fig. 12-4).

In "simple" Rb heterozygotes (Fig. 12-3), fertility is often near-normal in single or low multiple heterozygotes but generally highly reduced in those heterozygous for 6–10 chromosomes (Table 12-1). However, there is great variation between species and even between Rb rearrangements within species (e.g., humans). This variation may be due to: (1) structural differences, e.g., in *Akodon molinae* the rearranged chromosomes may include unusual small centric changes in addition to the Rb mutation, which may affect chromosome pairing (Bianchi and Merani, 1981); (2) differences in normal meiotic behavior between chromosomes and between species, e.g., in male humans acrocentric chromosomes tend to associate with the meiotic sex vesicle, perhaps predisposing autosome–sex chromosome interactions (Luciani et al., 1987); (3) the size and symmetry of the metacentric, which may affect position of chiasmata and orientation on the first meiotic spindle (White, 1973, p. 295); (4) genic effects. In humans and common shrews, female Rb heterozygotes have higher nondisjunction frequencies than males (Table 12-1), concordant with studies on laboratory mice (Gropp et al., 1981, 1982a). For female common shrews, the mean number of embryos is similar in Rb homozygotes and heterozygotes because the slightly greater nondisjunction frequency in heterozgotes is offset by a higher frequency of ovulation (fitness may still be lower in heterozygotes due to the more variable litter size) (Searle, 1990). An even higher nondisjunction frequency is offset by increased ovulation in one stock of house mice (Harris et al., 1986).

Slight differences in overall chiasma count have been recorded between simple Rb heterozygotes and homozygotes in male common shrews, marsh rats, and house mice (Searle, 1986b; Nachman, 1992b; Wallace et al., in press.). This point is of interest because abnormal chiasma behavior is an important feature of hybrid breakdown in a grasshopper hybrid zone (Coates and Shaw, 1982, 1984; see Ch. 7). Although the distribution and number of chiasmata are apparently rather firmly regulated in eutherians (Lyon, 1976; Maudlin and Evans, 1980; Laurie and Hultén, 1985a,b), there is no indication that slight deviations from the norm would deleteriously affect the genotype of gametes. Further comparative studies are needed.

Table 12-2 documents the fertility of "complex" Rb heterozygotes (Fig. 12-3), which form long-chain or ring configurations at meiotic prophase I (Fig. 12-4). Among house mice and lemurs, chain-forming heterozygous males suffer massive germ cell death and sterility, whereas chain-forming female mice and ring-forming males do not. These results reflect two general features of karyotypic heterozygotes (see above): that sterility due to germ cell death is more likely in males than in females (an example of Haldane's rule) and that chain-forming heterozygotes suffer greater germ cell death than those that form rings because chains have externally unpaired regions at pachytene (Fig. 12-4). Those long-chain-forming or ring-forming heterozygous house mice that are not sterile due to germ cell death may suffer high nondisjunction frequencies (Table 12-2).

Not all eutherian complex heterozygotes are so infertile. Even on the basis of studies of laboratory mice, it would be expected that individuals that produce short meiotic configurations (especially rings) should have a reasonable chance of producing offspring (Gropp and Winking, 1981). Indeed, chain-IV-forming complex heterozygotes among marsh rats, common shrews, and wild pigs have near-normal fertility. This point is of interest because it indicates that heterozygotes for single whole-arm reciprocal translocations need not be particularly unfit. Thus the possibility of whole-arm reciprocal translocations should no longer be ignored in the analysis of eutherian chromosomal evolution, as it has been up to now, with some notable exceptions (Winking, 1986; Wójcik, 1986).

Considering all types of karyotypic heterozygotes, a general problem is the discrepancy between data for wild-caught individuals and for heterozygotes generated by laboratory crosses between karyotypic races (often from geographically distant localities). Usually, feral heterozygotes have near-normal fertility whether collected from areas of within-race polymorphism (pericentric inversions in deer mice, Rb variants in marsh rats—but not *Akodon molinae*) or from chromosomal hybrid zones (common shrew and house mouse simple Rb heterozygotes—but not *Apodemus speciosus*). Karyotypic heterozygotes generated by laboratory interracial crosses, however, often have low fertility not only because the hybrids tend to be multiple heterozygotes, often with complex meiotic configurations, but also due to the contribution of genic factors to infertility [see Ratomponirina et al. (1982) on lemurs]. Genic hybridity alone may lead to meiotic aberrations similar to those associated with karyotypic heterozygosity and, furthermore, may cause a reduction in fertility by enhancing the specific effects of karyotypic heterozygosity (germ cell death associated with pachytene unpairing of heterozygous configurations and nondisjunction of those configurations). Both contentions are supported by data on "hybrid" house mice with a mixed laboratory mouse/wild mouse genetic background (Cattanach and Moseley, 1973; Gropp et al., 1974, 1981; Winking and Johannisson, 1980; Gropp and Winking, 1981; Winking, 1986; Searle, 1988a; Winking et al., 1988; Everett and Searle, 1990; Wallace et al., in press).

Genic factors may also contribute to the infertility of karyotypic heterozygotes from a long-standing hybrid zone. The fertility of such hybrids, however, may tend to be higher than those generated by laboratory interracial crosses because: (1) naturally hybridizing races do not tend to be as genically different as those hybridized in captivity; and (2) in a long-standing hybrid zone selection may be expected to reduce the frequency of alleles that cause karyotypic heterozygotes to suffer a high frequency of meiotic anomalies (Shaw, 1981).

A final point about the fertility data set is that it should be considered only a rough guide to reproductive fitness, because of the difficulty of: (1) detecting small, but biologically important, fertility differences; and (2) assessing all factors that may counteract increased embryo loss or reduced output of gametes, which I refer to collectively as "reproductive compensation" (see Bengtsson, 1980).

Within-Species Patterns of Karyotypic Variation. The clearest examples of widespread "high-level" karyotypic polymorphisms are those involving heterochromatic variation. For example, human populations appear to be characterized by a large number of differently sized variants for many of the centrometric blocks of heterochromatin, so that different individuals usually have a different karyotype (Craig-Holmes et al., 1973; Brown et al., 1980). The same is probably true of most other eutherian species, e.g., the house mouse (Forejt, 1973). There are also examples of widespread polymorphism for pericentric inversion of heterochromatin: chromosome 9 in humans (Bobrow, 1985; Kaiser, 1988); chromosome 9 in the black rat, *Rattus rattus* (Yosida, 1980a); and the woodrat *Neotoma micropus* (Warner, 1976). Heterozygosity for such heterochromatic variants causes little or no loss in fertility (see above); even the inversions would not cause meiotic aberrations because of the generalized inhibition of crossing-over in heterochromatic regions. Therefore these variants are probably selectively neutral.

In eutherians, widespread "high-level" polymorphism is much rarer for euchromatic markers. Perhaps the best example is again the black rat in which there is a polymorphism for a euchromatic pericentric inversion of chromosome 1 over much of Japan (Yosida, 1980a). Among several examples of widespread Rb polymorphisms, that in the gerbil *Gerbillus pyramidum* (chromosome 10) is found consistently in populations of both the northern and southern karyotypic races over a wide area in Israel and Sinai (Wahrman and Gourevitz, 1973). Such extensive polymorphisms are not particularly surprising given that heterozygotes for single Rb fusions and pericentric inversions tend to have near-normal fertility in eutherians (see Table 12-1); in some cases the heterozygotes may be as fit as homozygotes, either because meiotic aberrations are particularly rare or because the effect of any such aberrations is minimized owing to "reproductive compensation." These polymorphisms may reasonably be considered examples of neutral variation; but other cases where there is heterozygote excess in populations may be best considered "balanced" polymorphisms maintained by heterozygote advantage, e.g., the pocket gopher *Geomys bursarius* (J. C. Patton et al., 1980; Baker et al., 1983a); and the field mice *Akodon molinae* and *A. dolores* (chromosome 1) (Bianchi et al., 1979; Bianchi and Merani, 1981; see discussion in Searle, 1988a).

With regard to subdivision into karyotypic races, the best-studied eutherians are largely those included in the survey of hybrid zones (Table 12-3 and 12-4). The same range of chromosomal mutations as involved in karyotypic polymorphism predominate: Rb rearrangements, pericentric inversions, and heterochromatic variants. On the basis of available knowledge (see above), mutations of these types occur frequently, and heterozygotes tend to have near-normal fertility. Thus particularly for those species that tend to be subdivided into rather small populations (Bush et al., 1977), it is not difficult to envisage such variants occasionally becoming fixed within a population by genetic drift (Lande, 1979, 1984; Hedrick, 1981; Walsh, 1982; Coyne, 1984; but see Barton and Rouhani, 1991). Consistent with this mode of formation, the karyotypic races in some species occupy extremely limited areas, e.g., the house mouse (Capanna, 1982) and the pocket gopher *Thomomys talpoides* (Thaeler, 1985).

Generally, the recognizable contacts between eutherian karyotypic races are narrow hybrid zones with cline widths (Endler, 1977) of 50 km or less. For those species multiply subdivided into karyotypic races distinguished by euchromatic variants (the common shrew *Sorex araneus,* the house mouse *Mus musculus,* the mole-rat *Spalax ehrenbergi,* the pocket gopher *Thomomys talpoides*), there is little within-race polymorphism involving the same types of markers (Wahrman et al., 1969; White, 1978b; Thaeler, 1985; Searle, 1986a, 1988b), despite "high-level" enzyme polymorphism (Nevo and Shaw, 1972; Nevo et al., 1974; Wilson et al., 1974; Nevo and Cleve, 1978; Frykman et al., 1983; Searle, 1985; Britton-Davidian et al., 1989). This finding suggests that the chromosomal variants suffer a slight heterozygous disadvantage, i.e., sufficiently small to allow occasional fixation of a variant in a population and to permit formation of new karyotypic races but sufficiently large that any within-race polymorphism is short-lived and that a race characterized by a variant chromosome can form a narrow "tension zone" (Barton and Hewitt, 1985) on contact with the ancestral race or another derived karyotypic race.

SURVEY OF HYBRID ZONES

***Robertsonian Variation: Common Shrew* (Sorex araneus).** There are numerous karyotypic races over the wide Palaearctic range of the common shrew (Searle, 1984, 1988b) because the ancestral acrocentric karyotype (Wójcik and Searle, 1988; Volobouev, 1989) has been repeatedly modified by Rb fusions and possibly whole-arm reciprocal translocations (Fig. 12-2) (Searle et al., 1990). Interracial hybrid zones have been described in Britain, Sweden, Poland, and Siberia; there are some additional "contact zones" in Switzerland where races meet but are not known to hybridize (Table 12-3). The karyotypic differences between parapatric races may be large in this species, such that rings or substantial chain configurations are expected at meiosis I in hybrids (Table 12-3; Fig. 12-4). These complex heterozygous hybrids may, in general, be assumed to have lower fertility than single (or low multiple) simple heterozygotes, which are themselves probably slightly less fit than homozygotes (see Fig. 12-3, Tables 12-1 and 12-2, and associated text).

In Britain, there are three named karyotypic races of the common shrew: "Oxford," "Aberdeen," and "Hermitage" (Searle, 1984). Morphological and enzyme studies suggest no major genic differences between these races (Searle, 1985; Searle and Thorpe, 1987; Searle and Wilkinson, 1987). The Oxford and Aberdeen races make contact in central Scotland, where pure race samples have been collected within 10 km of each other along one transect (Table 12-3; Fig. 12-5A). It could indicate a narrow tension zone (Oxford-Aberdeen hybrids should form a ring VI meiotic configuration) (Table 12-3). Alternatively, the races may meet at a minor physical barrier (a river?), perhaps at the position of initial contact or (more likely) after the hybrid zone had moved there. Tension zones, whether (1) slowing down after moving over long distances (as suggested for the Oxford-Aberdeen hybrid zone (Searle and Wilkinson, 1987; Searle, 1988b, Searle et al., 1990) or (2) at formation on secondary contact, are likely to be "attracted" to a local geographical barrier or region of low density (Barton

Table 12-3. Contact Zones Between Robertsonian Races in the Common Shrew, House Mouse, and Other Species of Eutherian Mammal

| Species, Geographic Region | Races Making Contact[a] | | | | Contact at Geographical Barrier or Ecotone? |
| | First Race | | Second Race | | |
	Name	Karyotype	Name	Karyotype	
Common shrew *(Sorex araneus)*					
S. England	Oxford	*kq, no*	Hermitage	*ko, n, q*	High ground
C. Scotland	Oxford	*kq, no, pr*	Aberdeen	*ko, np, qr*	Marsh, river
C. Sweden	M	*kp, oq*	S^M	*ko, pq*	No
C. Sweden	M^N	*hi, n, p*	N	*hn, ip*	No
Switzerland	Valais	*gi, hj, kn, lo, r,m*	Vaud	*hi, jl, gm, kr, no*	Mountain passes
Switzerland	Valais	*gi, hj, kn, lo*	Acrocentric	*g,h,i,j,k,l,n,o*	Mountain passes
NE Poland	"VI"	*hq, mn, p*	"VII"	*hn, mp, q*	No
NE Poland	"II"	*hi, ko, gm, np, r*	"IV"	*hk, io, gr, mn, p*	No
Siberia	Tomsk	*hi, gk, mn, o,p*	Novosibirsk	*hn, go, ik, mp*	No
House mouse *(Mus musculus)*					
N Scotland	John o'Groats	4.10, 9.12, 6.13, 11.14	Standard	4,6,9,10, 11,12,13,14	No
Belgium	Namur	4.12, 5.10	Standard	4,5,10,12	No
Denmark	S Danish	2.5, 3.8, 6.9	Standard	2,3,5,6,8,9	No
S Germany	"II/III"	5.15, 13.14	"I"	5,13,14,15	No
N Italy	Poschiavo	8.12, 2,10	Upper Valtellina	2.8, 10.12	No
N Italy	Milano 1	2.4, 3.6, 7.8	Milano 2	2.8, 3.4, 6.7	No
N Italy	Milano 2	3.4, 2.8, 6.7, 5.15, 10.12, 9.14, 11.13, 16.17	Standard	2, 3, 4, 5, 6, 7, 8, 9, 10, 11, 12, 13, 14, 15, 16, 17	No
N Italy	Cremona	1.6, 3.4, 2.8, 5.15, 9.14, 10.12, 11.13, 7.18, 16.17	Standard	1,2,3,4,5,6, 7,8,9,10,11, 12,13,14,15, 16,17,18	No
C Italy	Cittaducale (CD)	1.7, 3.8, 6.13, 4.15, 10.11, 2.18, 5.17, 12.14, 9.16	Standard	1,2,3,4,5,6, 7,8,9,10,11, 12,13,14,15, 16,17,18	Mountain/ lowland contact
Tunisia	Monastir	4.6, 8.9, 1.11, 2.16, 3.12, 5.14, 7.18, 10.17, 13.15	Standard	1,2,3,4,5,6, 7,8,9,10,11, 12,13,14,15, 16,17,18	Urban/ rural contact

324

Tests for Genic Difference[b]	Hybrids[c] F₁	Hybrids[c] Obs	Structure of Hybrid Zone[d]	Absolute Width (km)[e] of Area Occupied by Simple Hets	Complex Hets	Chromosomal Cline Width (km)[e]	References[f]
m,e	CV	II, CIII-V	Acro peak; Cl 2, 1	⁓100	⁓10	10–20	1–5
m,e	RVI	None	Abrupt	—	—	—	2,4,6
m,e,d	RIV	RIV	Cl 4	—	>7	<10	7–10
m,E,d	CV	II, CIII-V	Acro peak; Cl 1,2?	?	?	?	7–9,11
M,E,D	CXI	None	Abrupt	—	—	—	12,13
M,E,d	4CIII	None	Abrupt	—	—	—	12,13
—	CVI	II,CIII	Acro peak; Cl 1,1,1,1	>150	<30	?	14
—	RIV+ CVI	II, RIV, CIII-VI, RIV+ CIII-VI	Recombinant peak, Acro peak; Cl 2,2,2,2	>10	⁓10	5–10	15
—	CIX	CVII, CIX	Some acros; Cl 9?	—	<20	?	16–18
—	4CIII	II, 1–2 CIII	Cl 2,1,1	>50[g]	—	10–30	19,20
—	2CIII	II, 1CIII	Cl 1,1	⁓75	–	?	21–23
E,D	3CIII	II, 1–2 CIII	Cl 1,1,1	>40	—	5–25	24
—	2CIII	II, 1–2 CIII	Cl 1,1	>40[g]	—	?	25
—	CV	II, 1–7 CIII,CV	Mottled	⁓15	⁓10	—	26
—	RVI	RVI	Unknown	—	?	?	27
e	8CIII	II, 1–3 CIII	Cl 1,7	>20	—	?	27,28
—	9CIII	1–6 CIII	Unknown	?	—	?	27
M,e	9CIII	II, 1–4 CIII	Cl 9 (mottled in parts?)	⁓10	—	⁓1	29–31
M,E	9CIII	1–9 CIII	Mosaic?	>20	—	⁓1	32,33

Table 12-3. Contact Zones Between Robertsonian Races in the Common Shrew, House Mouse, and Other Species of Eutherian Mammal (*Continued*)

Species, Geographic Region	Races Making Contact[a]				Contact at Geographical Barrier or Ecotone?
	First Race		Second Race		
	Name	Karyotype	Name	Karyotype	
Mole rat *(Spalax ehrenbergi)*					
N Israel	2n = 58	"B4" metacent	2n = 60	"B4" acrocent	No; river[h]
Spiny Mouse *(Acomys cahirinus)*					
E Sinai	2n = 36	1 metacent	2n = 38	2 acrocent	No
Gerbil *(Gerbillus pyramidum)*					
Israel	2n = 50/2	7 metacent	2n = 64/6	14 acrocent	Town
Mole-vole *(Ellobius tancrei)*					
Tadzhikstan	2n = 32	11 metacent	2n = 54	22 acrocent	No
Wild sheep *(Ovis ammon)*					
N Iran	2n = 54	2 metacent	2n = 58	4 acrocent	No

[a]The distinction between first and second race is arbitrary. Normally, the names for the races follow those in the publications cited. The $2n = 40$ karyotypic race of the house mouse is called the standard race. Only those chromosomes that differ between the hybridizing races are included in this table. For common shrews, chromosome kq is a metacentric composed of chromosome arms k and q, and chromosome n is an acrocentric. For house mice, chromosome 4.10 is a metacentric composed of the ancestral acrocentrics 4 and 10.

[b]M/m = morphological difference; E/e = difference at enzyme loci; D/d = difference in mt DNA sequence looked for and found/not found, respectively.

[c]II = homozygous karyotype(s) different from those of the pure races; CIII = single or multiple simple heterozygotes; nCIII = simple heterozygotes for n rearrangements; CIV, CV etc. = chain-forming complex heterozygotes; RIV, RVI = ring-forming complex heterozygotes (see Figure 12-3). With regard to "degree of hybrid formation" (see Table 12-4): where hybrids were observed, hybrid production may reasonably be assumed to be "extensive," although in some cases data are limited or only one type of hybrid can be produced in the zone. There is no evidence of assortative mating.

[d]Cl 4 = four coincident metacentric chromosome clines; Cl 1,2 = two metacentric chromosome clines coincident but staggered from a third metacentric chromosome cline within the hybrid zone; Acro peak = A high frequency of the chromosomes that differ among the races in an acrocentric state; Recombinant peak = high frequency of homozygotes within the hybrid zone whose karyotypes include metacentrics from both races; Abrupt = races known to occur within 10 km of each other but little or no intermixing and no known hybridization (thus some of these may prove to be separate species); Mottled = "mottled" hybrid zone with irregular distribution of populations characterized by different homozygous karyotypes; Mosaic = "mosaic" hybrid zone (Harrison and Rand, 1989).

and Hewitt, 1981), thereby, in the case of a sizable river, for example, effectively ceasing to exist as a zone of hybridization in that region. For eutherians, chromosomal races commonly meet at local geographic barriers, such as rivers (Tables 12-3 and 12-4).

The Hermitage race forms a 300 km long hybrid zone with the Oxford race between London and Liverpool; particularly detailed data are available from the vicinity of Oxford (Searle, 1986a). In this zone the frequency clines for the Oxford race metacentrics kq and no coincide with each other but are slightly staggered relative to the cline for the Hermitage race metacentric ko (Table 12-3; Fig. 12-5B,C). In addition to the kq and no Oxford race chromosomes and ko, n, and q of the Hermitage race, acrocentrics k and o are found in the hybrid zone. Interracial hybrids with metacentrics ko and kq (and/or no) do occur but rarely; the hybrid zone is dominated by the

| Tests for Genic Difference[b] | Hybrids[c] | | Structure of Hybrid Zone[d] | Absolute Width (km)[e] of Area Occupied by | | Chromosomal Cline Width (km)[e] | References[f] |
	F₁	Obs		Simple Hets	Complex Hets		
e	1CIII	1CIII	Cl 1	~3	—	?	34–36
e	1CIII	1CIII	Cl 1	~16	—	?	37,38
e	7CIII	1–2CIII[i]	?	>40	—	?	39,40
m,e	11CIII	II,CIII	Mottled	?	—	—	41
E	2CIII	1CIII[i]	?	>150	—	?	42,43

[e]Only for chromosomal cline widths is the standardized measure of Endler (1977) used; for distribution of heterozygotes, the absolute width of the area of their occurrence is recorded. ? = unknown; — = not relevant.

[f]Key to references: (1) S. J. Mercer and J. B. Searle, pers. obs.; (2) Searle, 1985; (3) Searle, 1986a; (4) Searle and Thorpe, 1987; (5) Searle, 1990; (6) Mercer and Searle, 1991; (7) Fredga, 1987; (8) K. Fredga and H. Tegelström, pers. obs.; (9) Sulkava et al., 1985; (10) Frykman et al., 1983; (11) Frykman and Bengtsson, 1984; (12) Hausser et al., 1991; (13) Taberlet et al., 1991; (14) Fedyk, 1986; (15) Fedyk et al., 1991; (16) Volobouev, 1983; (17) Aniskin and Lukianova, 1989; (18) V. T. Volobouev, pers. obs.; (19) Scriven and Brooker, 1990; (20) Searle, 1991; (21) Hübner, 1988; (22) Hübner and Koulischer, 1990; (23) Bauchau et al., 1990; (24) Nance et al., 1990; (25) Adolph and Klein, 1983; (26) Hauffe and Searle, in press; (27) Gropp et al., 1982b; (28) Britton-Davidian et al., 1989; (29) Spirito et al., 1980; (30) Corti et al., 1990; (31) Corti and Thorpe, 1989; (32) Said and Britton-Davidian, 1991; (33) Said et al., in press; (34) Nevo and Bar-El, 1976; (35) Nevo and Shaw, 1972; (36) Nevo and Cleve, 1978; (37) Wahrman and Goitein, 1972; (38) Nevo, 1985; (39) Wahrman and Gourevitz, 1973; (40) Nevo, 1982; (41) Lyapunova et al., 1980; (42) Nadler et al., 1971; (43) Valdez et al., 1978.

[g]In these cases one small race is surrounded by another much larger race, and polymorphism occurs throughout the smaller race. The distance given is the maximum radius of the area of polymorphism from the center of the local race.

[h]Information on contact from more than one place.

[i]Few reliable data.

acrocentrics *k, n, o, q*. In the nomenclature adopted here, this contact is a *staggered hybrid zone* (i.e., a zone comprising distinctly noncoincident chromosomal clines) with an *acrocentric peak*. In essence, instead of the expected "single" hybrid zone dominated by low-fertility complex heterozygous F₁-type hybrids (Table 12-3), there are two closely located "hybrid zones" each dominated by moderately high-fertility simple heterozygotes. The staggered structure and associated acrocentric peak can be explained in terms of selection in favor of individuals that have acrocentrics in their karyotypes. They produce fewer lower-fertility complex heterozygotes than individuals with metacentrics in their karytoypes (Searle, 1986a).

The Oxford-Hermitage hybrid zone is localized on high ground in the Oxford area. If, as is likely, the density of common shrews decreases with altitude, the high ground may have attracted the Oxford-Hermitage "tension" zone and could still influence the width of the individual clines of chromosomal polymorphism (Hatfield et al.,

Table 12-4. Contact Zones in Various Eutherian Mammals Among Races that Differ by Non-Robertsonian Rearrangements

Species or Species Complexes	Races, Subspecies, or "Species" Making Contact	Chromosomal Differences Between Races[a]	Contact at Geographic Barrier or Ecotone?	Tests for Genic Difference[b]	Degree of Hybrid Formation[c]	Absolute Width (km)[d] of Area with Hybrids	Chromosomal Cline Width (km)[d]	References[e]
Tent-making bat (Uroderma bilobatum)	b. convexum/b. davisii	2n = 38/44; 2 tan,1 Rb	No	M,e	Extensive	~400	?	1–4
White-footed mouse (Peromyscus leucopus)	northeastern/ southwestern	Both 2n = 48; 3 pii	Vegetation type; river[f]	E,D	Extensive	>250	30	5–7
Mole rat (Spalax ehrenbergi)	2n = 52/54	1 pii,1 Rb	River	E	?[g]	—	—	8–11
	2n = 54/58	3 pii,2 Rb	No	E	Almost all F₁'s	0.7	?	8–11
Pocket gopher Geomys bursarius species complex	2n = 52/58 attwateri F	4 pii,3 Rb	No; cliffs[f]	e	Extensive	0.3	?	8–11
	Both 2n = 70; 1 attwateri G	pai,1 hc	No	m	Extensive	~85	~35	12,13

attwateri/ breviceps	$2n = 70/74$; HC,Rb,?	No; river[f]	m	F₁'s only	1.5	0.8	12,13
bursarius/ breviceps	$2n = 70\text{-}2/74$; HC,?	Soil type; river[f]	M,E	Limited (F₁,Fₙ)	—	—	14,15
bursarius/ lutescens	Both $2n = 70\text{-}2$; HC	Soil type	M,E	Stable hybrid population	~2	1	16
Thomomys bottae species complex							
bottae/ umbrinus	$2n = 76/78$; HC,?	Soil type	M,E	Largely F₁'s	~4	?	17,18
b. actuosus/b. ruidosae	Both $2n = 76$; HC	Soil type	M,E	Extensive	>3	1	19

[a]tan = tandem fusion/fission; Rb = Robertsonian fusion/fission; pii = pericentric inversion; pai = paracentric inversion; hc = heterochromatic arm present in one race and absent in the other; HC = substantial differences in heterochromatin content and distribution; ? = further chromosomal difference, the exact nature of which is not established.

[b]See footnote b in Table 12-3.

[c]extensive = a variety of first and higher generation hybrids found in hybrid zone; limited = only occasionally are hybrids produced; Fₙ = higher generation hybrid or backcross. Note that some of these interacting forms may best be considered separate species. Assortative mating has been demonstrated in laboratory studies of *S. ehrenbergi* (see text) and may occur at the *T. bottae/umbrinus* contact (Patton, 1973).

[d]See footnote e in Table 12-3.

[e]Key to references: (1) Baker, 1981; (2) Greenbaum, 1981; (3) Baker et al., 1972, 1975; (4) Hafner, 1982; (5) Baker et al., 1983b; (6) Stangl, 1986; (7) Nelson et al., 1987; (8) Wahrman et al., 1969; (9) Nevo and Bar-El, 1976; (10) Nevo and Shaw, 1972; (11) Nevo and Cleve, 1978; (12) Tucker and Schmidly, 1981; (13) Dowler, 1989; (14) Bohlin and Zimmerman, 1982; (15) Cothran and Zimmerman, 1985; (16) Heaney and Timm, 1985; (17) Patton, 1973; (18) Patton et al., 1972; (19) Patton et al., 1979.

[f]Information on contact from more than one place.

[g]Hybrids were reported by Wahrman et al. (1969), but subsequent reports implied that hybridization does not occur because the river acts as a barrier between the races.

329

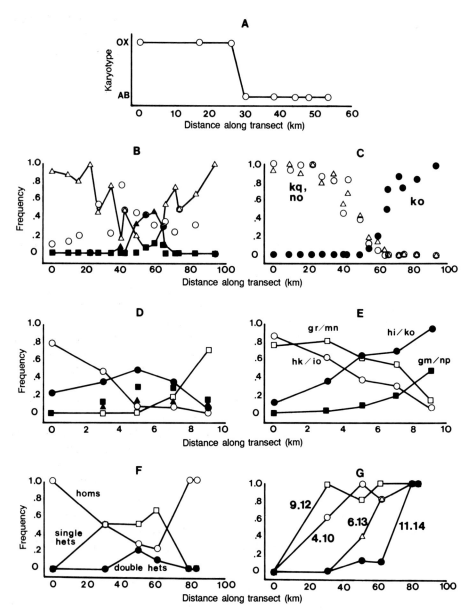

Figure 12-5. Transects through representative contact zones between karyotypic races of (**a–e**) common shrew and (**f, g**) house mouse (see Table 12-3 and text). (**a**) The "abrupt" contact zone between the Oxford and Aberdeen races in central Scotland in which no hybrids were detected, perhaps because a river formed a geographical barrier. (**b, c**) "Staggered hybrid zone" between the Oxford *(kq, no)* and Hermitage *(ko, n, q)* races in southern England, with an "acrocentric peak" (see Table 12-3). (**b**) Frequencies of acrocentric homozygotes *k, n, o, q* (closed triangles), other homozygotes (open triangles), simple heterozygotes (open circles), and complex heterozygotes (closed squares). (**c**) Frequencies of metacentric chromosomes *no* (triangles), *kq,* and *ko.* (**d,e**) "Staggered" hybrid zone between races II *(hi, ko, gm, np, r)* and IV *(hk, io, gr, mn, p)* in Poland, with a "recombinant peak" and an "acrocentric peak." (**d**) Frequencies of race IV homo-

1992). Both dispersal and the selective disadvantage associated with Robertsonian heterozygosity go to determine the width of these clines. The fact that they are some 10–20 km wide, despite the low vagility of common shrews (~ 100 meters per generation) (J. B. Searle and A. J. Reilly, pers. obs.) is consistent with the small unfitness of single, simple Robertsonian heterozygotes relative to homozygotes (Table 12-1 and associated text).

One unresolved problem is the origin of the acrocentrics k and o within the Oxford-Hermitage hybrid zone. There are three clear alternative explanations (see Searle, 1986a): (1) The acrocentrics could have already been present at the time of racial contact if the race-specific metacentrics spread through the ancestral acrocentric race in "stasipatric" fashion (White, 1978a,b). (2) There could have been chance infiltration of the hybrid zone by members of the ancestral acrocentric race. (3) There could have been Robertsonian fission (Fig. 12-2) of race-specific metacentrics generating the acrocentrics.

Despite uncertainty as to how its present structure arose, the Oxford-Hermitage zone is a clear demonstration of selection acting on different karyotypes (Bengtsson and Frykman, 1990). The acrocentric peak is maintained because of selection against the low-fertility complex heterozygotes. Because of the acrocentric peak, gene flow across the zone is presumably little impeded; a high frequency of complex heterozygotes would have created a much stronger genetic barrier.

Other hybrid zones for the common shrew have characteristics similar to the Oxford-Hermitage contact. In the five zones where F_1 hybrids are expected to form meiotic chain IV, V, or VI configurations (Table 12-3), a staggered clinal structure and acrocentric peak are found (including preliminary results from the Aberdeen-Hermitage zone) (Searle, 1988b). Thus the field data strongly endorse the notion that selection operates to minimize the occurrence of these types of hybrid.

The hybrid zone between races "II" and "IV" in Poland is one of those that has an acrocentric peak, but it also shows other remarkable features that are likewise consistent with selection minimizing low-fertility hybrids. For one of a variety of possible reasons, this zone and associated chromosomal clines are rather narrow in comparison with the Oxford-Hermitage contact (Fig. 12-5). The F_1 hybrids are complex heterozygotes expected to form both a ring IV and a chain VI configuration at prophase I of meiosis. Such hybrids are rare, first, because an acrocentric peak diminishes the occurrence of chain VI-forming complex heterozygotes, and, second, because of a high fre-

---

zygotes and simple heterozygotes (open circles), race II homozygotes and simple heterozygotes (open squares), recombinant homozygotes and simple heterozygotes *hi, ko, gr, mn* (closed squares), complex heterozygotes with a ring IV meiotic configuration *(ok/kh/hi/io)* but no chain configuration (closed circles) and complex heterozygotes with a chain IV–chain VI configuration (maximally *r/rg/gm/mn/np/p)* (closed triangles). (**e**) Frequencies of race-specific metacentrics (averaged between pairs). Note that the clines for the chromosomes involved in the meiotic ring IV configuration intersect at a frequency of 0.5, and that the intersection for those chromosomes involved in the meiotic chain VI configuration is at a frequency of 0.3–0.4 (due to the acrocentric peak) and is staggered by 3–4 km (due to the recombinant peak). (**f,g**) Hybrid zone between the John o'Groats ($2n = 32$: 4.10, 9.12, 6.13, 11.14) and standard ($2n = 40$) races in northern Scotland.

quency of homozygous recombinant individuals with metacentrics that derive from both races (*hi* and *ko* of race "II" and *gr* and *mn* of race "IV"). As a result of the independent segregation of the ring and chain configuration, two types of recombinant could be produced by F_1 hybrids. Selection, however, would be expected to favor only one type (presumably, whichever happened to be most common at the initial stages of the evolution of this zone). Thus considering the recombinant homozygotes sandwiched between the two races, contact between the recombinants and race "IV" should generate ring IV-forming individuals, and contact between the recombinants and race "II" individuals should generate chain VI-forming individuals; as already mentioned, though, the latter are rare because of the presence of an acrocentric peak.

An interesting feature of the hybrid zone between races "II" and "IV" in Poland and that between races M and S^M in Sweden is that ring IV-forming complex heterozygotes are common (Fig. 12-5D; Table 12-3); in neither case is there an acrocentric peak that would reduce their frequency. This type of complex heterozygote may suffer little fertility reduction (Fedyk et al., 1991), and therefore an acrocentric peak would not be selected for. The generation of such a peak may also be problematical in this case. In a zone where chain-forming complex heterozygotes are produced, it is expected that a fission is favored because it allows the production of more fertile heterozygotes with shorter chains. In a zone where ring IV-forming complex heterozygotes are produced, fission results in chain V-forming complex heterozygotes, which are probably less fertile. Complex heterozygotes that form meiotic chains suffer greater germ cell death than those with rings (Table 12-2 and associated text). Two specific fissions in close succession in one population would be necessary to generate an acrocentric peak and simple (chain III-forming) heterozygotes; this situation is perhaps unlikely to occur.

The contact zones in Siberia and Switzerland remain to be described (Table 12-3). The details of the Siberian zone are scanty, but complex heterozygous hybrids with a long meiotic chain configuration (nine elements) are expected and observed with no evidence of staggering of chromosomal clines (Table 12-3). This zone is interesting, but more information is required to contrast it properly with other interracial contacts among shrews. Shrew karyotypic races that differ so markedly in chromosome constitution are likely to belong to distinct lineages that have been geographically isolated in the past (Searle, 1984) and may have genic as well as chromosomal differences. Such genic differences are observed for two contact zones in Switzerland (Table 12-3) where no hybridization has been found and the forms making contact are probably best considered true species (Hausser et al., 1991). In one case there is a substantial karyotypic difference between the races, and in the other there is little difference; there is scant indication as to the contribution of the chromosomal difference to genetic isolation. The contact zones in Switzerland are located in areas of low shrew density; the same rules govern localization of a contact zone between sibling species with similar ecological requirements as for tension zones.

***Robertsonian Variation: House Mouse* (Mus musculus).** The house mouse is subdivided into two subspecies within Europe: *Mus musculus domesticus* to the west and *M. m. musculus* to the east. Both have a basic karyotype of 40 acrocentric chromosomes; but within *M. m. domesticus* numerous karyotypic races of limited distribution have been described, particularly in the Alps and around the Mediterranean basin

(Gropp and Winking, 1981; Capanna, 1982; Corti et al., 1986; Winking et al., 1988). The races are characterized by a complement of metacentrics formed by Rb fusion (Fig. 12-2) and sometimes probably modified by whole-arm reciprocal translocations (Winking, 1986; Searle, 1991). Thus laboratory hybrids between metacentric races of the house mouse often form large meiotic rings or chains (Table 12-2). Although many of these races are apparently parapatric, there are remarkably few data on contacts between metacentric races. Thus a ring VI-forming complex heterozygote (Fig. 12-3) was found at the contact between the Milano 1 and 2 races (Table 12-3), and chain IV-forming hybrids have been found in southern Switzerland (R. Hübner, pers. obs.). The only well-studied contact between metacentric races is that between the Upper Valtellina and Poschiavo races in northern Italy. Early studies based on only a few sites suggested distinct morphological, behavioral, and enzyme differences between the races and an absence of hybrids (Capanna and Corti, 1982; Thorpe, et al., 1982; Capanna et al., 1984, 1985), but an extensive study of the contact area indicated that the expected chain V-forming hybrids and an array of other karyotypes are found (Table 12-3), including the 40-chromosome standard race (Hauffe and Searle, in press). Populations consisting either of one of the pure races or a mixture of races are dotted along a 20-km stretch of valley in an irregular fashion. The patchy distribution of the races and hybrids within the contact area is reminiscent of the "mosaic" hybrid zone for the field cricket *Gryllus* (Harrison and Rand, 1989). However, although the patchiness of *Gryllus* can be related to a patchy distribution of habitats and differences in the ecological requirement of the hybridizing races, relevant ecological heterogeneity in this small region of the Upper Valtellina Valley is not obvious. Although little is known about the population dynamics of mice in this region, patchiness could evolve as a consequence of the following three common, but not universal, traits of commensal mouse populations (Berry, 1981): (1) little movement between neighboring populations, which, given heterozygote disadvantage, tends to lead to high frequencies of the most common homozygous karyotype within the various isolated populations; (2) rather frequent extinction of populations, enhancing the possibity of (3) occasional human-mediated long-distance colonization (e.g., to explain the unexpected occurrence of standard race mice in the region, perhaps brought with agricultural produce).

It may be appropriate to call the Upper Valtellina-Poschiavo contact a "mottled" hybrid zone to distinguish it from the "mosaic" hybrid zone of Harrison and Rand. This hybrid zone is unusual because three races occur in a small geographical area, with all three types of F_1 hybrid presumably highly unfit. However, other hybrid zones for the house mouse are also probably mottled as a result of inappropriate location of hybrid populations; that is, there may not be the expected sequence of populations across a hybrid zone from those more similar to one race to those more similar to others. Corti et al. (1990) described a likely example along one transect across the Cittaducale (CD)-standard race hybrid zone in central Italy.

A true "mosaic" hybrid zone may occur in Tunisia, where there is both a 22-chromosome highly metacentric form (the "Monastir" race) and the 2n = 40 all-acrocentric race. The distribution of the Monastir race is patchy and limited to urban centers, perhaps because of an adaptive advantage in that habitat (Said and Britton-Davidian, 1991). Although both F_1 and higher generation/backcross hybrids have been found, the transitions between the 22-chromosome and 40-chromosome populations are narrow ($\sim$1 km) and there is probably a strong genetic barrier between the races. F_1 hybrids have small litters (Said et al., in press), although the extent to which it is a result of karyotypic heterozygosity is unknown (Table 12-1). The Monastir race is genically distinct and shows low genic heterozygosity; these features may provide clues as to its mode of formation (Said and Britton-Davidian, 1991).

Further information is desirable to establish whether the Monastir–standard race contact is indeed a "mosaic" hybrid zone. For example, the pattern observed could also be explained assuming slow displacement of the Monastir race by the standard race, with urban habitat holding higher densities of house mice (any race) than rural areas.

All the remaining hybrid zones in the house mouse to be described are contacts between metacentric races with a continuous (though often small) distribution and widespread acrocentric races (usually the 2n = 40 standard race). As for the common shrew, there are examples of staggered hybrid zones that usually involve races with a rather small difference in karyotype, and cases of coincident or virtually coincident clines involving races with a large difference in karyotype. The transition from the CD race to standard race occurs over only 10 km along the transect collected by Spirito et al. (1980). The individual clines are narrow (Table 12-3), as may be expected when hybrids are unfit (Szymura and Barton, 1986). The clines appear to be exactly coincident, though coincidence is not easy to judge because of the small sample sizes and some indication of 'mottling'. In the Milano 2-standard race zone (Table 12-3), seven clines must run close to each other, whereas an eighth cline (metacentric 16.17: chromosome nomenclature in Table 12-3) is staggered by about 20 km. Over this area, individuals homozygous for metacentric 16.17 predominate. As a generality, staggered hybrid zones for the house mouse result in zonal populations dominated by individuals that are homozygous for some, but not all, of the metacentrics that characterize the metacentric race (i.e., individuals with a homozygous karyotype with intermediate features between the hybridizing races). In contrast to the common shrew, neither "complex heterozygotes" nor "acrocentric peaks" are expected or observed in these zones.

One of the clearest examples of a staggered hybrid zone in the house mouse is at the contact between the John o'Groats race (2n = 32: four pairs of metacentrics) and the standard race in northern Scotland (Fig. 12-5F,G; Table 12-3). The clines for metacentrics 4.10 and 9.12 may or may not be coincident, whereas these clines are themselves staggered from the 6.13 cline, which is also staggered from the 11.14 cline.

Other staggered hybrid zones occur in central Europe (Table 12-3). In Denmark a metacentric race occurs within *Mus musculus domesticus* close to the contact with *M. m. musculus.* The three metacentric chromosome clines lie on the *domesticus* side of the zone staggered both from each other and from the set of coincident clines for mtDNA, Y chromosome DNA, and enzyme variants, which mark the *musculus–domesticus* zone.

In the case of hybrid zones in Belgium and southern Germany, a small metacentric race is surrounded by a much more widespread acrocentric race. Once again there is staggering of chromosome clines, but in both cases there is a degree of asymmetry in this staggering. In Belgium the small metacentric race is characterized by chromosome 4.12 and 5.10; the clines for these metacentrics virtually coincide to the northwest, but metacentric 4.12 extends further to the east, south, and southwest than metacentric 5.10. In Germany the race "II/III" is characterized by metacentrics 5.15 and 13.14, but metacentric 13.14 is more widespread to the east and 5.15 more widespread to the west.

The frequent occurrence of staggered hybrid zones for the house mouse is of great interest and needs to be explained. This noncoincidence of character clines is unusual in hybrid zones (Barton and Hewitt, 1985), and the explanations involving complex heterozygotes used for the common shrew do not apply to the house mouse. I believe that a variety of factors may be important in the origin of staggered hybrid zones for the house mouse, and that different combinations of features may apply in different zones. These factors are as follows.

1. *Simple heterozygosity for one or a few metacentrics causes only a small fertility reduction in house mice* (Table 12-1). If hybrids are unfit, there is a tendency for those clines already coincident in a hybrid zone to remain together and for nearby clines to be "attracted" to the zone (Barton and Hewitt, 1985). This situation occurs because of the greater dispersal from the "pure" race populations into the zone than vice versa. In the case of a hybrid zone in the house mouse comprising a small number of chromosomal clines, such hybrids are never particularly unfit, so there is no strong "force" attracting together originally staggered clines or inhibiting originally coincident clines from separating. Contacts of this kind provide the best examples of staggered hybrid zones in the house mouse.

2. *The different chromosomal characters involved within a staggered hybrid zone are unlinked.* Clines for closely linked characters associate particularly readily in a tension zone (Barton and Hewitt, 1981). Thus lack of linkage is thought to contribute to the frequent noncoincidence of nuclear and mitochondrial character clines in animal hybrid zones (Barton and Jones, 1983; Harrison, 1990) and likewise, presumably, for the staggered metacentric clines for the house mouse.

3. *The distribution of the house mouse has been labile in recent times.* The species is thought to have entered Europe only 8000 years ago (Capanna, 1982); and as a commensal of humans, local distributions have been at least as fluid as those of our own species. Changes in distribution may cause hybrid zones to move (Barton and Hewitt, 1985). Under these circumstances coincident clines are particularly likely to become separated, albeit in an irregular fashion, if as argued in (1) and (2) above, the selective forces holding the clines together are not strong. For example, by chance, one of a set of moving clines could be held up for a longer time than the others by a minor geographical barrier in their path. Thus of an original set of coincident clines (e.g., formed as a result of secondary contact), some may remain coincident and others become staggered (as possibly occurs in the Scottish hybrid zone), and the degree of staggering may vary along the hybrid zone (as apparently occurs in the Belgian hybrid zone).

4. *Many of the metacentric races in the house mouse have a limited distribution and are surrounded by a much more substantial acrocentric race.* In this case, asymmetry of gene flow (or, more relevantly, "chromosome flow") across the zone may, in a fashion similar to that for zone movement, acts to separate originally coincident clines.

5. *House mice that are multiple simple Robertsonian heterozygotes may be more unfit than expected.* The staggering of chromosomal clines and formation of an acrocentric peak in the common shrew is thought to have originated because there is not a simple additive relation between the unfitness of simple and complex heterozygotes (Hatfield et al., 1992). Likewise, in the house mouse, if the unfitness of multiple simple heterozygotes is greater than the additive expectation based on single simple heterozygotes, it may cause originally coincident metacentric clines to separate (see Barton and Bengtsson, 1986). In the case of the Danish hybrid zone, the noncoincidence of chromosomal and genic characters may occur because the combination of *musculus–domesticus* genic hybridity and chromosomal heterozygosity causes greater unfitness than that expected from a combination of the unfitnesses associated with the characters in isolation. This theory fits with the evidence described above that meiotic aberration increases with genic hybridity in the house mouse. Further considerations are required to explain the data of Spirito et al. (1980) for the CD–standard race hybrid zone (Table 12-3) on this model. If there is a simple nonadditive relation between fitness of heterozygotes—such that the unfitness of double heterozygotes is greater than expected from that of single heterozygotes, and the unfitness of triple heterozygotes is greater than that expected of double heterozygotes, for example—it could be anticipated that the nine clines in the CD–standard race zone should separate. However, if it was a secondary contact and if the fitness of the highly heterozygous individuals at the center of the hybrid zone is low owing to meiotic aberrations, the "forces" keeping the clines together [see (1) above] may outweigh the tendency for the clines to separate (see Barton and Bengtsson, 1986).

6. *The metacentric races of the house mouse are thought to have originated since the colonization of Europe, i.e., within the last few thousand years* (Capanna, 1982; Searle et al., 1990). The structure of the chromosomal hybrid zones may reflect this recent race formation; and, in particular, the staggering of chromosomal clines could be a consequence of the pattern of accumulation of new metacentrics. It is often those metacentrics that in a regional sense are most widespread (and therefore, presumably, were the first formed) that have the widest distribution in the hybrid zone (16.17 in northern Italy, 4.10 and 9.12 in Scotland, 4.12 in Belgium). It may indicate that, even within a metacentric race of limited distribution, only a small part of that race is likely to become colonized by a new metacentric.

In this account, I have presented the different types of hybrid zone for the house mouse (staggered, "narrow coincident," mottled, mosaic) as separate alternatives. They could also be considered as extremes of a multidimensional continuum. Clearly, particular hybrid zones can demonstrate more than one of these features, either across the zone (e.g., Milano 2–standard: staggered, narrow coincident) or along it (e.g., CD–standard: narrow coincident, apparently mottled). Much more detailed surveys are needed to untangle these complexities.

Robertsonian Variation: Other Species. Simple Rb heterozygotes are found in hybrid zones of the mole-rat, spiny mouse, gerbil, and wild sheep (Table 12-3). The differences in width of these zones partly reflects vagility (compare the wide zone for the sheep and the narrow zone for the mole-rat); the extent to which heterozygote unfitness differs between these species is unknown. The hybrid zone for the mole-vole appears to have a mottled structure; the pure races and populations of intermediate homozygotes form a patchwork, and few definite heterozygotes have been described.

Whole-Arm Rearrangements: Tent-Making Bat (Uroderma bilobatum).

The two races of tent-making bat differ by one Rb rearrangement and two tandem rearrangements (Fig. 12-2; Table 12-4). Heterozygotes for tandem rearrangements should have a more substantially reduced fertility than simple Rb heterozygotes. In particular, heterozygotes for chromosome pair C are characterized by three acrocentrics (Baker, 1981) and are therefore expected to form 50% unbalanced gametes (see above). Although karyotype frequencies are compatible with this area being a tension zone (Barton, 1982), little is known about individual chromosome clines or about the relative contribution of chromosomal and genic heterozygosity to hybrid unfitness (there is evidence that the two races differ in morphology). The large area of occurrence of interracial hybrids ($\sim$ 400 km) is not surprising, even if there is substantial unfitness of hybrids, given the opportunity for long-range movements for this species.

Pericentric Inversions: White-Footed Mouse (Peromyscus leucopus).

The genus *Peromyscus* is well known for a high incidence of pericentric inversion (Robbins and Baker, 1981); Stangl (1986) described a hybrid zone between two races of *P. leucopus* that differs by three such inversions (Table 12-4). In a related species, inversion heterozygotes have near-normal fertility (Hale, 1986), and therefore *P. leucopus* is not expected to suffer substantial hybrid unfitness on chromosomal grounds. Indeed, the chromosome clines are wide (30 km), given the low vagility of this small mammal; and there is substantial introgression of variants (especially chromosome pairs 5 and 20). The three chromosome clines are coincident and lie close to an ecotone. Clines for enzyme and mtDNA variants lie at the same position, and there are some indications that the hybrid zone lies at the boundary of morphologically distinct subspecies (Stangl, 1986; but see Baker et al., 1983b). Further data are required: (1) to establish whether this area is a tension zone or another sort of contact between races adapted to different environments (Barton and Hewitt, 1985; Harrison, 1990); and (2) to explain the characteristics of the inversion clines.

Subterranean Rodents: Mole-Rat (Spalax ehrenbergi) and Pocket Gophers Geomys and Thomomys.

The mole-rat and pocket gophers are considered together, partly because their interracial chromosomal differences are rather heterogeneous and ill-defined and partly because of the distinct similarities in the features of hybrid zones among these subterranean rodents (Table 12-4) (Thaeler, 1974, 1985; see Ch. 11). The *Geomys bursarius, Thomomys bottae,* and *T. talpoides* species complexes are North American taxa, whereas *Spalax ehrenbergi* is found in the Near East and has been studied in particular detail in Israel. Most of the cytogenetic analysis of these species has been based on conventional chromosome preparations. Even when G-band and/or C-band preparations are available, the exact nature of the chromosomal differences between the hybridizing races is not always clear because of the small size of some of the variable chromosomes, e.g., those that are supposed to have undergone pericentric inversions in *Spalax ehrenbergi* (Wahrman et al., 1985). In no case among these subterranean forms is there detailed information on the degree of meiotic aberration associated with karyotypic heterozygosity.

Despite these reservations on cytogenetic interpretation, various generalizations may be made. The karyotypic races of *S. ehrenbergi* differ by Rb rearrangements involving some of the larger chromosomes and within-chromosome changes of some of the smallest elements. In the *T. bottae* complex, hybridizing forms usually differ substantially in heterochromatin content and distribution, with one race often having heterochromatic short arms not present in the other race. In the *G. bursarius* complex there may also be substantial differences in blocks of heterochromatin between races, but also other (e.g., Rb) rearrangements. In the *T. talpoides* complex, karyotypic differences predominantly result from Rb rearrangements and pericentric inversions (Thaeler, 1985).

Among these species, generally the hybridizing races show genic differentiation (Table 12-4). Thus in the *G. bursarius* and the *T. bottae* complexes, there are often differences in morphology, enzyme genotype, and soil preference. For *S. ehrenbergi* there is no obvious difference in gross morphology between the races, but all races show a degree of assortative mating (Nevo, 1991); and the 54-chromosome race, in particular, is found to be genetically distinct (Nevo and Shaw, 1972; Nevo and Cleve, 1978; Catzeflis et al., 1989). Numerous other genic, behavioral, morphological, physiological, and ecological differences have been found between widely separated samples of the four chromosomal races of *S. ehrenbergi* (for review see Nevo, 1991), but it is not clear if these features are truly race-specific (some clearly are not).

Among the hybrid zones described in Table 12-4, hybridization is rare in one *(G. bursarius/breviceps)* and largely limited to the production of F_1's in others (*G. attwateri/breviceps, T. bottae/umbrinus, S. ehrenbergi:* 2n = 54/58). Hybrid male sterility has been demonstrated in the *T. bottae/umbrinus* hybrid zone (Patton, 1973). Even when there is more extensive hybridization, the hybrid zones are narrow (<5 km absolute width; but see below for one exception) indicating strong selection against hybrids (given a vagility of 100 meters) (Hafner et al., 1983). Taking these narrow hybrid zones together, in some cases it is possible that chromosomal heterozygosity contributes substantially to hybrid unfitness (see discussion above on narrow coincident clines in the house mouse). However, given that there are genic differences between hybridizing races in all these zones and that the major differences between races are often heterochromatic changes, the chromosomes may more usually represent neutral or weakly selected markers at a strong genetic barrier (cf. enzyme variants in the hybrid zone of the toad *Bombina*) (Szymura and Barton, 1986). Although in most of the zones the hybrids are likely to be unfit relative to "pure" race individuals, at the contact between *G. bursarius* and *G. lutescens* the hybrids may be relatively fitter in the intermediate soil type.

In the *G. bursarius* species complex a wide hybrid zone (85 km absolute width) is observed between races F and G *(G. attwateri),* which differ only slightly in karyotype (Table 12-4). Genic differences have not been found, and the width of the zone may reflect merely a small unfitness associated with karyotypic heterozygosity. Results similar to those described for *G. bursarius* and *T. bottae* have been obtained for the *T. talpoides* species complex, which includes 38 karyotypic forms. Relatively broad hybrid zones (4–40 km absolute width) tend to involve morphologically identical karyotypic forms, whereas those contacts where there is little or no hybridization tend to be between forms that are both karyotypically and morphologically distinct (Thaeler, 1985).

CHROMOSOMAL HYBRID ZONES, RACIATION, AND SPECIATION

Staggered Hybrid Zones and Raciation. An interesting feature of some eutherian chromosomal hybrid zones is the occurrence of zonal populations dominated by individuals with a homozygous karyotype different from both of the hybridizing races. This situation has been demonstrated in the staggered and mottled hybrid zones of the house mouse and common shrew, between races that differ by Rb fusions (and possibly whole-arm reciprocal translocations). Through range changes, these small populations could come to occupy much larger areas and thereby become new widespread races. Such races could have (1) race-specific metacentrics from both races, (2) some but not all the metacentrics that characterize one race, and (3) acrocentrics in place of the race-specific metacentrics of both races. Races of type (1) could derive from a "recombinant peak" as observed at the contact between common shrew races "II" and "IV" in Poland; this process is similar to the "recombinational speciation" mechanism proposed for higher plants (Grant, 1981; for eutherians see Searle, 1984; Corti et al., 1986). Races of type (2) could derive from populations with high frequencies of intermediate homozygous karyotypes as observed in the John o'Groats race–standard race hybrid zone in the house mouse; and races of type (3) could derive from an "acrocentric peak," such as that between the Oxford and Hermitage races of the common shrew (Table 12-3 and associated text).

Searle et al. (1990) proposed that the numerous karyotypic races of common shrew in southern Britain (Fig. 12-6) may ultimately derive from populations formed in a hybrid zone between the Oxford and Aberdeen races (characterized by metacentrics *kq, no, pr* and *ko, np, qr,* respectively). These races are in direct contact in Scotland (Table 12-3) and may originally have been so in southern Britain. Once acrocentrics *k, n, o, p, q, r* arose in the zone, the metacentric clines could have become staggered as a result of selection against the complex heterozygous hybrids. The current racial distribution could then have been generated if the metacentric clines separated substantially in response to large population movements at the end of the last glaciation (Searle et al., 1990). If this hypothesis is correct, the Hermitage (hybrid) race would be classified as type (1) above, the Oxford and Aberdeen variant races and the Hermitage (standard) race would be type (2), and the "acrocentric peak" could ultimately develop into a race of type (3) (Fig. 12-6).

The transformation of populations within staggered hybrid zones into races can thus be viewed as a process of widening the separation of the clines. It is most likely to occur if the hybrid zone is moving, as it might during species range expansion. I term this process *zonal raciation* (Searle, 1991), and it would be expected to generate an approximately linear series of races, each differing by one or more Rb rearrangements from its neighbor. In the simplest case, chromosome number would increase or decrease along this series. This pattern has been described, e.g., in the mole-rat *Spalax ehrenbergi* (Wahrman et al., 1969) and the spiny rat *Proechimys guairae* (Reig et al., 1980), but is typically attributed to a process I call *invasive raciation* (after White, 1985). On this model, the linear series is supposed to represent sequential fixation of

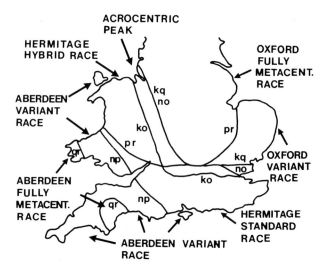

Figure 12-6. Location of chromosome clines for the common shrew in southern Britain, which may have originated by modification of the hybrid zone between the Oxford *(kq, no, pr)* and Aberdeen races *(ko, np, qr)*. In addition to the fully metacentric Aberdeen and Oxford races (which also have wide distributions in northern Britain), the following "minor" races (homozygous forms) can also be distinguished: Aberdeen "variant" race *(ko, np, q, r)*, Hermitage "standard" race *(ko, n, p, q, r)*, Hermitage "hybrid" race *(ko, pr, n, q)*, Oxford "variant" race *(kq, no, p, r)*. Also, there is an "acrocentric peak" between the Oxford and Hermitage races *(k, n, o, q* with arm combination *pr* present as a metacentric or acrocentric), and the Hermitage (standard) race acts as the "acrocentric peak" between the Hermitage (hybrid) race and the Aberdeen (variant) race. (For discussion of these "acrocentric peaks" see Table 12-3 and associated text.)

new Rb mutations at the periphery as a species invades new territory. A third model, *included raciation,* can also explain this pattern. This model involves sequential fixation of Rb mutations (as in invasive raciation), but in this case each new mutation arises within the "core" of the expanding distribution of previous mutations. Clinal movement could result from species range expansion (as in zonal raciation). White (1978b) has presented a "stasipatric" version of this model (the "chain process") in which clinal movement occurs as a result of selective advantage of each Rb mutation as it arises. However, as I reviewed above, there is no clear expectation that homozygotes for a Rb rearrangement will be more fit than the ancestral homozygote, and preferential transmission can be viewed only as a "possibly important process" for spread of Rb variants. Although all the models require multiple Rb mutations, the type of rearrangement involved differs. Thus for the mole-rat (Wahrman et al., 1969), the invasive model requires Rb fissions; the included model requires Rb fusions; and the zonal model could involve fissions or fusions.

Chromosomal Hybrid Zones as Genetic Barriers. Fertility studies have shown that some chromosomal variants are more likely to suffer heterozygous disadvantage than others (see above; King, 1987). Heterochromatic additions and deletions do not appear to cause fertility problems and polymorphisms involving these variants are common in eutherians. Although heterozygotes for some pericentric inversions may have reduced fertility in humans, so far there are no data to suggest a similar heterozygous disadvantage in wild eutherians, and again polymorphisms are well documented. Thus for hybrid zones that involve heterochromatic changes and pericentric inversions, e.g., in the pocket gopher *Thomomys bottae* and white-footed mouse *Peromyscus leucopus,* respectively (Table 12-4), it is likely that the chromosomal differences themselves are not creating a strong genetic barrier (but see Walsh, 1982). However, high fertility in inversion heterozygotes is associated with crossover suppression, and hence gene flow may be much reduced for loci in the region of the inversion.

A majority of the well-studied eutherian chromosomal hybrid zones differ by Rb variants (or possibly whole-arm reciprocal translocations in some cases). Commonly, single simple Rb heterozygotes (Fig. 12-3) have near-normal fertility in eutherians (Table 12-1 and associated text) and other organisms [the lizard *Sceloporus grammicus* (Porter and Sites, 1985, 1987) and the grasshopper *Dichroplus pratensis* (Bidau and Mirol, 1988)]; it is likewise true for single whole-arm reciprocal translocations, at least in some species (Table 12-2 and associated text). This high fertility enhances the probability of fixation of individual Rb variants or whole-arm reciprocal translocations. When such rearrangements continue to become fixed (one at a time) in a species range, races with increasingly different karyotypes develop. Such races are expected to form complex or multiple simple heterozygotes (Fig. 12-3) on contact, and it has long been realized that hybrid zones of this type may be substantial genetic barriers on cytological grounds alone (White, 1978b; Capanna, 1982; Baker and Bickham, 1986).

Not all hybrid zones where the F_1-type hybrids are complex or multiple simple Rb heterozygotes are strong genetic barriers, however. For both the house mouse and common shrew, numerous examples of "staggered hybrid zones" have been demonstrated. Such hybrid zones comprising staggered clines are intrinsically weaker genetic barriers than sets of coincident clines (Barton and Hewitt, 1981)—all the more so if they reduce the frequency of particularly unfit complex or multiple simple heterozygotes. Widening the taxonomic spectrum, a staggered hybrid zone has been described for the angiosperm *Clarkia speciosa* (Onagraceae) between races that differ by whole-arm reciprocal translocations (for reviews see Bloom and Lewis, 1972; Harrison, 1990). Again, this staggered hybrid zone is believed to be a much weaker genetic barrier than there would have been if the "pure" races were making direct contact. Nevertheless, although staggering is a possibility for races that differ by a moderate number of Rb rearrangements or whole-arm reciprocal translocations, it may not develop if the races differ such that there is particularly high initial hybrid unfitness on secondary contact (see discussion of CD–standard race zone). In this manner, the following hybrid zones may represent strong genetic barriers resistant to the forces promoting staggering of chromosomal clines because of the formation of highly unfit multiple simple or complex Rb heterozygotes (Table 12-3): for the common shrew—the Tomsk-Novosibirsk zone (chain IX-forming complex heterozygotes); for the house mouse—the Upper Valtellina-Poschiavo CD-standard race, Monastir-standard race, and Milano 2-standard race zones (chain V-forming complex heterozygotes, multiple simple heterozygotes for seven to nine arm combinations).

Speciation and Chromosomal Evolution. Speciation could occur as a sole consequence of karyotypic differences between races ("chromosomal speciation") in two ways: (1) if karyotypic differences between races accumulate such that hybrids are sterile purely as a result of karyotypic heterozygosity, i.e., any genic differences between the taxa do not contribute to the sterility; and (2) if the fertility of the hybrids is reduced purely as a result of the karyotypic heterozygosity, and this hybrid unfitness promotes genic changes leading to speciation. Karyotypic differences between races could also contribute to the speciation process without being the sole driving force. All three possibilities are plausible in eutherians and are considered in turn.

It is clear that complex Rb heterozygotes with long configurations and high multiple simple Rb heterozygotes may have very low fertility apparently because of the behavior of the heterozygous configurations at meiosis. Thus races that accumulate many Rb rearrangements or whole-arm reciprocal translocations may become species because they produce sterile F_1 hybrids on secondary contact with other races. These rearrangements may accumulate in races rapidly, apparently in the absence of major genic change, e.g., as postulated for the house mouse (Capanna, 1982; Britton-Davidian et al., 1989). Whatever the manner in which the divergent karyotypic races are formed, highly heterozygous hybrids are only likely to occur if the races become completely fixed for their chromosomal differences in allopatry. If multiple Rb differences build up between two races in parapatry by a "stasipatric" process (White, 1978b), even if a new fusion could spread to the hybrid zone, selection is likely to ensure that the cline for that fusion would remain staggered relative to other chromosomal clines.

Plausible though this first type of speciation process appears, there are no clear examples where it has been demonstrated. When two genically similar races that differ by multiple Rb rearrangements or whole-arm reciprocal translocations come into secondary contact, it is perhaps more likely that the hybrids will have greatly reduced fertility than complete sterility. In this case, selection may favor reinforcement (Dobzhansky, 1970; Sanderson, 1989; see Ch. 3), and the chromosomal rearrangements may thereby promote speciation, which occurs by genic changes. Reinforcement may be particularly likely in those cases where only a single type of (complex heterozygous) hybrid is expected. Reinforcement was proposed to explain the apparent reproductive isolation of the Upper Valtellina and Poschiavo chromosomal races of house mice (Capanna and Corti, 1982), but more recent work suggests that these races do actually randomly interbreed (Table 12-3).

Not all hybridizing chromosomal races are so different in karyotype that the hybrids are highly infertile. Frequently, races differ by only one or a few chromosomal rearrangements (Tables 12-3 and 12-4). Clearly, even in this case chromosomal speciation may occur because genic differences could accumulate between the races in response to even a small degree of unfitness and crossover suppression in karyotypic heterozygotes. However, this process would be slow, and the zone would most likely be disrupted (by range changes) before completion of speciation. More plausible is that karyotypic differences promote speciation acting *simultaneously* with genic factors. Genic hybridity may enhance the meiotic aberrations associated with karyotypic heterozygosity; for example, with certain genetic backgrounds, heterozygosity for only a single Rb fusion may lead to a anaphase I nondisjunction frequency of 39% in male and 61% in female house mice (Searle, 1988a). Genic hybridity may also contribute to a reduction in fertility in a more independent fashion.

This discussion on chromosomes and speciation has, so far, been based on observations of extant karyotypic races and hybrid zones. The hybrid zones currently seen

Table 12-5. Frequency Distribution of Rearrangements Identified in Chromosomal Phylogenies of Various Taxa of Eutherian Mammals[a]

Taxon	No. of Species Studied	Frequency of Rearrangements[b]									Ref[d]
		pii	pai	add/del	cms	Rb fus[c]	tan fus	Rb fis	tan fis	rcp	
Insectivora (*Sorex araneus* complex)	6				7	22	7				1
Chiroptera (Phyllostomatidae family)	12	3				24[e]	5[e]	1[f]			2
Rodentia											
Rattus genus	9	3				18		6			3
Peromyscus genus	30	41								6	4
Neotoma genus	10	8				9				6	4
Oryzomys genus	11	8	1			13				21	4
Arvicolidae family	22	18	12	26	17	38	20		8	1	5
Artiodactyla (Cervidae family)	14	2				12	29				6
Carnivora (Mustelidae family)	8	3	1		6	6	[g]	[h]	[h]	[g]	7
Primates (Primates order)	~60	43	6	[i]	[i]	51	[j]	[k]	[k]	1	8

[a]Chosen because they give the most precise information on the rearrangements of interest.

[b]pii = pericentric inversion; pai = paracentric inversion; add/del = addition/deletion of euchromatin; cms = centromeric shift; Rb fus = Robertsonian fusion; tan fus = tandem fusion; fis = fission; rcp = reciprocal translocation.

[c]This category may include some whole-arm reciprocal translocations.

[d]Key to references: (1) Volobouev, 1989 (Fig. 6); (2) Baker and Bickham, 1980 (Fig. 1); (3) Baverstock et al., 1983 (Figs. 8 and 9); (4) Baker et al., 1988 (Table 1); (5) Modi, 1987 (Table 1); (6) Neitzel, 1987 (Fig. 3); (7) Couturier and Dutrillaux, 1986 (Fig. 19); (8) Dutrillaux, 1979 (Table 4).

[e]Up to three additional examples.

[f]Up to one additional example.

[g]Two "non-Robertsonian" translocations.

[h]Five "fissions."

[i]Thirteen "complex intrachromosomal."

[j]Twenty-three "complex or unprecised [*sic*] interchromosomal."

[k]Ten "fissions."

are most likely to be long-term, stable entities. Comparisons of the karyotypic differences between species may provide information on shorter-lived hybrid zones, which may be the most significant with regard to speciation. These comparisons should also give a wider perspective on eutherian chromosomal evolution.

Data from eutherian chromosomal phylogenies (Table 12-5) suggest that the karyotypic differences between closely related species tend to be of the same type as between hybridizing karyotypic races (Tables 12-3 and 12-4) and between morphs in within-race karyotypic polymorphisms. Rb fusions (perhaps including whole-arm reciprocal translocations), pericentric inversions, and heterochromatic changes (not included in Table 12-5, but see, e.g., Patton and Sherwood, 1983) are the most common mutations in all three data sets. Pericentric inversions and heterochromatic changes are thought to have little effect on heterozygous fitness. Therefore there is some justification for considering that chromosomal speciation is most likely to occur between races that differ by multiple differences caused by Rb fusions or whole-arm reciprocal translocations (see above; Capanna, 1982; Baker and Bickham, 1986).

The high frequency of Rb fusions and pericentric inversions in species phylogenies also fits happily with the observation that single heterozygotes for either of these rearrangements tend to have only slightly reduced fertility. These rearrangements could relatively easily become fixed by genetic drift in small populations and hence come to characterize races and species.

However, the high proportion of Rb fusions and pericentric inversions in the species phylogenies casts doubt on the hypothesis that chromosomal rearrangements are an important evolutionary force in the creation of new genes or gene clusters (Wilson et al., 1974). Such genes can be created only if there are two euchromatic breaks; in Rb fusions, breaks are only in centromeric regions, which would be expected to be heterochromatin (Hsu and Arrighi, 1971), whereas for pericentric inversions one of the two breaks is also usually in the centromeric region, e.g., *Peromyscus* (Stangl and Baker, 1984). Paracentric inversions and reciprocal translocations, which would be the rearrangements most likely to create novel genes, are relatively rarely fixed in eutherian mammals (Table 12-5). Therefore although novel genes may occasionally be created by chromosomal rearrangements in this group of organisms, the data suggest that it is not a common phenomenon.

The data in Table 12-5 argue against another contentious hypothesis. Contrary to Todd (1970) and Imai (1988), Rb fusions are involved far more often in karyotypic change in eutherians than are Rb fissions. These mutations produce the same type of heterozygote, so presumably they have a similar probability of fixation once they have arisen. The inference is that viable Rb fissions (i.e., mitotically stable) (Searle, 1988a) are less frequent than viable Rb fusions, as may have been anticipated from mutation rates in humans, presented earlier. Doubt is cast on scenarios based on repeated Rb fissions (e.g., Wahrman et al., 1969, on mole-rats).

Those data on mutation rates also show that reciprocal translocations are the most common type of chromosomal rearrangement in humans (see above; Lande, 1979) and yet among primates they hardly feature in the chromosomal phylogeny (Table 12-5). The probability of fixation of a chromosomal rearrangement depends on both mutation rate and the degree of heterozygous disadvantage. The paucity of reciprocal translocations in the primate phylogeny is consistent with the low fertility of heterozygotes for reciprocal translocations in humans.

Following on from these arguments, the high frequency of tandem fusions and of reciprocal translocations in some lineages (Table 12-5) is likely to reflect a particularly high mutation rate for these variants in those lineages. It is expected that heterozygotes for reciprocal translocations and tandem fusions would have a high frequency of meiotic aberrations in all eutherians, especially when acrocentrics are involved, as, for example, in the Cervidae (Table 12-5). A high mutation rate plus such other factors as subdivision of the species into particularly small populations (Walsh, 1982) and reproductive compensation (see above; Bengtsson, 1980) could mean that even rearrangements that cause a high incidence of meiotic aberration in heterozygotes sometimes become fixed by genetic drift. Clearly, once one or several of such rearrangements characterize a karyotypic race, a strong genetic barrier may form on hybridization with other races, and speciation could ensue.

In conclusion, "chromosomal speciation" in eutherians is most likely to occur where two races differ by multiple Rb fusions (perhaps including whole-arm reciprocal translocations) or by one or more tandem fusions or reciprocal translocations. For such karyotypic differences to develop, there must be high mutation rates for the variants in question. However, chromosomal speciation is not inevitable under these circumstances and may be rare. The karyotypic races may form a long-lasting hybrid zone permitting gene flow. Population movements and selection may induce a weakening of the genetic barrier by promoting a staggering of chromosomal clines.

Probably of more general significance to speciation is that karyotypic heterozygosity and genic hybridity may act simultaneously to generate hybrid unfitness. Therefore, in contrast to White (1968, 1978a), I believe that a chromosomal difference between races is most likely to have a major influence on hybrid infertility if the races have been geographically isolated. Clearly, in allopatry, populations diverge genically and, given periods of low population size, have the opportunity to accumulate chromosomal rearrangements that are disadvantagous in the heterozygous state. Once such populations come into contact, the combination and interaction of genic and chromosomal difference may cause substantial hybrid unfitness.

ACKNOWLEDGEMENTS

I am grateful to the following for kindly providing me with unpublished data or manuscripts: H. C. Hauffe, S. Fedyk, J. Britton-Davidian, K. Fredga, Y. Rumpler, M. W. Nachman, V. T. Volobouev, P. Taberlet, H. Tegelström, R. Hübner, C. A. Redi, and H. Winking. Also I warmly thank R. G. Harrison, N. H. Barton, G. L. Bush, A. E. Douglas, B. O. Bengtsson, P. de Boer, B. M. N. Wallace, H. Wallace, H. C. Hauffe, S. J. Mercer, and R. Hübner and an anonymous reviewer for many invaluable comments on previous versions of the chapter. I acknowledge the financial support provided by the Royal Society of London.

REFERENCES

Adolph, S., and Klein, J. 1983. Genetic variation of wild mouse populations in souther Germany. I. Cytogenetic study. Genet. Res. 41:117–134.

Aniskin, V. M., and Lukianova, I. V. 1989. A new chromosome race and the analysis of hybridization zone of two karyoforms of *Sorex araneus* (Insectivora, Soricidae). Doklady Akad. Nauk SSR 309:1260–1262 [in Russian].

Baker, R. J. 1981. Chromosome flow between chromosomally characterized taxa of a volant mammal, *Uroderma bilobatum* (Chiroptera: Phyllostomatidae). Evolution 35:296–305.

Baker, R. J., and Bickham, J. W. 1980. Karyotypic evolution in bats: evidence of extensive and conservative chromosomal evolution in closely related taxa. Syst. Zool. 29:239–253.

Baker, R. J., and Bickham, J. W. 1986. Speciation by monobrachial centric fusions. Proc. Natl. Acad. Sci. USA 83:8245–8248.

Baker, R. J., Atchley, W. R., and McDaniel, V. R. 1972. Karyology and morphometrics of Peters' tent-making bat, *Uroderma bilobatum* Peters (Chiroptera, Phyllostomatidae). Syst. Zool. 21:414–429.

Baker, R. J., Bleier, W. J., and Atchley, W. R. 1975. A contact zone between karyotypically characterized taxa of *Uroderma bilobatum* (Mammalia: Chiroptera). Syst. Zool. 24:133–142.

Baker, R. J., Qumsiyeh, M. B., and Rautenbach, I. L. 1988. Evidence for eight tandem and five centric fusions in the evolution of the karotype of *Aethomys namaquensis* A. Smith (Rodentia: Muridae). Genetica 76:161–169.

Baker, R. J., Chesser, R. K., Koop, B. F., and Hoyt, R. A. 1983a. Adaptive nature of chromosomal rearrangement: differential fitness in pocket gophers. Genetica 61:161–164.

Baker, R. J., Robbins, L. W., Stangl, F. B., and Birney, E. C. 1983b. Chromosomal evidence for a major subdivision in *Peromyscus leucopus*. J. Mamm. 64:356–359.

Barton, N. H. 1982. The structure of the hybrid zone in *Uroderma bilobatum*. Evolution 36:863–866.

Barton, N. H., and Bengtsson, B. O. 1986. The barrier to genetic exchange between hybridising populations. Heredity 56:357–376.

Barton, N. H., and Hewitt, G. M. 1981. Hybrid zones and speciation. In W. R. Atchley and D. S. Woodruff, eds. Evolution and Speciation: Essays in Honour of M. J. D. White. Cambridge: Cambridge University Press, pp. 109–145.

Barton, N. H., and Hewitt, G. M. 1985. Analysis of hybrid zones. Ann. Rev. Ecol. Syst. 16:113–148.

Barton, N. H., and Jones, J. S. 1983. Mitochondrial DNA: new clues about evolution. Nature 306:317–318.

Barton, N. H., and Rouhani, S. 1991. The probability of fixation of a new karyotype in a continuous population. Evolution 45:499–517.

Bauchau, V., Smets, S., Viroux, M-C, Nootens, D., and de Caritat, A-K. 1990. Robertsonian translocations in free-living populations of the house mouse in Belgium. Biol. J. Linn. Soc. 41:193–201.

Baverstock, P. R., Gelder, M., and Jahnke, A. 1982. Cytogenetic studies of the Australian rodent, Uromys caudimaculatus, a species showing extensive heterochromatin variation. Chromosoma 84:517–533.

Baverstock, P. R., Gelder, M., and Jahnke, A. 1983. Chromosome evolution in Australian Rattus—G-banding and hybrid meiosis. Genetica 60:93–103.

Bengtsson, B. O. 1980. Rates of karyotype evolution in placental mammals. Hereditas 92:37–47.

Bengtsson, B. O., and Frykman, I. 1990. Karyotype evolution: evidence from the common shrew (Sorex araneus L.). J. Evol. Biol. 3:85–101.

Berry, R. J. 1981. Population dynamics of the house mouse. Symp. Zool. Soc. Lond. 47:395–425.

Bianchi, N. O., and Merani, S. 1981. Chromosome 1 polymorphism and viability in Akodon (Rodentia, Cricetidae). In O. A. Reig, ed. Ecología y Genetica de la Especiación Animal. Caracas: Equinoccio Press, University Simón Bolivar, pp. 211–253.

Bianchi, N. O., Merani, S., and Lizarralde, M. 1979. Cytogenetics of South-American akodon rodents (Cricetidae). VI. Polymorphism in Akodon dolores (Thomas). Genetica 50:99–104.

Bidau, C. J., and Mirol, P. M. 1988. Orientation and segregation of Robertsonian trivalents in Dichroplus pratensis (Acrididae). Genome 30:947–955.

Bloom, W., and Lewis, H. 1972. Interchanges and interpopulational gene exchange in Clarkia speciosa. Chromosomes Today 3:268–284.

Bobrow, M. 1985. Heterochromatic chromosome variation and reproductive failure. Exp. Clin. Immunogenet. 2:97–105.

Bohlin, R. G., and Zimmerman, E. G. 1982. Genic differentiation of two chromosome races of the Geomys bursarius complex. J. Mamm. 63:218–228.

Bond, D. J., and Chandley, A. C. 1983. Aneuploidy. Oxford: Oxford University Press.

Britton-Davidian, J., Nadeau, J. H., Croset, H., and Thaler, L. 1989. Genic differentiation and origin of Robertsonian populations of the house mouse (Mus musculus domesticus Rutty). Genet. Res. 53:29–44.

Britton-Davidian, J., Sonjaya, H., Catalan, J., and Cattaneo-Berrebi, G. 1990. Robertsonian heterozygosity in wild mice: fertility and transmission rates in Rb (16.17) translocation heterozygotes. Genetica 80:171–174.

Brown, T., Robertson, F. W., Dawson, B. M., Hanlin, S. J., and Page, B. M. 1980. Individual variation of centric heterochromatin in man. Hum. Genet. 55:367–373.

Burgoyne, P. S., and Baker, T. G. 1984. Meiotic pairing and gametogenic failure. Symp. Soc. Exp. Biol. 38:349–362.

Bush, G. L., Case, S. M., Wilson, A. C., and Patton, J. L. 1977. Rapid speciation and chromosomal evolution in mammals. Proc. Natl. Acad. Sci. USA 74:3942–3946.

Capanna, E. 1982. Robertsonian numerical variation in animal speciation: Mus musculus, an emblematic model. In C. Barigozzi, ed. Mechanisms of Speciation. New York: Alan R. Liss, pp. 155–177.

Capanna, E., and Corti, M. 1982. Reproductive isolation between two chromosomal races of Mus musculus in the Rhaetian Alps (northern Italy). Mammalia 46:107–109.

Capanna, E., Corti, M., and Nascetti, G. 1985. Role of contact areas in chromosomal speciation of the European long-tailed house mouse (Mus musculus domesticus). Boll. Zool. 52:97–119.

Capanna, E., Corti, M., Mainardi, D., Parmigiani, S., and Brain, P. F. 1984. Karyotype and intermale aggression in wild house mice: ecology and speciation. Behav. Genet. 14:195–208.

Capanna, E., Gropp, A., Winking, H., Noack, G., and Civitelli, M-V. 1976. Robertsonian metacentrics in the mouse. Chromosoma 58:341–353.

Carothers, A. D., Buckton, K. E., Collyer, S., de Mey, R., Frackiewicz, A., Piper, J., and Smith, L. 1982. The effect of variant chromosomes on reproductive fitness in man. Clin. Genet. 21:280–289.

Cattanach, B. M. 1978. Crossover supression in mice heterozygous for tobacco mouse metacentrics. Cytogenet. Cell Genet. 20:264–281.

Cattanach, B. M., and Moseley, H. 1973. Nondisjunction and reduced fertility caused by the tobacco mouse metacentric chromosomes. Cytogenet. Cell Genet. 12:264–287.

Catzeflis, F. M., Nevo, E., Ahlquist, J. E., and Sibley, C. G. 1989. Relationships of the chromosomal species in the eurasian mole rats of the Spalax ehrenbergi group as determined by

DNA-DNA hybridization, and an estimate of the Spalacid-Murid divergence time. J. Mol. Evol. 29:223–232.

Chandley, A. C. 1984. Infertility and chromosome abnormality. Oxford Rev. Reprod. Biol. 8:1–46.

Chandley, A. C. 1988. Meiotic studies and fertility in human translocation carriers. In A. Daniel, ed. The Cytogenetics of Mammalian Autosomal Rearrangements. New York, Alan R. Liss, pp. 361–382.

Coates, D. J., and Shaw, D D. 1982. The chromosomal component of reproductive isolation in the grasshopper Caledia captiva. I. Meiotic analysis of chiasma distribution patterns in two chromosomal taxa and their F₁ hybrids. Chromosoma 86:509–531.

Coates, D. J., and Shaw, D. D. 1984. The chromosomal component of reproductive isolation in the grasshopper Caledia captiva. III. Chiasma distribution patterns in a new chromosomal taxon. Heredity 53:85–100.

Corti, M., and Thorpe, R. S. 1989. Morphological clines across a karyotypic zone of house mice in Central Italy. J. Evol. Biol. 2:253–264.

Corti, M., Capanna, E., and Estabrook, G. F. 1986. Microevolutionary sequences in house mouse chromosomal speciation. Syst. Zool. 35:163–175.

Corti, M., Ciabatti, C. M., and Capanna, E. 1990. Parapatric hybridization in the chromosomal speciation of the house mouse. Biol. J. Linn. Soc. 41:203–214.

Cothran, E. G., and Zimmerman, E. G. 1985. Electrophoretic analysis of the contact zone between Geomys breviceps and Geomys bursarius. J. Mamm. 66:489–497.

Couturier, J., and Dutrillaux, B. 1986. Evolution chromosomique chez les Carnivores. Mammalia 50:124–163.

Coyne, J. A. 1984. Correlation between heterozygosity and rate of chromosome evolution in animals. Am. Naturalist 123:725–729.

Craig-Holmes, A. P., Moore, F. B., and Shaw, M. W. 1973. Polymorphism of human C-band heterochromatin. I. Frequency of variants. Am. J. Hum. Genet. 25:181–192.

Daniel, A., ed. 1988. The Cytogenetics of Mammalian Autosomal Rearrangements. New York: Alan R. Liss.

De Boer, P. 1986. Chromosomal causes for fertility reduction in mammals. In F. J. de Serres, ed. Chemical Mutagens. Vol. 10. New York: Plenum Press, pp. 427–467.

DeBoer, P., and De Jong, J. H. 1989. Chromosome pairing and fertility in mice. In C. B. Gillies, ed. Fertility and Chromosome Pairing: Recent Studies in Animals and Plants. Boca Raton, FL: CRC Press, pp. 37–76.

De Boer, P., and Searle, A. G. 1980. Summary and synthesis: workshop on chromosomal

aspects of male sterility in mammals. J. Reprod. Fertil. 60:259–265.

Dobzhansky, T. 1970. Genetics of the Evolutionary Process. New York: Columbia University Press.

Dowler, R. C. 1989. Cytogenetic studies of three chromosomal races of pocket gophers (Geomys bursarius complex) at hybrid zones. J. Mamm. 70:253–266.

Dutrillaux, B. 1979. Chromosomal evolution in primates: tentative phylogeny from Microcebus murinus (Prosimian) to Man. Hum. Genet. 48:251–314.

Dutrillaux, B., and Rumpler, Y. 1977. Chromosomal evolution in Malagasy lemurs. II. Meiosis in intra- and interspecific hybrids in the genus Lemur. Cytogenet. Cell Genet. 18:197–211.

Dutrillaux, B., and Rumpler, Y. 1987. The role of chromosomes in speciation: a new interpretation. Chromosomes Today 9:75–90.

Eichenlaub-Ritter, U., and Winking, H. 1990. Nondisjunction, disturbances in spindle structure, and characteristics of chromosome alignment in maturing oocytes of mice heterozygous for Robertsonian translocations. Cytogenet. Cell Genet. 54:47–54.

Endler, J. A. 1977. Geographic Variation, Speciation, and Clines. Princeton, NJ: Princeton University Press.

Epstein, C. J. 1986. The Consequences of Chromosome Imbalance. Cambridge: Cambridge University Press.

Everett, C. A., and Searle, J. B. 1990. Infertility in male house mice heterozygous and homozygous for Robertsonian translocations. J. Reprod. Fertil. Abstr. Ser. 6:27.

Fedyk, S. 1986. Genetic differentiation of Polish populations of Sorex araneus L. II. Possibilities of gene flow between chromosome races. Bull. Pol. Acad. Sci. Biol. Sci. 34:161–171.

Fedyk, S., Chetnicki, W., and Banaszek, A. 1991. Genetic differentiation of Polish populations of Sorex araneus L. III. Interchromosomal recombination in a hybrid zone. Evolution 45:1384–1392.

Ferrier, D. K., Searle, J. B., and Everett, C. A. 1990. The influence of genic background on infertility in male feral house mice that are multiple heterozygotes for Robertsonian translocations. J. Reprod. Fertil. Abst. Ser. 6:28.

Ford, C. E. 1972. Gross genome unbalance in mouse spermatozoa: does it influence the capacity to fertilise? In R. A. Beatty and S. Gluecksohn-Waelsch, eds. The Genetics of the Spermatozoon. Edinburgh: Beatty and Gluecksohn-Waelsch, pp. 359–369.

Forejt, J. 1973. Centromeric heterochromatin polymorphism in the house mouse: evidence from inbred strains and natural populations. Chromosoma 43:187–201.

Forejt, J. 1982. X-Y involvement in male sterility caused by autosome translocations—a hypothesis. In P. G. Crosignani and B. L. Rubin, eds. Genetic Control of Gamate Production and Function. Orlando FL: Grune & Stratton, pp. 135–151.

Forejt, J. 1984. X-inactivation and its role in male sterility. Chromosomes Today 8:117–127.

Fredga, K. 1987. (1) The distribution of chromosome races in Sweden and Denmark. (2) The northern hybrid zone in Sweden. Abstracts to the 1st International Meeting on the Population and Evolutionary Cytogenetics of *Sorex araneus*, Oxford, U. K.

Frykman, I., and Bengtsson, B. O. 1984. Genetic differentiation in *Sorex*. III. Electrophoretic analysis of a hybrid zone between two karyotypic races in *Sorex araneus*. Hereditas 100:259–270.

Frykman, I., Simonsen, V., and Bengtsson, B. O. 1983. Genetic differentiation in *Sorex*. I. Electrophoretic analysis of the karyotypic races of *Sorex araneus* in Sweden. Hereditas 99:279–292.

Futuyma, D. J., and Mayer, G. C. 1980. Nonallopatric speciation in animals. Syst. Zool. 29:254–271.

Garagna, S., Zuccotti, M., Searle, J. B., Redi, C. A., and Wilkinson, P. J. 1989. Spermatogenesis in heterozygotes for Robertsonian chromosomal rearrangements from natural populations of the common shrew, *Sorex araneus*. J. Reprod. Fertil. 87:431–438.

Garagna, S., Redi, C. A., Zuccotti, M., Britton-Davidian, J., and Winking, H. 1990. Kinetics of oogenesis in mice heterozygous for Robertsonian translocations. Differentiation 42:167–171.

Grant, V. 1981. Plant Speciation. New York: Columbia University Press.

Greenbaum, I. F. 1981. Genetic interactions between hybridizing cytotypes of the tent-making bat *(Uroderma bilobatum)*. Evolution 35:306–321.

Greenbaum, I. F., and Reed, M. J. 1984. Evidence for heterosynaptic pairing of the inverted segment in pericentric inversion heterozygotes of the deer mouse *(Peromyscus maniculatus)*. Cytogenet. Cell Genet. 38:106–111.

Gropp, A., Giers, D., and Kolbus, U. 1974. Trisomy in the fetal backcross progeny of male and female metacentric heterozygotes of the mouse. I. Cytogenet. Cell Genet. 13:511–535.

Gropp, A., and Winking, H. 1981. Robertsonian translocations: cytology, meiosis, segregation patterns and biological consequences of heterozygosity. Symp. Zool. Soc. Lond. 47:141–181.

Gropp, A., Winking, H., and Redi, C. 1982a. Consequences of Robertsonian heterozygosity: segregational impairment of fertility versus male-limited sterility. In P. G. Crosignani and B. L. Rubin, eds. Genetic Control of Gamete Production and Function. Orlando, FL: Grune & Stratton, pp. 115–134.

Gropp, A., Winking, H., Redi, C., Capanna, E., Britton-Davidian, J., and Noack, G. 1982b. Robertsonian karyotype variation in wild house mice from Rhaeto-Lombardia. Cytogenet. Cell Genet. 34:67–77.

Gropp, A., Winking, H., Redi, C., Noack, G., Kolbus, U., and Louton, T. 1981. Consequences of Robertsonian heterozygosity in the mouse—meiotic segregation and post-zygotic abnormality in male versus female carriers of Rb chromosomes. Unpublished manuscript.

Hafner, J. C. 1982. Genetic interactions at a contact zone of *Uroderma bilobatum* (Chiroptera: Phyllostomatidae). Evolution 36:852–862.

Hafner, J. C., Hafner, D. J., Patton, J. L., and Smith, M. F. 1983. Contact zones and genetics of differentiation in the pocket gopher *Thomomys bottae* (Rodentia: Geomyidae). Syst. Zool. 32:1–20.

Hale, D. W. 1986. Heterosynapsis and suppression of chiasmata within heterozygous pericentric inversions of the Sitka deer mouse. Chromosoma 94:425–432.

Hale, D. W., and Greenbaum, I. F. 1988a. Chromosome pairing in deer mice heterozygous for the presence of heterochromatic short arms. Genome 30:44–47.

Hale, D. W., and Greenbaum, I. F. 1988b. Synapsis of a chromosomal pair heterozygous for a pericentric inversion and the presence of a heterochromatic short arm. Cytogenet. Cell Genet. 48:55–57.

Harris, M. J., Wallace, M. E., and Evans, E. P. 1986. Aneuploidy in the embryonic progeny of females heterozygous for the Robertsonian chromosome (9.12) in genetically wild Peru-Coppock mice (Mus musculus). J. Reprod. Fertil. 76:193–203.

Harrison, R. G. 1990. Hybrid zones: windows on the evolutionary process. Oxford Surv. Evol. Biol. 7:69–128.

Harrison, R. G., and Rand, D. M. 1989. Mosaic hybrid zones and the nature of species boundaries. In D. Otte and J. A. Endler, eds. Speciation and Its Consequences. Sunderland, MA: Sinauer, pp. 111–133.

Hatfield, T., Barton, N., and Searle, J. B. 1992. A model of a hybrid zone between two chromosomal races of the common shrew *(Sorex araneus)*. Evolution 46:1129–1145.

Hauffe, H. C. and Searle, J. B., in press. Extreme karyotype variation found in a mosaic hybrid zone between five chromosome races of *Mus musculus domesticus* in Upper Valtellina, northern Italy: The tobacco mouse story revisited. Evolution.

Hausser, J., Bosshard, F., Taberlet, P., and

Wójcik, J. 1991. Relationships between chromosome races and species of *Sorex* of the araneus group in the western Alps. Mem. Soc. Vaud. Sci. Nat. 19:79–95.

Heaney, L. R., and Timm, R. M. 1985. Morphology, genetics, and ecology of pocket gophers (genus *Geomys*) in a narrow hybrid zone. Biol. J. Linn. Soc. 25:301–317.

Hedrick, P. W. 1981. The establishment of chromosomal variants. Evolution 35:322–332.

Hsu, T. C., and Arrighi, F. E. 1971. Distribution of constitutive heterochromatin in mammalian chromosomes. Chromosoma 34:243–253.

Hübner, R. 1988. Populations robertsoniennes chez le souris "sauvage" (*Mus domesticus* Rutty 1772) en Belgique. Annls. Soc. R. Zool. Belg. 118:69–75.

Hübner, R., and Koulischer, L. 1990. Cytogenetic studies on wild house mice from Belgium. Genetica 80:93–100.

Imai, H. T. 1988. Centric fission in man and other mammals. In A. Daniel, ed. The Cytogenetics of mammalian Autosomal Rearrangements. New York: Alan R. Liss, pp. 551–582.

Kaiser, P. 1988. Pericentric inversions: their problems and clinical significance. In A. Daniel, ed. The Cytogenetics of Mammalian Autosomal Rearrangements. New York: Alan R. Liss, pp. 163–247.

King, M. 1987. Chromosomal rearrangements, speciation and the theoretical approach. Heredity 59:1–6.

Kirk, K. M., and Searle, A. G. 1988. Phenotypic consequences of chromosome imbalance in the mouse. In A. Daniel, ed. The Cytogenetics of Mammalian Autosomal Rearrangements. New York: Alan R. Liss, pp. 739–768.

Kurnit, D. M. 1979. Satellite DNA and heterochromatin variants: the case for unequal mitotic crossing-over. Hum. Genet. 47:169–186.

Lande, R. 1979. Effective deme sizes during long-term evolution estimated from rates of chromosomal rearrangement. Evolution 33:234–251.

Lande, R. 1984. The expected fixation rate of chromosomal inversions. Evolution 38:743–752.

Laurie, D. A., and Hultén, M. A. 1985a. Further studies on bivalent chiasma frequency in human males with normal karyotype. Ann. Hum. Genet. 49:189–201.

Laurie, D. A., and Hultén, M. A. 1985b. Further studies on chiasma distribution and interference in the human male. Ann. Hum. Genet. 49:203–214.

Lewis, K. R., and John, B. 1963. Chromosome Marker. London: Churchill.

Loidl, J. 1991. Coming to grips with a complex matter: a multidisciplinary approach to the synaptonemal complex. Chromosoma 100: 289–292.

Luciani, J. M., Guichaoua, M. R., Delafontaine, D., North, M. O., Gabriel-Robez, O., and Rumpler, Y. 1987. Pachytene analysis in a 17;21 reciprocal translocation carrier: role of the acrocentric chromosomes in male sterility. Hum. Genet. 77:246–250.

Lyapunova, E. A., and Yakimenko, L. V. 1985. Genetics of *Ellobius* (Ellobius, Rodentia). IV. Decrease in the fertility of hybrids between the forms of *Ellobius talpinus* subspecies with different chromosome numbers. Genetika 21:1960–1969 [in Russian].

Lyapunova, E. A., Vorontsov, N. N., Korobitsyna, K. V., Ivanitskaya, E. Y., Borisov, Y. M., Yakimenko, L. V., and Dovgal, V. Y. 1980. A Robertsonian fan in Ellobius talpinus. Genetica 52/53:239–247.

Lyon, M. F. 1976. Distribution of crossing-over in mouse chromosomes. Genet. Res. 28:291–299.

Lyon, M. F. 1991. The genetic basis of transmission-ratio distortion and male sterility due to the t complex. Am. Naturalist 137:349–358.

Lyon, M. F., and Searle, A. G., eds. 1989. Genetic Variants and Strains of the Laboratory Mouse. 2nd Ed. Oxford: Oxford University Press.

Madan, K. 1988. Paracentric inversions and their clinical implications. In A. Daniel, ed. The Cytogenetics of Mammalian Autosomal Rearrangements. New York: Alan R. Liss, pp. 249–266.

Malorni, W., Capanna, E., Cristaldi, M., and de Martino, C. 1982. Changes of seminiferous epithelium in hybrids of mice carrying Robertsonian karyotype. Arch. Androl. 9:333–341.

Mandahl, N. 1978. Variation in C-stained chromosome regions in European hedgehogs (Insectivora, Mammalia). Hereditas 89:107–128.

Martin, R. H. 1988. Abnormal spermatozoa in human translocation and inversion carriers. In A. Daniel, ed. The Cytogenetics of Mammalian Autosomal Rearrangements. New York: Alan R. Liss, pp. 397–417.

Maudlin, I., and Evans, E. P. 1980. Chiasma distribution in mouse oocytes during diakinesis. Chromosoma 80:49–56.

Merani, M. S., Capanna, E., and Bianchi, N. O. 1980. Cytogenetics of South American akodont rodents. VI. Segregation of the polymorphic chromosomes 1 in the testicular meiosis of *Akodon molinae*. Nucleus 23:226–233.

Mercer, S. J., and Searle, J. B. 1991. Preliminary analysis of a contact zone between karyotypic races of the common shrew *(Sorex araneus)* in Scotland. Mem Soc. Vaud. Sci. Nat. 19:73–78.

Mercer, S. J., and Searle, J. B. in press. Captive breeding of the common shrew *(Sorex araneus)* for chromosomal analysis. J. Mamm.

Mercer, S. J., Searle, J. B., and Wallace, B. M. N. 1991. Meiotic studies of karyotypically homozygous and heterozygous male common shrews. Mem. Soc. Vaud. Sci. Nat. 19:33–43.

Mercer, S. J., Wallace, B. M. N., and Searle, J. B. 1992. Male common shrews *(Sorex araneus)* with long meiotic chain configurations can be fertile: implications for chromosomal models of speciation. Cytogenet. Cell Genet. 60:68–73.

Miklos, G. L. G., and John, B. 1979. Heterochromatin and satellite DNA in man: properties and perspectives. Am. J. Hum. Genet. 31:264–280.

Modi, W. S. 1987. Phylogenetic analyses of chromosomal banding patterns among the Nearctic Arvicolidae (Mammalia: Rodentia). Syst. Zool. 36:109–136.

Nachman, M. W. 1992a. Geographic patterns of chromosomal variation in South American marsh rats *Itolochilus brasiliensis* and *Holochilus vulpinus.* Cytogenet Cell Genet. 61:10–16

Nachman, M. W. 1992b. Meiotic studies of Robertsonian polymorphisms in the South American marsh rat *Holochilus brasiliensis.* Cytogenet. Cell Genet. 61:17–24.

Nachman, M. W., and Myers, P. 1989. Exceptional chromosomal mutations in a rodent population are not strongly underdominant. Proc. Natl. Acad. Sci. USA 86:6666–6670.

Nadler, C. F., Lay, D. M., and Hassinger, J. D. 1971. Cytogenetic analyses of wild sheep populations in northern Iran. Cytogenetics 10:137–152.

Nance, V., Vanlerberghe, F., Nielsen, J. T., Bonhomme, F., and Britton-Davidian, J. 1990. Chromosomal introgression in house mice from the hybrid zone between *M. m. domesticus* and *M. m. musculus* in Denmark. Biol. J. Linn. Soc. 41:215–227.

Neitzel, H. 1987. Chromosome evolution of Cervidae: karyotypic and molecular aspects. In G. Obe and A. Basler, eds. Cytogenetics. Berlin: Springer-Verlag, pp. 90–112.

Nelson, K., Baker, R. J., and Honeycutt, R. L. 1987. Mitochondrial DNA and protein differentiation between hybridizing cytotypes of the white-footed mouse, *Peromyscus leucopus.* Evolution 41:864–872.

Nevo, E. 1982. Genetic structure and differentiation during speciation in fossorial gerbil rodents. Mammalia 46:523–530.

Nevo, E. 1985. Genetic differentiation and speciation in spiny mice. *Acomys.* Acta Zool. Fenn. 170:131–136.

Nevo, E. 1991. Evolutionary theory and processes of active speciation and adaptive radiation in subterranean mole rats, *Spalax ehrenbergi* superspecies, in Israel. Evol. Biol. 25:1–125.

Nevo, E., and Bar-El, H. 1976. Hybridization and speciation in fossorial mole rats. Evolution 30:831–840.

Nevo, E., and Cleve, H. 1978. Genetic differentiation during speciation. Nature 275:125–126.

Nevo, E., and Shaw, C. R. 1972. Genetic variation in a subterranean mammal, *Spalax ehrenbergi.* Biochem. Genet. 7:235–241.

Nevo, E., Kim, Y. J., Shaw, C. R., and Thaeler, C. S. 1974. Genetic variation, selection and speciation in *Thomomys talpoides* pocket gophers. Evolution 28:1–23.

Patton, J. C., Baker, R. J., and Genoways, H. H. 1980. Apparent chromosomal heterosis in a fossorial mammal. Am. Naturalist 116:143–146.

Patton, J. L. 1973. An analysis of natural hybridization between the pocket gophers, *Thomomys bottae* and *Thomomys umbrinus,* in Arizona. J. Mamm. 54:561–584.

Patton, J. L., Hafner, J. C., Hafner, M. S., and Smith, M. F. 1979. Hybrid zones in Thomomys bottae pocket gophers: genetic, phenetic, and ecologic concordance patterns. Evolution 33:860–876.

Patton, J. L., Selander, R. K., and Smith, M. H. 1972. Genic variation in hybridizing populations of gophers (genus *Thomomys*). Syst. Zool. 21:263–270.

Patton, J. L., and Sherwood, S. W. 1983. Chromosome evolution and speciation in rodents. Ann. Rev. Ecol. Syst. 14:139–158.

Porter, C. A., and Sites, J. W. 1985. Normal disjunction in Robertsonian heterozygotes from a highly polymorphic lizard population. Cytogenet. Cell Genet. 39:250–257.

Porter, C. A., and Sites, J. W. 1987. Evolution of *Sceloporus grammicus* complex (Sauria: Iguanidae) in central Mexico. II. Studies on rates of nondisjunction and the occurrence of spontaneous chromosomal mutations. Genetica 75:131–144.

Ratomponirina, C., Andrianivo, J., and Rumpler, Y. 1982. Spermatogenesis in several intra- and interspecific hybrids of the lemur *(Lemur).* J. Reprod. Fertil. 66:717–721.

Redi, C. A., Garagna, S., Merani, M. S., Capanna, E., Bianchi, N. O., and Manfredi Romanini, M. G. 1982. Microdensitometric evolution of the DNA content, as ploidy parameter, of spermatozoa in the polymorphic chromosomal system of *Akodon molinae* Cabrera (Rodentia, Cricetidae). Gamete Res. 5:345–354.

Redi, C. A., Garagna, S., Pellicciari, C., Manfredi Romanini, M. G., Capanna, E., Winking, H., and Gropp, A. 1984. Spermatozoa of chromosomally heterozygous mice and their fate in male and female genital tracts. Gamete Res. 9:273–286.

Reig, O. A., Aguilera, M., Barros, M. A., and

Useche, M. 1980. Chromosomal speciation in a Rassenkreiss of Venezuelen spiny rats (genus *Proechimys,* Rodentia, Echimyidae). Genetica 52/53:291–312.

Richler, C., Uliel, E., Rosenmann, A., and Wahrman, J. 1989. Chromosomally derived sterile mice have a 'fertile' active XY chromatin conformation but no XY body. Chromsoma 97:465–474.

Robbins, L. W., and Baker, R. J. 1981. An assessment of the nature of chromosomal rearrangements in 18 species of Peromyscus. Cytogenet. Cell Genet. 31:194–202.

Rumpler, Y., and Warter, S. 1990. Remaniements chromosomiques et sterilité chez les Lemuriens. Pathol. Biol. 38:170–172.

Rumpler, Y., Warter, S., Ratomponirina, C., Hauwy, M., and Ishak, B. 1985. Cytogenetic study of complex hybrids in the genus *Lemur* (Primates Prosimians). Folia Primatol. 44:108–116.

Rumpler, Y., Warter, S., Hauwy, M., and Brun, B. 1990. Stillborn triploid complex hybrid of lemurs (*Eulemur macaco* X *E. coronatus*). Cytogenet. Cell Genet. 53:213–215.

Rutledge, J. C., Cain, K. T., Cachiero, N. L. A., Cornett, C. V., Wright, C. G., and Generoso, W. M. 1986. A balanced translocation in mice with a neurological defect. Science 231:395–397.

Said, K., and Britton-Davidian, J. 1991. Genetic differentiation and habitat partition of Robertsonian house mouse populations *(Mus musculus domesticus)* of Tunisia. J. Evol. Biol. 4:409–427.

Said, K., Auffray, J-C., and Britton-Davidian, J. In press. Are there two chromosomal species of the house mouse in Tunisia? Actes Coll. Rongeurs Espace.

Saitoh, M., and Obara, Y. 1988. Meiotic studies of interracial hybrids from the wild population of the large Japanese field mouse, Apodemus speciosus speciosus. Zool. Sci. 5:815–822.

Sanderson, N. 1989. Can gene flow prevent reinforcement? Evolution 43:1223–1235.

Scriven, P. N., and Brooker, P. C. 1990. Caithness revisited: Robertsonian chromosome polymorphism in Caithness house mice. Heredity 64:25–27.

Searle, J. B. 1984. Three new karyotypic races of the common shrew Sorex araneus (Mammalia, Insectivora) and a phylogeny. Syst. Zool. 33:184–194.

Searle, J. B. 1985. Isoenzyme variation in the common shrew *(Sorex araneus)* in Britain, in relation to karyotype. Heredity 55:175–180.

Searle, J. B. 1986a. Factors responsible for a karyotypic polymorphism in the common shrew, *Sorex araneus.* Proc. R. Soc. Lond. B229:277–298.

Searle, J. B. 1986b. Meiotic studies of Robertsonian heterozygotes from natural populations of the common shrew, *Sorex araneus* L. Cytogenet. Cell Genet. 41:154–162.

Searle, J. B. 1988a. Selection and Robertsonian variation in nature: the case of the common shrew. In A. Daniel, ed. The Cytogenetics of Mammalian Autosomal Rearrangements. New York: Alan R. Liss, pp. 507–531.

Searle, J. B. 1988b. Karyotypic variation and evolution in the common shrew, *Sorex araneus.* In P. E. Brandham, ed. Kew Chromosome Conference III. London: HMSO, pp. 97–107.

Searle, J. B. 1989. An autosomal trisomic cell line in a wild common shrew *(Sorex araneus).* Hereditas 110:183–184.

Searle, J. B. 1990. A cytogenetic analysis of reproduction in common shrews *(Sorex araneus)* from a karyotypic hybrid zone. Hereditas 113:121–132.

Searle, J. B. 1991. A hybrid zone comprising staggered chromosomal clines in the house mouse *(Mus musculus domesticus).* Proc. R. Soc. Lond. B246:47–52.

Searle, J. B. and Thorpe, R. S. 1987. Morphometric variation of the common shrew *(Sorex araneus)* in Britain, in relation to karyotype and geography. J. Zool. Lond. 212:373–377.

Searle, J. B., and Wilkinson, P. J. 1987. Karyotypic variation in the common shrew *(Sorex araneus)* in Britain—a "Celtic fringe." Heredity 59:345–351.

Searle, J. B., Hübner, R., Wallace, B. M. N., and Garagna, S. 1990. Robertsonian variation in wild mice and shrews. Chromosomes Today 10:253–263.

Shaw, D. D. 1981. Chromosomal hybrid zones in orthopteroid insects. In W. R. Atchley and D. S. Woodruff, eds. Evolution and Speciation: Essays in Honour of M. J. D. White. Cambridge: Cambridge University Press, pp. 146–170.

Sites, J. W., and Moritz, C. 1987. Chromosomal evolution and speciation revisited. Syst. Zool. 36:153–174.

Speed, R. M. 1989. Heterologous pairing and fertility in humans. In C. B. Gillies, ed. Fertility and Chromosome Pairing: Recent Studies in Plants and Animals. Boca Raton, FL: CRC Press, pp. 1–35.

Spirito, F., Modesti, A., Perticone, P., Cristaldi, M., Federici, R., and Rizzoni, M. 1980. Mechanisms of fixation and accumulation of centric fusions in natural populations of *Mus musculus* L. I. Karyological analysis of a hybrid zone between two populations in the central Appenines. Evolution 34:453–466.

Stack, S. M. 1984. Heterochromatin, the synaptonemal complex and crossing over. J. Cell Sci. 71:159–176.

Stangl, F. B. 1986. Analysis of a contact zone between two chromosomal races of *Peromyscus leucopus* (Rodentia: Cricetidae). J. Mamm. 67:465–473.

Stangl, F. B., and Baker, R. J. 1984. Evolutionary relationships in *Peromyscus:* congruence in chromosomal, genic, and classical data sets. J. Mamm. 65:643–654.

Stene, J., and Stengel-Rutkowski, S. 1988. Genetic risks of familial reciprocal and Robertsonian translocation carriers. In A. Daniel, ed. The Cytogenetics of Mammalian Autosomal Rearrangements. New York: Alan R. Liss, pp. 3–72.

Stewart-Scott, I. A., and Bruère, A. N. 1987. Distribution of heterozygous translocations and aneuploid spermatocyte frequency in domestic sheep. J. Hered. 78:37–40.

Sudman, P. D., Greenbaum, I. F., Hale, D. W., and Smith, S. A. 1989. Synaptic adjustment in *Peromyscus beatae* (Rodentia: Cricetidae) heterozygous for interstitial heterochromatin. Cytogenet. Cell Genet. 50:1–5.

Sulkava, S., Vahtola, M., and Fredga, K. 1985. Structure of the upper tooth-row of *Sorex araneus* in Scandinavia. Acta Zool. Fenn. 173:237–239.

Szymura, J. M., and Barton, N. H. 1986. Genetic analysis of a hybrid zone between the fire-bellied toads, *Bombina bombina* and *B. variegata*, near Cracow in southern Poland. Evolution 40:1141–1159.

Taberlet, P., Fumagalli, L., and Hausser, J. 1991. mtDNA comparison of the Alpine chromosomal races and species of the *Sorex araneus* group: preliminary results. Mem. Soc. Vaud. Sci. Nat. 19:107–118.

Thaeler, C. S. 1974. A study of four contacts between the ranges of different chromosome forms of the *Thomomys talpoides* complex (Rodentia: Geomyidae). Syst. Zool. 23:343–354.

Thaeler, C. S. 1985. Chromosome variation in the *Thomomys talpoides* complex. Acta Zool. Fenn. 170:15–18.

Therman, E. 1986. Human Chromosomes. Structure, Behavior, Effects. 2nd Ed. New York: Springer-Verlag.

Thorpe, R. S., Corti, M., and Capanna, E. 1982. Morphometric divergence of Robertsonian populations/species of *Mus:* a multivariate analysis of size and shape. Experientia 38:920–923.

Todd, N. B. 1970. Karyotypic fissioning and canid phylogeny. J. Theor. Biol. 26:445–480.

Troshina, A., Gustavsson, I., and Tikhonov, V. N. 1985. Investigations of two centric fusion translocations of wild pigs by different banding techniques. Hereditas 102:155–158.

Tucker, P. K., and Schmidly, D. J. 1981. Studies of a contact zone among three chromosomal races of *Geomys bursarius* in East Texas. J. Mamm. 62:258–272.

Valdez, R., Nadler, C. F., and Bunch, T. D. 1978. Evolution of wild sheep in Iran. Evolution 32:56–72.

Viroux, M.-C. and Bauchau, V. 1992. Segregation and fertility in *Mus musculus domesticus* (wild mice) heterozygous for the Rb (4.12) translocation. Heredity 68:131–134.

Volobouev, V. T. 1983. Les types du polymorphisme chromosomique et leur rôle evolutif chez les mammifères (Insectivora, Rodentia et Carnivora). Thesis, University of Pierre and Marie Curie, Paris.

Volobouev, V. T. 1989. Phylogenetic relationships of the *Sorex araneus-arcticus* species complex (Insectivora, Soricidae) based on high-resolution chromosome analysis. J. Hered. 80:284–290.

Wahrman, J., and Goitein, R. 1972. Hybridization in nature between two chromosome forms of spiny mice. Chromosomes Today 3:228–237.

Wahrman, J., and Gourevitz, P. 1973. Extreme chromosome variability in a colonising rodent. Chromosomes Today 4:399–424.

Wahrman, J., Goitein, R., and Nevo, E. 1969. Mole rat *Spalax:* evolutionary significance of chromosome variation. Science 164:82–83.

Wahrman, J., Richler, C., Gamperl, R., and Nevo, E. 1985. Revisiting *Spalax:* mitotic and meiotic chromosome variability. Isr. J. Zool. 33:15–38.

Wallace, B. M. N., Searle, J. B., and Everett, C. A. In press. Male meiosis and gametogenesis in wild house mice *(Mus musculus domesticus)* from a chromosomal hybrid zone: a comparison between "simple" Robertsonian heterozygotes and homozygotes. Cytogenet. Cell. Genet.

Wallace, B. M. N., Searle, J. B., and Garagna, S. 1991. Oogenesis in common shrews homozygous and heterozygous for Robertsonian rearrangements. Mem. Soc. Vaud. Sci. Nat. 19:23–31.

Walsh, J. B. 1982. Rate of accumulation of reproductive isolation by chromosome rearrangements. Am. Naturalist 120:510–532.

Warner, J. W. 1976. Chromosomal variation in the Plains woodrat: geographic distribution of three chromosomal morphs. Evolution 30:593–598.

White, M. J. D. 1968. Models of speciation. Science 159:1065–1070.

White, M. J. D. 1973. Animal Cytology and Evolution. 3rd Ed. Cambridge: Cambridge University Press.

White, M. J. D. 1978a. Modes of Speciation. San Francisco: W. H. Freeman.

White, M. J. D. 1978b. Chain processes in chromosomal speciation. Syst. Zool. 27:285–298.

White, M. J. D. 1985. Types of hybrid zones. Boll. Zool. 52:1–20.

Wilson, A. C., Sarich, V. M., and Maxson, L. R. 1974. The importance of gene rearrangement in evolution: evidence from studies on rates of chromosomal, protein, and anatomical evolution. Proc. Natl. Acad. Sci. USA 71:3028–3030.

Winking, H. 1986. Some aspects of Robertsonian karyotype variation in European wild mice. Curr. Top. Microbiol. Immunol. 127:68–74.

Winking, H., and Johannisson, R. 1980. Pattern of pachytene pairing in mouse hybrids with chain and ring multivalents. Clin. Genet. 17:94.

Winking, H., Dulíc, B., and Bulfield, G. 1988. Robertsonian karyotype variation in the European house mouse, *Mus musculus:* survey of present knowledge and new observations. Z. Säugetierk. 53:148–161.

Wójcik, J. M. 1986. Karyotypic races of the common shrew (*Sorex araneus* L.) from northern Poland. Experientia 42:960–962.

Wójcik, J. M., and Searle, J. B. 1988. The chromosome complement of *Sorex granarius*—the ancestral karyotype of the common shrew *(Sorex araneus)*? Heredity 61:225–229.

Wu, C-I., True, J. R., and Johnson, N. 1989. Fitness reduction associated with the deletion of a satellite DNA array. Nature 341:248–251.

Yosida, T. H. 1976. Karyotypes and meiotic segregation of hybrids between Asian and Oceanian type black rats. Proc. Jpn. Acad. B52:304–307.

Yosida, T. H. 1980a. Cytogenetics of the Black Rat. Karyotype Evolution and Species Differentiation. Tokyo: University of Tokyo Press.

Yosida, T. H. 1980b. Segregation of karyotypes in the F$_2$ generation of the hybrids between Mauritius and Oceanian type black rats with a note on their litter-size. Proc. Jpn. Acad. B56:557–561.

Index